Anna-Maria Gasser

Form and Clarity in Euclid's *Elements*

Science, Technology, and Medicine in Ancient Cultures

Volume 15

Anna-Maria Gasser

Form and Clarity in Euclid's *Elements*

DE GRUYTER

ISBN 978-3-11-067033-2
e-ISBN (PDF) 978-3-11-067046-2
e-ISBN (EPUB) 978-3-11-067059-2
ISSN 2194-976X

Library of Congress Control Number: 2024933416

Bibliographic information published by the Deutsche Nationalbibliothek
The Deutsche Nationalbibliothek lists this publication in the Deutsche Nationalbibliografie; detailed bibliographic data are available on the internet at http://dnb.dnb.de.

Typesetting: Integra Software Services Pvt. Ltd.
Printing and binding: CPI books GmbH, Leck

www.degruyter.com

MIX
Papier | Fördert gute Waldnutzung
FSC® C083411

Meinen Eltern und meinen Kindern

Acknowledgements

This book is the revised version of my doctoral dissertation that was accepted by the Faculty of Language, Literature and Humanities of Humboldt University Berlin in 2018. Most of the literature published since then could not be considered in this revision.

First and foremost, I would like to thank my supervisor Markus Asper, who sparked my interest in Greek mathematics, for his constant encouragement, countless advice, and unfailing support over many years up to the present day. I am profoundly grateful also to Bernd Seidensticker for his guidance since my days as a graduate student at the CRC Aesthetic Experience and the Dissolution of Artistic Limits, especially for his supervision in the Posidippus project, and I would like to thank him and Jochen Althoff for reading my dissertation and providing invaluable comments and suggestions.

I am indebted to Markus Asper, Christoph Markschies as well as the Caroline von Humboldt Scholarship Program for giving me time and space to write this study, and to Susanne Gödde for her understanding and support while it was prepared for publication.

For all sorts of help and advice, I extend my heartfelt thanks to Roland Baumgarten and Thomas Poiss, for their loyal support along the way to Eva María Mateo Decabo and Antje Wessels. For reading parts of the book at different stages and for discussing particular problems, I would like to express my gratitude to Martin Stöckinger, Orly Lewis, Renate Burri, Chiara Thumiger, Oliver Overwien, Katharina Kanthak, and Carlos Gasser. For greatly improving the language and style of the book, thanks are due to Stev Talarman. All remaining errors are, of course, my own.

Finally, I am grateful to the editors of STMAC for accepting my manuscript, to Torben Behm of De Gruyter for his patient and competent help during the whole publication process, to Anne Stroka of the De Gruyter delivery unit at Integra Software Services for her efficient and flexible organization of the typesetting, and to the typesetters and graphic artists of Integra for many corrections, a lot of last minute work, and the beautiful diagrams.

https://doi.org/10.1515/9783110670462-202

Contents

List of figures

https://doi.org/10.1515/9783110670462-204

1 Introduction

1.1 Subject and aim

In this study, I analyze the textual form of Euclid's *Elements* and its relation to clarity.[1] The *Elements* (Στοιχεῖα) is Euclid's main work and the most famous one of the genre with the same name.[2] The origin of the generic term στοιχεῖα is unknown.[3] It refers to mathematical "axiomatic-deductive" texts, i.e., texts which consistently deduce propositions from a number of unprovable basic assumptions (axioms) and terms.[4] The function of such texts would have been to accumulate and codify mathematical knowledge for later use.[5] The Euclidean *Elements* itself has long been dated to around 300 BC, but more recently to the second half of the third century BC.[6] Its 13 books focus on different elementary subjects of geometry and arithmetic.[7] In the main, each book is a sequence of propositions. The discrete unit of the proposition consists of a claim (πρότασις) and its proof. The propositions are deductively arranged, with more complex propositions being based on simpler ones and on the unprovable assumptions which are listed at the beginning of some of the books. The *Elements* is generally recognized to be one of the most widely received and most influential works from antiquity, and it has decisively shaped the form of modern scientific texts in terms of its standardized language and exposition and its use of diagrams.

My study of the relation between form and clarity in the *Elements* focuses on the features of the 'literary' (or 'textual') form of the work.[8] I regard as the decisive fea-

1 On the notions of 'textual form' and 'clarity' see below, pp. 1 f., 27 f., and 54 f.

2 In addition to στοιχεῖα, στοιχείωσις was also in use (Asper 2007, 99, 104). On the paradigmatic position of Euclid and specifically of his *Elements* within the genre cf. ibid., 106.

3 Asper 2007, 99; on possible origins and ancient uses and explanations of the term ibid., 98–104.

4 For this definition, cf. Asper 2007, 97 and 98, n. 36.

5 Asper 2007, 145 and *passim.*

6 In this regard, the relation of Euclid to Archimedes has been deemed decisive: whether Archimedes (ca. 287–212 BC) used the Euclidean work, which would suggest an early date, or not, which might point to a later date; cf. Asper 2007, 97 with reference to Knorr 1978 and Schneider 1979, 47 f., who have first doubted the standard early dating, and Netz 2009, 232, n. 5, who tends to retain the traditional dating on the basis of what might be a reference to Euclid in the Archimedes Palimpsest (see his Netz 2004b, 44), but above all on "the coincidence between Euclid's *Elements* and the toolbox of previous results used by Hellenistic mathematicians."

7 In addition to this, two non-Euclidean books have been passed down as part of the *Elements*: book XIV is originally by Hypsicles and has been added to the work only in late ancient times; book XV has been added to the text only after the sixth century AD; part of it has been ascribed to Isidore of Miletus (Acerbi 2007, 400).

8 Throughout my study, I shall use the term 'textual form' synonymously with 'literary form' in order to distinguish the kind of form which I analyze from the 'logical form,' that is, above all the axiomatic-deductive structure (on this distinction, see Section 1.2).

https://doi.org/10.1515/9783110670462-001

ture of the literary form its 'standardization,'[9] by which I mean a tendency toward linguistic, stylistic, and structural unification. In the *Elements*, such a tendency of using the same formal features for the same textual units is recognizable on all levels of the text: lexicon, syntax, structure, and also diagram.[10] By 'clarity,' I understand textual 'unambiguousness' which guarantees the 'comprehensibility' and 'unmistakability' of a text. That is, provisionally, I conceive of it on the one hand as a feature of the text and on the other hand in terms of its cognitive-communicative effect. I shall employ the English terms largely interchangeably throughout my study, but shall tend to use 'unambiguousness' for referring to clarity as a quality of the text form (both literary and logical) and the other terms for emphasizing the communicative dimension of clarity.[11] Both aspects, the textual-logical and the communicative one, can be connected with the Greek logical and rhetorical term σαφήνεια (lat. *perspicuitas*),[12] which is usually translated as 'clarity' or 'distinctness.'[13] However, as I shall show shortly, the concept to which the term refers cannot simply be applied to the Euclidean *Elements*, above all because it was developed neither by the authors of 'elements texts'[14] nor with respect to elements texts. Therefore, it is, if at all, applicable only in terms of its overall meaning but not in terms of specific semantic nuances of the term. Neither can it be applied in terms of the textual features nor of the causes, aims, and effects which are distinguished by the different ancient theoretical accounts. This means that we can analyze the relation between textual form and clarity in the *Elements* only as an outside observer, that is, from an 'etic' perspective, even if we make recourse to the ancient concept. 'Etic' concepts are liable to anachronism; nevertheless, applying them is potentially productive, and in any case to some extent inevitable, if one wishes to investigate a notion which is insufficiently reflected from

---

9 Asper 2007, 113 distinguishes three "literarische[] Formmerkmale[]:" "*Diese Texte sind 'diskret,' standardisiert und unpersönlich*" (his italics). By contrast, I treat the "discrete" and "impersonal" features of the style as direct consequences of the specific standardization of the language and style of the *Elements* (while, of course, not all standardized texts are necessarily "discrete," that is, consisting of separate textual units, and "impersonal").

10 Cf. the definition given by Asper 2007, 29, n. 35: "Unter 'Standardisierung' verstehe ich hier das Bestreben, Syntax und evtl. Wortwahl ähnlicher Einheiten eines Textes möglichst ähnlich zu halten, d.h. also einem formalen und/oder semantischen Standard zu unterwerfen [. . .]." I prefer to recognize standardization first of all as a 'tendency' of unification, before considering whether there is an 'effort.'

11 I explain my uses of the terms 'clarity' and 'unambiguousness' in more detail in Section 1.4, 54 f.

12 Asmuth 2003, 814 f.

13 These translations, however, as well as the corresponding German translations 'Klarheit' and 'Deutlichkeit,' are themselves in need of explanation. The ancient terms comprise a variety of semantic nuances, amongst them 'unambiguousness' and 'comprehensibility.' Cf. below, pp. 51–53, on the uses of the terms in ancient logic and rhetoric and pp. 42–51 on their uses in the commentaries on scientific texts.

14 I shall use the term 'elements texts' throughout to refer to texts of the genre of 'elements' (i.e., in the sense of the German 'Elementa' or 'Elementaschriften') because 'elements' alone is ambiguous.

an 'emic' perspective, that is, from inside a specific group in the extant contemporary literature.[15]

My aim is to show how the standardized form of the *Elements* relates to unambiguousness and clarity. For this purpose, I shall first explore in which ways standardization characterizes the text form, and whether standardization leads to textual unambiguousness. Then I shall ask to what extent textual standardization is necessitated, on the one hand, by the mathematical content, and, on the other hand, by the communicative situation. That is, I will consider first to what extent textual standardization serves the mathematical aim of the text, i.e., the unambiguousness and consistency of the proofs; second, I will ask to whom the text, if at all, conveys clarity, whether to the general readership or only to a specific subset of specialist readers; and how much standardization the intended readership requires in order to understand the proofs. Finally, I shall determine whether and how far standardization is motivated by the effort to be clear, and if not, or if not entirely, what other motivation there might be for it and what its function could be. I shall argue that the textual form of the *Elements* is extremely standardized; that it is, in fact, more standardized than would be necessary in order for the proofs to be unambiguous and consistent; that its standardization, nevertheless, does not convey clarity universally, but only to a specific readership; and that cognitive-communicative clarity is, in any case, not the (only) function of the standardization of the *Elements*.

More broadly speaking, my analysis of the relation between form and clarity in Euclid's *Elements* is a study of the function of the textual form of an ancient Greek scientific text from a literary-aesthetic perspective. 'Science,' like 'clarity,' is an 'etic' concept.[16] I regard as 'scientific texts' or 'science texts' such that, amongst other things, have the basic function of transmitting knowledge.[17] Neither in ancient nor in

---

15 For the distinction of 'emic' and 'etic,' and the implications of applying etic concepts cf. Asper 2013b, 4 f.

16 Note, however, that it is not really possible to talk about science from an emic perspective, not even as a scientist about contemporary 'science' from a supposedly inside perspective, because 'science' is a construction (Groebner 2012, 49, who argues that there are merely different scientific disciplines). "Mathematics," too, is an "etic" concept, but its application is less problematic than that of "science" because of the strong overlap of the ancient and modern usage (Asper 2013b, 4).

17 Cf. the functional definitions by van der Eijk 1997, 77: "As for 'scientific,' this is used throughout this article in a non-sophisticated, non-evaluative manner to refer to any serious attempt at studying and understanding the nature of things (or part thereof), and – as far as texts are concerned – to any verbal expression intended to communicate about this with an audience;" and Asper 2007, 13: "Als 'Wissenschaftstexte' sollen hier Texte gelten, die in der Intention ihrer Autoren gesichertes, proportionalisierbares, nicht-narratives, d.h. keiner Zeitstruktur unterworfenes Wissen weiterzugeben bemüht sind, dessen Inhalt oder auch Validisierungsweisen in der Regel einer bestimmten Gruppe als konsensfähig gelten. [. . .] Ich betrachte also 'Wissenschaftstexte' als eine primär funktional, nicht oder erst sekundär formal, abgegrenzte Gruppe von Texten." The basic function of 'transmitting' or 'transferring' knowledge may be taken to include the two other basic cognitive-communicative functions of science texts, that is, to discover or acquire and to store knowledge (as listed by Asper 2007, 21 f.) be-modern times have scientific texts been part of the traditional literary canon; accordingly, they have received little attention by both ancient and modern literary criticism.[18] Modern literary studies have long neglected them because they cherished a non-functional concept of literature in terms of an 'aesthetics of autonomy.'[19] Such a concept does not cover texts which serve a specific purpose as do scientific texts. Classical philology, however, could not have afforded to ignore scientific texts which happen to account for a large part of the transmitted corpus of ancient Greek and Latin writings. However, until recently, the discipline has considered scientific texts on the one hand mostly anachronistically with respect to the truth and validity of the knowledge and theories which they contain and on the other hand with respect to their textual constitution. It is indeed still largely focused on editing the vast amount of scientific texts extant, although more recently, cultural-comparative approaches to ancient science have come to be important.[20] Only lately, the 'rhetoric' of scientific texts has been taken into view; this approach starts from the assumption that communicating knowledge to readers requires that these readers be persuaded first, and it analyzes the textual devices which scientific authors apply for doing so.[21] In addition, scientific texts have been analyzed increasingly by means of linguistic, narratological, and sociological approaches. These have been applied to investigate specific aspects of the form of scientific texts (e.g., "terminology," "narrative," and "authorship") with respect to social-communicative functions of these texts,[22] such as authority, which are closely related to the basic purposes of transmitting knowledge and persuading the reader.

By contrast with such approaches, which consider the function of individual preconceived parts and aspects of the literary form of science texts in relation to the overall functions of the texts, "literary-aesthetic" approaches to ancient scientific texts are still largely missing.[23] By this, I mean perspectives which take into view the textual

cause both the documentation of newly discovered knowledge and the storage of knowledge imply that this knowledge will be available for later access by later readers (including the author) and will be communicated at some point.

18 Asper 2013b, 1 f.

19 Asper 2007, 46.

20 Cf. Asper 2013b, 2.

21 The need for such an approach, methodological requirements, and possible strands of research has been outlined programmatically by van der Eijk 1997.

22 Asper 2013b, 3 f. Cf. Asper 2003a on "authorship" in Greek medical and mathematical texts (see Asper 2013b, 3 f. for references to other recent studies).

23 Asper 2013b, 2. For the very few literary-aesthetic studies of Greek theoretical mathematics see my short literature review Section 1.5, 56 f. Aesthetic approaches to Greek scientific texts have even been criticized as being overstated: Cf. van der Eijk 1997, 82 on Schütrumpf's 'aesthetic' (rather than 'rhetorical') approach to the form and style of Aristotle's treatises in his 1997 article "Form und Stil aristotelischer Pragmatien:" "[. . .] one need not go so far as to make strained efforts to ascribe an aesthetic dimension to texts which have traditionally been despised because of their supposed 'lack of care' and 'clumsy style.'" Unlike Aristotle's writings, the Euclidean *Elements* do not have the reputation of

form of science writing *as textual form*, that is, which are concerned with its language, style, and structure from an aesthetic point of view. My own study presents precisely such an attempt of a literary-aesthetic approach toward the textual form of Euclid's *Elements*: I am interested in what makes the *Elements* a literary text, in what is 'literary' about the form of the *Elements*. Accordingly, I shall first analyze the text form comprehensively and thoroughly with respect to its most prominent characteristic (i.e., standardization). Then I shall ask whether and to what extent the standardized text form serves the logic and the communication of the proofs. I shall do so by considering the relation between the textual form and clarity as a 'textual-logical' and as a communicative quality: with the help of this twofold distinction of dimensions of clarity, I ask to what extent the form serves the function of clarity, that is, logical-communicative purposes, and to what extent the text is 'formalized,' that is, 'has literary form,' beyond this function, and is 'aesthetic' in this very sense. Finally, I shall consider the function or point of this aesthetic 'excess' of textual form, that is, of the amount of form which is non-functional from the point of view of the transmission of knowledge.

My first point of interest is, thus, in the meaning of the literary form of Euclid's *Elements* and its literariness. In order to determine the functional and (in terms of communication) non-functional dimension, I employ the concept of clarity. My second point of interest is in the role of 'clarity' itself. After analyzing the form of the *Elements* with the help of the concept of 'clarity,' I shall reevaluate the relevance of this concept for the form of the *Elements*. Finally, I shall suggest some broader conclusions about the relation of form and clarity in (Greek) scientific texts.

The origin of the 'Euclidean' features

The formal features of Euclid's *Elements* are highly peculiar, but they are, for the most part, not idiosyncratic.[24] Rather, they are generic features, that is, of elements and texts which go under the name of Greek 'theoretical mathematics.' This is to distinguish the genre from the longstanding tradition of 'practical' mathematics against the backdrop of which Greek theoretical mathematics probably emerged, even though its origins mostly lie in the dark.[25] The decisive generic features seem to be already

being carelessly written; however, today, they are the epitome of a clearly, but dryly written text, and on the face of it an aesthetic approach toward it might suggest itself just as little as in the case of Aristotle.

24 There may be exceptions: Netz 2009, 123 claims that the concluding formula of the propositions ὅπερ ἔδει δεῖξαι (the famous 'Q.E.D.') is "idiosyncratic" to Euclid, but this is doubtful.

25 Cf. Asper 2007, 166–171. I shall compare briefly the context of practical mathematics with that of elements texts at the end of my study (pp. 430–432).

attested for Hippocrates of Chios around 430 BC.[26] The extant works of the elements genre mostly stem roughly from the third century BC; these include the works of Archimedes and Apollonius of Perge, the lesser known writings of Euclid, and the works of Autolycus of Pitane and Aristarchus of Samos.[27] Within ancient Greek literature, the form of the genre is indeed unique; it is confined to the few works which represent this small genre;[28] only individual features of it recur in other genres which draw on it, mostly for reasons of authority or application in other fields of science, above all philosophy and mechanics.

One may try to determine certain subgenres of elements texts, but the respective categories which arise from the different criteria applied overlap. One possible differentiation is that of elements proper, as it were, which expose comprehensively and systematically the basic knowledge (of their predecessors) of a field of knowledge (in the case of Euclid's *Elements*, geometry and arithmetic),[29] and secondary elements texts which, beyond summarizing previous knowledge, aim to present an innovative mathematical discovery that is disclosed at the end of the text, such as most of the extant books of Archimedes.[30] The second(ary) type seems to have developed from the first: the Euclidean *Elements* became a standard work of reference which provided the basis for the presentation of new results in elements texts which built on and generically developed from elements texts of the Euclidean type. However, it is not clear whether the *Elements* indeed predated the works of the second type; especially the chronological relation of Euclid and Archimedes is unclear.[31] It is equally unclear whether works of what I call the second(ary) type existed all along, and whether the impression that it developed from the Euclidean type is due to the state of transmission by which only first-type texts survived as supposedly early works of the genre.[32] In any

---

26 The so-called 'lune of Hippocrates' is the earliest testimony for elements texts; however, it is not possible to retrieve the original form of Hippocrates' text and determine how much it resembled that of Euclid's *Elements* since the text has only been transmitted by Simplicius (sixth century AD; Rudio 1907, 30–37) who provides a more detailed rendering of the abridged version which he finds in Eudemus (fourth century BC; fr. 140 Wehrli, Becker 1936b, 417–419). Cf. Netz 2004a.

27 For an overview cf. Asper 2007, 97.

28 There have always been very few theoretical mathematicians; it was an elitist occupation (cf. Netz 2002b for an estimation of the – extremely small – number of Greek theoretical mathematicians around at a particular time).

29 For an overview of the subjects of the individual books of the *Elements* cf. Asper 2014, 463.

30 Asper 2014, 456.

31 Netz 2009, 232, n. 5; he either classifies Euclid (with hesitation) as Hellenistic mathematics (109) or as "early" and, more precisely, as "pre-Archimedean Greek mathematics" (which he distinguishes again from the "early Greek mathematics of Hippocrates and Archytas") as against "Hellenistic," Archimedean mathematics (233 f.). Cf. n. p. 1, n. 6 on the dating of Euclid (in relation to Archimedes).

32 Netz 2009, 233. Apart from Euclid's *Elements*, he counts among "the pre-Archimedean works" of this type also the works of Autolycus and "the fragments extant in the Peripatetic corpus." To him, this elementary type is characterized by its ("pedagogic") function and its ("sedate") style (on these criteria see below, pp. 7 f.).

case, at some point, the two types co-existed and even blended; for the authors of this secondary type of elements texts also wrote their own first-type elements in the manner of Euclid's main work: Archimedes quotes his own, now lost, Κωνικὰ στοιχεῖα,[33] and Apollonius' *Conics* I–IV are an introductory exposition of the knowledge on conic sections,[34] which integrates new knowledge in its third book and provides the basis for the presentation of more advanced and new material in the subsequent books V–VIII. In this way, the latter work presents a hybrid of the two aforementioned types of elements texts.[35] Compared with the strict impersonality of the Euclidean *Elements*, which is not addressed to any particular reader,[36] the personal address of a specific reader – a fellow mathematician – in the introductory letters and forewords of Archimedes and of Apollonius is conspicuous. However, the degree of (im)personality cannot serve generally to distinguish two types of elements texts; for three of the first four books of Apollonius' *Conics* (I, II, and IV), which otherwise resemble the *Elements* as regards their form and function, are headed by prologues. In many works of the non-Euclidean, 'Archimedean' type, as it were, Netz has identified a playful element; among this "ludic" type of treatise, i.e., "work[s] based on obtaining results in surprising, intricate ways," Netz counts Archimedes' books, Aristarchus' *Sizes and Distances of the Sun*, Eratosthenes' *Duplication of the Cube*, and Diocles' *On Burning Mirrors*.[37] However, he also characterizes as "ludic" Apollonius' *Conics* III (and VII) and Hypsicles' 'book XIV' of the *Elements* whose form differs considerably from that of the work's Euclidean books; and he even holds that *Elements* XIII, the last book of the work which Netz classifies as "pedagogic," "already suggests the ludic."[38] That is, according to Netz, it marks the transition to the "ludic" which characterizes Hellenistic mathematics as a part of the Hellenistic literature.[39] While what Netz calls "the ludic" overlaps roughly with our second(ary), Archimedean type, he has *two* different categories to cover what our first type – that which exposes comprehensively and systematically the basics of a field of knowledge – comprises: apart from "the pedagogic," represented mainly by the *Elements*, also the "survey" ("a work where a certain field is exhaustively treated"), represented by Euclid's

33 *Con. sph.* 3 (vol. 1, p. 270.23 f. Heiberg), cf. Asper 2007, 147, n. 368.

34 Asper 2014, 469.

35 Regarding the function of 'elements proper' which accumulate ("thesaurieren") consensual knowledge, the function of these elements texts could be described as "Parallel- bzw. Entlastungsthesaurierung" (Asper 2007, 147).

36 The absence of the reader from the *Elements* goes along with that of the author: Euclid is "the most self-effacing of Greek mathematicians" (Netz 2009, 92). On the communicative context of the Euclid's *Elements* cf. Section 5.2, 423–426.

37 Netz 2009, 108.

38 Netz 2009, 108. On Hypsicles' book cf. n. 7. On its form see Netz 2009, 92–99.

39 Netz 2009, 115. Cf. Section 1.5, 56 f. and Section 5.3, 434, n. 35 on the hypothesis of Netz 2009.

Data, *Elements* X, Apollonius' *Cutting off of a Ratio* and *Conics* IV.[40] However, the border between the latter two types is somewhat fuzzy, for the supposedly "pedagogic" *Elements* likewise covers a field, or even several fields exhaustively (satisfying Netz' definition of the "survey" category), while it is uncertain whether the work originated from a didactic context at all;[41] and the *Data*, which Netz classifies as "survey," might have had an accumulating function like the *Elements*[42] or even a pedagogic aim.[43] In sum, one may distinguish cautiously between two types of elements texts – the more comprehensive and formally more stringent treatment of a subject and the often playful kind which leads to specific mathematical results; but there is considerable overlap between the two – as regards chronology, form, and probably also function. In any case, the Euclidean *Elements* was not the first exponent of its type, the first of the two which I have just sketched out. The work was canonized early, and all previous elements texts of its type were lost. Euclid's Neoplatonist commentator Proclus, in his 'catalogue of geometers' which goes back to the Peripatetic Eudemus, lists 14 writers of slimmer predecessor versions, of which Hippocrates is the first one.[44] Euclid has collected and restructured their propositions and has made the earlier exponents of the genre obsolete because of the greater scope of his *Elements* and because he appears to have perfected the generic formal features.[45]

The extent of Euclid's own contribution in the standardized form of his *Elements*, as it presents itself in the modern editions, is uncertain. On the one hand, his predecessors were in principle already characterized by the same formal features. This can be assumed from Proclus' classifying them as authors of elements, i.e., as texts of the same genre,[46] and from some quotations of them in later authors. Aristotle often gives mathematical examples which show features of the Euclidean language and therefore prove themselves to be quotations from Euclid's predecessors, probably Leon or Theu-

40 Netz 2009, 108 f. The "three types of Greek mathematical treatises" which Netz identifies indicate by their names that they actually pertain to different categories: "ludic" refers to the particular style of a certain type, "survey" to the content, and "the pedagogic" to the aim and communicative context.
41 Cf. Section 5.2, 423–426 for my discussion of the original function of the work; it only seems to have been used for didactic purposes, and reshaped accordingly, in post-Euclidean times (cf. 425 f.).
42 Thaer 1970, 64.
43 Cf. Asper 2014, 464 on the "purely pedagogic aim" ("rein pädagogische Zielsetzung") of the *Data*, and more cautiously 2007, 141 with reference to the Neoplatonic commentator of the *Data* ("Die Intention der *Data* ist aber durchaus didaktisch, wie Marinos feststellt"), who, however, should have had his own reasons for ascribing a didactic function to the work (cf. Section 2.3.2, 103 for Proclus' insinuation of a didactic function of the *Elements*).
44 Proclus *In Eucl.* 65.21–68.6 Friedlein = Eudemos Fr. 133 Wehrli. Cf. Asper 2007, 109.
45 Asper 2014, 464. The genre, in any case, entails that Euclid did not invent his *Elements* from scratch – which has promoted some critics to see in him 'merely' a compiler rather than an author. The derogative connotation of the former betrays a certain concept of authorship (marked by the 'genius aesthetic') rather than saying something about Euclid's actual achievements (Asper 2007, 106 f.).
46 Asper 2007, 109.

dius.[47] Other fragments, such as Simplicius' excerpt from Eudemus who cites Hippocrates of Chios, can only provide limited information as to the form of the original text which is mediated several times.[48] From Euclid's text itself, it is impossible to conclude much about the formal features of individual predecessors.[49] Nevertheless, one may go as far to say that the changes which Euclid has made to the textual form are not qualitative in nature, but only quantitative,[50] that is, he seems to have applied the generic features more consistently, and to have thus increased the standardization of the text. The exact degree of this quantitative development of the form, however, is difficult to determine. Proclus credits Euclid with significant formal improvements of the foregoing versions, but we have to take into account the teleological coloring of his catalog; Proclus describes the development of elements texts as one of linear progress which culminates in Euclid, who, therefore, has to stand out positively from those before him.[51]

On the other hand, the 'original' text of Euclid itself, whatever it looked like, has undergone considerable additional standardization in the course of its transmission and reception by late ancient editors and commentators.[52] It was canonized and institutionalized early, came under the influence of the philosophical schools and was used therein as a textbook for didactic purposes.[53] The text was changed most fundamentally by Theon. The *Elements* has been mainly transmitted in six Greek manu-

47 Asper 2014, 462. Above all, the characteristic third-person imperatives indicate the underlying presence of an elements text.

48 Asper 2007, 110.

49 If at all, it is only possible to reconstruct parts of the contents of some of the predecessors such as the origin of particular terms or proof techniques (Asper 2007, 109).

50 Cf. Asper 2007, 113.

51 Proclus, *In Eucl.* 65.21–67.1 Friedlein, cf. Asper 2013c, 416, who points out that Proclus imagines the development of elements texts as the result of a continuous group effort, and that this projection could well be in accord with the self-image of the writers of elements texts, lacking, however, the teleological aspect of Proclus' account. By contrast, Netz 1998, 283 argues that the mathematicians "were pursuing their own projects (which are only vaguely understood by us)," whereas Proclus' description of the supposed development up to Euclid reflects his own textual practices as a commentator which aim at establishing the ideal text; Netz even goes as far as to say that it is "far from clear that anyone prior to Euclid has tried to put together a book of Elements," which however does seem to be indicated in view of our pre-Euclidean evidence for a common genre of *stoicheia*. – As regards my own line of inquiry, it is not clear from Proclus' appraisal of Euclid whether the latter's supposed achievements regarding the form really concern the *textual* form, which I am interested in, or rather the 'logical' form (on this distinction see n. 8 above and more extensively Section 1.2); I shall discuss this question in Chapter 2.

52 Cf. Netz 1998, 268–270 on "standardization" as a common practice of "deuteronomic texts," i.e., "texts depending fundamentally on earlier texts" (ibid., 261).

53 Asper 2007, 211. The turning-point was the second half of the third century BC when the *Elements* became canonical ("nacheuklidische Phase").

scripts,[54] and all but one of the manuscripts date back to Theon's version of the text, which has decisively shaped the text of the standard edition by Heiberg & Stamatis (1969–1973).[55] And even the text of the one non-Theonine manuscript, the Vatican manuscript P was changed by pre-Theonine editors.[56] Although Heiberg & Stamatis take P into account, it is still heavily influenced by Theon, which is indicated by the Arabic tradition of the text.[57] The Arabic manuscripts sometimes provide a better and generally shorter text than all Greek manuscripts and prove that P, too, has "suffered later [pre-Theonine] editorial work."[58]

Some features of the transmitted text can be identified with some certainty as later inventions, as they contradict the generic rules of elements texts, e.g., the explicit standardization by means of enumerating propositions or subdivision, "obvious intermediate arguments,"[59] or the explication of proofs presupposed by the deductive structure of explicitly recalling arguments implicitly presupposed by Euclid.[60] However, with regard to implicit textual standardization, it is difficult, and in many cases impossible, to determine who is responsible for the respective degree of standardization, whether Euclid (if not one of his predecessors) or one of his editors. For such changes to the Euclidean text are only a matter of degree, with the authorial/generic and the "deuteronomic"[61] standardization adding to the same style.

It is not my aim in this study to pinpoint Euclid's personal contribution to the textual standardization of the *Elements* by means of an analysis of Euclid's sources or by means of work on the text of Heiberg & Stamatis. As regards the question how much the *Elements* owes to its predecessors in terms of standardization, it is not decisive for my line of inquiry exactly how much the formal features were already fixed before Euclid and who exactly brought about the standardization in one or another respect. My interest lies in the extreme manifestation of the *generic* features in the *Elements*

54 The Greek manuscript transmission (P, and the Theonine F, b, B, p, and V) is supplemented by a few papyrus fragments and the Arabic-Latin tradition of the text. The textual evidence of the papyrus fragments is very small and so does not add much to the reconstruction of the text; I shall examine their diagrams as for indications of the original ones and present them in Section 4.1.

55 Asper 2014, 463.

56 Vat. gr. 190.1 and Vat. gr. 190.2.

57 Cf. Asper 2014, 463.

58 Saito 2009, 816 f. mentions the example of the manuscript b, which belongs to the Theonine tradition, but, in the case of book XII, turns out to be very close to the Arabic one; he points out, however, that none of the manuscript traditions can in general be considered superior to the other ones because different manuscripts provide a better text for different books and parts of the work. The first Arabic translation was produced in the eighth century, but at least two more were to follow subsequently; the first Latin translations from the Arabic originate from the twelfth and thirteenth centuries (Morrow 1992, xlvii).

59 These are omitted in P (Saito 2009, 815).

60 They run counter the deductive principle of "never go[ing] backwards;" such additions can be found in all Greek manuscripts, including P, but not in the Arabic (Saito 2009, 815 f.).

61 Cf. p. 9, n. 52.

and how they relate to cognitive-communicative clarity, regardless of how much exactly the standardization owes to that of earlier elements texts. I take 'Euclid' to refer to the authorship of the generic standardization which can be observed in the *Elements*, without indicating that the standardization is the original work of a single historical author. Conversely, I sometimes refer to the accomplishments of the text as those of 'the writer' or 'the author' in order to indicate that they do not necessarily have to be Euclid's own.

As far as the post-Euclidean standardization of the *Elements* is concerned, it does make a difference for my inquiry how much of the standardization was imposed on the text only later, for I attempt to determine the degree of the generic standardization in the case of the *Elements*. While in Euclid's own time the text seems to have had the same function as its predecessors, it continued to be used and transmitted, its form being variously modified and adapted to the exigencies of its later recipients. However, the examination of the standard-edition text as to its validity in every single case of a feature's standardization would by far exceed the scope of my study. Therefore, I shall base my analysis on the text of Heiberg & Stamatis and reconsider the tradition of the text only in qualitative terms: that is, regarding the question whether a feature was (standardly) present *at all* in the Euclidean text.[62] In any case, I shall discuss the implications of a reinforcement of the standardization of the *Elements* in the aftermath of Euclid.

There are two (partly interrelated) reasons why I focus specifically on Euclid's *Elements* of all extant works of the genre, even though I maintain that its formal features are merely generic: the first is that the work seems to apply these generic features more consistently than all other extant works of Greek theoretical mathematics and thus offers the best material basis for observing standardization in relation to clarity. To be sure, the other works of the genre, most of which are of the second 'non-elementary' type which I have distinguished above,[63] especially those of Archimedes and Apollonius, are overall similarly standardized as the *Elements* as regards their language and style.[64] However, the standardization of many other works as regards their textual form differs from that of the *Elements* in two ways which seem to have something to do with the communicative context and function of the respective writings; they are directed at specific individual fellow mathematicians to which they present new discoveries instead of basic knowledge. On the one hand, many of them are headed by personal introductory letters due to their communicative situation: unlike the *Ele-*

---

62 E.g., regarding the Q.E.D., I shall discuss the results of Acerbi 2003 and Netz 2009 (see p. 23, n. 135); regarding the preexistence of the diagram in the text, I shall make my own inquiry (see Chapter 4).
63 Cf. above, pp. 6–8.
64 The statistical evaluations of Netz 1999a overall do not show any significant differences in the use of the language between the mathematicians (cf. below Section 3.1.1, 113, n. 21).

ments, the works by Archimedes and Apollonius are sent as letters to their addressees.[65] In terms of textual form, these paratexts add a personal, non-standardized element to text by which the works deviate from the *Elements* and its uncompromising standardization and impersonality.[66] On the other hand, and more importantly, the *Elements*, and especially book I, on which I am going to focus, are more standardized in terms of structure (both as regards the structure of the proposition and the macrostructure of the work) than other works of the genre. Reasons for this include the following: (1) Many elements texts are not as standardized as the *Elements* in terms of macrostructure, that is, they do not present their propositions as distinct units, which has consequences for the internal structure of the proposition, namely, by rendering it less standardized (cf. Apollonius' *Conics* I 52–53).[67] (2) Different kinds of structure result from different kinds of proof: a different *kind* of structural standardization characterizes those geometrical propositions which are written in what Netz calls the "analysis and synthesis mode of proof" (cf. Archimedes' *Sphere and Cylinder* II); in itself, this type is as standardized as the proposition type of the *Elements*, but works which feature both types are, as a result, less standardized in terms of structure.[68] (3) The "playful" character of many elements texts of the second type, identified and investigated by Netz, results in some structural non-standardization.[69] These deviances from the structural form of the *Elements* may again simply be due to the specific function and context of the extant works; that is, potential 'elementary' works by Archimedes might have been just as stringently standardized as Euclid's *Elements*.[70] As we have pointed out, the first four books of Apollonius' *Conics* seem to have a similar introductory function as the *Elements*; and it is precisely in the third book, which itself also presents new knowledge, that the style becomes more "playful" and in this way less standardized.[71] However, as it is, the most salient exposition of the generic features of Greek theoretical mathematics is Euclid's.

65 Cf. Asper 2007, 117, n. 166, who argues that the letters are to prevent misunderstanding of the mathematical content in view of decontextualization.

66 On the relation of impersonality and standardization in the *Elements* cf. Section 1.2, 29 my overview of the formal features. The *Elements* as it has come down to us contains a paratextual element, too: i.e., the axiomatic lists; Netz regards the introductory letters and the axiomatic lists as part of the same tendency to "start with some piece of prose preceding the sequence of proved results" (1999a, 94). Euclid's axiomatic lists also differ in their form from the standardized main body of the work (without, however, adding a personal element to the text), but their originality is (at least partly) unclear (see pp. 22 f.). On the strong deviation from the impersonality of the *Elements* in the non-Euclidean, Hypsiclean *Elements* XIV cf. Netz 2009, 92 f.

67 Netz 1999b, 294.

68 Netz 1999b, 293.

69 Cf. Netz 2009, 117–136.

70 Netz 2009, 233 suggests this *mutatis mutandis* for what he calls "pre-Archimedean" mathematics, for which only works of the style of the *Elements* survive (cf. above p. 6). See Section 5.3, 434 for my own interpretation of the "play."

71 Cf. above p. 7.

The second reason why I focus on Euclid is that his name is identified with those features even to the present day. This is partly because he used them most stringently, but above all it is the result of his canonization and identification with the genre and its form from early on.[72] He has come to be the most famous and most influential exponent of the form which has decisively shaped the form of modern science and our idea of scientificity and scientific clarity.[73] In the light of an analysis of the form of the *Elements* it will be possible to reconsider and question this idea. It is only toward the end of my study that I shall briefly consider the specific shape and position of Euclid's *Elements within* the genre by comparing its features briefly to that of the surviving elements texts of the 'Archimedean type.'[74]

1.2 The form of Euclid's *Elements*

The characteristic features of the form of the *Elements* are lettered diagrams, definitions, terms, assumptions, propositions, proofs, generality, discreteness, deductiveness, systematicity, increasing complexity, formalization of arguments, third-person imperatives, standardized language and structure, and impersonality. This jumbled list reads like a description of the (ideal) form of texts of the contemporary 'hard' sciences. The reason is that the form of the *Elements* has decisively influenced the form of today's science writing, both in terms of the argumentative form, to which I shall refer as 'logical' form, and in terms of its textual form.[75]

By 'logical' form, I mean those aspects of the form of the *Elements* which are part of the proof method. I put 'logical' in inverted commas whenever using it with reference to the form of the *Elements* because the axiomatic-deductive proofs, while relying on logical principles, are not formalized according to (Aristotelian) logic.[76] The proof method is the method by which the proof is presented to be done – and effectively by which it is done in writing. That is, I do not mean by it the components of

72 Asper 2007, 106. Archimedes, by contrast, has always been associated with his discoveries and anecdotes about them; see Asper 2014, 465.

73 Asper 2014, 462.

74 Cf. n. 30 on Netz' distinction of "pre-Archimedean" and, accordingly, Archimedean, "Hellenistic" mathematics (2009, 233 f.).

75 Cf. Asper 2014, 462 and 464 on the influential text form of Greek mathematics; on Euclid's own contribution to it see above, pp. 8–10.

76 Mueller 1969, 297 f. uses a similar distinction, claiming that the Euclidean proofs are "logically correct" but "not based on logic." However, the question is precisely whether it can be determined that they are logically correct without the application of formal logic (cf., e.g., below pp. 23–27 for the problem of generalizing the result of a particular proof). This is why I prefer to say that the proofs rely on logic, in the nontechnical sense of drawing inferences from given statements. As far as Aristotle is concerned, he develops his scientific theory and proof method with the help of mathematics (not the

whatever method was used originally to discover the proof,[77] but rather the proof method which is followed in the written text.

For in the case of the *Elements*, discovering the proof and 'writing up' the proof are two different processes, even though the discovery will have involved writing, and even though the act of 'writing up' implies presenting the proof in a valid way and, in doing so, accomplishing it in the first place.[78] The discovery of the proof will have been based on the same 'logic' as are the written proofs; it will have entailed a diagram[79] and thinking about how the proposition can be deduced in connection to what is already proved. However, the cognitive process which preceded the writing up is no longer accessible to us. This is because the text of the *Elements* bears no traces at all of the process which led up to the proof discovery; the text fixates only the result: the proof itself. The *Elements* does not document in writing Euclid's or his predecessor's cognitive processes ("thought experiments")[80] by means of which they arrived at the proofs nor does it reflect how proving the propositions was taught orally to a learner. The *Elements* is not the result of putting into writing any oral processes of knowledge acquisition and communication; it is, from the first, a written enterprise, predicated by writing.[81] The resulting text does not imitate oral lectures for the sake of knowledge communication;[82] its own strategy of knowledge communication consists in presenting only the acquired knowledge.

---

other way around, Asper 2011c, 307 f.), however, by doing violence to the mathematical examples in order to fit them to his theory (Kullmann 1974, 348 f. and 1981 *passim*).

77 I.e., by Euclid or one of his predecessors. As we have seen in the preceding section (p. 8 with n. 45), his work with the *Elements* presumably only partly consisted in finding new proofs but also in rearranging and rewriting the proposition corpus already available to him.

78 This has a cognitive aspect and a social one: on the one hand, only by accomplishing the proof in writing will one know whether it really works out and realize possible inconsistencies (cf. Asper 2007, 121 with reference to Holmes 1987, 226 f. and Heintz 2000, 169). On the other hand, only the written version may be made available to others who will be able to confirm the validity of the proof: according to consensus theory, it is only then that a proposition can be considered proved (for the role of consensus in modern mathematics cf. Heintz 2000, 178).

79 Though probably not the one which would have been part of the later text (if at all); cf. Asper 2007, 120 with reference to Arist. *Metaph.* Θ 9.1051a21–33 for the discovery of the proof by means of a diagram.

80 Mueller 1969, 295.

81 This emerges not only from the presentation of the proof rather than of how it was reached but also from the extreme textual standardization, to which I will immediately turn. Cf. Asper 2007, 119–122.

82 Cf. Asper 2007, 121 f., for such kind of decontextualized knowledge communication in the Pap. Rhind.

The *Elements* is not just a *collection* of theorems and problems;[83] the result of the kind of knowledge acquisition which the *Elements* fixates entails in itself a kind of process too: that of arriving at the conclusion from the πρότασις. For the result is not the truth of a claim but its proof. Therefore, the method of the proof (how it is applied in the individual case of the proposition) is part of the result; the proof method is at the same time the object of the description and the (practice of the) description itself.[84] In this way, the proof method seems to be part both of the content and form of the *Elements* – however, it is part of the content only *as form*, that is, as the 'logical' form which is standardly applied, but not because it is an intrinsic part of geometry. The 'logical' form only seems to be inextricably linked with the geometrical content because it has been canonized and has come to be regarded as the only possible way not only to talk about mathematics, especially geometry, but also to 'do' mathematics. But in fact, the 'logical' form is contingent because technically geometry does not have to be done and described in terms of it (even if this is hard to imagine from our own pre-shaped 'Euclidean' point of view). Only (some of) the *individual* features of the 'logical' form seem to be indispensable, as they are a necessary part of the chosen method; changing them means changing the whole method.

I regard as features of the 'logical' form the characteristics of the text as it can be described from the point of view of the proof method. By this I mean that these features can be pointed to in the text, that is, as words, sentences, or parts of the proof, but that their designations describe the 'logical'-methodical function which those parts of the text have as part of the proof. This means that one cannot relate their designations to those features unless one is familiar with the method and its (later) terminology. By contrast, by literary or 'textual' features I mean the characteristics of the text as regarded from a linguistic, stylistic, or grammatical point of view.[85] Those features can be described without making reference to the proof at all, but simply by drawing on conventional literary categories of describing texts of whatever genre. The textual features are wholly contingent,[86] that is, they do not bear any intrinsic

83 The papyrological evidence for the *Elements* shows that such collections of the propositions of the *Elements* (along with the respective diagrams) once existed, too, but these must have been accomplished on the basis of the *Elements*, which contained the complete proofs, maybe for the sake of training the proof method or memorizing the proofs (cf. Section 4.1). Another case is that of mathematicians who appear to have sent only propositions to fellow mathematicians in order for them to prove the propositions first, before sending the proof as well.

84 Cf. Asper 2007, 121: "die Praxis des Beweisens besteht aus der *Beschreibung* eines Beweises."

85 The same kind of distinction underlies Netz's threefold distinction of "the level of mathematical contents," "the level of logical form," and "the level of form, of presentation [. . .] style" (1999b, 282). A similar distinction between the linguistic and the 'logical' side is made by Acerbi's differentiation of the "levels" on which the structure may be described: "linguistic, logical, mathematical" (2010, 88). Again, the logical level takes a middle position between the textual form and the content, but it is not clear whether Acerbi regards it as part of the 'form' as well.

86 Asper 2007, 211.

relation to geometry or the 'logic' of axiomatic-deductive proofs. However, again, as in the case of the 'logical' features of the form, they might *seem* to be a natural part of doing geometry because they have proved so successful that they have continued to be used until today. I shall now first describe in some detail the features of the 'logical' form, before I briefly outline the 'textual' or 'literary' form which I am mainly interested in, and which I analyze comprehensively in Chapter 3. Here I shall concentrate on the geometrical propositions of the *Elements*.[87]

Features of the 'logical' form

Methodically, the proof is presented, on the one hand, so as to be achieved by reasoning from unprovable principles and proved claims, that is, it is axiomatic-deductive; on the other hand, it is based on a diagram. I consider first the ramification of the axiomatic-deductive method for the 'logical' form of the *Elements*. The work standardly applies the method in order to prove general propositions logically. The main components of this method are certain evident and unprovable assumptions (axioms)[88] and propositions: these consist of the proposition proper (πρότασις), which is either a theorem (θεώρημα) or a problem (πρόβλημα) of construction,[89] and of a proof along with a lettered diagram on the basis of which the theorem is proved or the problem is solved.[90] The proof is accomplished axiomatic-deductively, i.e., by means of reasoning from certain premises to the proposition. It deduces the proposition from the presupposed 'axioms' and from previously proved propositions.[91]

The proof section as a whole is not exclusively deductive. This is because the proof is accomplished with respect to a particular diagrammatic case; for this pur-

87 This is because the main focus of my study will be on book I of Euclid's *Elements*, which comprises only geometrical propositions. The form of arithmetical propositions of books VII–IX largely coincides with that of the geometrical ones, but differs in a few important ways.

88 I use the term 'axioms' here in a wide sense as comprising all of the three types of principles which may appear at the beginning of the books of the *Elements* ('definitions,' 'postulates,' and 'axioms' proper). This is indicated by the fact that all of them can function as axioms (for which see below p. 20). I explain their names, originality, and function a little further below (pp. 19, n. 108 and 20–23), and, for the time being, refer to them invariantly as 'axioms,' '(unprovable) assumptions,' and '(first/unprovable) principles.'

89 'Proposition' has these two senses: the πρότασις which is to be proven, and the unit which comprises the πρότασις and the proof. I refer to propositions whose *protasis* is a theorem as 'theorem propositions,' and to propositions whose *protasis* is a problem as 'problem propositions.'

90 By 'proof,' I mean here the whole part of the proposition by which the πρότασις is proved, not only the 'proof proper,' that is, the section of the proposition text which accomplishes the proof (cf. Netz 1999a, 255 f.).

91 The *Elements* is, however, not axiomatic-deductive in the modern, Hilbertian sense of the term (Asper 2007, 98), and has received much criticism by Hilbert and others for the logical gaps which it has according to their criteria.

pose, the text first has to set up the diagram and formulate a certain aim with respect to it. The corresponding parts of the text preceding the 'proof proper' contain some of the assumptions from which the proposition is subsequently deduced; further relevant assumptions and previously proved propositions are mentioned as part of the deduction itself. In fact, the 'proof proper,' which performs the deduction of the proposition,[92] consists only of assumptions and propositions which are deductively related to one another. The logical relations are indicated by specific standardized particles which work as "logical connectors."[93] The proof is directed almost exclusively forward, that is, the text always justifies what comes next, not what comes before.[94] Thus, the propositions are strictly focused on preparing and accomplishing the proof. They do not contain any metamathematical remarks or contextual information as to the position of the proposition within the work, the choice of proof (which is often only one of several possible ones), or earlier handlings of the propositions, additional explanations, qualifications, explicit cross-references to the preceding propositions which are used in the deduction, or (partial) renditions of preceding proofs.[95] Everything seems to be subordinated to the 'logical' structure.

The axiomatic-deductive method (and the concomitant microstructure of the propositions) has consequences for the 'logical' macrostructure of the work: the *Elements* is a "Satzsystem,"[96] that is, the axioms and propositions are not merely listed but arranged systematically "according to their logical status:" the axioms come first because they form the basis of all subsequent propositions. Then, the propositions are deduced in such a way that the succeeding propositions presuppose the preceding ones.[97] Thus, the propositions each form "discrete" 'logical' units,[98] which, at the same time, are chained by means of the deductive proofs and form part of a larger deductive structure. From the implicitness of the individual propositions regarding the axioms and propositions from which they are deduced, it follows that the 'logical' macrostructure is likewise an implicit one. The underlying assumptions and propositions are 'quoted,' but the fact that they have already been stated or proved, and where, is not mentioned.[99]

---

92 Strictly speaking, it deduces the previously formulated *n*, that is, the concrete formulation of the claim of the πρότασις (see further below pp. 23–27 on the influence of the diagram on the 'logical' structure of the proposition and on the problem of the generalizability of the concrete, diagrammatic proof).

93 Netz 1999a, 115–118.

94 Netz 1999a, 203.

95 The commentators of the text readily use the proof-centeredness and resulting implicitness of the work to supply information on all of these subjects (cf. Section 2.3.1).

96 Asper 2007, 98.

97 Asper 2007, 95 and 99.

98 On the feature of discreteness from a textual point of view Asper 2007, 114–116.

99 That is, 'quoted' only in the ancient sense of the practice, which did not formally indicate the quotation nor name its source.

The deductive “implication structure”[100] entails that the propositions tend to be ordered according to increasing complexity and degree of implication. The first proposition of the work (*Elem.* I 1) is only deduced from unprovable principles; *Elem.* I 2 is deduced from I 1 (and unprovable principles), *Elem.* I 3 from I 2 (and unprovable principles); in this way, the subsequent propositions become more complex by necessitating themselves propositions of increasing complexity.[101] That is, the propositions are presented largely in the inverted order of the deductive chain: from what is evident to simple to complex propositions.[102] However, the deductive macrostructure is not as linear as the example of the first three propositions of the work suggests. Not only the immediately preceding proposition is presupposed (and in this way implicitly all which are presupposed by that one). But just as the unprovable assumptions are at the disposal of the proofs of all propositions, the proved propositions, too, are explicitly applied again and again in the work.[103] In this way, the propositions overall tend to become ever more complex by presupposing more and more of the foregoing propositions (in addition to several first principles), but this development is only palpable in the synopsis of the whole work; immediately succeeding propositions do not regularly presuppose more propositions than those before, but just as often the same number or fewer ones. Moreover, the propositions do not always build on the immediately foregoing one; they may go back to a much earlier proposition, though there is a tendency to deduce a proposition from *one of* those which have just been proved.

These deviations from a strictly linear deductive structure are due in some cases to inconsistencies, but above all, they are the result of the thematic organization of the work. The *Elements* does not aim at one single complex construction or theorem, to which each proposition would contribute; rather, they present the available knowledge in the two main fields of geometry and arithmetic according to certain subthemes. This means that, besides the deductive structure, there is also a “competing” thematic structural principle.[104] On the one hand, individual books or groups of books have common subjects: books I–VI are on surface geometry, VII–IX on arithmetic, X on incommensurability and irrationality, and XI–XIII on solid geometry; on the other hand, within and across the boundaries of the individual books, propositions are thematically grouped together. The beginning of these thematic units mostly interrupts the overall deductive chain to varying degrees either by switching to the treatment of a completely different subject (e.g., the change from geometry to arithmetic) or by beginning to focus on a different geometrical object with a basic proposition, which presupposes little, of the respective thematic field (e.g., *Elem.* I 27, starting a sequence on

100 Asper 2007, 105.
101 Cf. Acerbi 2007, 2543–2559.
102 Asper 2007, 115.
103 By ‘explicitly’ I mean that the text of a proved proposition is used as part of the proof; the reference itself is however always implicit.
104 Asper 2007, 99, n. 37 with reference to Mueller 1981, 204.

pairs of straight lines and relying only on *Elem.* I 16). The thematic group may constitute a deductive structure, a "Satzsystem," which fits into the overall "Satzsystem," even though it is not connected to the immediately preceding propositions: but it presupposes some of the earlier material and is itself presupposed by what comes after it. However, not each group of propositions is in itself deductive in a strictly linear way: sometimes, the propositions present variations of a theme and are thus based on the same preceding proposition rather than forming a deductive structure of rising complexity (e.g., *Elem.* I 33 and I 34 which prove different theorems by means of the same diagram and are both deduced from *Elem.* I 29 amongst others, rather than from *Elem.* I 32 and 33, respectively). Thus, the "Satzsystem" of the *Elements* itself consists of several differently sized and overlapping "Satzsysteme"[105] which are determined by the overlap and mutual subversion of the thematic and 'logical'-deductive principle. The overarching structure is in any case supposed to be axiomatic-deductive: every proof is accomplished by deduction from unprovable assumptions or propositions proved earlier in the work (no matter at what point), and whatever is stated (as assumption) or proved is presupposed at a later stage of the deduction.

However, critique in modern times, above all by Hilbert, has been of the "unarticulated assumptions," that is, assumptions which are necessitated by the proof but have not been included in the preliminary axiomatic lists; they have to be read from the diagram (I will explain its role for the proof shortly).[106] This brings us to the form and status of the list of axioms in the 'logical' structure of the *Elements*. There are three types of axiomatic principles (ἀρχαί) which are listed at the beginning of some of the books of the *Elements*: definitions (ὅροι), specifically of the geometrical terms, which are supplied at the beginning of nine of the 13 books: I, II, III, IV, V, VI, VII, X, and XI, that is, whenever books on different subjects require further definitions;[107] and postulates (αἰτήματα) and axioms (κοιναὶ ἔννοιαι), each provided only at the beginning of book I, that is, of the whole work.[108]

105 Asper 2014, 463.

106 Cf., e.g., already I 1 for such an "implicit assumption," i.e., that the two circles meet (Netz 1999a, 27). It is precisely the aim of Hilberts *Grundlagen der Geometrie* (1899) "to rewrite geometry without any unarticulated assumptions" (ibid., 26). He assumes 20 in total compared with nine which have been transmitted as part of Euclid's *Elements* (cf. below for issues of authenticity).

107 Book X in fact contains two further lists of definitions in addition to the one at the beginning. The lists of books VII and XI do not just precede respectively these very books themselves, but also those which follow upon each of them and are *not* introduced by introductory lists; that is, VII–IX are headed by the list at the beginning of VII and the list of XI introduces XI–XIII. This means that the lists indicate the (thematic) structure of the work which I have just outlined above (p. 18, see Waschkies 1995, 105).

108 The Greek names were probably coined later (Waschkies 1995, 105–111 and Acerbi 2013, 680 for the term αἰτήματα). On the relation of the terms κοιναὶ ἔννοιαι and ἀξιώματα cf. von Fritz 1955, 43–63.

The borders between the three types are fuzzy.[109] Axioms are certain general unprovable assumptions, that is to say they are not related to a specific geometrical object, e.g., axiom 1: "Things which are equal to the same thing are also equal to one another."[110] What remains as 'postulates' falls into two main groups: the first three of them are "rules licensing basic constructions" (e.g., postulate 1: "Let it be postulated to draw a straight line from any point to any point") and "the fourth and the fifth are assertions stating properties of particular geometric objects" (e.g., postulate 4: "That all right angles are equal to one another").[111] In a way, they work as axioms as well because they assume that objects are or can be constructed, and that particular objects have certain properties. Particularly with regard to the second-type postulates, the distinction between the two types is not clear; some statements have been classified once as axioms, another time as postulates, and some vice versa.[112] The definitions, too, can function as axioms because they "do not settle linguistic usage but geometrical propriety: they set out when a property, independently understood, is considered to hold."[113] For example, the first definition "A point is that which has no part" (Σημεῖόν ἐστιν, οὗ μέρος οὐθέν)[114] is of the geometrical concept of the object 'point,' not σημεῖον, the term used for it, and it assumes that the property of 'having no part' holds for the object 'point.'

In themselves, the lists are 'logically' arranged; their structure in fact resembles that of the following main body of the work: they are ordered systematically according to increasing complexity and thematic substructures within which logical relations are indicated by means of particles.[115] Otherwise, they stand out from the rest of the work by their form because they explicitly standardize a number of terms and assumptions. By contrast, the terms and *Sätze* of the proposition corpus are only implicitly standardized, that is, by their being repeatedly used in textually standardized

---

109 Cf. Netz 1999a, 95 on the "[f]uzziness between 'definition' and 'axiom,'" Thaer 1970, 419 notes: "die Grenze zwischen Axiomen und Postulaten fließt."

110 Τὰ τῷ αὐτῷ ἴσα καὶ ἀλλήλοις ἐστὶν ἴσα (*Elem.* I, Ax. 1 [vol. 1, p. 5.9 Heiberg & Stamatis]).

111 Acerbi 2013, 680 with reference to Procl. *In Eucl.* 182–184 and 188–193 Friedlein. Postulate 1: Ἠιτήσθω ἀπὸ παντὸς σημείου ἐπὶ πᾶν σημεῖον εὐθεῖαν γραμμὴν ἀγαγεῖν (*Elem.* I, Post. 1 [vol. 1, p. 4.14 f. Heiberg & Stamatis]). Postulate 4: (Sc. Ἠιτήσθω) Καὶ πάσας τὰς ὀρθὰς γωνίας ἴσας ἀλλήλαις εἶναι (*Elem.* I, Post. 4 [vol. 1, p. 5.3 Heiberg & Stamatis]).

112 Thaer 1970, 419.

113 Cf. Netz 1999a, 102 f. with reference to Mueller 1991.

114 *Elem.* I, Def. 1 (vol. 1, p. 1.1 Heiberg & Stamatis).

115 Asper 2007, 117 with nn. 164 and 167. In view of this use Netz 1999a, 94 emphasizes that the definitions are actually not a list (as is suggested by the anachronistic numbering in the modern editions), but a continuous piece of prose. However, the definitions can still be regarded as a list (the items of which are in part connected by recurring particles) because they move from object to object, and the statements about the different objects are, except by particles, unconnected (on the characteristics of literary lists cf. Mainberger 2003, 6–13). The same holds for the other principles.

form.[116] One might suppose that the explicitness of the first principles is due to their presumed 'logical' function at the beginning of the work (and subsequent books): they seem to work as explicit starting points of the deduction which are to provide what the proofs presuppose.

Interestingly, however, the axiomatic lists do not really provide what the proofs presuppose. The definitions of the geometrical terms and concepts hardly delimit all recurring terms systematically, but rather only cover a very small part of the terms which are actually in use;[117] and these few definitions even include such terms of all things which are not presupposed by the propositions.[118] Moreover, the example of the first definition of the "point" as "that which has no part" shows that many of the definitions do not really define. They do not disambiguate the concepts – neither for readers who do not know them nor as a 'logical' basis for the proofs. The term 'point' is not commonly understood and does not seem to be presupposed specifically as having the meaning "that which has no part."[119] Thus, it appears that a least a certain number of the definitions do not actually have a 'logical' (axiomatic) function with regard to the subsequent deduction.[120] Accordingly, Euclid does not, for the most part, refer to his definitions.[121] The information about geometrical terms which is necessitated by the proofs is either taken from the diagram or presupposed as evident. Similarly, the axioms and postulates explicated at the beginning of book I cover only a few

116 Conversely, the introductory lists, which standardize explicitly, are themselves not standardized by their language and style (cf. below, p. 22).

117 Netz 1999a, 96, 99, and 103.

118 E.g., of *Elem.* I, Def. 22, ἑτερόμηκες ("oblong"), ῥόμβος ("rhombus"), and ῥομβοειδές ("rhomboid"), cf. Heath 1926, 151.

119 Cf. Thaer 1970, 417 on this definition: "Euklid will höchstens abgrenzen, was an sich bereits existiert. Aber auch dies tut er meist nur so, daß er die Anschauung schon voraussetzt, bloß einzelne Merkmale hervorhebt, wie in Def. 1 vom Punkt die Unteilbarkeit. Wer nicht weiß, was ein Punkt ist, wird es aus Euklids Definition nicht lernen." It is for this reason that the purpose of the definitions does not seem to be "disambiguation" and not because they do not aim toward completeness (Netz 1999a, 102). Nevertheless, the definitions' statements about the properties of geometrical *concepts* may have a disambiguating effect on the language of the propositions (Netz 1999a, 103), as they connect one term to the concept which they define prominently at the beginning of the work (cf. the end of Detel's remark in the following note).

120 Cf. Detel 1993, 217: "In der Tat kann man sich nur schwer vorstellen, was man in einem euklidischen Beweis etwa mit der Definition 'Dreieck: = zweidimensionale Figur, die durch drei Geraden begrenzt ist' anfangen sollte. In einem fundamentalen Sinne muß der beweisende Geometer natürlich immer schon *wissen*, daß ein Dreieck eine derartige Figur ist; aber diese dihairetische Definition ist zu informationsarm, als daß sich aus ihr vieles folgern ließe: sie expliziert ja wirklich nicht viel mehr als gerade den *Namen Dreieck* – das, was sozusagen jede Person darunter versteht." Netz 1999a, 96 and 103 thinks they are part of second-order discourse originating from an interest in speaking about mathematics.

121 In *Elem.* I, Euclid uses I, Def. 10 and 15 twice and I, Def. 23 once. Detel 1993, 216 with reference to Heath; cf. also Thaer 1970, 418 f.

of the assumptions which serve as "starting points" for the proofs of the *Elements*,[122] and the proofs, again, rarely refer to them.[123]

In view of the little overlap between the *Elements*' axiomatic principles and what the propositions actually presuppose, and of their explicitness as well as their linguistic non-standardization, one might question their originality. Indeed, it is unlikely that they had been part of the generic features from the beginning.[124] As regards the axioms proper, four out of the nine transmitted in the manuscripts have been commonly deemed unoriginal, as they explicate some of the many implicit assumptions of the work – which apparently were not assumed as explicit starting points. Of the remaining five, I, Ax. 1–3 are recognized as Euclidean by Hero of Alexandria (first century AD)[125] and I, Ax. 4 and 5 (originally 7 and 8) only by Proclus.[126] But it is possible that even those had not been part of the Euclidean text;[127] it could be argued that their explication at the beginning was not necessary because they are, by definition, evident[128] and for this reason merely assumed without being made explicit. In view of the general tendency of 'deuteronomic' texts to explicate,[129] the axioms might have been made explicit on the basis of the propositions only in the aftermath, which is possible precisely because they are explicitly mentioned by the propositions.[130] The same could apply to the postulates because of their proximity to the axioms. As far as the definitions are concerned, the first 10 are attested as early as the third century AD on papyrus.[131]

122 Netz 1999a, 96.
123 Netz 1999a, 182 f.
124 Asper 2007, 116, n. 163. The proofs were, in any case, developed prior to the definitions (ibid., 175).
125 Heath 1926, 222 thinks that the first three axioms were already known to Aristotle and therefore part of pre-Euclidean elements on the grounds that Aristotle several times quotes the third axiom and refers to 'axioms' in the plural. However, although Aristotle distinguishes kinds of ἀρχαί in different passages of his *Posterior Analytics* (*An. post.* I 2.72a14–22, I 7.75a39–b2, I 10.76b11–77a4), his distinctions do not compare to the mathematical ones. Cf. Mueller 1979, 104 f., Kullmann 1981 and Kahn 1981, 388–393, referred to by Detel 1993, 223 as opponents of the "traditional view" put forward by Heath 1926, 117–124 that Aristotle's terms ὅρος/ὁρισμός and ἀξίωμα match the Euclidean ones, and that only the mathematical postulates are not matched by the Aristotelian ὑπόθεσις; cf. also von Fritz 1955; cf. more recently Acerbi 2013 with a new argument for the mismatch of the mathematical and Aristotelian distinction.
126 Thaer 1970, 420.
127 Acerbi 2013, 685 holds that "none of the [i.e., axiomatic] rules was made explicit as a principle in any mathematical writing," that is, despite Aristotle's reference to what is transmitted as Euclid's I, Ax. 3.
128 Cf. Thaer's translation of the later term κοιναὶ ἔννοιαι: "allgemein Eingesehenes," and his explication which emphasizes the evidential character: "ein allgemein logischer Grundsatz, den kein Vernünftiger, auch wenn er von Geometrie nichts weiß, bestreitet" (1970, 420).
129 Netz 1998, 263–268; for the term cf. p. 9, n. 52.
130 Thaer 1970, 420 notes on the axioms: "ja es besteht die Möglichkeit, daß Euklid solche überhaupt nicht aufgestellt hat."
131 P. Mich. III 143 (cf. Section 4.1). Heath 1926, 151 argues that those of Euclid's definitions which do not recur in the propositions stem from pre-Euclidean elements texts.

The occupation with scientific definitions in general is attested – outside mathematics – only from the end of the second half of the fourth century BC, that is, they were still rather new in Euclid's own time.[132]

The second methodical aspect which determines the 'logical' form of the work is the diagram. Each proof is based on a diagram which is attached to the text through letters which designate points in the diagram and feature in the references to the geometrical objects in the text. The diagram is intended to present a particular case of the general proposition; it is this particular case which is proved and then generalized. Thus, the general proposition must be concretized in terms of the particular case of the diagram in order to be proved, and, after the proposition has been proved in particular terms, generalized back again. This twofold transition is accomplished by means of a number of highly formalized proof steps:[133] after the πρότασις, which gives the general proposition ('if this is the situation, then this is true'),[134] the ἔκθεσις sets out the particular situation, and the διορισμός asserts the truth of the general claim in the particular situation which has been set out. The διορισμός is then proved by the construction (κατασκευή) and proof proper (ἀπόδειξις) of the particular case, at the end of which the truth of the claim in the particular situation is established. Finally, the particular conclusion is generalized (συμπέρασμα), and the concluding formula *quod erat demonstrandum* confirms that the συμπέρασμα is exactly the same as the πρότασις.[135]

---

132 Asper 2007, 175 with reference also to Aristotle's discussion of definitions in *An. post.* II 3–13.
133 Cf. Section 3.4.1, 176–178.
134 Cf. Netz 1999a, 253–257. The terms for the parts of the proof are derived from Proclus' systematization of the parts of the geometrical (theorem) proposition, which, he claims, applies to "every problem and theorem" (*In Eucl.* 203.1–2 Friedlein), but actually presents an ideal developed from the theorem propositions of *Elements* I (Netz 1999b, 282); that is, it does "not exactly" fit the problem propositions of *Elements* I, and only "roughly" the whole work of the *Elements*; but it cannot be used to describe other elements texts adequately; for deviations of non-Euclidean elements texts from the Proclean structural scheme cf. Netz 1999b, 292–294 and 302. While I use Proclus' terminology here to describe the 'logical' form of the proposition, I shall discuss the scheme and its usefulness in Section 3.4.1, 178–182 from the point of view of *textual* (rather than 'logical') structure. Netz in his 1999b classifies it as part of the "stylistic" (i.e., in our terminology, "textual") rather than 'logical' form to begin with. It certainly can be used to describe the structural form from either perspective, but the terminology is 'logical' rather than textual, as it refers to the 'logic' of the proof.
135 The συμπέρασμα is rare outside of Euclid, and even in the *Elements*, it is not applied systematically; for this reason, Acerbi 2003, 199–201 believes that the fact that it nevertheless features extensively in the *Elements* is a consequence of later revisions. It is in any case conspicuous that the συμπέρασμα is much rarer in problem than in theorem propositions. This could be because their πρότασις, which orders a general construction problem, is already 'proved' if a particular case of it has been constructed. By contrast, in theorem propositions, the πρότασις can only be considered proved if it is proved to be generally true. Surely, the proof of problem propositions was considered to be generalizable, too (which can be seen from its being characterized by the very formal features which, as I will show, are used to suggest generalizability); only it seems to have been less obvious to

This means that the claim of the proposition recurs in total four times, twice as the *demonstrandum* (first in general form as the πρότασις, second in concrete form as the 'ἔκθεσις and διορισμός'), twice as the conclusion of the proof (first in concrete form at the end of the ἀπόδειξις, second in general form as the συμπέρασμα). The concrete conclusion exactly repeats the concrete *demonstrandum* (the διορισμός), and the general conclusion (συμπέρασμα) the general claim (πρότασις).[136] The purpose of this multiple repetition which results in a double ring structure appears to be bridging the gap between the concrete and the general.[137] For the main problem is how the particular conclusion can be generalized.[138] The 'logical' structure alone, however, does not account for the generalizability of the particular case.[139] This is also clear from the fact that the structure is more varied than the systematization suggest; it would have to be strictly standardized if generalizability depended on the structure.

The two features of the 'logical' form by which the text appears to establish the generalizability of the proved cases and which occur without exception are the generality of the terms ('line') and operations ('to draw a line') of the proposition and the use of diagrammatic letters for describing the particular case.

On the one hand, the same kind of general terms and operations are applied invariantly in both the abstract and the concrete parts of the proposition, and those abstract and concrete parts which are repetitions of one another even feature the exact same general terms. In fact, the repetition of the πρότασις and conclusion is only enabled by the use of general terms. In this respect, the use of standardized language plays a decisive role because it makes for the generality of terms and operations.[140] The particular

claim its generalizability on the basis of the proof. It seems to be for similar reasons that the συμπέρασμα is completely absent from arithmetical propositions (cf. Netz 1999a, 169).

136 Netz 1999a, 253.

137 Asper 2007, 118.

138 Mueller 1981, 13: "How can one move from an argument based upon a particular example to a general conclusion, from an argument about the straight line AB to a conclusion about any straight line?" Cf. also Netz 1999a, 255.

139 Netz claims precisely that, based on a formalization of the proposition structure in logical terms (1999a, 255–257): his main argument is that the διορισμός is shown to be provable on the basis of the ἔκθεσις alone, on the one hand because the affirmative formulation of the διορισμός itself indicated this, and because the following parts κατασκευή and proof are only a "footnote" to the διορισμός (rather than a proof of the πρότασις, which is "taken to be proved by the combination of *ekthesis and diorismos*," 256), proving that its claim is true and that it is indeed provable on the basis of the ἔκθεσις alone. According to Netz, the generalization of the particular proof (not of the particular proof result) is guaranteed by the proof's being based on the ἔκθεσις alone: "The same proof must be repeatable for any other object as long as the same ἔκθεσις applies to that object" (256 f.). However, one must add to this that the repeatability of the proof crucially depends on the use of general terms in the whole proposition (cf. my above ensuing argument).

140 Acerbi 2010, 21–56. One of the main questions of my book is precisely to what extent the standardization of the language exhausts itself in the function of establishing generality and guaranteeing generalizability.

case is established precisely through the use of the general terms and operations of the proposition; the invariant repetition of the general terms and operations of the πρότασις constitutes the particular case in the first place – and constitutes it as one case of the general proposition. Conversely, the use of general terms and operations throughout the proof of the particular case and the formulation of the particular conclusion by means of the general terms of the πρότασις suggest that this particular conclusion can be generalized. Finally, the repetition of the general terms of the particular conclusion in the general conclusion, which, in turn, repeats the general terms of the πρότασις, confirms the generalizability of the particular conclusion. Further, the use of general terms and operations throughout suggests the generalizability of the particular case through its repeatability: in the proof of the particular case, everything except for the letters is put in the same general terms of the propositions; therefore, if the particulars, the letters, are changed, the proof remains valid; it is repeatable and therefore generalizable.[141]

The other way in which the propositions indicate the generalizability of their result is by means of their use of letters for referring to the particular case. In fact, the particular case which is proved is not really particular to begin with; for the use of the letters as placeholders for exemplary particular numbers (rather than numbers themselves) already implies the generalization of the seemingly particular case which is proved.[142]

However, while the letters themselves suggest generality rather than particularity, that to which the letters refer to, the implementation of the diagram, is inevitably concrete. This is the case regardless of whether the diagram actually accompanied the text or whether the reader had to draw the diagram from the text:[143] if it came with the text, it would present a single one of all the particular cases to which the letters could refer (e.g., a particular line AB with a particular length) and prompt the question whether there was something particular about it.[144] If it had to be drawn subsequently, a line AB could be drawn as any line AB, but once the diagram was drawn, and the proposition proved on its basis, the question arises whether another implementation of the diagram according to the proposition text also presents a case covered by the proof.

This shows that, despite the generality of the terms and letters, the case which is proved is in any case a particular one which is (merely) "assumed to be generalizable.

141 Netz 1999a, 270: "Greek generality derives from repeatability."

142 Asper 2007, 158 f.

143 This is the central question of my fourth chapter. The ancient diagrams for the *Elements* are not extant (except for very few papyrus fragments whose relation to the *Elements* is unclear). There are strong indications that the diagrams were indeed included in the text, but since I analyze the standardization of all textual features, I do not simply presuppose their existence.

144 I shall discuss drawing conventions and relevance of the diagrams and their style for the proof in Chapter 4.

In the absence of something like the rules of logic there is no uniform procedure for checking the correctness of this assumption in individual cases."[145] This is not to question the validity of proofs of the *Elements*: the text suggests their generalizability not only by the generality of its formal features but also by deducing from propositions (which are proved with the aid of particular proofs and particular diagrams) other propositions whose particular proofs use different diagrams.[146] The generalizability of the proofs was apparently accepted in the context of Greek mathematics; therefore, they are in any case generalizable and true by convention. Anyhow, there remains a certain tension between the particularity of the proof case and the generality of the claim, and this tension is felt precisely in the strong effort to produce generality in the text by means of the general terms and their repetition.

In view of the difficulty of proving a general proposition by means of a particular case of it, the question simply is why the proof refers to a concrete (that is, lettered) diagram at all. Of course, in order to be able to *discover* the proof, a diagram is indispensable,[147] but this does not necessarily mean that *writing* the proof requires referring to one as well. For discovering and writing the proof are two completely different acts, and the *Elements* in any case does not reproduce the process of finding the proof – this would probably have involved even more material which is not part of the work: trial diagrams, notes, and dead ends. Rather, it codifies (the result of) the successful proof itself.[148] Neither does the work originate in (semi-)oral teaching of geometry with the help of diagrams because the strong stylistic standardization indicates a written conception of the work.[149] From the point of view of proof 'logic,' it is technically not necessary to refer to a particular lettered diagram. The proofs would be just as valid and even more straightforward if they were put in general terms without having to revert to a concrete level. It is, however, from a communicative point of view that it is necessary to make use of diagrammatic letters. For while it is conceivable, though somewhat cumbersome, to write some of the easier propositions of the *Elements* without referring to a lettered diagram, more complex propositions do not seem to be communicable in writing without using letters. Only by the letters is the work able to be textually autonomous.[150] Therefore, one might assume that the 'logical' proposition structure which is due to referencing a lettered diagram does not have any intrinsically logical motivation, but rather serves communicative ends.

Whatever its original motivation, the reference to a lettered diagram came to be standardized, and the diagram appears to have assumed both 'logical' and communicative functions. It visualizes the relations of the abstract terms of the proposition

---

145 Mueller 1981, 11.
146 Cf. Netz 1999a, 269.
147 Netz 1999a, 13.
148 Cf. p. 14 above.
149 Asper 2007, 118.
150 Cf. Asper 2007, 103 for the letters as guaranteeing "textuelle Autonomie."

and the abstract geometrical operations performed with the general objects and, in this way, makes it possible in the first place to deduce general propositions from other propositions in writing.[151] Some of the assumptions and conclusions even appear to be visible *only* in the diagram.[152] This suggests that the diagram itself has 'logical' functions – that it does not just provide a basis for the textual proof.

Features of the textual form

The textual form is the concern of the present study.[153] No ancient study of the textual form is extant, and none seems to have existed.[154] I will focus on the *propositions* of the *Elements* (and here on the geometrical ones)[155] and not on the axiomatic lists which precede the propositions at the beginning of the work and some of the following books. The first reason for leaving them out is that they do not share in the characteristic textual standardization of the main body of the work, in which I am mainly interested.[156] Second, I do not consider them in my analysis of the form because their origin is (at least partly) unclear.

By 'textual' or 'literary' form, I mean all matters of language and style. These comprise all aspects of the text which can be described independently of the geometrical content and the axiomatic-deductive proof method, even though they seem to be inseparably connected with the mathematical subject.[157] If considered within the context of (Greek) mathematics, the stylistic features of the *Elements* do not astonish; they are generic features. What is more, they do not even suggest themselves in terms of style, that is, as a matter which can be considered independently from the geometrical proofs of which they seem to be an inevitable part. If, however, consciously regarded *as form* from a more general, literary point of view, within the scope of (Greek) literature, the language and style of the work appear to be extremely strange and unparalleled outside the genre. I focus on the standardization of the text as the main source of the

151 Cf. Asper 2007, 158 for the diagram as "überpersönliche Evidenztechnik."
152 Cf. Netz 1999a, 20–35 for the "mutual interdependence" of text and diagram and especially 26–29 for the role of the diagram for derivations.
153 For analyses of the textual features cf. Asper 2007, 114–135, Acerbi 2010, 26–55, Acerbi 2012, 169–189, Netz 1999a, and, by way of overview, Acerbi 2018, 269–271 (cf. Section 1.5).
154 Cf. Netz 1999a, 124.
155 Acerbi emphasizes that originally there has been not only one single style of Greek theoretical mathematics, but three "stylistic codes" ("the demonstrative, the procedural, and the algorithmic codes"), of which the "demonstrative code, in which all geometrical texts are written," only became the dominant one (2018, 269). He offers a detailed description of the features of each in his 2012, but in his 2018 overview of "The Styles of Hellenistic Mathematics" (269–271) confines himself to mentioning the three and describes only the demonstrative one.
156 Cf. below, p. 30.
157 Cf. pp. 15 f. above.

peculiarity of the *Elements*' form. I explain why this is the case by considering first the most apparent characteristics of the language and style and second their relation to standardization.

For the most part, the *Elements* makes use of ordinary words and simple syntax. Even the words used for referring to geometrical objects are mostly known from other contexts as well. But two *linguistic* features are unusual within the realm of Greek literature – at least their extensive use is: on the one hand, the characteristic third-person imperatives: ἔστω ("let there be") and the otherwise unusual third-person perfect passive imperatives such as προσεκβεβλήσθωσαν ("let there be drawn out further," e.g., in *Elem.* I 5); especially by their repetition, they appear ritualized and add to the strangeness of the text.[158] On the other hand, the Euclidean language stands out by its use of letters: single letters or combinations of letters form part of the formulaic phrases which, in the core section of the proposition, refer to particular geometrical objects in a (thereby invoked) lettered diagram. As the text constantly revolves around geometrical objects, their construction, and relation, its core section is full of letters and is distinct from all other sorts of text merely by its appearance. Moreover, the noun phrases which denote geometrical objects in a general way at the beginning and end of the proposition often become elliptic by losing the actual noun as a result of the insertion of the letters: the resulting phrases (e.g., τὸ Α [sc. σημεῖον], "the point A") are incomprehensible for those who encounter them for the first time.

The most conspicuous *stylistic* features are repetition and impersonality. First, repetition: every element of the text is repeated constantly throughout the text: letters, words, verb endings, phrases, parts of sentences, sentence constructions, parts of the proposition – both within a single proposition and across propositions – and even the whole proposition structure itself. No other text outside the elements genre is as repetitive. The constant repetition entails other figures of speech which are based on repetition: ring composition (of the structure of the proposition), homoioteleuton (in the case of verb endings such as that of the imperatives), and anaphora (in the case of particles which connect the statements in the proof). Second, impersonality: except for one first-person singular form (λέγω) which occurs in virtually all theorem propositions in the standardized phrase λέγω ὅτι (and occasional personal forms of δείκνυμι, likewise used in a standardized way), the text does not contain any 'I' or 'we,' nor any 'you,' that is, no personal remarks or address of any kind. It cultivates an absolutely impersonal, objectifying style.

Both the repetitive and the impersonal character of the text as well as the prominence of the above linguistic features (the imperatives and the letters) have to do

158 Cf. Asper 2007, 178 on the "solemn" third-person (often perfect) imperatives ("feierliche Imperative") of laws, the only other context in which comparable imperatives are used independently of mathematics (they are only a "Scheinparallele"). "Ritualized" is not meant in a religious sense; rather, I use it to describe the imperatives from a sociological perspective on the communication of the mathematicians.

with the extreme standardization of the text. Repetition is both the result and the means of standardization: the text does not simply repeat its features, but it almost invariantly repeats certain features *instead of others*. That is, it repeats those features which are to be standardized, and, in doing so, standardizes them.

Similarly, impersonality is both the means and the result of standardization:[159] on the one hand, the proposition text almost only applies third-person verb forms, even where it could use a form of personal address, most notably when demanding the construction of geometrical objects by means of the above-mentioned third-person passive imperatives; it exclusively talks about its objects, not about its author or addressee; thus, the text standardizes impersonality. On the other hand, the text becomes impersonal because of its very standardization: standardization by definition excludes the (however stylized) contingency, variance, and exceptionality which any self-representation of the author or personal coloring presuppose. Of course there are scientific texts which have a personal tone throughout, but this presupposes that the authorial voice is not completely standardized. Accordingly, even the λέγω ὅτι does not carry any personal meaning but rather also contributes to standardization and impersonality because it standardly appears in the same form and function. The standardization of the *Elements* makes its author vanish behind the text.[160]

In the same way, the conspicuous linguistic characterization of the textual form by the two above-named features is due to their standardization. Other genres have adopted them, that is, technical, astronomical, and geographical texts which apply mathematics, and even before Euclid's own time, Aristotle took them over as part of the 'mathematization' of his scientific theory. But the third-person imperatives and the diagrammatic letters occur nowhere near as standardly as in the 'Euclidean' propositions. What makes the rare imperatives even more unusual is their standardized and therefore repeated use at standardized places within almost each proposition. The same is true for the letters. Moreover, the geometrical nouns can be elided from the phrases in which the letters occur precisely because of the letters' standardized use, which is how the 'letter phrases' receive their remarkable appearance.[161] I have provided these examples of the characteristics of the *Elements*' language and style in order to show why I focus on standardization for determining the role and meaning of the textual form of the work: standardization is arguably the most characteristic quality of the Euclidean textual form which gives it its peculiar shape. The uniqueness of this text form within Greek literature is due most of all to the degree to which the form is standardized and, as we have seen, also to the kind of features which are standardized.

159 Cf. p. 2, n. 9 on the relation of standardization and impersonality as features of Greek theoretical mathematics.

160 Asper 2014, 456.

161 For the term cf. Section 3.2.2, 124 f.

By standardization, I mean, as defined above,[162] a comprehensive tendency toward formal unification. This unification is practiced by the text and takes the form of the exact repetition of the same textual features rather than of alternative ones.[163] Accordingly, I use the term 'standardization' to refer both to the process and to its result. The textual standardization entails that every feature tends to retain its meaning and textual function throughout the text, and conversely every function always tends to be fulfilled by the same feature. The standard correspondences are not stated explicitly anywhere in the text; the standardization of the textual form of the propositions is exclusively implicit. By contrast, the lists of definitions at the beginning of nine of the books provide an element of explicit standardization, as they explicitly connect some of the main geometrical terms with a definition of their meaning. This explicit standardization does not constitute a feature of the textual, but rather of the 'logical' form.[164] The definitions, in turn, are themselves not standardized, neither regarding their own existence nor regarding their own textual form, and, accordingly, do not standardize the text by their form (but by means of explicit statements *about* the terms, if at all).[165] Therefore, they are not relevant for my question of the meaning of the textual form,[166] except in the way that the existence of the definitions might have further enhanced the tendency of standardization.

The implicit, textual standardization of the *Elements* as well as of all other works of the genre is all-encompassing. It pertains to all levels of the textual form: to the lexical, grammatical (above all, morphological and syntactical), and to the structural level (both to the microstructural one of the proposition and to the macrostructural one of the books and of the whole work). Every item of the text tends to be standardized, and larger parts of the text are standardized by means of the standardization of subordinate features. Standardization also pertains to what may be called the diagrammatic level of the text form, insofar as a diagram is regularly evoked by diagrammatic letters, which themselves seem to be subject to standardization as far as their application is concerned.

Whether the 'textual' standardization also included that of the diagram itself and if so, how far its form was standardized, is uncertain, although the standardization of the diagram suggests itself in view of the standardization of all other parts of the text. As has been said in the preceding part on the 'logical' form of the *Elements*, the diagrams to which the letters refer enable accomplishing the proof in writing in the first place, and they have certain 'logical' functions. The question is whether the diagrams

162 P. 2.

163 Cf. more thoroughly my discussion of the concept of 'standardization' pp. 64 f. and 108–110.

164 Cf. above, pp. 20 f.

165 Cf. above, p. 27.

166 Even though I am not going to compare the axiomatic lists with the propositions in terms of standardization, the axioms usefully provide evidence for linguistic and stylistic non-standardization in mathematics against which the extreme standardization of the main body stands out.

do so simply by being implied by the letters which enable their construction, or whether they need to be present as part of the proposition in the form of lettered drawings from the beginning. In other words: does the text presuppose the presence of a diagram or does it not?

This prompts the question of how the proofs came into being (as distinct from the question of how they were conceived and drafted). This is what either possibility suggests in terms of how the propositions were written and when the diagram became part of the text: if, on the one hand, the text presupposes a diagram, that is, if it does not work or if it (or any feature of it) cannot be understood without it being available from the beginning,[167] it can be assumed that the text was written with the finished, lettered diagram at hand as a necessary visual reference.[168] In this case, it is likely that the diagram was included in the proposition text by the writer because he must have been aware that he relied on the diagram and that any reader will need to do so, too. This is true unless the writer deliberately tried to make the text not depend on the presence of the diagram, even though writing on the basis of it, but cognitively this is nearly impossible.[169] If, on the other hand, the text does not presuppose a diagram, i.e., if it is self-sufficient, this suggests that the writer drew and lettered the diagram simultaneously with writing the text. In this case, he could have included the diagram in the text once it was accomplished; or he could have left it out, e.g., for pedagogical reasons: whoever reads the text has to accomplish the diagram himself. This means that only in the first case, if at all, can we draw conclusions about whether the diagram was included in the text. Indeed, there is indication that the text presupposes the diagram, and its presence is generally assumed. However, the question is how standardly this is the case. Moreover, the same considerations apply to the role of the form of the diagram: whether a certain kind of form was generally necessitated and therefore can be assumed to be standardized. To be clear: all we can determine by considering the 'logical' form, if at all, is whether the text was written on the basis of a diagram, not whether this diagram was also included in the text: the *textual* standardization of the diagram (that is, its standardization as a feature of the proposition text) does not necessarily follow from its regular presupposition by the text. But it is quite likely that if the diagram was presupposed it was also included. In any case, the only way for us to know if it was likely to be normally included as a textual feature is

167 To be sure, whether the text can be understood does not only depend on the information which the text provides but also on the reader. I am going to take account of this in Chapter 4 by discussing the need for the diagram both from the point of view of proof 'logic' and from a communicative point of view. The distinction between these two viewpoints is also the basis of my interpretation of the textual form of the *Elements* in general in Chapter 5.

168 I assume that the diagrams which were used for finding the proof were not lettered, and that these were in any case not the ones which were included in the text, if any were included at all.

169 Cf. Netz 1999a, 22 on the difficulty of abstracting from our knowledge of a diagram in the reverse process of trying to read the proof as if there was no diagram (precisely for the purpose of testing whether the text presupposes it).

by considering the 'logical' form. By contrast, the standardization of the features of the text proper can be told from the textual form itself.

The standardized form of the *Elements* appears as highly peculiar when compared to other literary forms because it is extremely repetitive and seems to be automatized – to the extent that it is hardly even recognizable as literary form. Ironically, it is precisely the extreme degree of literary form which disguises its literariness. The textual form appears to be a natural part of the proof and the way in which it is done.

Overlap of the textual with the 'logical' form

I have tried to describe the logical and textual features separately from each other in order to single out the part of the form which is not intrinsically linked to the proof method and content. Now I shall briefly consider with the help of two examples how the two aspects of form relate in the proposition. First, the διορισμός: the term refers to the 'logical' form; it is the part of the proof which follows upon the ἔκθεσις of the main geometrical objects and which defines the goal of the proposition in particular terms which has been stated as part of the πρότασις in general terms. From a 'literary' point of view, the part referred to as διορισμός is a sentence which standardly begins with the phrase λέγω ὅτι ("I say that") or with δεῖ δή ("Thus it is required to")[170] and goes on by repeating the second part of the sentence with which the proposition starts off only by using letters with the nouns and phrases which refer to the geometrical objects (whereas the letters do not appear in the first sentence). The letters themselves present another example: on a 'logical' level, they mark the section of the proposition which performs the proof of the particular case and are added to general geometrical terms for referring to particular objects in the diagram. On a purely textual level, they form part of noun phrases referring to geometrical objects in the whole proposition except for its beginning and ending; the order in which the letters of the alphabet are applied to the noun phrases seems to follow certain rules.

The two examples show that there is considerable overlap between the two types of form. In fact, almost all parts of the text can be described from either perspective. What is more, either type can be used to say something about the other: the 'logical' parts of the proof such as the διορισμός can be described by the standardized textual features which constitute it, and the conventionalized names for the 'logical' parts can be used to describe the textual features (indeed, I will do so for the sake of convenience). This suggests that the textual form is at least partly in service of the 'logical' one; it is standardized so as to standardize the 'logical' form as well.[171] The extent of the resulting overlap of 'logical' and 'textual form' is crucial for the present study of

170 Acerbi 2012, 174.
171 That is, the textual form is *used* to some degree to serve the logic; it is, however, contingent.

the 'textual' form. I will investigate how much and which of the textual standardization is necessitated by the framework of the 'logical' form, and whether less or different standardization will do as well from a 'logical' viewpoint. That the texts are not strictly formalized according to the rules of logic is what leaves room for interpreting the 'textual' form in the first place.

1.3 The form of Euclid's *Elements* and clarity

The overall impression is that the standardized form makes the text extremely clear: the proofs and their deductive connection are, at least for the most part, unambiguous, if the rules of the genre are applied rather than modern standards for axiomatic-deductive proofs. Technically, they are universally comprehensible. This is indicated by the long-term clarity of the work: the *Elements* is still, even today, comprehensible without any prior knowledge because it itself, precisely by its standardization, contains everything necessary for being understood. From observing the standardized features in a certain number of proofs, the rules of the genre can be acquired. This is facilitated by the "discreteness" of the proofs; they form small recognizable units which can be worked through one by one.[172] Therefore, the proofs still can and, arguably, could be understood in ancient times by anyone who is able to read and willing to understand them. The long-term clarity of the text is particularly striking if the text is compared to other works of ancient Greek science, many of which are far from clear today. It is possible, however, that we still perceive the *Elements* as clear because we have all been trained in Euclidean geometry in school and are familiar with its form, and because modern science, which still cherishes the formal ideal of the genre of elements, connects it with scientific clarity. Of course, the form of elements texts might have been chosen as a medium for modern science communication precisely because of its clarity.[173]

Nevertheless, many readers, both in ancient times and today, might object to the impression that Euclid is clear and find him rather obscure. This could be because the text of the *Elements* is quite inaccessible to those who do not know it. This is due to its uninviting impersonality and lack of contextual information; and it makes a highly unusual impression: outside the genre, no other text is stylized like it. Moreover, the uncompromising standardization entails that the text is written in a strictly linear way without referring explicitly back to previous proofs.[174] In fact, the text is implicit in every way because of its standardization. There is no room for any explanations of content, method or form, additional details, cross-references, or addressing the reader. In

172 Asper 2007, 115.
173 Asper 2014, 456.
174 Asper 2007, 115.

this way, it necessitates a lot of work on the part of the reader who is not trained in the standardized form of the proofs. This suggests that the *Elements* would have been immediately clear only to expert readers, but not to lay readers or students.

The ancient reception of the form of Euclid and the elements genre seems to testify to both views: that the form was especially clear and that it was rather obscure. The two afore mentioned commentators provide the only extant explicit remarks on the question with regard to Euclid's main work itself: both praise specifically Euclid's clarity, and Proclus relates the characterization explicitly to Euclid's form.[175] The commentators' view is a late and, as we shall see shortly, a biased one, but it is noticeable in view of the large commentary tradition of ascribing obscurity rather than clarity to one's source authors.[176]

Outside of the commentaries, there are only a few scattered and more general remarks on the mathematical language and form which could be interpreted in terms of clarity and obscurity, although they do not mention the technical terminology. First, Aristotle: although he makes ample use of mathematical examples and methods, there are only few passages talking explicitly about 'mathematics,' 'the mathematical sciences,' or 'the geometers,' which could be interpreted with regard to clarity and obscurity. Two passages seem to indicate the unambiguousness of the mathematical language and objects on account of the definitions, but do not mention any terminology of clarity.[177] Two other passages address the role of the diagram in enabling the clarity and comprehension of the proof. One of them is concerned with the question of the relevance of the diagram for the proof:[178] it states with regard to the natural imprecision of diagrammatic drawings that "the geometer does not conclude anything from there being this line which he himself has described, but from what is made clear through them (τὰ διὰ τούτων δηλούμενα)."[179] In the rhetorical and logical context, the adjective δῆλος is often used synonymously with σαφής ('clear,' 'distinct'), also by Aristotle himself, but it is also used to express that something is 'evident.' Here, δηλούμενα indicates the latter sense: it is precisely the point that the diagrammatic drawings of mathematics do *not* unambiguously and distinctly represent the object qualities and relations; rather, their function is – according to Aristotle – to make 'evident' or 'visible' those things – i.e., presumably, constructions, objects, and object relations – from which the geometer deduces the proposition.[180] The other pas-

---

175 Cf. Chapter 2 for quotations of the respective passages and a detailed account on the question whether Proclus meant the textual or 'logical' form.

176 See the following section on clarity and (Greek) science.

177 *An. post.* I 12.78a10–13 and *Metaph.* M 3.1078a36–b2 (cf. Asper 2007, 149).

178 It is, therefore, central for my argument regarding the proof-relevance of the diagram in Chapter 4.

179 ὁ δὲ γεωμέτρης οὐδὲν συμπεραίνεται τῷ τήνδε εἶναι γραμμὴν ἣν αὐτὸς ἔφθεγκται, ἀλλὰ τὰ διὰ τούτων δηλούμενα (*An. post.* I 10.77a1–3, transl. Barnes).

180 The function of the diagrams, as it is indicated by the text of the *Elements* and by the Euclidean diagram, will be discussed in Chapter 4. I shall explain further below in Section 1.4, which is devoted

sage about the mathematical use of diagrams does not use any terminology of clarity, but claims that the construction of the diagram enhances understanding by presenting the figure as coming into being.[181]

Two more Aristotelian passages indicate that the mathematical language was considered unusual by Aristotle's readers. In *Rhetoric* III 1, Aristotle debates the role of 'language and style' (λέξις) in making a speech. Aristotle claims that language and style do have some importance "in every kind of instruction" (ἐν πάσῃ διδασκαλίᾳ)[182] because the way in which something is said makes a difference "with regard to clarity" (πρὸς τὸ δηλῶσαι), but that the λέξις is not so much important because it is only φαντασία ("appearance") and used for the sake of the hearer. He concludes: "Therefore, no one teaches geometry in this way."[183] According to Aristotle, geometry dispenses with (and *can* dispense with) a rhetorical style which is simply used for pleasing the reader and, in doing so, being clear to them. Aristotle seems to presuppose that his (non-mathematical) readers considered the mathematical text form to be wholly non-rhetorical and therefore highly unusual because it is different from the style in which all of them were trained. This also suggests that non-mathematical readers would have considered the mathematical 'style' as unclear because, according to Aristotle, (the rhetorical) style is recommended for use precisely for the purpose of being clear to one's readers.

Something similar is suggested by a passage from *Metaphysics* α about the ἀκριβολογία μαθηματική, the "precision of mathematical speech." Aristotle claims that hearers find language to which they are unaccustomed "somewhat unintelligible and foreign" (ἀγνωστότερα καὶ ξενικώτερα), whereas they find customary language "intelligible" (γνώριμον), which he explains using the example of mathematical language:[184]

> Some people do not listen to a speaker unless he speaks mathematically (μαθηματικῶς λέγῃ), others unless he gives instances, while others expect him to cite a poet as witness. And some

---

to the concept of clarity, why I do not consider ἐνάργεια as a function of the diagrams, a concept which is closely related to σαφήνεια in the rhetorical treatises (p. 54).

181 *Cael.* I 10.279b33–280a10. Here, Aristotle does not mention mathematicians explicitly, but they must be meant by the formulation οἱ τὰ διαγράμματα γράφουσι ("those who draw diagrams").

182 On the relation of language/style and *philosophical* clarity in Aristotle cf. Rapp 2013. For a discussion of the present passage from the *Rhetoric* and the following one from the *Metaphysics* in the context of Aristotle's reflections on the form of terms and its relation to clarity cf. Gasser 2024.

183 τὸ μὲν οὖν τῆς λέξεως ὅμως ἔχει τι μικρὸν ἀναγκαῖον ἐν πάσῃ διδασκαλίᾳ· διαφέρει γάρ τι πρὸς τὸ δηλῶσαι ὡδὶ ἢ ὡδὶ εἰπεῖν, οὐ μέντοι τοσοῦτον, ἀλλ' ἅπαντα φαντασία ταῦτ' ἐστί, καὶ πρὸς τὸν ἀκροατήν· διὸ οὐδεὶς οὕτω γεωμετρεῖν διδάσκει (*Rhet.* III 1.1404a10–12; cf. Asper 2007, 116, n. 159).

184 *Metaph.* α 3.995a8–16: οἱ μὲν οὖν ἐὰν μὴ μαθηματικῶς λέγῃ τις οὐκ ἀποδέχονται τῶν λεγόντων, οἱ δ' ἂν μὴ παραδειγματικῶς, οἱ δὲ μάρτυρα ἀξιοῦσιν ἐπάγεσθαι ποιητήν. καὶ οἱ μὲν πάντα ἀκριβῶς, τοὺς δὲ λυπεῖ τὸ ἀκριβὲς ἢ διὰ τὸ μὴ δύνασθαι συνείρειν ἢ διὰ τὴν μικρολογίαν· ἔχει γάρ τι τὸ ἀκριβὲς τοιοῦτον, ὥστε, καθάπερ ἐπὶ τῶν συμβολαίων, καὶ ἐπὶ τῶν λόγων ἀνελεύθερον εἶναί τισι δοκεῖ. [. . .] τὴν δ' ἀκριβολογίαν τὴν μαθηματικὴν οὐκ ἐν ἅπασιν ἀπαιτητέον, ἀλλ' ἐν τοῖς μὴ ἔχουσιν ὕλην. Transl. Ross.

want to have everything done accurately (ἀκριβῶς), while others are annoyed by accuracy, either because they cannot follow the connection of thought or because they regard it as pettiness (μικρολογίαν).[185] For accuracy has something of this character, so that as in trade so in argument some people think it mean (ἀνελεύθερον).[186] [. . .] The minute accuracy of mathematics (τὴν δ' ἀκριβολογίαν τὴν μαθηματικὴν) is not to be demanded in all cases, but only in the case of things which have no matter. Therefore its method is not that of natural science; for presumably all nature has matter.

Ἀκρίβεια can be understood as referring to mathematical standardization; this passage is therefore a reflection that is directly relevant to our discussion. It suggests that any readers who are not familiar with the standardized form, above all, most of Aristotle's own readers, would have found it unintelligible, while mathematicians themselves would have considered it very intelligible.[187] The strangeness of the mathematical language and its unintelligibility for non-mathematical readers are pointed out by Galen too: "Wenn man z.B. von einer waagerechten Fläche spricht, so versteht das von denen, die es hören, keiner außer dem Mathematiker [. . .]."[188] The passages quoted seem to indicate that outside the context of mathematics, the textual form of elements texts was perceived as highly unusual and therefore could have been considered unclear.

From ancient Greek mathematics, no study of the *textual* form of elements texts is extant; accordingly, no ancient theoretical reflection from an 'emic' perspective on the relation of the textual form of elements texts and clarity survives. Therefore, I now broaden my focus and turn to the larger context of the relation of clarity and form in (Greek) scientific texts. I consider whether the ancient discourse on clarity in Greek scientific texts and beyond is at all useful for analyzing the textual form of the *Elements* with regard to clarity.

1.4 Clarity and/in (Greek) scientific texts

The aim of my study – to analyze the relation of form and 'clarity' in Euclid's *Elements* – relies on a set of assumptions: the main assumption is that the textual form of the *Elements* has to do with the purpose of being unambiguous and comprehensible. This seems to be indicated by the presumed function of the *Elements*. As a scientific text, the work should have the basic function of written knowledge communication. This means that the text must communicate precisely a kind of content which is relatively difficult.

---

185 Ross translates "pettifoggery," but the simpler "pettiness" seems to be more apt.

186 Cf. Asper 2007, 116 for the social connotations of ἀκρίβεια.

187 In the rhetorical treatises, ἀκρίβεια is a concept which is closely connected to σαφήνεια (Vatri 2016, 102–104). I return to this passage below p. 52 when discussing the applicability of rhetorical terms to the *Elements*.

188 Galen, *De nominibus medicis*, translated from Arabic by Meyerhof & Schacht 1931, 1 (ar.)/ 8.16 f. (dt.).

However, any communication – even oral everyday communication – is "contingent:" it constantly runs the risk of failing and is even "improbable" in the sense that the information to be communicated is unlikely to be understood in the intended way.[189] Communicating in writing should be particularly difficult because it faces the problem of decontextualization: unlike oral communication in which speakers may make sure constantly that they are being understood correctly and in which they may adjust their communicative strategy ad hoc accordingly, written texts are received out of the context in which they have emerged. By the time readers receive the texts, authors no longer control how they are being understood. Therefore, writers must aim in advance at unambiguousness and unmistakability in order for the communication of knowledge to be successful. Accordingly, one may expect the text form to be in service of 'clarity' in the above sense of 'unambiguousness' and 'unmistakability.'

Ancient scientific writers on textual clarity and knowledge communication

Ancient scientific writers themselves do not explicitly refer to the clarity of form as a solution to the issue of decontextualization in written knowledge communication. Some, however, do discuss the problems to be encountered when communicating knowledge in writing, while others reflect on the relation of clarity and written knowledge communication.

The difficulty of conveying 'truth' and 'trustworthy evidence' to someone else's mind by means of words is reflected as early as Empedocles: "Friends, I know that truth is in the words/ That I shall speak out; but it is very irksome/ For men, and causes distrust, the impulse of argument to the mind."[190] He, however, does not talk about the specific problems of *written* knowledge communication.

The ramifications of decontextualization for texts which communicate knowledge in writing have been famously outlined in the 'critique of writing' of Plato's *Phaidros*: the written account cannot provide additional explanations in response to questions which a reader may have but remains silent; it cannot choose its readers but can fall into the hands of those who will not understand it, and it cannot defend nor help itself against undeserved criticism.[191] Because of these drawbacks of writing, the explicit recommendation which Socrates makes in the *Phaidros* is not to commit any philosophical knowledge or insights to writing at all; one should only write for the sake of aide-

---

189 Asper 2007, 11 f. with reference to Iser 1994, 257 f. and Luhmann 1981, 26–30 on the "Unwahrscheinlichkeit der Kommunikation."

190 ὦ φίλοι, οἶδα μὲν οὕνεκ' ἀληθείη πάρα μύθοις,/ οὓς ἐγὼ ἐξερέω · μάλα δ' ἀργαλέη γε τέτυκται/ ἀνδράσι καὶ δύσζηλος ἐπὶ φρένα πίστιος ὁρμή (DK 31 B 114), translation Laks & Most 2016, 365 (modified), who classify the fragment under "Communication of the truth and access to divinity."

191 *Phaidr.* 275d4–e5.

mémoire or play.[192] The only suitable way to communicate (philosophical) knowledge according to Plato's Socrates is by means of the (oral) art of dialectic because only λόγοι of this kind "are able to help themselves."[193]

In light of this, the existence of Plato's own writings presents a well-known paradox. In order to solve it, some scholars have assumed either an 'unwritten doctrine' of Plato which contains his true philosophy whereas the extant dialogues only hint at the dialectic method; others, alternatively, have argued that the dialogues do contain Plato's philosophy, however not explicitly, but by imitating oral dialectic and thus by revealing the knowledge only if read many times (i.e., the text answers only if it is asked again and again). This would mean that the dialogues can only be understood by some: those who make an effort at rereading the texts and are able to grasp the answers which the text only gives indirectly.[194] Thus, the solution to the problem of decontextualization which the *Phaidros* suggests is to withhold knowledge from one's readers, at least to a certain extent: either (according to the first interpretation) by refraining from writing and publishing any knowledge texts at all or (according to the second) by writing dialogues which obscure the knowledge from certain readers and do not give it away immediately even to readers who will be able to understand eventually.

The Platonic evidence displays awareness of the problem of decontextualization for written knowledge communication even before Euclid's own time. The 'critique of writing' even links the notions of decontextualization and clarity. However, clarity of form is not presented as a potential solution to the decontextualized nature of writing; rather, writing cannot produce anything clear and certain (τι σαφὲς καὶ βέβαιον) precisely because it becomes decontextualized.[195]

The only other ancient author who testifies to an awareness of the problem of decontextualization is the medical author and commentator Galen in the second century AD.[196] In his treatise *De dignoscendis pulsibus*, he criticizes Praxagoras' account (λόγος) of the relationship between the pulse, the state of the arteries, and the types of humors for not introducing (ἐπεισάγει) a "common perceptible sensation,"

---

192 *Phaidr.* 276d.

193 *Phaidr.* 276e4–277a4.

194 According to the 'Tübinger Platonschule,' the dialogues do not contain Plato's philosophical doctrine, but by their form only hint to the esoteric 'unwritten doctrine' which Plato taught in the Academy by means of the dialectic method. Advocates of the other main line of interpretation which goes back to Friedrich Schleiermacher hold that the dialogues did present the core of Plato's philosophy, and that the 'critique of writing' (although it is explicitly referred to γραφή) did not pertain to writing in general, but only to the systematic treatise (σύγγραμμα), while the Platonic dialogues, by imitating oral dialectic, were the only possible way to write philosophy (cf. the overview of the controversy by Szlezák 2001, 180–190 from the Tübingen point of view).

195 The whole critique of (the decontextualization of) writing (*Phaidr.* 275d4–e5) serves as an explanation (γάρ) of why writing cannot be clear (*Phaidr.* 275c5–d2).

196 Asper 2007, 16 f. with n. 41, and 331 with n. 85.

i.e., "a perceptible aspect of the pulse [. . .] which can serve as a criterion for distinguishing different pulses."[197] Galen argues that

> in this way, he risks to render his account (τὸν λόγον) a riddle, so that someone reading it and making no sense of it, will think that something deep and marvelous is hidden in it and he will thus be clearly wearing himself out, since searching through it all he will find nothing. For how could he find that which is not at all manifest?[198]

It is true that in this passage Galen does not explicitly talk about the risks of written knowledge communication in general; nor does he imply that Praxagoras' theory had actually contained that which is missing and that Praxagoras simply did not mention it for whatever reason. What he criticizes is the theory (and according account of it) rather than primarily the account itself.[199] Galen's point is that a diagnostic criterion for distinguishing different pulses should be part of the theory and that not only himself, but readers in general ("someone reading it") will expect it to be mentioned and search the account for it. As they will do so in vain, the account will be a "riddle" to them. This means that while Galen raises a particular problem of content rather than of the form of writing, he judges the Praxagorean *logos* with regard to its reception by a reader. In so doing, he implies the more general problem of decontextualization which comes with written knowledge transmission, especially if a text is received a long time after its composition; for if Praxagoras was present in person, he would be able to confirm immediately that a "common perceptible sensation" was indeed missing from his theory. By contrast with the Platonic 'critique of writing,' Galen describes how decontextualization affects the reader rather than the author. However, a reader who keeps looking for something which is not there will misunderstand the text, and the author's communication of knowledge eventually will be unsuccessful.

Galen does not explicitly give a solution to the problem of decontextualized knowledge communication; the one which is implied most obviously by his critique is to im-

---

197 *Dign. puls.* VIII,4, 941.20–941.19 Kühn. Cf. Lewis 2017 for translation (59–61) and commentary (147–149, here 149).

198 αἴνιγμα γὰρ ἂν οὕτως τὸν λόγον ποιήσειεν, ἵνα τις ἀναγινώσκων, εἶτα μὴ νοῶν, οἴηταί τι βύθιον ἐγκεκρύφθαι τῷ λόγῳ καὶ θαυμαστόν, εἶτα κατατρίβηται δηλαδή, ζητῶν μὲν διὰ παντός, εὑρίσκων δ' οὐδέν. πῶς γὰρ ἂν εὕροι τὸ μηδ' ὅλως φαινόμενον; (*Dign. puls.* VIII,4, 941.20–942.4 Kühn, transl. Lewis 2017, 61, modified). I take κατατρίβηται to mean "he will be wearing himself out" (cf. LSJ s.v. 2 and Asper 2007, 331, n. 85) as referring to the effort which the reader will be putting into looking for the missing element rather than to the time he will be losing in looking for it, which Lewis' translation "he will be wasting his time" suggests. For ἵνα equaling ὥστε, i.e., in a non-final sense which it may have in later authors cf. LSJ s.v. B II 1. A final sense would change the whole thrust of the passage, suggesting that Praxagoras intentionally turned his *logos* into a riddle in order for the reader to get lost. While such assumptions are topical in Galen and other commentators of Greek scientific texts (see below p. 51), Galen in the present passage seems to point to the danger of Praxagoras' written communication rather than to any intentional obscurity on the part of Praxagoras.

199 Conveniently, λόγος refers to both the theory and the account of it.

prove the content and add what is missing. Alternatively, Galen seems to suggest another measure for preventing misunderstanding, that is, not to improve the theory itself, but its *presentation*. If Praxagoras mentioned the absence of a diagnostic criterion and explained why it is missing from his theory, the reader would not have to look for it and would not puzzle over not finding it. Galen's critique is essentially of the theory: (part of) it does not make sense to him; but the language of obscurity which he uses to describe the account and its impact suggests that a reader will not even be able to decide whether Praxagoras is right or wrong because his *logos* is so unclear and incomprehensible (αἴνιγμα, βύθιον ἐγκεκρύφθαι, τὸ μηδ' ὅλως φαινόμενον: it will appear as a "riddle" and as if something "deep is hidden in it [. . .] which is not at all apparent").[200] The Galenic passage implies, in any case, that in order to respond to the danger of misunderstanding in decontextualized knowledge communication one should take action to prevent misunderstanding – rather than to not write at all or to write obscurely and understandably only for few people. Galen does not explicitly recommend clarity as a measure for preventing misunderstanding the written account; but the description of Praxagoras' account in terms of 'obscurity' *ex negativo* suggests that 'clarity' would be a quality of the kind of written text which does not lead its readers astray but facilitates its own reading reception; in this way, it implies that clarity would reduce the risks of written knowledge communication. By implicitly ascribing Praxagoras's obscurity to his (textual) presentation, Galen suggests that clarity too would be effected by the presentation of scientific knowledge, that is, by the form of a scientific text.

In two other passages from *De alimentorum facultatibus* and *De compositione medicamentorum secundum locos* Galen explicitly mentions clarity as that quality by which written texts may communicate knowledge successfully. Both times Galen emphasizes that the "living voice" (ζῶσα φωνή) is to be preferred over a book as a medium for teaching (i.e., transferring knowledge to students) but admits that it may be adequately

200 That Galen would consider *explaining* the absence of a diagnostic criterion enough to make the account comprehensible could be argued from comparison with other instances where he criticizes a certain point in the writings of authors of medical writings. Of course, even if Praxagoras had explicitly rejected and justified the element which Galen is missing, Galen would criticize Praxagoras anyway at this point, as the passage is part of a polemic against Archigenes and his followers (Lewis 2017, 147). For criticizing the authoritative author (cf. Lewis 2017, 27–29 on Galen's ambivalent stance toward Praxagoras), the notion of obscurity comes in handy. Like most other ancient commentators of scientific authors, Galen too topically ascribes obscurity to his source texts, which provides him with ample opportunity to variously explain them. At present, he uses the topical language of obscurity, which is not restricted to the word itself, but also includes vocabulary such as the above. Even though the 'comment' on Praxagoras is not from a commentary work, it inscribes itself into the tradition of the topos of obscurity, and the Galenic critique of Praxagoras would have to be reconsidered in light of this tradition. I am going to explain shortly how the commentators' use of the language of clarity and obscurity is relevant for our subject; the present passage is unique in that it uses the obscurity topos to refer to the risks of written knowledge communication.

replaced by that which is "clearly written" (σαφῶς γεγραμμένα).[201] The use of the adverb with respect to 'writing' could suggest that the quality of clarity describes the *way* of writing, i.e., the form of the text, not its theory or content. Typically, Galen refers to his own current writing (or to his writings in general) as being able to transfer knowledge. That is, what at first appears to be a general reflection about the preconditions of communicating knowledge to students turns out to be a self-praise which may serve Galen to enhance his authority as a teacher or to provide against contrary, unfavorable perceptions of his work.

Whatever his motivation in bringing up the subject of clarity in the two passages, Galen explicitly states that 'writing clearly' is necessary for communicating knowledge through a book. However, he does not link 'clarity' to the problem of decontextualization, i.e., to the problem raised in the aforementioned Galenic passage. There could be other reasons why Galen presupposes that it is more rewarding to learn from a real person rather than from a book unless it is written clearly (e.g., psychological ones); so he does not necessarily recommend 'writing clearly' as a remedy to decontextualization. One could read into σαφῶς (sc. γεγραμμένοις, "clearly written") some of the meanings which the term has in the contexts of logic and rhetoric: 'unambiguously,' 'comprehensibly,' or 'unmistakably;' then, the present passages could be understood to point to the danger of misunderstanding entailed by committing knowledge to decontextualized writing. This is quite likely also because the preference of the "living voice" over writing is evocative of the Platonic 'critique of writing' whose crux is the decontextualized nature of writing. However, I am looking for an explicit link between textual 'clarity' and the problem of potential misunderstanding posed by written, decontextualized knowledge communication precisely in the context of science itself.

---

201 *Alim. fac.* Corp. Med. Graec. V 4.2, 216.19–27: "[. . .] I think that the majority are correct who say that the best instruction is through personal contact, and that it is impossible for anyone to become either a helmsman or an expert in any other craft from a book. These are reminders for those who have previously studied and understood, not complete instruction for the ignorant. Just the same, if any of the latter who lack instructors are willing to attend carefully to what has been written clearly and in detail as I am doing, they will profit greatly, especially if they do not hesitate to read it over and over again." ([. . .] μοι δοκοῦσι καλῶς οἱ πολλοὶ λέγειν ἀρίστην εἶναι διδασκαλίαν τὴν παρὰ τῆς ζώσης φωνῆς γιγνομένην, ἐκ βιβλίου δὲ μήτε κυβερνήτην τινὰ δύνασθαι γενέσθαι μήτ' ἄλλης τέχνης ἐργάτην· ὑπομνήματα γάρ ἐστι ταῦτα τῶν προμεμαθηκότων καὶ προεγνωκότων, οὐ διδασκαλία τελεία τῶν ἀγνοούντων. εἴ γε μὴν ἐθέλοιέν τινες καὶ τούτων, ὅσοι διδασκάλων ἀποροῦσιν, ἐντυγχάνειν ἐπιμελῆς τοῖς σαφῶς τε καὶ κατὰ διέξοδον, ὁποίαν ἡμεῖς ποιοῦμεθα, γεγραμμένοις, ὀνήσονται μεγάλως, καὶ μάλιστα ἐὰν πολλάκις ἀναγιγνώσκειν αὐτὰ μὴ ὀκνῶσιν. Transl. Powell 2003, 39). Alongside 'clarity,' 'detailedness' is mentioned as a second quality in virtue of which books may substitute oral teaching. The second passage to link 'clarity' and written transfer of knowledge is *Comp. med. sec. loc.* XII,2, 894.1–6 Kühn: ἀληθὴς [. . .] ὁ λεγόμενος ὑπὸ τῶν πλείστων τεχνιτῶν ἐστι λόγος. ὡς οὐκ ἴσον οὐδ' ὅμοιον εἴη παρὰ ζώσης φωνῆς μαθεῖν ἢ ἐκ συγγράμματος ἀναλέξασθαι. τοῖς γε μὴν φιλοπόνοις καὶ φύσει συνετοῖς οὐχ ἡ τυχοῦσα πολλάκις ὠφέλεια γίνεται βιβλίοις ἐντυγχάνουσι σαφῶς γεγραμμένοις.

There is not any such explicit link: the passages by Empedocles, Plato, and Galen which we have first surveyed show awareness of the risk of misunderstanding in written knowledge communication but do not, or only implicitly, suggest clarity (of the textual form) as a possible solution. The latter two passages by Galen mention clarity in a normative way, but without making direct reference to the problem of misunderstanding, and thus without making explicit in what sense clarity is necessary for knowledge communication: in the sense of the avoidance of misunderstanding, i.e., unambiguousness, or in another sense. Moreover, the two passages in which 'clarity' appears in the form of the adverb σαφῶς do not explicitly define it as a quality of the textual form; the reference to the textual form, if there is one at all, is only implicit.

'Clarity' in the ancient commentators of scientific texts

The absence of any explicit reflection about the relation of clarity as a quality of the text form and decontextualized knowledge communication could be due to the lack of an extant theory of the (literary) form of science or handbooks on how to write science. For a modern pendant, one could refer to manuals which give instructions on how to write science. There are various subtypes of this genre: some address students, undergraduate or postgraduate, of either the humanities or the (hard) sciences, others researchers at different stages of their careers; accordingly, they offer stylistic and structural guidance for writing works of scientific genres as different as student essays, master's and doctoral theses, peer-reviewed papers, and books. Mostly, however, they are directed to a larger readership of students and researchers in general and not confined to a single scientific discipline.[202] What the manuals have in common amongst other things is that clarity features among the main virtues of structure and style. Interestingly, this is true of English as well as, e.g., German publications, although the respective academic languages differ considerably in terms of style.

From ancient times, no comparable normative-theoretical works on the form of scientific writing in general survive. None is even known from the time of Euclid – and accordingly none which would have dealt with clarity as a quality of the form and with its benefits for knowledge communication. We are only aware of the title of

202 E.g., in German, Schnur 2010 (*Schreiben. Eine lebensnahe Anleitung für die Geistes- und Sozialwissenschaften*), which is directed at authors of all kinds of scientific texts (8), or Groebner 2012 (*Wissenschaftssprache. Eine Gebrauchsanweisung*), which addresses "Autorinnen und Autoren [. . .] die [. . .] an einem größeren Text für ein wissenschaftliches Publikum arbeiten – an einer Dissertation zum Beispiel" (16), especially in the humanities (17). The English classic *The Elements of Style* (Strunk & White 2000) seems to be directed to "writers" in the sense of everyone who communicates in writing, and thus includes, but is not limited to, authors of scientific texts (foreword by Roger Angell, xi).

a much later Galenic treatise *On Clarity and Obscurity*, which displays the respective ancient technical rhetorical terms (Περὶ σαφηνείας καὶ ἀσάφειας).[203] As Galen mentions it only once,[204] it could be just as well an alternative name for his lost treatise Περὶ ἐξηγήσεως (*On Exegesis*) which dealt with the principles of the 'interpretation' or 'explanation' of (scientific) texts by means of commentaries. The main principles of this treatise are summarized in the proem to Galen's commentary on the Hippocratic *De fracturis*. The summary also contains a distinction of forms of obscurity.[205] If the *On Clarity and Obscurity* is indeed identical with the *On Exegesis*, it will have dealt with the two notions systematically and theoretically. However, it will have done so not from the point of view of scientific authors and their addressees, but from a commentator's perspective. And even if the two titles do not refer to the same treatise, the *On Clarity and Obscurity* will have tackled the problem from a commentator's viewpoint, that is, it is presumably about the clarity or obscurity of scientific texts *as* source texts of commentaries, not of scientific texts in general. This is because the obscurity of the source text (Greek ἀσάφεια and Latin *obscuritas*) is topical in Galen's commentaries and in other ancient commentators of scientific writing.[206]

Galen appears to be the only commentator to have written a whole treatise on the subject. But many of the extant commentaries, above all Galen's commentaries on the Hippocratic writings and the Neoplatonic commentaries on Aristotle, contain semi-theoretical discursions on obscurity of varying length and abound with references to the supposed obscurity of a word, passage, author, or work and to their own clarification of the allegedly obscure text.[207] Indeed, while Heraclitus *is* 'the obscure one,' and while the obscurity of Hippocrates and Aristotle, the two scientific authors

---

203 The English translation cannot imitate the etymological relation of the Greek σαφήνεια and ἀσάφεια, as the word 'unclarity' does not exist. Ἀσάφεια is commonly translated as 'obscurity' in the tradition of *obscuritas*, the Latin translation of ἀσάφεια.

204 *Lib. prop.* 124.16 Müller.

205 Mansfeld 1994, 151, see below pp. 45–47 with n. 217. Ἐξήγησις in general means 'explanation' or 'interpretation' (LSJ s.v. A II) but seems to be used here to refer to 'commenting' in the form of 'commentary writing' in general (cf. ἐξηγητής LSJ A II 3 'commentator' with reference to Galen, *In Hipp. Victu acut. comm.* XV,4, 518.16 f. Kühn). That the work was concerned with commenting on *scientific* texts (not poetry or scripture, for example) is suggested by the summarizing passage from *In Hipp. Fract. comm.* in which Galen first gives a definition of exegesis in general but in explaining it switches to talking about his own commentary work, which is concerned with medical (and philosophical) texts.

206 The following section on the topos of obscurity in the ancient commentators of science relies on my Kanthak 2013.

207 The explicit treatment of obscurity in the commentaries oscillates between topical references to and attempts to theorize the phenomenon, which, however, become topical themselves, hence "semi-theoretical" (Kanthak 2013, 178).

most heavily commented upon, is likewise notorious,[208] almost no ancient writer of science escapes the verdict of obscurity by their commentator(s).[209]

The omnipresence of ascriptions of obscurity in the commentaries on scientific writing by means of the rhetorical term ἀσάφεια implies a paradox.[210] For as we have stated above, scientific texts can be assumed to aim for transmitting knowledge and for being comprehensible in order to do so. Ἀσάφεια, however, denotes a lack of σαφήνεια. Without going into the implications of the originally rhetorical term, the term σαφήνεια covers a range of nuances amongst which 'unambiguousness' and 'comprehensibility' are prominent. Accordingly, ἀσάφεια refers to a lack of unambiguousness or comprehensibility.[211] Therefore, if indeed ἀσάφεια is used in this sense, its topical ascription to texts which are meant to communicate knowledge comprehensibly would call for explanation – especially because it is precisely the prominent scientific authors who are called obscure, so the reason for their obscurity cannot be their inability to write clearly.[212]

The paradoxical obscurity of scientific writing in general is also implied by how Galen defines the task and function of commentaries of scientific texts.[213] I shall now take a closer look at this definition and its context, as it implicates an explanation for the alleged omnipresence of obscurity. In two of his proems to Hippocratic commentaries, Galen states that the ἴδιον ἔργον of the exegesis, and its δύναμις, is 'to clarify (σαφηνίζειν/ἐργάσασθαι σαφῆ) what is unclear (ἀσαφές/ἀσαφῆ).'[214] In light of this, the vast quantity of ancient commentaries on scientific texts, including a great number

208 For Heraclitus' epithet, e.g., Cic. *Fin.* II 15: *Heraclitus, cognomento qui* σκοτεινός *perhibetur*. The ancient topos of obscurity centered around the rhetorical term ἀσάφεια, but was not limited to it. – According to Soranus' *Life of Hippocrates*, Hippocrates has usually been depicted with a veiled head, which Soranus links with his well-known obscurity; Aristotle's (intentional) obscurity is already addressed as early as the first century – in the alleged, probably spurious letter from Aristotle to Alexander the Great which may have been forged by Andronicus – and from then on by various authors and commentators (cf. Kanthak 2013, 157 and 169 f.).

209 For a list of authors cf. Kanthak 2013, 157.

210 Kanthak 2013, 166 f.

211 Ἀσάφεια and the corresponding Latin *obscuritas* do not refer to the opposite of σαφήνεια / *perspicuitas* but only to a partial absence of clarity (Fuhrmann 1966, 47).

212 Kanthak 2013, 163.

213 More precisely, Galen defines the task of "exegesis," but what he means seems to be the task of the commentator of scientific texts.

214 The first appearance of this formula is *In Hipp. Aphor. comm.* XVIIB,2, 561.4–7 Kühn: μάλιστα μὲν οὖν ὅσον ἐν αὐτοῖς ἀσαφές ἐστι σαφηνίζοντες, ἔργον γὰρ τοῦτο ἴδιον ἐξηγήσεως, οὐ μὴν ἀλλὰ καὶ τὴν ἀπόδειξιν ἑκάστου τῶν ὀρθῶς εἰρημένων προστιθέντες, ἐπειδὴ καὶ τοῦτ' ἔστιν ἐν τοῖς ὑπομνήμασιν γίνεσθαι. The second one is the proem to the commentary to *De fracturis* already mentioned above: *In Hipp. Fract. comm.* XVIIIB,2, 318.1–4 Kühn): Πρὸ τῆς τῶν κατὰ μέρος ἐξηγήσεως ἄμεινον ἀκηκοέναι καθόλου περὶ πάσης ἐξηγήσεως, ὡς ἔστιν ἡ δύναμις αὐτῆς, ὅσα τῶν ἐν τοῖς συγγράμμασίν ἐστιν ἀσαφῆ, ταῦτ' ἐργάσασθαι σαφῆ ("Before I begin the detailed interpretation, it is worth saying in general about any interpretation that its purpose is to render clear whatever is unclear in the text itself." Transl. Barnes 1992, 271).

by Galen alone, indicates that ancient scientific writing was commonly perceived as largely obscure. According to Galen's definition, unclear is precisely everything which is commented upon.[215] Galen's insinuation of the obscurity of the source texts of commentaries is especially noticeable as his pointedly antithetical formula itself suggests that obscurity is a defective state of (scientific) texts, while clarity is the ideal one:[216] an ideal state, however, which is regularly not achieved by the (scientific) writers themselves but has to be accomplished by the commentators' clarification. One wonders then why the authors of the texts would not have made them clear in the first place but would instead have left it for later commentators to do so. Galen takes the ideal of clarity and the according aim of clarification for granted and does not mention why the source would need to be clear at all. But from his subsequent remarks it seems that clarity is meant to enable the readers' understanding. Accordingly, the transmission of knowledge would not be accomplished by Hippocrates but only afterward by Galen's clarification.

Galen does not mention that Hippocrates' supposed obscurity is paradoxical. It is probably in answer to the question why Hippocrates' text is not clear to begin with that Galen, in the proem of his commentary on the Hippocratic *De fracturis* after his definition of the function of the exegesis, goes on to differentiate two types of exegesis, and accordingly two types of obscurity according to the origin of the obscurity: "that which is genuinely unclear and is so itself in itself" (τὸ μὲν ὄντως ἀσαφὲς αὐτὸ δι' ἑαυτὸ τοιοῦτον ὑπάρχον) and "that which did not have its origin in it in the first place" (τὸ δὲ ἐν αὐτῷ πρότερον τὴν γένεσιν οὐκ ἔχον), i.e., that which was not originally obscure. Subsequently, he explains the second type by the heterogeneous readership which differed in educational background and practice as well as disposition: there are experienced and inexperienced, sharp and obtuse, wise and witless recipients. Apparently, the second type of obscurity (which was not originally there) pertains only to the stupid and untrained type of readers. One must conclude that the obscurity of the first type is thought to be there (even) for the best kind of readers.[217]

---

215 Galen reaffirms his view that clarification is the main aim of exegesis by pointing out that other things which are usually done by commentators actually should not be part of the exegesis: *In Hipp. Fract. comm.* XVIIIB,2, 318.5–319.3: τὸ δ' ἀποδεῖξαί τι τῶν γεγραμμένων ὡς ἀληθὲς ἢ ὡς ψεῦδος ἐλέγξαι, καὶ εἰ κατηγόρησέ τις σοφιστικῶς ἀπολογήσασθαι, κεχώρισται μὲν ἐξηγήσεως, εἴθισται δὲ γίγνεσθαι πρὸς ἁπάντων ὡς εἰπεῖν τῶν γραφόντων ὑπομνήματα. καὶ νὴ Δία οὐδὲν κωλύει καὶ τούτου μετρίως ἅπτεσθαι τὸν ἐξηγητήν. τὸ δ' ἀγωνίζεσθαι τελέως ὑπὲρ τῶν τοῦ γράφοντος δογμάτων ἐκπέπτωκε τὸν ὅρον τῆς ἐξηγήσεως ("To prove that what is written in the text is true, or to refute it as false, or to defend it against captious criticisms, these are no part of interpretation, although they are customarily done by pretty much everyone who writes commentaries. There is indeed no reason why an interpreter should not touch lightly on these matters; but a thorough examination of the author's doctrines falls outside the boundary of interpretation." Transl. Barnes 1992, 271).

216 Kanthak 2013, 164 f.

217 *In Hipp. Fract. comm.* XVIIIB,2, 319.6–17 Kühn: οὔσης μέντοι καὶ κατὰ ταύτην τὴν ἐξήγησιν διαφορᾶς διττῆς, ὅτι καὶ τὸ ἀσαφὲς αὐτὸ διττό ἐστιν. [. . .] δέδεικται δὲ ἐν ἐκείνῳ τὸ μὲν ὄντως ἀσαφὲς

Galen seems to be talking about the differences among his own potential readers for whom he is going to clarify Hippocrates' text and not (or not specifically) about Hippocrates' original readership. Thus, the temporal aspect of the definition of the second type (πρότερον) refers to the original absence of obscurity from the text, in contradistinction to the first type, which is in the text and therefore present from the start. The second kind originates only later: in the perception of its readers, that is, a certain type of readers. While Galen is talking primarily about part of his own readers, πρότερον does not imply that the original readership of Hippocrates would have understood what is obscure to (part of) Galen's audience. Galen simply does not reflect on the influence of time and context on the perception of obscurity: that is, on whether his own readers would be more likely to receive the second-type obscurity than Hippocrates' original readers because of the greater decontextualization.

From the negative characterization of the type of reader who would also perceive the second type of obscurity – they are described as ἀγύμναστοι ("untrained"), ἀμβλεῖς ("dumb"), and ἀσύνετοι ("stupid") – it does not seem like Galen would be interested in explaining that which is obscure for them, but that his commentaries are concerned with the "genuine" type of obscurity. However, his subsequent examples and description of his addressees are not clear on this point. First, Galen gives an example for a "formulation" which "is obscure in itself" (τὴν ἀσάφειαν ἔχει αὐτὴ δι' ἑαυτὴν ἡ λέξις) from Hippocrates' treatise *De fracturis* on which he is about to comment. Then, he introduces a second example by saying τοιοῦτόν ἐστι κἀκεῖνο ("such is also that one"), which indicates that it is again an example of something which is "obscure in itself." But after explaining the sentence, Galen notes that the λέξις would "not be obscure" (οὐδεμίαν ἀσάφειαν ἔχει) to someone who saw the bones of the arm, but "fairly obscure" (ἀσαφὴς εἰκότως) to someone who did not know the anatomy of

---

αὐτὸ δι' ἑαυτὸ τοιοῦτον ὑπάρχον, τὸ δὲ ἐν αὐτῷ πρότερον τὴν γένεσιν οὐκ ἔχον, ἐπειδὴ τῶν ἀκουόντων τοῦ λόγου διαφοραὶ πάμπολλαι τυγχάνουσιν οὖσαι κατά τε τὸ προπαιδεύεσθαι, καὶ γεγύμνασθαι περὶ λόγους ἢ παντάπασί γε ἀγυμνάστους ὑπάρχειν, εἶναί τε φύσει τοὺς μὲν ὀξεῖς τε καὶ συνετούς, τοὺς δὲ ἀμβλεῖς καὶ ἀσυνέτους. ("Since even regarding this [sc. kind of] exegesis, there is a twofold difference, because the unclear itself is twofold too [. . .]. For it has been shown in that one [sc. my book on exegesis] that what is genuinely unclear is so in and by itself, that the other sort however is not originally obscure, because there are after all numerous differences among those who read the argument as to their having either received a preliminary education and training in relation to arguments, or being entirely untrained. And as regards their natural disposition, some are sharp and intelligent, others dumb and stupid." Transl. Mansfeld 1994, 150 f.). Galen here uses οἱ ἀκούοντες in the sense of 'readers of a book' (LSJ s.v. A.4); this becomes clear when he subsequently describes the communicative situation as a decontextualized one. The causal subclause (ἐπειδὴ . . .) is strangely added to the distinction of the two forms. I take it to be an explanation of the second type of obscurity which is recipient-dependent, while the first type is presented as being not. Alternatively, it could also refer to the distinction as a whole, indicating that some obscurity (the first type) pertains (even) to the best recipients at the one end of the scale, while a different obscurity is only perceived by those at the other end. Either way, the sense is the same, except that the second interpretation implies understanding the first 'objective' type in subjective terms too (cf. below p. 48 on the difficulty of doing so).

the arm.[218] Accordingly, the example seems to illustrate the second rather than the first type of obscurity. Galen adds that he "think[s] it is better to explain *all such things too*, as most of those who read the book are without knowledge of anatomy."[219] This suggests that his concern was originally the "genuine" obscurity, but that he decided to include the other type too. Hence the two types of clarification-exegesis. Notably, Galen characterizes the recipients of the second type of obscurity more neutrally than before as "those who do not know" (τῷ δ' ἀγνοοῦντι) or those who "have no knowledge of" (ἀμαθεῖς) something.[220] The reason seems to be that Galen has in mind different kinds of recipients even of the second type of obscurity and that he intends to clarify some but not all obscurity of this kind:

> For if I read a book together with someone present, I can aim at the (sc. right) measure of the explanation, paying attention each time to the state of the learner. When I write for all, I aim neither at the best nor at the worst prepared person. For one thing will be unclear to the greatest number of people, the other will distress those who spend time on the clear material. I think it is best to aim at those who are in the middle.[221]

Alternatively, the two types of obscurity which Galen distinguishes here with regard to its recipients could correspond to the two types distinguished before, with the "genuine" obscurity pertaining "to the greatest number of people" and "that which was not originally obscure" to those who are grieved by that which is (actually) clear. However, at the beginning of the proem, Galen mentions that there are two distinct types of exegesis according to two essentially different types of obscurity.[222] Presently, by contrast, the two forms of obscurity are at two ends of a scale, of which Galen is concerned with the middle.

In any case, by orientating the scope of his clarification toward the disposition and prior knowledge of his recipients, Galen suggests that he is aware of the subjectiv-

218 *In Hipp. Fract. comm.* XVIIIB,2, 320.12–16 Kühn: εἰ μέντοι τις ἑώρακεν ὁποῖόν ἐστι τῶν ὀστῶν ἑκάτερον, ὑπὲρ ὧν ὁ λόγος ἐστὶν, οὐδεμίαν ἀσάφειαν ἔχει. τῷ δ' ἀγνοοῦντι τῆς κατ' ἀγκῶνα διαρθρώσεως τὴν φύσιν ἀσαφὴς εἰκότως ἡ λέξις φαίνεται.

219 *In Hipp. Fract. comm.* XVIIIB,2, 320.16–20 Kühn: δοκεῖ δέ μοι βέλτιον εἶναι καὶ τὰ τοιαῦτα πάντα ἐξηγεῖσθαι, διὰ τὸ τοὺς πλείστους τῶν ἀναγινωσκόντων τὸ βιβλίον ἀμαθεῖς ἀνατομῆς εἶναι (my italics).

220 *In Hipp. Fract. comm.* XVIIIB,2, 320.14 and 19 Kühn.

221 *In Hipp. Fract. comm.* XVIIIB,2, 321.7–14 Kühn: ἐγὼ γὰρ ὅταν μὲν παρὼν παρόντι συναναγινώσκω τι βιβλίον, ἀκριβῶς στοχάζεσθαι δύναμαι τοῦ μέτρου τῆς ἐξηγήσεως, ἀποβλέπων ἑκάστοτε πρὸς τὴν τοῦ μανθάνοντος ἕξιν. ὅταν δὲ γράφω πᾶσιν, οὔτε τοῦ ἄριστα παρεσκευασμένου οὔτε τοῦ χείριστα στοχάζομαι. τὸ μὲν γὰρ τοῖς πλείστοις ἀσαφὲς ἔσται, τὸ δὲ ἀνιᾶται χρονίζοντας ἐν τοῖς σαφέσιν. ἄριστον οὖν ἡγοῦμαι τῶν μέσην ἕξιν ἐχόντων στοχάζεσθαι· (my own translation). Galen adds that while having in mind the average recipients, he is inclined to write for "the more capable ones" (τοὺς ἐκτικωτέρους) since those below average do not tend to be readers of commentaries anyway (*In Hipp. Fract. comm.* XVIIIB,2, 321.14 f. Kühn).

222 Cf. p. 45 with n. 217.

ity and relativity of clarity: what is obscure to some is not necessarily obscure to others. In the face of this, it is strange that he postulates an essential, 'objective' type of obscurity besides the 'subjective' one.[223] Moreover, the two types do not really compare; hence the attempt to make them comparable: "An item is obscure 'in itself' if it is obscure to *every* reader."[224] However, obscurity of this type in Hippocrates of necessity cannot pertain to Galen himself, at the very least; otherwise he would not be able to clarify it.[225] Accordingly, it pertains to most people, if at all, and the difference of this type from the subjective form would be no more than a gradual one.[226] But what "will be unclear to the greatest number of people" is hardly an adequate rendition of "that which is genuinely obscure itself in itself." It could be that Galen formulates the first type in subjective terms by analogy with the second – especially because he emphasizes the subjectivity of obscurity shortly afterwards. But he seems to have a distinct interest in positing a true obscurity *in the text*: the point of the phrase ὄντως ἀσαφὲς αὐτό δι' ἑαυτό is precisely that it locates the obscurity essentialistically in the text itself and not in the perception of certain types of readers.

Galen's initial principle that commentators are to "clarify what is unclear"[227] implicitly provides an explanation for why it insinuates that the source texts are obscure, in part even objectively so: if the clarification of obscurity is the supposed main task of the commentators, they must persuasively convey the impression that their source texts are substantially obscure because the obscurity of the source text is their *raison d'être*. If they admit that the obscurity which they clarify is purely subjective, the need for their clarification is, too: they risk not being of interest to expert readers and to those who do not think of themselves as ἀμαθεῖς. By contrast, a "genuine" kind of textual obscurity enables the commentator to make a real contribution in terms of genuine clarification. Moreover, the postulation of an objective type of obscurity opens up various possibilities of 'clarification:' if the reader depends completely on the commentator's clarification of what is "genuinely obscure," the commentator can interpret the alleged obscurity in any way and do things other than clarification without the reader being aware of it.[228] In light of this interpretation of the postulation of an objective type, Galen's discussion of the subjectivity of obscurity and his concession to discuss the subjective type too could, in turn, be understood as a safeguard against those readers who call the identified obscurity into question. The introduction of a subjective type does not relativize the implications of the formula "clarifying

223 See Mansfeld 1994, 150 for labelling Galen's types of obscurity thus.

224 Barnes 1992, 172.

225 However, note that he does not explain the example of "genuine" obscurity which he provides (by contrast with the example for the second type which he does explain).

226 Kanthak 2013, 167.

227 P. 44, n. 214.

228 Mansfeld 1994, 155 f. has coined the term "creative exegesis" for the kind of clarification enabled by obscurity.

what is unclear." Rather, it implies a possibility to 'clarify' even where the text is not really obscure and to insinuate additional obscurity in so doing.

Thus, Galen's proem to his commentary on the Hippocratic *De fracturis* indicates that the commentators fundamentally rely on the obscurity of their source authors because their self-understanding, *qua* commentators, is predicated upon their ability to produce clarity where it was ostensibly absent in the original. This has some implications for interpreting the paradoxical topos of obscurity: as obscurity is necessary for the commentators, they have to discover it in their texts all the time; they even have to construct it wherever they would like to perceive it – this could explain the commentators' eagerness to point out obscurity. Further, they have to explain the paradoxical (omni)presence of obscurity – this could be why they frequently have semi-theoretical discussions of the supposed phenomenon, some of which are quite lengthy. If, however, the commentators construct a good deal of the obscurity which they purport to have identified, their true aim cannot (only) be clarity. Rather, the alleged obscurity of the source texts seems to allow the commentators to interpret them according to their own purposes for the supposed sake of clarification – hence, the ideal of clarity at the center of Galen's description of the commentator's work.

How relevant is then the discussion of obscurity in Galen and other commentators on Greek science for our inquiry on the role of clarity in scientific texts? On the one hand, Galen in the above-mentioned proem presents obscurity as that which prevents the reader's understanding and thus the transmission of knowledge, and clarity (or clarification) as that which enables it. In so doing, he suggests that his own utterances about obscurity and those of his fellow commentators add to the question of the relation of clarity to the (written) transfer of knowledge. Galen even links the problem of the decontextualization of written knowledge communication with the subject of clarity, though not with respect to his scientific source texts, but rather to his own commentaries: he says that he can adjust the scope of his clarification ad hoc in oral teaching but has to prepare his clarification in advance for an average reader when writing a commentary. This could imply that non-commentator authors of scientific texts, too, use clarity for ensuring the comprehension of their readers in the face of the texts' decontextualization, although Galen does not explicitly say so, because in the passage we have been looking at he is focused on the obscurity of the source texts from a commentator's point of view. As for the place of obscurity, Galen in his proem always locates it in the λέξις, even in the case of the recipient-dependent obscurity: the term could simply refer to the source 'text' as opposed to the commentary,[229] or part of it, that is, to a 'sentence' or 'phrase,' but it could also mean specifically the 'phrasing,' 'wording,' or 'diction' of it. In his second example, the obscurity seems to be due to the technicality of the content because it does not exist for those who possess the respective anatomical knowledge. This notwithstanding it could be the wording which enables anatomical ex-

229 LSJ s.v. II 4.

perts to understand and which simultaneously excludes laypeople. Thus, Galen presents obscurity and, by implication, clarity, as qualities of the text and its language and style by which the text is 'incomprehensible' or 'comprehensible,' respectively, to some or all of its readers. That is, obscurity appears as what prevents the transmission of knowledge, clarity as what guarantees it. In this way, Galen's summary treatment of clarity and obscurity seems to be directly relevant for our subject, that is, the relation of clarity and form with regard to the communication of knowledge in ancient scientific texts. Insofar as Galen's discourse forms part of a larger tradition, manifestations of the topos in other commentators, too, can be considered pertinent to the subject of clarity and form in the *Elements*.

On the other hand, Galen's account reveals that the commentators are fundamentally biased toward the subject of clarity. Although Galen has written a whole treatise on it in which he distinguishes types of "the obscure," he does not really seem to be interested in the notion itself. Rather, he seems to apply and discuss it for the sake of something else, that is, supposedly, for the sake of interpreting the text according to his own purposes. This means that he indeed seems to apply the terms 'obscurity' and 'clarity' (as in 'clari-fication') to scientific texts in their original technical sense. However, it must be assumed from Galen's account that for him and his fellow commentators, these terms have a different function: 'clarification' will stand for virtually all things which commentators do to the text, while 'obscurity' will signpost the places of possible commentary intervention in the source text. Because of their special need for the obscurity of their source texts, the commentators' explicit remarks and debates on the subject are inconclusive as to the role of clarity in scientific writing: they cannot serve as a testimony for actual obscurity or clarity in science writing nor can they tell us anything about the place and shape of clarity in the self-understanding of scientific writers and in the expectations of their primary non-commentator readers. Thus, though the contributions of the commentators to the subject are ample – indeed they present the most detailed treatments of obscurity in the context of ancient science writing – they cannot be applied straightforwardly to the analysis of clarity in scientific writing.

Of course, any account of obscurity will be subjective and determined by its function and the communicative context in which it takes place. In the present case, the way in which Galen and other commentators of scientific texts use the notions of clarity and obscurity is presumably shaped by the whole tradition of the concept and is therefore different from how their source authors and their own readers would have applied it many centuries before. Moreover, the addressees of the commentaries would have had different levels of prior knowledge and of experience with the textual conventions than the primary readers of commentaries' source texts and thus would have found different things obscure. Any analysis of the concept of obscurity in a particular commentator must take these factors into account. However, what most fundamentally shapes any given commentator's perspective on obscurity is that clarity is at the heart of the commentators' self-image. With regard to this, the commentators have their own agenda with obscurity, and it is this agenda by which their perspective on obscurity differs essentially from the point of view of the author and the addressees of scientific texts.

It is noticeable that the topos of obscurity is not limited to commentaries of scientific texts but also features in the other great traditions of ancient commentary writing: in the commentaries on biblical texts and in the poetic scholia, amongst others. The reason could be that poetical, biblical, and scientific texts are similarly obscure, but it is more likely that 'clarity' is part of the self-image not only of scientific commentators rather than that the source texts are actually obscure. This, too, suggests that the obscurity topos in commentaries on scientific texts tells us more about the agenda of the commentator than about the comprehensibility of ancient scientific texts.

The manifestations of the obscurity topos outside of the Galenic passage which we have discussed in detail confirm that the commentators use it primarily for their own purposes.[230] The commentators treat 'obscurity' selectively and only as far as it comes in handy. For the most part, they debate *why* an author is unclear and often diagnose 'intentional obscurity.'[231] This category explains the above-mentioned paradox of the obscurity of scientific writing; it allows the commentators to reconcile their own insinuation of obscurity with the canonical status of the scientific source text which indicates that is largely clear[232] and to provide 'clarifications.' Less frequently do the commentators systematically discuss what constitutes obscurity from a formal point of view – even though they use the term for the rhetorical virtue of style – and the means by which the text can be clarified.[233] This gives rise to doubts as to whether they indeed perceive obscurity and attempt to remove it.

The ancient contemporary discourse on clarity: Aristotle

The topos of obscurity forms part of a larger ancient discourse on clarity and obscurity. Terminologically, the commentators stand in the tradition of the classical rhetorical treatises in which σαφήνεια is a virtue of style. From pre-Euclidean times, the most comprehensive discussion of σαφήνεια comes from Aristotle on whom I focus

230 See for an overview of the main passages my Kanthak 2013.

231 The term by which they do so is ἀσάφειαν ἐπιτηδεύειν ('to practice obscurity [sc. deliberately'], e.g., Galen, *In Hipp. Prorrh.* Corp. Med. Graec. V 9.2, 100.9–10).

232 Cf. Sluiter 2016 on the paradox of the obscurity of canonical texts. The most detailed discussions of 'intentional obscurity' can be found in the commentators of Aristotle (Kanthak 2013, 171–177); Galen does not only apply this category to Hippocrates, but also pejoratively to his opponents and 'non-Hippocratic' writings (ibid., 168).

233 The only distinction of textual manifestations of obscurity can be found in the commentators of Aristotle (Kanthak 2013, 174–177).

briefly in order to debate the relevance of the ancient discourse for our subject. He deals with σαφήνεια not only in his *Rhetoric* but also in the *Poetics* and in his logical treatises.[234] The treatments in the different writings bear resemblance to one another and seem to form part of an overall discourse in Aristotle; he names similar means for achieving clarity in each.[235] But they also differ from one another, which points to the underlying assumption that what is deemed clear depends on the generic context.

The logical treatment is arguably the most relevant: Aristotle develops his scientific theory with recourse to mathematics, using both features of the logical form of elements texts (definition, proof) as well as examples from mathematics. It is both with regard to definitions (ὅροι) and to the proof (ἀπόδειξις) that he discusses the need for clarity and the means to achieve it. However, Aristotle's logical treatises, while using mathematics, are not *about* mathematics. Therefore, his remarks are not straightforwardly applicable.[236] This can be seen from the fact that the features of the form of elements texts do not conform to the Aristotelian prescriptions.[237] In the passage about ἀκριβολογία from the *Metaphysics* which has been quoted in Section 1.3,[238] Aristotle himself argues that understanding depends on the use of the language to which one is accustomed and uses precisely the example of the mathematical language which is a commonly accepted 'jargon' amongst some readers yet is unfamiliar and irksome to others. In this way, Aristotle implies that comprehension and comprehensibility depend on the recipient.

The focus of the *Rhetoric* is rather narrowly on public speech, even though large parts are written in such a general way that they seem to be applicable to other genres.[239] The question is whether they are also applicable to scientific writing and, in particular, to elements texts. Science and rhetoric have in common that they are concerned with speakers and writers who must make themselves unmistakably understood to their audience: as Quintilian puts it with regard to speeches at court: "We should not speak so that it is possible [sc. for the judge] to understand us, but so that it is impossible to misunderstand us."[240] And in any case, ancient scientific writers are trained in rhetoric so that scientific writing will have been influenced by the art of rhetoric.[241] However, the category of "appropriateness" (τὸ πρέπον) which Aristotle

234 I focus on the rhetorical and logical treatments in my brief account.

235 For a comparison of Aristotle's discussions of form and clarity that is focused on the relation of clarity and the *terminological* form cf. Gasser 2024.

236 Netz 1999a, 45: "[. . .] Aristotle – as ever – has his own, non-mathematical project, which makes him a difficult guide." Cf. also Kullmann 1981 *passim* for Aristotle's distortion of mathematical examples.

237 Cf. Section 1.2 (p. 22, n. 125) for the relation of the Aristotelian first principles to the Euclidean ones.

238 Pp. 35 f.

239 Rapp 2013, 285–287.

240 *non ut intellegere possit (sc. iudex), sed ne omnino possit non intellegere curandum* (Quint. *Inst. or.* VIII 2.24).

241 van der Eijk 1997, 80.

defines as "that which is 'analogous' to the subject matter" (τοῖς ὑποκειμένοις πράγμασιν ἀνάλογον)[242] seems to suggest that the rhetorical account of σαφήνεια cannot be used for writing and interpreting elements texts. The notion also occurs in the definition of "good prose style" in *Rhetoric* III 2: "[. . .] the excellence of the stylistic form should be defined as being clear (for since a speech is a sign, it will not fulfill its proper work if it is not clear), and neither banal nor exceedingly sublime, but appropriate."[243] Christof Rapp has argued persuasively that Aristotle does not mention two separate virtues here (clarity and appropriateness) but, since he uses the singular, only one (clarity), and that the criterion of 'appropriateness' refers to the opposition 'banal/sublime:' good prose style should be appropriately sublime. This criterion is connected with the virtue of clarity insofar as too much clarity produces banality. Therefore, appropriate sublimity is needed to counterbalance excessive clarity, e.g., by an appropriate quantity of uncommon words. The appropriate degree of sublimity is determined by the object of prose speech by contrast with that of poetry.[244] This implies that the appropriate degree of sublimity would be a different one in another prose genre. Indeed, in the above-discussed passage, Aristotle himself points out the difference between the mathematical and rhetorical style, claiming that "no one teaches geometry in this (i.e., the rhetorical) way."[245] As I have argued in a recent paper, mathematical and logical texts occupy one end of a scale in Aristotle's discussions of clarity and form, that is, the end at which clarity does not have to be counterbalanced.[246] Indeed, the main textual feature of elements texts, standardization, is not reflected explicitly by the rhetorical accounts, even though Aristotle, both in the rhetorical and in the logical accounts, recommends lexical standardization by saying that homonyms must be avoided for the sake of clarity.[247] The recommendation to use ordinary words that Aristotle makes in all his treatments of the concept could be referred to the lexicon of the *Elements*; but neither does the work seem to be clear specifically because of its use of ordinary words – they are used within otherwise unusual phrases and verb forms – nor will Aristotle have made the recommendation with regard to the mathematical λέξις.

---

242 *Rhet.* III 7.1408a10 f. (Porter 1995, 125).

243 [. . .] ὡρίσθω λέξεως ἀρετὴ σαφῆ εἶναι (σημεῖον γάρ τι ὁ λόγος ὤν, ἐὰν μὴ δηλοῖ οὐ ποιήσει τὸ ἑαυτοῦ ἔργον), καὶ μήτε ταπεινὴν μήτε ὑπὲρ τὸ ἀξίωμα, ἀλλὰ πρέπουσαν (*Rhet.* III 2.1404b1–4). My translation, which I quote from Gasser 2024, 147, follows the German translation by Rapp 2013, 290.

244 Rapp 2013, 289–294.

245 *Rhet.* III 1.1404a12 (the passage has already been quoted in Section 1.3, p. 35 with n. 183 above).

246 Gasser 2024, 161 f. and 175.

247 E.g., *Top.* VI 2.139b–140a.

The use(s) of 'clarity' in the present study

As it is not certain how far, if at all, the logical-rhetorical concept of σαφήνεια is applicable to elements texts, I shall not base the analysis of the form of the *Elements* on it – neither in terms of the stylistic means which are recommended for achieving it nor in terms of the semantic nuances of the term whose meaning is notoriously difficult to pin down. I shall, however, broadly define my own non-technical usage of 'clarity' in a double sense with regard to the two main areas of the ancient contemporary discourse of the concept, i.e., the logical and the rhetorical ones: on the one hand, as 'unambiguousness' and, on the other hand, as 'comprehensibility' or 'unmistakability'. I shall *not* consider 'clarity' in the sense of 'clearness' or 'vividness' with regard to the diagram and its possible function of 'visualization' (that is to say, in a sense which is also present in σαφήνεια but more often connected with ἐνάργεια). This is because as regards the diagram, my interest is in its standardization which cannot be proved on the basis of considering how it visualizes.

This is how I shall use the notion of 'clarity' in the different parts of the study: first, I will meticulously analyze the textual standardization regarding the kind and number of features which are standardized and the degree to which each of the features is standardized. Here, unambiguousness suggests itself as a benchmark by which to analyze standardization because it is, in the proper meaning of the word, an immediate consequence of complete standardization. Therefore, I shall assess both the degree of the standardization of the individual features and the nature of the features which are standardized with respect to unambiguousness. Second, I shall interpret the standardization of the individual features of the text by considering the overall function and meaning of the standardization of the textual form. Here, clarity suggests itself as a function of standardization: both in view of the mathematical content and in view of the general problem of how to avoid misunderstanding in knowledge communication. On the one hand, the proofs need to be accomplished and described consistently, irrefutably, and unequivocally and are therefore in need of unambiguousness; on the other hand, the propositions need to be communicated unmistakably. Therefore, I examine the relation between standardization and clarity as 'unambiguousness of proof' and as a function of knowledge communication.

Thus, it is in the following three respects that the relation between standardization and clarity-as-unambiguousness will be tested: (i) the standardization of all individual textual elements, (ii) the textual requirements of the proof 'logic,' and (iii) the comprehensibility of the propositions. Both the assessment of the standardization of the individual textual features and the interpretation of their standardization with regard to proof 'logic' are focused on the text and consider clarity/unambiguousness as a quality of it. Only the assessment of the communicative dimension of the textual standardization takes the perspective of the readers of the *Elements* into account. Different readers will be considered, but the focus is on the original context of reception. With regard to this communicative situation, I will consider clarity as a function of standardization and eventually identify the meaning and function of that part of the standardized form which cannot be explained by the effort to be clear.

1.5 Review of the literature

As of yet, there is no study of Euclid's *Elements* which focuses on its textual form *as textual* (or *literary*) *form*. Neither is there any study of the relation of the textual form and clarity in the *Elements*; the link has been noted but not systematically investigated.

Studies on the *Elements* and its genre are largely focused on questions of the work's textual transmission,[248] on the origin and implications of the name of the genre,[249] the prehistory and development of the *Elements* and the reconstruction of Euclid's predecessors from the text of the *Elements*,[250] on issues of date,[251] on the *Elements* in the history of Greek mathematics,[252] on Euclid's method (what I call the 'logical' form), also more generally from the perspective of the philosophy of mathematics,[253] the translatability of the verbalism of Euclidean mathematics into the formalism of modern mathematics,[254] on the social context,[255] on the relation of philosophy and mathematics,[256] on the late ancient reception,[257] and more recently, on the transmission and logical function of the diagrams.[258] Some of these explicitly or implicitly touch upon the issue of textual standardization: e.g., by claiming that the standardization of the text is due to some extent to its commentators and editors;[259] by concluding from the linguistic and stylistic standardization the existence of a common source of certain books as part of the reconstruction of pre-Euclidean mathematics;[260] by considering the substitution of words and phrases of the mathematical text with modern mathematical

---

248 E.g., Heiberg 1885, 1900, 1903a, 1903b, 1903c, Knorr 1996, Saito 2009, Vitrac 2001a, 2001b, Acerbi 2016.

249 Burkert 1959.

250 E.g., Knorr 1975, Neuenschwander 1972/73, Artmann 1985, 1991, 1999, Fowler 1999, Vitrac 2004, 2008. With a focus on the origin of the first principles: Szábo 1962, van der Waerden 1977/78, Knorr 1981, Acerbi 2013.

251 E.g., Knorr 1978.

252 E.g., Heiberg 1882, Becker 1936a, Seidenberg 1977/78, Waschkies 1989, 1998, Cuomo 2001.

253 E.g., von Fritz 1955 (on first principles), Neuenschwander 1972/73, Mueller 1969, 1981, Cambiano 1992, Artmann 1999, Waschkies 1995, Vitrac 2012 (on 'proofs by contradiction').

254 Cf., diametrically opposed, Unguru 1975 and van der Waerden 1976.

255 E.g., Asper 2007, 2009, Netz 2002a, Netz 2002b, Fowler 1999, Hoyrup 1990.

256 Asper 2003, Burnyeat 1987, Burkert 1982, Carter 1986, Detel 1993, Cleary 1995, Knorr 1981, 1982.

257 E.g., Netz 2004b.

258 Overview in Mumma & Panza 2012. On the manuscript transmission: Saito 2006, 2009, 2012, Saito & Sidoli 2012. On the deductive role: Manders 2008, Allwein & Barwise 1996, Mumma 2010, 2012, Panza 2012, Shin 2012.

259 Netz 2004b.

260 Neuenschwander 1972/73.

formulae (which is only possible on the basis of the complete standardization of the verbal units which are to be substituted); or by noting the implicitness of the references to previous propositions and constructions (which is accomplished precisely by their verbatim repetition, which, in turn, is possible because of the standardization of the text).[261] However, in such contexts, the standardization of the text is merely presupposed, but not itself considered from a literary(-aesthetic) perspective. Individual linguistic features have been mostly regarded with regard to their deductive function,[262] and most of the observations which have supposedly been made regarding 'form' and 'style' actually refer to the 'logical' rather than to the literary form.[263]

Only most recently, studies of specific traits of the textual form of the *Elements* and other elements texts from a literary perspective have emerged: Asper has analyzed narratives and narrative structures in Greek mathematics (and in Greek science more generally): on the one hand, anecdotes, which cannot be found in Euclid (only *about* Euclid), but only in the introductory letters of contemporary elements texts, e.g., by Archimedes, and in later mathematics, mostly in commentaries;[264] on the other hand, he has argued that the structure of the (written) proof ("proof-as-published"), though not being a narrative one, nevertheless contains formal elements of narratives; and that the proposition text, in this way, narrates the proof as the reader receives it (as opposed to how the author found the proof: this is the narration of the "proof-as-discovered" which is obliterated by the actual proof 'narration').[265] Netz has shown in his 2009 study *Ludic Proof* that Archimedes and Apollonius even use special narrative strategies in structuring some of their books in order to play with the expectations of the readers and create effects of confusion, suspense, and finally, surprise.[266] While Asper merely notes that (ordinary) proofs and literary narratives use the same linguistic means, Netz goes as far as to interpret the special narrativity of the Archimedean proofs within "the Alexandrian aesthetic." He claims that it characterizes Hellenistic poetic and mathematical works alike in terms of "a certain carnivalesque play of detail, with its ironic self-undercutting; mosaic composition; narrative surprise; indeed a certain

261 Asper 2007, 129 with references.
262 Federspiel 1992, 1995, Acerbi 2008, 2011a, 2011b.
263 Szábo 1994, 365 is an exception in suggesting an aesthetic interest in linguistic formalization.
264 Asper 2013b on "accounts of progress" in Archimedes, Hero, Pappus, and Proclus; 2011b on "frame tales" in Greek mathematics (94–99), and 2011a, 27–31 on anecdotes about the attitude of mathematicians toward money.
265 2011a, 12 f.
266 2009, 91. Asper 2011a, 16 f. notes that "unexpectedness" and "surprise" are among the "'purely aesthetic' qualities" which the mathematician Hardy ascribes to *Elements* X 9 and IX 20 (1940, 113) which have been chosen as the most beautiful theorems in a later survey. The notion of 'mathematical beauty' has not been discussed with respect to the Greek theoretical mathematics, except in Netz's study (Asper 2013, 2). In antiquity, the concept of 'mathematical beauty' is a Platonist, i.e., a different one (Asper 2011a, 15).

tendency to experiment with one's very generic boundaries."[267] On a similar note, Doxiadis & Sialaros argue that the textual structure of the mathematical proposition resembles the "standard form" of forensic speeches and the "chiastic and ring-compositional patterns" of rhetorical πίστεις.[268]

Thus, these recent literary-aesthetic studies of Euclid and other elements texts focus on certain aspects of the form, particularly the 'narrative' and 'stylistic' ones, and interpret them as the influence of other genres or the zeitgeist or as the result of using the same tools as other, 'more' literary genres. They do not, however, make reference to what according to the present study is the most important characteristic of the literary form, i.e., standardization, although it seems to underlie the stylistic traits on which these studies focus: as I have pointed out above, the rhetorical figures which can be observed in the *Elements* (e.g., ring composition and chiasmus) all come about as variations of repetition, which, in turn, is the result of standardization. The Archimedean play with the structure, in turn, presupposes first of all a standardized type of structure to play with, such as can be found in the *Elements*.

The standardization of the textual features of the *Elements* is at the center of three larger studies, though it is not considered *as* textual standardization in either of them and not systematically with respect to clarity.

First, Netz's *The Shaping of Deduction. A Study in Cognitive History*. This seminal book provides an extremely detailed and comprehensive analysis of the language and style of elements texts according to the following aspects: the lettered diagram (chapter 1), the letters (chapter 2), the lexicon (chapter 3), and "formulae" (chapter 4), by which Netz means "groups of words" (but also single words) which recur in set meanings and functions as building blocks of the text.[269] The "formulae" thus account for the form of the proposition structure.[270] The analysis of the proposition structure is complemented by a more detailed assessment of the value of the Proclean scheme of the parts of the proposition, published in the same year.[271] Using samples not only from Euclid's *Elements* (including Hypsicles' book which survives as *Elements* XIV)[272] but also from Euclid's second main work, the *Data*, as well as from all other main authors of Greek elements texts, Netz makes a great number of quantitative assessments of the standardized features of the text.[273] In so doing, he effectively assesses

267 Netz 2009, 227. Cf. Asper 2013d, 77 and Section 5.3, 434, n. 35 of the present study on the hypothesis that the "ludic" is due to a Hellenistic zeitgeist.

268 2013, 390–403.

269 Cf. below pp. 127–130 for criteria for what counts as formulaic according to Netz.

270 The analysis of the proposition structure as part of his chapter 6 ("The shaping of generality") is focused on the 'logical form' (see above pp. 16–27 my outline of the features of the 'logical' form).

271 1999b.

272 Cf. p. 1, n. 7.

273 As in the present study, Netz's focus is on the geometrical propositions of these authors (Lehoux 2000, 1 f.).

their standardization, even though he does not conceive of the results in terms of textual 'standardization,' a term which accordingly does not recur at all in his study.

While his 2009 study of the "playful" manifestations of the (otherwise standardized) textual form is a literary-aesthetic one, he takes a *cognitive* approach in analyzing the regular features of the proposition form in his 1999a.[274] Here, Netz's interest is not in the features of the textual form themselves, but in their function as "cognitive tools" shaping the deduction, as "*prerequisites* of deduction."[275] He claims to offer both "a description of the practices of Greek mathematics" and "a theory of the emergence of the deductive method,"[276] but it turns out that the former happens precisely with regard to the latter. Netz describes the practices with respect to how they help bring about the proof – however, not only in writing but also conceptually in the mind of the mathematician. The underlying assumption of the cognitive approach is that the text form reflects how the mathematician first of all conceives the proof.[277] However, as a matter of fact, the form is solely a matter of *writing* the proof.[278] Indeed, Netz's sequence of steps of how the proof came into being has 'writing the final version' as a well-defined distinct step; but he does not present the difference between the preliminary conception and the final version as a qualitative one.[279]

The cognitive approach determines the choice, conceptualization, and scope of the features according to which the form is assessed, and in so doing, also to some extent the outcome of the analysis (as well as its interpretation): e.g., Netz assumes that the mathematicians think and write in "formulae," and accordingly, analyzes the texts with

274 By "regular" I mean that Netz in his 1999a study, rather than focusing specifically on the deviations from standardization as in his 2009, looks at the characteristic standardization of the propositions by means of his assessment of larger samples. I shall argue toward the end of the present study (Section 5.3, 434 with n. 35) that Archimedes' and Apollonius' occasional playfulness is to be explained precisely against the backdrop of the otherwise extremely standardized form (which is, as I shall argue, no less aesthetic and, in any case, equally deserves a consideration *as form*).

275 1999a, 89. Accordingly, his chapters 5 and 6 investigate how "necessity" and "generality," as logical preconditions of the proof, are "shaped."

276 1999a, 1. The third aim is to provide – "implicit[ly]" – a "case-study for a general view on the history of science."

277 Cf. Netz 1999a, 311: "Throughout the book, I have stressed form rather than content, partly as a method of getting at the cognitive reality behind texts [. . .]."

278 Asper 2003b, 9 with regard to "formulae," which are, if at all (see my discussion of the concept in Section 3.3.3), only a feature of the written proof. Cf. above pp. 14 and 26 on the vital difference of discovering the proof and 'writing up' the proposition.

279 Netz 1999a, 85 does point out a major difference: "Finally, a text, which assumes this lettered diagram, is written down. Because this text assumes this lettered diagram, it will be different from the dress rehearsal [. . .]." However, as the lettered diagram crucially determines the form of the proposition text, this text will be *completely* different from any version which does not presuppose the diagram and no longer reflect previous cognitive processes.

the help of the concept only to find it to be shaped indeed formulaically.[280] Overall, the quantitative assessments reveal that the features are "conventionalised." Netz concludes from the significant deviations from the conventions[281] that the "conventionality" is "neither random nor regulated" through "explicit codification, something like a style manual,"[282] but "self-regulating;"[283] he imagines that in a professionalized context, a group of writers[284] tends to use the same language for writing about the same "subject matter," by which "self-perpetuating" process "conventions [. . .] will necessarily emerge." Though Netz admits that "the conventionality is not just the operation of blind, general cognitive forces [but] does have a component of intentional agency,"[285] he does not say why the mathematicians repeat the features they do, why they repeat them in such an extreme way, and why they started repeating them in the first place.[286] Netz is vague on this: he calls the driving force of repetition a "natural tendency"[287] and describes the emergence of a convention thus: "What starts as the raw material of conventions, what is merely the work of natural forces, becomes, by being tinkered with, a meaningful tool, a convention."[288] If, however, the conventions of the form are thought to be the result of "natural" or "cognitive forces,"[289] this suggests a natural connection between the formal features and the proof content; the form appears to be the most intuitive and useful way of coming to terms with the content. But what does it then mean to say that "Greek mathematics stresses form," that it is "a cultural practice in which the dominant was the form?"[290] How can it be more important than the content if it is shaped by its cognition (and shapes the deduction of the propositions in turn)? In my view, the question rather is what is more important for shaping the form: cognitive, logical, or communicative efforts or an

280 In Section 3.3.3, I shall challenge the formulaic nature of the texts (in the way in which Netz suggests it).

281 "The overall structure [sc. "of the regularity"] is 'usually – but not always' [. . .]" (1999a, 78).

282 1999a, 74, and see 90, 98, 123, 126, 154 f., 253.

283 1999a, 74, 78 f., 88, 113, 126, 239.

284 Tellingly, Netz writes "group of speakers," whereas the extreme conventionality which he observes is certainly only possible in *writing*.

285 1999a, 78.

286 As Asper 2007, 211 argues, "self-regulating conventionality" cannot account for the emergence of a convention: "So lässt sich allerdings nur das Fortwirken einer solchen Konvention erklären, nicht ihre Entstehung." I shall argue that it does not even account for its perpetuation.

287 1999a, 74. The deviation, in turn, is to be explained by "inexact[ness]" of "memory" (ibid.).

288 1999a, 78. This and the previous remarks are made with regard to the conventions of letter use, but Netz assumes "self-regulating conventionality" to be at work throughout.

289 Although Netz does not distinguish explicitly in his book between 'logical' and 'stylistic' (i.e., in my terminology 'textual' or 'literary') form, as he does in his 1999b where he regards the Proclean parts of the proposition as "stylistic" rather than "logical" form (cf. p. 23, n. 134), it is clear that he thinks that it is the stylistic form which 'shapes the deduction;' he mentions several times the "style" as the object of his discussion and only once, in a different sense, the "logical form" of the *protasis* (259).

290 1999a, 311.

interest in the form itself (either for aesthetic reasons or for the sake of endeavors which do not have to do with the proof content and its unambiguous communication)? Netz does not consider unambiguousness/clarity as a function of standardization;[291] he does not discuss any communicative function of standardization at all. Neither does he debate any aesthetic functions of the standardized form.

The second main contribution on the standardization of the textual form is Acerbi's 2010 study *Il silenzio delle sirene. La matematica greca antica*; a large part of its first chapter is concerned with the style of Greek theoretical mathematics,[292] more precisely, with what Acerbi calls "the demonstrative code, in which all geometrical texts are written."[293] This study is complemented by his 2012 comparative analysis of the three "stylistic codes" which he recognizes in Greek mathematics (demonstrative, procedural, algorithmic), in which he shows how the demonstrative one became the canonical one.[294] Acerbi provides a detailed and comprehensive analysis of the standardized textual features of the Euclidean geometrical propositions. Contrary to Netz who grasps the style of the proposition parts in terms of "formulae," Acerbi analyzes style according to traditional linguistic, above all, grammatical categories, such as tenses and, above all, particles. In this way, he accounts for the standardization of the *Elements* and other texts of the genre more fittingly, in my view. However, Acerbi only deals with the results of standardization, that is, with the standardized form as it is. But he does not enquire into how the standardization initially came about nor does he assess its exact degree.[295] This is because his interest is not in the form of the text as such, and in the variety of functions it could have; he regards it exclusively as serving the logic of the proof, more precisely with regard to the above-mentioned problem of the generalizability of the particular proof. He argues that the textual features ensure the "generality" of the proof. Thus, while Netz exclusively considers cognitive functions of the standardized text form, Acerbi presumes that the function of the text

291 He only considers it once *ex negativo* when he points out that "ambiguity," that is, of the lexicon, is common outside mathematics, e.g., in the Hippocratic corpus (1999, 121); otherwise, he only explicitly dismisses "disambiguation" as a function of the explicit standardization of the definitions (1999a, 99–103).

292 "come è scritta la matematica greca antica" (2010, 21–56).

293 Acerbi 2018, 269.

294 Acerbi 2012 ("I codici stilistichi della matematica greca: dimostrazioni, procedure, algoritmi"). Another major contribution by Acerbi was published in 2021, i.e., his monograph *The Logical Syntax of Greek Mathematics*. I was only able to consider it in Section 3.3.1 on sentence constructions.

295 His 2012 paper, however, considers the standardization of mathematical texts in the course of their transmission. Acerbi argues that even texts which were written in one of the other two styles were gradually revised according to the "demonstrative code." One might consider whether the preference for the demonstrative style has anything to do with its standardizability, and if so, why standardization was preferred: whether for logical-communicative or for other reasons, and whether this betrays something about the original driving force for the standardization in those texts which were written in the "demonstrative code" to begin with.

form is a logical one.[296] Like Netz, he does not consider whether the meaning of the form exhausts itself in the function which he attributes to it.

The third major contribution to the study of the standardized form of Euclid's *Elements* and elements texts in general is the large, monographic chapter on axiomatic-deductive texts of Asper's *Griechische Wissenschaftstexte*.[297] Asper mentions "linguistic standardization" as one of the main features of the form of elements texts.[298] He analyzes the standardization of elements texts comprehensively, but, for the most part, not linguistically. That is, he does not himself make quantitative assessments as to the exact degree of standardization, or close analyses of the standardized features, except in select cases (e.g., that of the third-person perfect passive imperatives).[299] He concludes from the review of the textual form that the main function of the standardized form is unambiguousness for the sake of facilitating the reception of the text,[300] without investigating the exact relation between the formal features and communicative clarity. This is because his interest is in the emergence and function of standardization as it can be explained by the literary and socio-communicative context of the writers of elements texts. Asper focuses on the "pre-Euclidean period," which he defines as the time before Euclid became canonical, i.e., the time from the second half of the third century onward (that is, he considers the composition, as opposed to the reception, of Euclid's *Elements* as belonging to the "pre-Euclidean" period). The reason is that in the "post-Euclidean period," the communicative context of elements texts changed: the *Elements* itself became canonical and was used in institutionalized teaching and in philosophical contexts.[301] However, as regards the "pre-Euclidean period," the emergence and the socio-communicative function of the textual form cannot be concluded from Euclid and his predecessors directly. The *Elements* itself does not betray anything explicitly about its communicative context, and all earlier elements texts are lost, so the development of the features, for the most part, cannot be traced. What is more, all information about the context and function of Euclid's predecessors stems from the later philosophical reception and can be used only cautiously. Asper uses the later reception mostly *e contrario*; in addition, he takes into account a great number of sources and comparative textual traditions, both Greek and from other cultures. From these, he draws conclusions about the circumstances of knowledge transmission under which Greek theoretical mathematics emerged and the ac-

---

296 Cf. above Section 1.2, p. 24 for 'generality' as a feature of the 'logical' form. In Section 5.1, I try to determine how far the textual features intend to serve the aim of logical unambiguousness, e.g., by creating generality, and to what extent they have different functions beyond the logical one.

297 2007, 94–212.

298 2007, 113. Cf. above p. 2, n. 9.

299 2007, 116–125.

300 Asper 2007, 124 f.

301 Asper 2007, 210.

cording cognitive-communicative function of the form, about the "milieu"[302] in which the mathematicians first applied the textual form, about earlier models of the form, and about its social functions:[303] the standardized form appears to have compensated for the lack of institutions in which knowledge would have been transmitted; it served the functions of building consensus and securing unmistakable communication in written, decontextualized circumstances of knowledge transmission;[304] it is indicative of the lack of competition among mathematicians;[305] it could have been modeled analogically on that of early law texts, which uses similar features for similar purposes;[306] and apart from cognitive-communicative purposes, it served to distinguish the practicians of such standardization from other mathematicians of the "non-Euclidean," "practical" tradition of mathematics.[307]

The present approach to the standardized form of the *Elements* differs from those of all three preceding major works on the subject, but it builds on the methods, achievements, and insights of all of them. The interest of the present study is in the meaning and function of the extreme manifestation of the generically standardized form of the *Elements*.[308] Unlike previous studies, this analysis starts from and is focused on the form itself in order to retrieve its function. This is partly because we lack information about Euclid's communicative context[309] and partly because the form is so conspicuous that I expect it to yield clues as to its functions.

As this study is focused on the form of the text, it is concerned with the written product, the "proof-as-published."[310] In this way, it does not consider any processes prior to writing the proof which led to the proof discovery because the text form will not have had any function in them.[311] Further, it regards the textual form as contingent; by this I mean that it does not naturally or necessarily emerge from the proof content and that it is not the only or mathematically most convenient form for presenting the proof. I question precisely any logical or cognitive-communicative functions of the form in order to find out to what extent the form serves these purposes at all. Therefore, I analyze the textual features with regard to non-mathematical, above all, grammatical categories (for which cf. also Acerbi 2010), which I take as indicators

302 Asper 2007, 157.
303 Asper 2007, 94–212.
304 Asper 2007, 147–156.
305 Asper 2007, 157–173.
306 Asper 2007, 173–196.
307 Asper 2007, 197–210. Cf. Sections 5.2, 428 f. and 5.3, 430–432 for my discussion of his hypotheses.
308 Cf. above p. 10 f. and below p. 66.
309 Conversely, the focus of the chapter on elements texts in Asper 2007, 94–212 has been the original motivation for standardization, that is, in the first texts of the genre previously to Euclid (p. 61 f.); in the case of these, we do have context information available but lack the works themselves.
310 For Asper's term, cf. above p. 56.
311 Cf. above, pp. 14, 26, and 58 on the distinct processes of discovering the proof and 'writing up' the proposition.

of the degree and kind of standardization rather than regarding the form from a functional point of view from the outset. For assessing the standardization according to these textual categories, I adopt the quantitative method from Netz 1999a. While I disagree with his main assumption regarding the origin of the standardized form and accordingly deviate from certain resulting methodological decisions as well as the interpretation of the results of the statistical evaluations, I will nevertheless use his study as my main point of reference in the analytical part (Chapters 3 and 4). In addition to applying the quantitative method, I will consider some of the original concepts according to which Netz analyzes the form, but will challenge and either substitute or readjust them so as to fit my own purposes.

It is only when I subsequently ask what motivates the form that I consider the logical and socio-communicative functions suggested by the earlier studies (Asper, Netz) with regard to the amount of standardization for which they account. For this purpose I use the notion of (logical and communicative) clarity which has not been at the center of any of the previous studies on Euclid.

1.6 Method and plan

Overview

The study consists of four chapters: Chapter 2 traces the terminology of 'clarity' in the commentators Pappus and Proclus. It is focused on the reception of the form of the Euclidean *Elements* in terms of clarity. In this way, the chapter stands apart from the subsequent main part (Chapters 3 and 4) which is concerned with the literary form of the Euclidean text itself. Chapter 3 analyzes comprehensively the standardization of the *text* of the proposition, Chapter 4 the standardization of the *diagrams* and of the diagrammatic letters in the text. Chapter 5 discusses the results of the analysis of textual and diagrammatic standardization with respect to the relation of standardization and cognitive-communicative clarity and offers a conclusion on the relation of form and clarity in the *Elements*.

Chapter 2

This chapter frames the study of the relation of form and clarity in Euclid's *Elements* with an analysis of the concept of clarity in the two extant commentaries of the work, that is, by Pappus (early fourth century AD) and Proclus (fifth century AD). Additionally, one could consider the excerpts of the lost commentaries by Hero of Alexandria und Simplicius in al-Nayrīzī's Arabic commentary which has survived in a Latin translation by Gerard of Cremona. However, I have decided to concentrate on the two commentaries which survive in substantial parts or as a whole, respectively, and which are not mediated in such a complex way as the two lost Greek commentaries. For I will analyze the language of the commentators and the context in which they use certain terms.

The focus of this chapter will be mainly on Proclus' commentary of *Elements* I because, of Pappus' commentary of the *Elements*, only that on the 10th book survives (in Arabic). There are two reasons why I place the study of the perspective of the commentators before my own analysis of the *Elements*: first, Proclus and Pappus both attribute clarity to Euclid contrary to the usual practice of the ancient commentators of scientific texts who perpetuate the topos of the obscurity of their source authors; second, Proclus is the only extant author to use specifically the term σαφήνεια ('clarity,' 'distinctness')[312] with respect to Euclid's *Elements*, the *terminus technicus* for the rhetorical virtue of style. In this way, the attribution of 'clarity' to Euclid in the two extant commentaries of the work is directly relevant to our study of the form and clarity in the *Elements*. The aim of the chapter is to consider whether the commentators, especially Proclus, provide (however late) ancient evidence for linking the textual form of the *Elements* with cognitive-communicative clarity, and in which way. For this purpose, I will take into account the whole language of clarity in Proclus, which includes, but is not limited to, cognates of σαφήνεια. As this chapter will be concerned with the reception of Euclid's *Elements*, it will stand apart somewhat from the subsequent chapters which will be based on the text of the *Elements* itself.

Chapters 3 and 4

These two chapters form the bulk of the study. The aim of the comprehensive analysis of the standardization is to show how much and in which way the textual and diagrammatic features are standardized individually. I now confine myself to making only a few remarks on the method and the corpus and will elaborate on them in the general introduction to the two chapters at the beginning of Section 3.1 below.[313]

The main method which I use in the analytical part (Chapters 3 and 4) is the quantitative assessment of the individual features of the textual and diagrammatic form. My choice of using this method instead of more conventional philological ones is due to the subject matter: standardization can only be measured and analyzed in a statistical way. I assess the number and kind of different textual components resulting from standardization and the actual degree of their standardization.[314] This includes making various counts of the textual features on all levels of the text. In evaluating these counts, I regard as 'standardization' any tendency of using the same formal ele-

312 For the problem of translating the term cf. above p. 2, n. 13.

313 Cf. Section 3.1.1, 108–113.

314 'Standardization' may refer both to the process and to the result of this process (cf. above p. 30).

ments in the same textual parts or function, that is, as a tendency of formal unification, toward formal uniformity or sameness. I regard this tendency as a result of a preferred application of certain formal elements in the place of alternatively available elements which the text applies less frequently or in a less unified way or not all.[315] This means that I do not conceive of 'standardization' as perfect and complete uniformity, but only as a certain degree of sameness. However, for reasons which I shall explain below,[316] I do not predefine a minimum degree, i.e., a certain required percentage of unification which I count as 'standardization.' Nor do I attempt to determine systematically whether the standardization of a feature is indeed the result of its preferred application, let alone whether it suggests an effort to standardize.[317] I leave this question for Chapter 5; in the analytical Chapters 3 and 4, I merely consider for each textual feature how strong the tendency of textual uniformity is with regard to the virtual top-end of the scale in the case of each textual feature, i.e., 'full,' 100% standardization. For referring to this benchmark of complete standardization, I use the term 'unambiguousness.' Apart from quantitatively assessing the textual standardization, I also analyze it qualitatively: I assess what *kind* of formal features are standardized, how standardizable and unambiguous these features are *per se*, and how far they, if at all, add to the unambiguousness of the text. Here, now, I use the term 'unambiguousness' in a different, looser sense: for referring to the effect of standardization or as a characterization of certain formal features. Both the quantitative and the qualitative analysis will be focused on the text as far as possible, but I take into account the perception of different kinds of readers wherever it is instructive.

As far as the text corpus is concerned, I shall limit the analysis to *part* of the *Elements* rather than taking the whole work into account and shall limit it to the *Elements only* rather than comparing other works. As explained above, the focus will be on the textual form of the *geometrical propositions* rather than on the arithmetical propositions[318] and the axiomatic lists.[319] Instead of taking all geometrical books into account, I shall use book I of the *Elements* as a convenient sample wherever I need to statistically evaluate a larger portion of text: the book is the most famous and most widely received one; it is the book on which Proclus' commentary is based, which I discuss in Chapter 2 and which contains influential conceptualizations of the form of the *Elements*; most of the scarce ancient diagrammatic evidence which I review in

---

315 Cf. above pp. 2 and 30.
316 Cf. Section 3.1.1, pp. 108–110.
317 Cf. p. 2, n. 10.
318 P. 16.
319 P. 27.

Section 4.1 survives only for book I; and book I is the most standardized and at the same time the least mathematically complex book of the whole work.[320] In this way, it may be expected to show the workings of standardization most visibly, and furthermore, in view of its relatively easy geometrical contents, it raises the question of the motivation for standardization – that is, if one proceeds on the assumption that the standardization serves above all cognitive-communicative purposes.[321] This focus on book I does not exclude occasional references to other books of the *Elements* wherever necessary.[322] However, I shall not extend my analysis beyond the *Elements* neither to other elements texts nor to scientific texts beyond the elements genre. Among texts of the genre, I focus on the example of the *Elements* because I am interested in the meaning and function of the *generic* standardization of elements texts, which seems to be at its greatest and most clearly visible in this work.[323] That is, it is not my purpose to show how unique the exact degree of the standardization of the *Elements* is *within* the genre.[324] Quantitative comparisons with texts *outside* the elements genre could prove useful in certain cases, but they would exceed the scope of this study, and the choice of the texts would run the danger of being arbitrary.[325]

In the first chapter of the analytical part (Chapter 3), I assess the standardization of the *text* of the propositions of *Elements* I. I do so on the basis of the standard text by Heiberg & Stamatis and do not take issues of textual transmission into account, first, because this would be beyond the scope of the present study, and second, because individual controversial cases do not make any crucial difference to the overall statistical results. The quantitative and also qualitative analysis of the text will have three main parts: 'lexicon' (Section 3.2), 'grammar' (Section 3.3), and 'structure' (Section 3.4). With the help of these categories, I will consider all parts of the text, except for the diagrammatic letters which I consider in Chapter 4. The analysis will include the grammatical and structural features which the text uses in order to construct the diagram; but I will consider these aspects more thoroughly in Chapter 4.

The second, much larger analytical chapter is focused on the standardization of the *diagram* in *Elements* I (Chapter 4). Unlike the textual standardization proper, the diagrammatic standardization cannot be analyzed straightforwardly, but only indirectly from different angles. This is because we do not know whether diagrams were

---

320 On the representativeness of book I in terms of standardization as well as on possible reasons for the special standardization of book I and the role of the comparative mathematical simplicity therein cf. Section 3.1.1, pp. 110 f.

321 For this assumption, see, for example, p. 36.

322 Cf. Section 3.1.1, p. 112.

323 Cf. pp. 10 f.

324 The exact degree of Euclid's own contribution to the original standardization of the *Elements* is also impossible to find out (cf. pp. 8–10). On potential candidates for intra-generic comparisons see Section 3.1.1, pp. 112 f.

325 I elaborate on the possibility and problems of such comparisons Section 3.1.1, p. 113.

originally included in the text at all, let alone in what form, or whether the diagrams had to be constructed on the basis of the text; the transmitted diagrams might have been added to the text only after Euclid's own time.[326] I shall examine the issue from scratch: the aim of Chapter 4 is to analyze both the transmitted diagrams and the proposition texts for indications of the standardized preexistence and form of the diagram and to assess these indications with regard to their conclusiveness. As has been pointed out above, we can only, if at all, find out whether propositions were written on the basis of a diagram, that is, if a diagram is 'logically' presupposed by the text, in which case it is likely that this diagram was indeed a material part of the text.[327] By referring to the ('material') presence of' the diagram, its 'preexistence,' 'underlying,' or 'the text's coming with the diagram,' I mean thus the likelihood of the material presence of the diagram on the grounds of its logical presupposition.

After briefly reviewing the few explicit references to diagrams in the Greekmathematicians as part of the introduction to Chapter 4, I shall start by reviewing the *diagrammatic* evidence (Sections 4.1 and 4.2).[328] First, I examine six ancient ostraca and papyrus fragments which present the only ancient evidence of diagrams which we have for the *Elements* (Section 4.1). This includes, on the one hand, a review of the fragments regarding their function and trustworthiness, on the other hand, an analysis of the fragments' diagrams with respect to what they betray about the preexistence and the aesthetics of the diagrams in Euclid. Then I turn to the Byzantine manuscripts of the *Elements* which standardly include diagrams (Section 4.2). These are liable to reflect the diagrammatic conventions of their own time or, if at all, late ancient drawing conventions.[329] Nevertheless I will consider them as potential evidence of the use and form of diagrams in the Euclidean text. The analysis of the diagrams and their standardization within and across manuscripts will focus, on the one hand, on how the geometrical configuration of the proposition is realized diagram-

---

326 This applies to the diagrams of all ancient mathematical authors, cf. Netz 1999a, 12: "I talk about a void, an absent object, for the diagrams of antiquity are not extant [. . .]." To be sure, we cannot trust the *textual* transmission either as to the exact degree of textual standardization (cf. pp. 9 f.), but the proposition *texts* are transmitted comparatively reliably.

327 Cf. Section 1.2, pp. 30–32, also for the respective underlying scenarios of how the text was written.

328 By 'diagrammatic evidence,' I mean throughout the evidence in diagrammatic form which has come down to us for Euclid's *Elements* on ostraca, papyrus, and manuscripts. I also call it 'material evidence' because I look at the precise form of how the diagrams have been transmitted on these materials. Rather than implying by the terms 'diagrammatic' and 'material evidence' that the transmitted diagrams *are evidence* of diagrams in the original text of Euclid's *Elements*, it is precisely my aim to find out whether, and if so, what kind of evidence they provide of original diagrams.

329 Cf. Netz 1999a, xvii, who claims that the "understanding of what a diagram should look like" is "culturally variable" when criticizing modern editors of Euclid for reconstructing the diagrams according to their own understanding of them rather than drawing them, like himself, on the basis of the Byzantine diagrams. However, the Byzantine manuscripts, too, are culturally determined.

matically and, on the other hand, on the style of the hand-drawn diagrams. In order to take account of the possible differences in the two main textual traditions, I choose two manuscripts for closer inspection: the only one non-Theonine manuscript (P) and one of the Theonine ones (B).[330]

Subsequently, I shall examine the *textual* evidence for indications of whether the *Elements* originally contained diagrams and what they looked like (Sections 4.3 and 4.4). I shall focus on the 'diagrammatization' of the text, that is, on how the diagram as an object of the proposition has implicitly shaped it. I shall examine the diagrammatization of the text above all with regard to the question of the standardization of the diagram as a literal part of the proposition, but also with regard to how it relates to *textual* standardization, which I shall, thus, review once more.

The first of the sections on diagrammatization is about the 'textual construction,' by which I mean 'the way(s) in which the text constructs the diagram' (Section 4.3). Accordingly, I shall focus here on the parts and features of the proposition text which are devoted to the construction of the diagram. I shall try to determine how 'constructional' the text is, i.e., whether it is indeed focused on constructing a diagram, and what kind of diagram, if any, it enables, and how this diagram is related to the proof. In the first part, I shall investigate the temporal implications of the structure of the proposition and of the imperative forms which assign the construction of geometrical objects. In the second part, I shall first consider the practicality of the construction imperatives, i.e., how doable a construction of the diagram is on the basis of the information which they provide, and what kind of drawing in terms of precision and technicality they enable. In the third part, I shall consider the functionality of the diagram which results from the textual construction. This includes asking, on the one hand, how conclusive the diagram is and whether the proof can be accomplished on the basis of it; on the other hand, how proof-relevant the diagram is, that is, whether the proof requires it at all. Here, I shall focus specifically on the proof-relevance of the diagrammatic *form*, that is, on the extent to which the proof requires the depiction of the diagram to be technical and precise; all issues involving the letters will be left for the following section.

The second of the two chapters on the diagrammatization of the text (Section 4.4) scrutinizes the letters by means of which the text refers to points of geometrical objects in the diagram, such as A in τὸ A σημεῖον ('the point A').[331] This section shall ignore the technical side of the diagrammatic construction and regard the diagram solely as a 'letter configuration;' it observes whether the use of the letters reflects an underlying 'diagram-as-letter configuration.' This part thus aims to look for evidence only of whether diagrams were regularly part of the text, not of whether and how

330 Cf. p. 9 f. above for an outline of the textual transmission of the *Elements*.

331 On my focus on the geometrical propositions cf. pp. 16 and 65; on the meaning of the letters in arithmetical contexts cf. p. 283, n. 175.

they were formally standardized. For this purpose, I shall explore some of the letter-related practices in great detail, above all the (non-)alphabetical order of introducing the letters into the text, the act of linking letters with points in the diagram (i.e., what Netz has called the "specification" of the letters),[332] and the ordering of the letters in words which include several letters, such as τὸ ΑΒΓ τρίγωνον ('the triangle ΑΒΓ').

Chapter 5

The first two parts of the final chapter (Sections 5.1 and 5.2) discuss the results of the foregoing analysis of textual and diagrammatic standardization with respect to 'logical' and cognitive-communicative clarity. While the textual analysis in Chapters 3 and 4 will have considered the relation of the standardization of all textual features individually to unambiguousness, Sections 5.1 and 5.2 will problematize the overall relationship of standardization and unambiguousness/clarity in the *Elements*. First, I ask how standardization is related overall to the 'logical' unambiguousness of the proofs (Section 5.1) and second, how it relates to the 'comprehensibility' of the proofs (Section 5.2). That is, I use the notion of clarity in the two senses defined above in order to determine the relation of the standardization to proof logic and communication.[333] The guiding question of these two sections will be whether the textual standardization of the *Elements* is logically or communicatively 'necessary' or 'unnecessary,' or which part of it is necessary or unnecessary with regard to logical and communicative clarity. The method which I will use for distinguishing 'necessary' from 'unnecessary' standardization is what I call the 'method of subtraction.' I shall 'subtract' the standardization of textual features from the text by way of thought experiment and consider, on the one hand, whether or not the proof remains consistent[334] even after the subtraction, and on the other hand, whether it stays comprehensible for the intended readership of Euclid (which I will attempt to determine beforehand). On the basis of this discussion, I shall attempt in the third part of the chapter (Section 5.3) to identify the function of standardization and, with regard to this, the role of clarity in the *Elements*. Finally, I will briefly consider the specificity of Euclidean clarity within the context of the genre and beyond.

---

332 Netz 1999a, 19 and *passim*.

333 As I have explained in Section 1.4, p. 54, it is not my purpose to apply the ancient theories of σαφήνεια in terms of the features which they connected to either of the two sides of clarity, nor to enlarge the arsenal of these features with respect to the genre of elements texts, but I only consider the notion of clarity in terms of two dimensions which were present also in the ancient term (without meaning to imply that the ancient and modern notion are congruent).

334 I regard the notions of unambiguousness and consistency as largely synonymous with regard to mathematical proof logic.

The photographs of the ancient diagrams and their reconstructions in Section 4.1 are taken from Fowler's book *The Mathematics of Plato's Academy* and, in the case of P. Berol. [inv.] 17469, from Brashear's edition of the fragment.[335] The schematic diagrams printed in Sections 4.1–4.4 have kindly been drawn by de Gruyter on the one hand according to the diagrams of the standard edition Heiberg & Stamatis 1969–1973 and on the other hand to those of the Vatican manuscript P and the Bodlein manuscript B[336]

335 Fowler 1999, Plates 1–3 between pp. 6 and 7; Brashear 1994, 30 and Plate 6.
336 Cf. p. 68 on my choice of these two manuscripts and p. 217, n. 81 on their digital editions.

2 'Clarity' in the commentators of Euclid

2.1 Plan of the chapter

In the present chapter, I shall investigate the notion of clarity as it is used and applied to Euclid by his own extant commentators, that is, the fourth-century mathematician Pappus and the Neoplatonic philosopher Proclus (fifth century). My aim is to show how the originally rhetorical concept has been applied to Euclid within (late) science writing itself, especially whether the commentators have referred it to Euclid's *textual* form. Considering the evidence of Pappus and Proclus is worthwhile because the prominent rhetoric of clarity in both commentators could mirror how Euclid was generally perceived among late ancient mathematicians and philosophers and, in so doing, perhaps also reflect the perception of earlier mathematical authors or even a tradition of perceiving Euclid as clear which may have a long history.

I proceed in the following way: first, I review the language of clarity in both commentators but especially in Proclus because only his commentary survives in Greek.[1] This will be the largest part of the chapter. I start by analyzing the emphatic attribution of 'clarity'/σαφήνεια to Euclid in both commentators; I consider the prominent place of the attribution in each, the qualities to which clarity is explicitly referred, and the provenance of other terms mentioned alongside clarity in these passages. Subsequently, I review the nonemphatic uses of cognates of σαφήνεια in Proclus, that is, the many times in which he states that something is or becomes clear, that someone says something clearly or makes something clear; I consider what these uses, above all of the adjective σαφής and the adverb σαφῶς, add to and reveal about the aforementioned emphatic use of the noun. Then I explore the terminology of clarity employed besides that of σαφήνεια in the Proclean commentary, especially ἐνάργεια, which is also used prominently, and its relation to σαφήνεια. In addition to this, I consider terms which are related to the language of clarity proper, above all ἀκρίβεια. Second, and only by way of a brief interpretation of the language of clarity, I compare the explicit evidence for 'clarity' with the commentary practices of Proclus and Pappus beyond the passages about clarity. In so doing, I aim to draw conclusions on the function which the concept serves the commentators and what, in light of this, its use can further divulge about the perception of Euclid.

1 Cf. above, p. 64.

https://doi.org/10.1515/9783110670462-002

2.2 Review of the language of 'clarity'

2.2.1 The emphatic attribution of 'clarity'/σαφήνεια to Euclid in Pappus and Proclus

The place of the attribution

I begin by analyzing the explicit and emphatic attribution of 'clarity' to Euclid in both commentators. Both Pappus und Proclus mention 'clarity' as a laudable quality of Euclid in a prominent passage of their commentaries. Pappus greatly applauds Euclid's "wonderful clarity" toward the beginning of the surviving part of his commentary in a passage in which he reflects on the "good" of the *Elements*.[2] Since the commentary only survives in an Arabic translation of the tenth century by Abu Uthman Al-Dimishqi, we cannot be sure which Greek term Pappus has used for what Thomson translates as "clarity."[3] A case can be made that it was σαφήνεια, which Proclus one century later, in the second part of the prologue to his commentary, names as one of several qualities which an elements text should have and in which "one should find Euclid's *Elements* to stand out from all others."[4] For, as has been pointed out, the prologues and beginnings of commentaries are the very place where we often find a topical reference to (and mostly justification of) the ἀσάφεια, that is, the 'lack of σαφήνεια,' of the author commented upon.[5] Indeed, the explicit mention of "clarity" in this particular place in Pappus' work is not made perchance randomly, but in implicit contradistinction to the ἀσάφεια of other scientific writers. Pappus even shows himself aware of the topos of obscurity and its regular place in the *prolegomena* since he explicitly sets the clarity of Euclid's account of irrationals in *Elements* X in contrast to the Pythagoreans' rather drastic recommendation not to disclose this knowledge to "the common herd."[6]

---

2 *In Eucl. X* I 2–3 (p. 64 Thomson & Junge).

3 *In Eucl. X* I 2 (p. 64 Thomson & Junge). The Arabic term is al-stiṣfā'u, but we cannot say which Greek term corresponds to it (Oliver Overwien via email).

4 εὕροι τις ἂν τὴν < Εὐκλείδου > στοιχείωσιν τῶν ἄλλων διαφέρουσαν (*In Eucl.* 74.9–11 Friedlein, my translation). Morrow 1992, 61 translates τῶν ἄλλων simply as "others," but οἱ ἄλλοι equals the Latin *ceteri* and means "the/all others" (Smyth §1188).

5 See above Section 1.4, 43 f.

6 Pappus relates the Pythagorean anecdote of the man who first revealed the knowledge of irrationals and, as a consequence, "perished by drowning" (*In Eucl. X* I 2 [Thomson & Junge 1930, 64]), which he reads both as a request to conceal such knowledge and as a warning as to what happens to whoever does not do so. The transition from here to the praise of Euclid's clarity is far from clear: "This was the consideration which Pythagoreans and the Athenian Stranger held to be an incentive to particular care and concern for these things and to imply of necessity the grossest foolishness in him who imagined these things to be of no account. Such being the case, he of us who has resolved to banish from his soul such a disgrace as this, will assuredly seek to learn from Plato, the distinguisher of accidents, those things that merit shame, and to grasp those propositions which we have endeavoured to explain, and to examine carefully the wonderful clarity with which Euclid has investigated each of

The immediate verbal context of the attribution

Since Pappus does not elaborate on Euclid's "wonderful clarity" in this passage, we do not know in which respect he thinks Euclid is clear. There is reason to assume, however, that it is Euclid's form of presentation that Pappus has in mind because he distinguishes 'clarity (of investigation)' from 'content:' "he [. . .] will assuredly seek [. . .] to examine carefully the wonderful clarity with which Euclid has investigated each of the ideas (or definitions) of this treatise (i.e., Book X.)."[7] To be sure, Pappus, according to Thomson's translation, says "investigated," not "presented," but what we can see is only how he investigates them in front of the reader and, thus, *presents* them.[8]

Proclus is more explicit. In his discussion of the necessary qualities of "elementary treatises"[9] he betrays his opinion concerning how the "clarity" of the *Elements* comes about: "its method of proceeding from simpler to more complex matters and its laying the foundations of the science on the 'common notions' produce clarity and articulateness (τὸ δὲ σαφὲς καὶ διηρθρωμένον)."[10] With this remark Proclus ascribes Euclid's clarity to the two components of the axiomatic-deductive method, suggesting that it lies primarily in the logical form of the work or its individual books, especially book I, the object of Proclus' commentary. However, the quality of "articulateness/distinctness" (τὸ διηρθρωμένον), which Proclus here mentions alongside "clarity" cannot but refer to the stylistic form. Like σαφήνεια/τὸ σαφές, διάρθρωσις/τὸ διηρθρωμένον are technical

---

the ideas (or definitions) of this treatise (i.e., Book X.)" (*In Eucl. X* I 2-3 [Thomson & Junge 1930, 64]). The "consideration" which Pythagoras and Plato, according to Pappus, have deemed an "incentive to particular care and concern" should be the aforementioned parable on the discloser of knowledge of the irrationals; the "disgrace" which one should "banish from [one's] soul" before turning to Plato and Euclid for irrationals should refer to the "grossest foolishness" of him "who imagined these things [i.e., the topic of irrationals] of no account." This would mean Pappus recommended to treat Plato and Euclid, who has been especially clear about the subject, with special care, when turning to them for the knowledge of irrationals. However, since we do not have the Greek text from which we could reconstruct the reference of the pronouns "this [consideration]" and "these [things]," we cannot be sure whether this is what Pappus is saying. In any case, while he seems to share the opinion that the topic needs to be treated with care on the part of the reader, he praises the clarity of Euclid who seems to have done just what the Pythagoreans wanted to avoid: to unveil the knowledge of irrationals to the "common herd."

7 *In Eucl. X* I 3 (p. 64 Thomson & Junge).

8 Nevertheless, the translation from the Arabic remains difficult, for it is unclear which part of the content exactly is presented with clarity in Pappus' view; Thomson seems unsure whether to translate "ideas" which is rather general and could refer to the content of the whole book including introduction and all propositions, or "definitions" (*In Eucl. X* I 3 [p. 64 Thomson & Junge]), which would refer only to a small part of *Elements* X, namely the three lists of definitions heading and contained in it, respectively (cf. p. 19).

9 στοιχειώσεις, *In Eucl.* 73.14 und 25 Friedlein. For the generic term, cf. above p. 1, n. 2.

10 τὸ δὲ σαφὲς καὶ διηρθρωμένον ἡ ἀπὸ τῶν ἁπλουστέρων ἐπὶ τὰ ποικιλώτερα μετάβασις ἀπεργάζεται καὶ ἡ ἀπὸ τῶν κοινῶν ἐννοιῶν καταβολὴ τῆς θεωρίας (*In Eucl.* 74.13–16 Friedlein, transl. Morrow). Cf. Morrow's comment (1992, 61): "Κοιναὶ ἔννοιαι occurs frequently in Proclus but is nowhere defined as a technical term. Cf. also κοιναὶ ἐπίνοιαι (*In Eucl.* 188.12 Friedlein)."

terms from classical Greek rhetoric;[11] and the present close conjunction of σαφές and διηρθρωμένον, which are governed by one and the same nominalizing article, implies at least that τὸ σαφές, too, is understood as a stylistic quality, if not that the two qualities form a hendiadys. Similarly, the first mention of σαφήνεια in this passage is in combination with συντομία ("conciseness") which, like σαφήνεια, is one of the stylistic virtues in classical rhetoric.[12] Again, apart from their joint function in the latter, the close juncture of the qualities by means of ἅμα καί[13] suggests that σαφηνείας, too, is referred to with regard to style. In any case, σαφήνεια/τὸ σαφές appears as a quality of the text to be produced by the author, not as something which is inherent in the content.

The concept of σαφήνεια emerging from the attribution to Euclid

Considering the verbal context of Proclus' use of σαφήνεια in the second part of the prologue to his commentary, it seems appropriate to assume for the term the meaning which it has in the original rhetorical context, i.e., 'clarity' in the sense of 'unambiguousness' or 'unmistakability' as a quality of the stylistic form of the text.[14] It is precisely by alluding to its rhetorical origins, but not redefining the term that Proclus refers the reader back to the classical (logical-)rhetorical treatments of the concept. The reference, however, seems to be to the general gist of the term rather than to the arsenal of formal means by which clarity is to be achieved, given the extreme differences between the Euclidean *Elements* and the texts to which the rhetorical recommendations pertain regarding their contents, form, and (insinuated) function. For Proclus, who himself comments on the *Elements* (at least partly) in a teaching context and uses it as a textbook, suggests that not only his own commentary but also the *Elements* itself was written for didactic purposes.[15] Here, he presents clarity as one of several qualities of elements treatises which are desirable from this didactic perspective.[16] He even gives clarity (alongside conciseness) some prominence among the other qualities, for he claims that an elements treatise "must devote great attention

11 Cf. Longinus' use of the term with respect to λόγος (Porph. *Vita Plot.* 20.68).

12 *In Eucl.* 74.4 Friedlein. Cf., e.g., Arist. *Rhet.* III 5.1407b28.

13 *In Eucl.* 74.4 f. Friedlein: σαφηνείας δ' ἅμα καὶ συντομίας πολλὴν πεποιῆσθαι πρόνοιαν ("it must devote great attention *both* to clarity *and* to conciseness;" Morrow's transl., my italics).

14 Cf. pp. 2, 41, and 54 above.

15 Cf. below p. 102–104 for Proclus' dual readership and Section 5.2, 423–426 on the original function(s) of Euclid's *Elements*.

16 The whole passage *In Eucl.* 74 Friedlein abounds with words denoting or having to do with the teaching and acquisition of knowledge: μάθησιν ("learning," 74.1), ἐπιστήμην ("knowledge," 74.3), διάνοιαν ("understanding," 74.6), περιλήψεως ("comprehension," 74.7), διδασκαλίαν ("teaching," 74.8), γνῶσιν ("knowledge," 74.9), θεωρίαν ("study," 74.13; "science," 74.16), γίγνεται γνώριμα ("can be learned," 74.19; Morrow's translations).

(πολλὴν πρόνοιαν)" to both of these qualities or else "confuses our understanding" (ἐπιθολοῖ τὴν διάνοιαν ἡμῶν).[17] Thus, clarity emerges as an important prerequisite for the activities of the διάνοια, that is, understanding and learning.

2.2.2 Nonemphatic occurrences of cognates of σαφήνεια

Apart from the prominent passages in which Pappus and Proclus commend Euclid's clarity, they refer to it also nonemphatically in the course of their actual commentaries of (part of) the *Elements*, that is, without seeming to emphasize the point of clarity. There are but a few such nonemphatic instances of 'clarity' in Pappus' commentary (as per the English translation of the Arabic),[18] in which he simply notes that a proof is clear[19] or that he makes something clear by means of an example.[20] Going through Proclus' nonemphatic mentions of clarity seems to be more rewarding because his commentary abounds by cognates of σαφήνεια.

Three types

First, the adjective σαφής is used with εἶναι ('to be') or γίγνεσθαι ('to become') in order to claim that something 'is' or 'becomes clear' (σαφές).[21] That which is or becomes clear is "what has been said" (τὸ λεγόμενον)[22] or simply "this" (ταῦτα),[23] that is, either a theorem by Euclid[24] or an interpretation of a Euclidean statement or passage by Proclus[25] or somebody else.[26] As the means from or by which something is or becomes clear Pro-

17 *In Eucl.* 74.4–6 Friedlein.

18 Besides, there seems to be one more emphatic passage in which he insinuates that Euclid performs an operation because he is "zealous to be clear," but here the Arabic does not seem to be certain; cf. Thomson & Junge 1930, 166, n. 8: "I read Ka-l-Mujiddi (like one who is zealous) instead of Ka-l-Muḥiddun [. . .]."

19 *In Eucl. X* I 31, II 9 and 17 (pp. 90, 126, and 138 Thomson & Junge).

20 *In Eucl. X* I 17 (p. 82 Thomson & Junge).

21 *In Eucl.* 273.15, 358.11, 374.5, 385.16, 428.7 Friedlein.

22 *In Eucl.* 273.15, 358.11, 374.5, 428.7 Friedlein.

23 *In Eucl.* 385.16 Friedlein.

24 I 30 (*In Eucl.* 374.2–8 Friedlein) and I 47 (*In Eucl.* 428.7 Friedlein). Toward the beginning of his commentary on I 33, after Proclus has outlined the position and importance of this proposition (that it is "a kind of boundary between the study of parallels and the study of parallelograms," *In Eucl.* 385.5–7 Friedlein), which he explains with regard to the construction of I 33 and to the subsequent theorem, he states that "This is evident;" (Ταῦτα μὲν οὖν σαφῆ, *In Eucl.* 385.16 Friedlein, Morrow's translation). "This" refers to (Proclus' remark on) the position and the meaning of theorem I 33.

25 *In Eucl.* 358.10 f. Friedlein.

26 Proclus cites potential critics of Euclid's construction in I 9: Ἴσως δ' ἄν τινες ἐνσταῖεν τῷ γεωμέτρῃ [. . .] γίγνεσθαι δὲ σαφὲς τὸ λεγόμενον διὰ τῶν στοιχείων ("Some may perhaps object to our geometer [. . .] and that this becomes clear from the *Elements*," *In Eucl.* 273.11–15 Friedlein, transl. Morrow).

clus mentions an ad-hoc diagram (τῶν ἀριθμῶν καταγραφῆς) drawn by himself,[27] "numbers" by which he illustrates the Pythagorean theorem,[28] and a "common notion" (κοινῆς ἐννοίας).[29] By the latter, Proclus does not seem to mean one of the axioms from the beginning of *Elements* I, but rather an idea[30] which is evident to everyone and based on the diagrammatic arrangement of the three parallel lines of I 30 ("Straight lines parallel to the same straight line are also parallel to one another."):[31]

> Our author himself takes the lines that are parallel to the same line as extremes and the line to which they have a similar relation as lying between them, *so that* the assertion may also be evident to us from a common notion. *For* if the outer lines meet one another, most certainly they will intersect the line lying between them and no longer be parallel to it.[32]

That is, it is actually the diagram of I 30 by which the theorem becomes "clear" or "evident,"[33] as Morrow translates σαφές here because Proclus seems to refer to the 'immediate clarity' of the diagram.

This is in fact the only passage of this group which refers to a particular part of the text for the clarity of Euclid. The remaining passages all directly or indirectly refer to Proclus' interpretation of or dealings with the text. The use of σαφής in them remains inconspicuous because they do not explain what clarity is or how exactly it comes about. Proclus seems to say that 'it is (unmistakably) recognizable or understandable from something else' or that 'it is such and such, as one can (obviously) see

27 I.e., a diagram which Proclus makes up in order to explain something and which is not identical with the Euclidean one (of I 27), contrarily to what he claims: γιγνέσθω δὲ ἐπὶ τῆς αὐτῆς καταγραφῆς σαφὲς τὸ λεγόμενον (*In Eucl.* 358.10 f. Friedlein, "An identical diagram will make clear what I mean," transl. Morrow).

28 Σαφὲς οὖν τὸ λεγόμενον ἐπὶ τῶν ἀριθμῶν· (*In Eucl.* 428.7 Friedlein, "The statement, then, is clear for numbers," transl. Morrow).

29 Besides, "the *Elements*" is mentioned as the source of clarity, but only in the report of Euclid's potential critics (*In Eucl.* 273.11–15 Friedlein). In *In Eucl.* 385.16 Friedlein, Proclus simply states that "this is clear" (Ταῦτα μὲν οὖν σαφῆ, my transl.), without explicitly specifying why (cf. p. 75, n. 24).

30 Proclus uses κοινὴ ἔννοια in both ways: in the sense of "axiom," synonymously with ἀξίωμα (either with reference to one of the Euclidean axioms transmitted as κοιναὶ ἔννοιαι at the beginning of *Elements* I – "Axioms" with a capital "A" in Morrow's translation, e.g., 1992, 293 – or more generally with regard to this kind of first principles), or in the even more general sense of 'common notion' (cf. p. 73, n. 10). This comment of Proclus' on *Elem.* I 30 does not refer to any particular axiom, although the beginning of the commentary (*In Eucl.* 373.8 f. Friedlein) makes reference to *Elem.* I, Ax. 1: "Things equal to the same thing are equal to each other," Τὰ τῷ αὐτῷ ἴσα καὶ ἀλλήλοις ἐστὶν ἴσα.

31 Αἱ τῇ αὐτῇ εὐθείᾳ παράλληλοι καὶ ἀλλήλαις εἰσὶ παράλληλοι (*Elem.* I 30 [vol. 1, p. 42.17 f. Heiberg & Stamatis]).

32 καὶ αὐτὸς μὲν τὰς τῇ αὐτῇ παραλλήλους ἄκρας ἔλαβεν καὶ μέσην, πρὸς ἣν αὗται τὴν ὁμοίαν ἔχουσι σχέσιν, ἵνα καὶ ἀπὸ κοινῆς ἐννοίας ἡμῖν γένηται σαφὲς τὸ λεγόμενον. εἰ γὰρ αἱ ἐφ' ἑκάτερα συμπίπτουσιν ἀλλήλαις, πάντως καὶ τῇ μεταξὺ κειμένῃ συμπεσοῦνται καὶ οὐκέτι πρὸς αὐτὴν ἔσονται παράλληλοι (*In Eucl.* 374.2–8 Friedlein, transl. Morrow).

33 For the overlap of the meaning of cognates of σαφήνεια with that of ἐνάργεια, the technical term for 'immediate clarity,' cf. below p. 97.

from . . . .' In doing so, he uses the rhetoric of 'clarity' in order to underscore (the truth of) his point.

The second category of instances are those which point out that someone 'says' or 'shows something clearly (σαφῶς).' In three cases, Euclid is the one to do so (μόνον οὐχὶ λέγει σαφῶς,[34] σαφῶς ἐδήλωσεν αὐτὸς ὁ στοιχειωτής,[35] δεικνύει σαφῶς[36]), once Proclus (δείκνυται σαφῶς),[37] and once Plato (σαφῶς ἀποφαίνεται).[38] That which is being said or shown clearly is always the content of an adjunct ὅτι or ὡς clause. In the case of Euclid and Proclus this is a mathematical statement, in the case of Plato a philosophical one. The three passages referring to Euclid pertain to his definitions I, Def. 1 and 17 and to his postulate I, Post. 2; they affirm that what they say is unmistakably clear, without explicitly specifying by virtue of what they say it clearly. The clarity seems to be due to the unambiguous (formulation of the) contents of the principles, although Proclus does not explicate the aspect of their 'formulation' in terms of 'textual form' or 'style.' His emphasis appears to be on *what* the principles say clearly; the paraphrase of I, Def. 1 and the reference to I, Post. 2 serve Proclus to justify what he has just argued (in the case of I, Def. 1, about this very definition).[39] If at all, Proclus' comment on definition I, Def. 17 refers to the way in which this definition is said: "The author of the *Elements* himself made it clear that he is defining not every diameter, but the diameter of the circle."[40] That is, Proclus points out that the definition unambiguously specifies its subject by confining the general subject "diameter" to "diameter of the circle," but this is not a matter of the kind of formulation, but merely of the act of mentioning. Moreover, Proclus' emphasis on the unambiguousness of Euclid's definition above all gives him opportunity to elaborate on different kinds of diameters and their terminology;[41] that is, the emphasis of the passage is not so much on Euclid's clarity but on the subject of 'diameters.' Similarly, Proclus' reference to a "clearly demonstrable" statement (the falsity of the converse of Euclid's I, Post. 4) only affirms the fact that it is proved – and unmistakably so, rather than elaborating on the reasons for the clarity of

34 *In Eucl.* 94.4 Friedlein; "for he all but clearly says" (my translation; cf. Morrow 1992, 77 for having "it," i.e., *Elem.* I, Def. 1, which Proclus' statement is about, as the subject).

35 *In Eucl.* 156.11 Friedlein: "the author of the *Elements* himself made it clear" (transl. Morrow, modified).

36 *In Eucl.* 215.24–26 Friedlein: "[sc. the postulate] shows clearly," of *Elem.* I, Post. 2.

37 *In Eucl.* 184.3 Friedlein: "it is [. . .] clearly demonstrable" (transl. Morrow). Proclus does not refer the clarity of the demonstration to himself explicitly, but he supplies his demonstration a little further on with reference to Pappus (*In Eucl.* 189.11–191.4 Friedlein, cf. Morrow 1992, 144, n. 9).

38 *In Eucl.* 29.26–30 Friedlein: "(Plato himself) clearly affirms."

39 *In Eucl.* 94.4 Friedlein: μόνον γὰρ οὐχὶ λέγει σαφῶς ("For he all but clearly says [. . .]," Morrow's transl., modified; my italics); *In Eucl.* 215.24–26 Friedlein: τὸ γὰρ καὶ πεπερασμένην εὐθεῖαν ἐπ' εὐθείας ἐκβάλλειν δεικνύει σαφῶς ("*For* this principle that a finite straight line may be extended in a straight line shows clearly that [. . .]," transl. Morrow, my italics).

40 Σαφῶς ἐδήλωσεν αὐτὸς ὁ στοιχειωτής, ὅτι οὐ πᾶσαν ὁρίζεται διάμετρον, ἀλλὰ τὴν κυκλικήν (*In Eucl.* 156.11 f. Friedlein, transl. Morrow, modified).

41 *In Eucl.* 156.11–19 Friedlein.

the proof. Proclus' comment on this statement seems to be first and foremost about the clarity of his own ensuing demonstration of it to which he implicitly draws attention.[42] One may conclude that whenever Proclus himself claims that someone 'says' or 'shows' something 'clearly,' he simply seems to mean that they (unmistakably) say it, or show it. The emphasis of these passages is on *what* is said clearly rather than *in which way* it is said clearly, because the precise accomplishments are not specified. Notably, more instances of this second (as opposed to those of the first) category are indeed related to Euclid's text. They seem to put more emphasis on the Euclidean authorship of the clarity of the statements than those of the first group because clarity is referred to the authorial act of 'saying' or 'showing.' However, as these references to Euclid's clarity at least partly serve Proclus for making his point, Euclid's authority simply might have come in handy.

Thirdly and finally, Proclus refers to a statement or an author who 'makes something clear' or 'clarifies something' (three times by σαφὲς/-ῆ ποιεῖν,[43] twice by means of σαφηνίζειν,[44] twice by σαφήνεια in the sense of 'clarification,'[45] and once by διασαφεῖν[46]). As agents of clarification, i.e., as that which or those who make something clear, Proclus mentions himself twice[47] and Euclid twice, once indirectly and once explicitly.[48] In the first case (the indirect reference to Euclid as clarifier), he draws attention to the clarifying potential of the 'definition of the goal' (ὁ δὲ διορισμός) as a paradigmatic part of the Euclidean proposition: it "takes separately the thing that is

42 The "clearly demonstrable" statement to which Proclus refers *In Eucl.* 184.3 and which he proves *In Eucl.* 189.11–191.4 Friedlein (cf. p. 77, n. 37) is the falsity of the converse of Euclid's postulate I, Post. 4. – The above-quoted reference to Plato's "clearly affirm[ing] that mathematics purifies and elevates the soul" (p. 77, n. 38) does not shed further light on Proclus' notion of clarity; it is a non-mathematical statement and not a direct quotation from Plato, but an interpretation of *Resp.* 527e, so the point of reference of 'clarity' can only be Plato's own words. – Another instance of Proclus' mentioning that someone 'says something clearly' which I have not cited so far is a quotation from *Phaedo* 73b (*In Eucl.* 45.15–17 Friedlein), where Plato himself makes use of the phrase: ἐὰν γάρ τις ἐπὶ τὰ διαγράμματα ἄγῃ, φησὶν ἐκεῖνος, ἐνταῦθα ἄρα σαφέστατα κατηγορεῖ, ὡς ἔστιν ἡ μάθησις ἀνάμνησις. ("'If you take a person to a diagram,' he says, 'then you can show most clearly that learning is recollection.'"). At first glance, it appears to provide an answer to our question by virtue of what are (mathematical) statements considered to be clear according to Proclus, as it is about diagrams in the context of μάθησις ("learning"), but the Platonic remark is again a philosophical one and does not really address the role of diagrams in mathematics.

43 *In Eucl.* 144.2, 210.18, 376.12 Friedlein.

44 *In Eucl.* 38.24, 428.22 Friedlein.

45 *In Eucl.* 238.25, 418.19 Friedlein. While we tend to regard 'clarity' as a quality of a text produced by its author and 'clarification' as the work of the commentator, the Greek σαφήνεια comprises both meanings and may be applied to both types of authors.

46 *In Eucl.* 203.10 Friedlein.

47 *In Eucl.* 210.17 f., 238.25 f. Friedlein.

48 In addition to himself and Euclid, Proclus mentions "doctors" (οἱ ἰατροί, *In Eucl.* 38.23 f. Friedlein) and "the ancients" (οἱ παλαιοί, *In Eucl.* 144.1–5 Friedlein) as clarifiers in non-mathematical remarks.

sought and makes clear precisely (διασαφεῖ) what it is."[49] It is not certain in what sense Proclus thinks that the διορισμός makes the goal of the proposition clear. Perhaps he just means to say that it states the goal unambiguously,[50] but considering that Proclus admits shortly afterwards that the 'definition of the goal' may as well be missing from a proposition, the use of the language of clarity here seems to contribute above all to the schematic and simplifying presentation of the parts of a proposition.[51]

In the latter case (the explicit mention of Euclid as agent of clarification in the commentary on problem I 31), he adduces I 12, "To a given infinite straight line, from a given point which is not on it, to draw a perpendicular," in which Euclid "made clear [. . .] by adding" (διὰ τοῦ προσθεῖναι σαφὲς ἐποίησεν) the phrase "which is not on it" (ὅ μή ἐστιν ἐπ' αὐτῆς) that the point is not on the given straight line.[52] However, the reason why Proclus mentions Euclid's unambiguous formulation in I 12 actually is to show a supposed lack of unambiguousness in the analogously formulated I 31 which does not contain the disambiguating qualification of the point as to its position ("Through a given point to draw a straight line parallel to a given straight line."). Thus, Proclus' comment about Euclid's clarity is at least as much about his own present clarification. Notably, both theorems, I 12 and 31, are logically unambiguous even without the specification of the point, as both the perpendicular and the parallel line which are to be drawn from and through the given point, respectively, already imply that the point is not on the given straight line.[53] The reason why Euclid does not qualify the point in I 31 seems to be that there, by contrast with I 12, he mentions the point first, so it does not make sense to say "Through a given point which is not on it" with-

---

49 ὁ δὲ διορισμὸς χωρὶς τὸ ζητούμενον, ὅτι ποτέ ἐστιν, διασαφεῖ (*In Eucl.* 203.9 f. Friedlein; transl. Morrow, modified). This passage stems from the famous outline of the parts of the Euclidean proposition which, although it claims to be, is not universally valid but oversimplifying (cf. p. 23, n. 134).

50 The actual function of the 'definition of the goal' is to state the goal – which is formulated in the 'enunciation' in general terms – in specific terms, i.e., with regard to the diagram of the proposition (cf. p. 23), but Proclus, in his scheme of the proposition parts, does not refer to the processes of concretizing the general proposition in terms of the particular case and generalizing the proof of the particular case (cf. p. 23 f.).

51 *In Eucl.* 203.23–204.2 Friedlein. The reason why it can be missing is that it does not belong to the most essential proposition parts (τὰ δὲ ἀναγκαιότατα, *In Eucl.* 203.17 Friedlein).

52 *In Eucl.* 376.11–14 Friedlein. ὅπερ ἐπὶ τῆς καθέτου καὶ διὰ τοῦ προσθεῖναι σαφὲς ἐποίησεν, ἐπὶ τὴν δοθεῖσαν εὐθεῖαν ἄπειρον ἀπὸ τοῦ δοθέντος σημείου, ὅ μή ἐστιν ἐπ' αὐτῆς, κάθετον ἀγαγεῖν. ("This is just what he made clear in the case of the perpendicular by adding a qualification. 'To a given straight line, from a given point which is not on it, to draw a perpendicular,'" transl. Morrow).

53 Proclus himself admits that the qualification is actually not necessary in his commentary to I 31 immediately before the passage quoted above, as Euclid, in I 31, showed (ἐδήλωσεν) implicitly "by mentioning separately the point and the straight line" "that the point must be taken outside the straight line" (*In Eucl.* 376.5–11 Friedlein). As Proclus nevertheless mentions the clarity produced by the additional relative clause in I 12, we must assume that he does so in order to draw attention to the 'unmistakability' of the formulation of I 12 and to how he himself made I 31 likewise unmistakable. On logical non-necessity in Euclid, cf. Section 5.1, 416–421.

out having mentioned "it," i.e., the given straight line first. In I 12, Euclid mentions first the given straight line, then the given point as the starting point (ἀπό, "from") of the perpendicular, and finally the infinitive to draw a perpendicular. Here, the qualification of the point "which is not on it" is made logically superfluous immediately by the subsequent infinitive, but at the very moment of its mention, the position of the point is unclear if it is not specified.[54] This could indicate how Proclus understands the term σαφές in the present passage: that is, with regard to his didactic approach, as "easily comprehensible" or "unmistakable" rather than as "logically unambiguous."

In another passage, about why Euclid did not include "the term 'complements'" (τὸ δὲ ὄνομα τῶν παραπληρωμάτων, *In Eucl.* 418.15 Friedlein) in his definitions, Proclus describes what "would have been necessary" (ἔδει) "for clarity" (πρὸς τὴν σαφήνειαν) but what Euclid has chosen not to provide because it would been (too) "complex."[55] Proclus purports to approve of Euclid's dispensing with the complicated clarification of the term which he outlines. By mentioning the "complexity" of the clarification, he implies that the text is simpler and more readable (i.e., clearer) without it. At the same time, by outlining it, he stages himself as clarifier of what he only insinuates to be (but not explicitly calls) obscurity on Euclid's part. The present passage shows the close connection, or even overlap, of σαφήνεια and 'understanding' which I have noted with regard to the preceding passage: ποικιλίας γὰρ ἔδει πρὸς τὴν σαφήνειαν, ἵνα γνῶμεν ("For complexity would have been necessary for clarity, so that we understand") and τούτων γὰρ σαφηνισθέντων [. . .] ἐγένετο γνώριμον ("after these had been explained [. . .] would have become clear").[56]

In the passages which refer to Proclus himself or to Euclid and his text, clarity and clarification are either of specific mathematical issues – such as the term (ὄνομα) "complements" which we have just seen, or a theorem (τοῦ θεωρήματος) –[57] or of metamathematical ones such as the goal (τὸ ζητούμενον, "the thing which is sought") and the structure of the proof.[58] In each case, Proclus indicates how clarity and clarification have or could have come about. As his own means, he mentions the giving of an example, such as the very first proposition, which serves him for the clarification

54 I discuss the relevance of the order of the text for its unambiguousness with regard to the 'textual construction' in Section 4.3, 243–245, 270 f.

55 ποικιλίας γὰρ ἔδει πρὸς τὴν σαφήνειαν, ἵνα γνῶμεν, τί παραλληλόγραμμον, καὶ τίνα τὰ περὶ τὴν αὐτὴν διάμετρον τῷ ὅλῳ. τούτων γὰρ σαφηνισθέντων καὶ τὸ παραπλήρωμα μόνως ἂν ἐγένετο γνώριμον ("It would have required a complex explanation to make us understand what a parallelogram is and what are the parallelograms that are constructed about the same diameter as the whole; for only after these had been explained would the meaning of 'complement' have become clear." *In Eucl.* 418.19–23 Friedlein, Morrow's transl.).

56 Cf. the preceding note for Morrow's more elegant and less literal translation.

57 *In Eucl.* 376.11–14 Friedlein on I 12 (cf. p. 79, n. 52), and 238.25 f. Friedlein on I 4 (cf. p. 78, n. 47 and p. 81, n. 60).

58 *In Eucl.* 203.9 f. Friedlein on the (definition of the) goal of the proposition (cf. p. 79, n. 49); 210.17 f. Friedlein on the structure of the proof (cf. p. 78, n. 47 and p. 81, n. 59).

of the standard structure of all of the books' propositions,[59] and providing "preliminary explanations" before going through Euclid's actual proposition;[60] and as Euclid's own means, he names, as we have seen, the "definition of the goal"[61] and an "addition" (τοῦ προσθεῖναι) in the form of a relative clause which disambiguates a more general term.[62] As a potential means of clarification (but not employed in the respective passage by Euclid), Proclus mentions a complex set of definitions or explanations of terms which are necessary for understanding another term.[63]

Accordingly, 'making something clear' may comprise the commentatory practices of systematization, illustration (by using examples), and anticipation (of necessary information), and the authorial practices of definition (of the goal of a proposition and of mathematical terms), specification (of something more general), and explicit mention (of what is already logically presupposed). But even though Proclus in these passages finally seems to be more explicit about how clarity is achieved, σαφηνίζειν and σαφὲς ποιεῖν can just as well be understood as 'to explain'[64] something so that it is unmistakably understood.[65] The emphasis again seems to be on *what* is made clear and *that* it is made clear rather than on the phenomenon of clarity, although the use of cognates of σαφήνεια is more diverse and frequent here than in the passages reviewed before and includes the noun which Proclus also uses in his emphatic praise of Euclid's clarity.[66] This is because the passages – even those which state that Euclid makes something clear – are mostly about Proclus' 'explanations' of Euclid. Remarkably, while the talk of 'clarification' or 'making something clear' invites ascriptions of obscurity, Proclus does not take the opportunity to call Euclid 'obscure.'

---

59 Ἡμεῖς μὲν οὖν ἐφ' ἑνὸς τοῦ πρώτου προβλήματος ἐγυμνάσαμεν πάντα ταῦτα καὶ σαφῆ πεποιήκαμεν ("We have thus exercised ourselves and clarified all these distinctions by applying them to a single case, the first problem," *In Eucl.* 210.17 f. Friedlein, transl. Morrow); cf. p. 23, n. 134 on the claim and scope of the Proclean scheme of the structure of the propositions.

60 Πρὸς μὲν οὖν τὴν σαφήνειαν τοῦ θεωρήματος τοσαῦτα προειλήφθω (*In Eucl.* 238.25 f. Friedlein, transl. Morrow). Literally, Proclus says: "For the clarity of the theorem, so much should be anticipated (προειλήφθω)."

61 ὁ δὲ διορισμὸς (*In Eucl.* 203.9 f. Friedlein, cf. p. 79, n. 49 and p. 80, n. 58).

62 διὰ τοῦ προσθεῖναι, *In Eucl.* 376.12 Friedlein.

63 Literally, "For complexity would have been necessary for clarity," ποικιλίας γὰρ ἔδει πρὸς τὴν σαφήνειαν, *In Eucl.* 418.19 Friedlein (my transl.).

64 Just like the English verb, the Greek phrases retain the original reference to clarity, but can be used without highlighting it.

65 The emphasis on understanding is particularly strong in these passages, and above all in *In Eucl.* 418.19–23 Friedlein (cf. above, p. 80, n. 55).

66 In one passage (*In Eucl.* 418.19–23 Friedlein), Proclus even uses two different terms of clarification (σαφήνειαν, σαφηνισθέντων). Cf. above p. 80, n. 55.

The relation of the nonemphatic occurrences to the initial praise of σαφήνεια

Even though the passages from these three aforementioned groups are not particularly conspicuous as regards the use of cognates of σαφήνεια, their terminology recalls, nonetheless, the initial passage quoted at the beginning about Euclid's surpassing all others in σαφήνεια.[67] It is in the light of that passage that the sheer number of occurrences (24) of cognates of σαφήνεια is remarkable. For it confirms the importance of the notion as regards Euclid's text, especially because a considerable number of these 24 instances[68] have Euclid himself as the subject of clarity and clarification.[69] By contrast, the opposite term ἀ-σάφεια ('obscurity') or its cognates hardly appear in the commentary and, in any case, are never applied to Euclid.[70] As we have just seen, Proclus does not even call Euclid 'obscure' when he purports to clarify him. Thus, the regular appearance of cognates of σαφήνεια (and the absence of ἀσάφεια) throughout the commentary corresponds with the emphasis which is put on it in the prominent metamathematical passage.

Also the meaning of σαφήνεια and its cognates in the nonemphatic appearances seems to be easily reconcilable with the emphatic occurrence of the word in the prologue: by and large, it can be translated in the way of the rhetorical term as 'unmistakability.' The implied reference of this translation to readers or learners who cannot misunderstand Euclid and Proclus (the two main agents of clarity and clarification) indicates the didactic context of the commentary. More rarely, one encounters two other meanings: that of 'unambiguousness,' which occurs only once with regard to the subject of a definition,[71] and that of 'evidential, immediate clarity,' which is used with regard to a theorem which becomes immediately understandable from its diagram[72] and with regard to the meaning of a theorem within the chain of propositions.[73]

67 Cf. p. 72 with n. 4.

68 Eight passages have Euclid has the (implied) subject of clarity (*In Eucl.* 94.4, 156.11 f., 203.9 f., 215.24–26, 374.2–8, 376.11–14, 385.16, and 428.7 Friedlein), one more implies that Proclus' clarity is based on Euclid's own (184.3 Friedlein), and yet another one suggests that Euclid chooses to be more readable, and thus clear, by dispensing with complex clarifications of terms (418.19–23 Friedlein).

69 Cf. p. 78, n. 45 for the double sense of σαφήνεια.

70 It is used only twice in a paraphrase of Carpus the engineer (called Carpus of Antioch at *In Eucl.* 125.25 Friedlein, maybe a contemporary of Hero, cf. Morrow 1992, 110) who applies it twice to problems (as opposed to theorems), but uses it, in effect, only *ex negativo* because he thinks that problems are not obscure at all: in the first case (242.9–13 Friedlein), he asks rhetorically: "What is unclear or difficult about these?" (τί γὰρ τούτων ἀσαφὲς καὶ περιειργασμένον;), i.e., problems, implying by his subsequent remarks about theorems that the latter can be obscure; 242.9–13 Friedlein, cf. below p. 94, n. 144); in the second case, it is stated that even "the most obscure problems" (τὰ ἀσαφέστερα [. . .] τῶν προβλημάτων) can be solved by the "method of analysis" (242.14–17 Friedlein, transl. Morrow). The only other time it appears is in the discussion of the definitions of axioms and postulates where it is applied to Apollonius' middle terms, that is, not to the author in question (195.10 Friedlein).

71 *In Eucl.* 156.11 f. Friedlein, cf. above p. 77 with nn. 40 f.

72 *In Eucl.* 374.2–8 Friedlein, cf. above p. 76 for Morrow's translation of σαφές as "evident."

73 *In Eucl.* 385.16 Friedlein, cf. above p. 75, n. 24 for Morrow's translation of σαφῆ as "evident."

However, as far as the means are concerned by which clarity is achieved, the passages which we have surveyed hardly relate to the metamathematical remarks about Euclid's clarity in the prologue. To remind ourselves: the latter link σαφήνεια with the 'logical' form, as it is produced, as Proclus says, by the "method of proceeding from simpler to complex matters and [. . .] laying the foundations of knowledge on the 'common notions.'"[74] Of those 10 passages which explicitly state or imply Euclid's clarity,[75] only one touches upon the subject of "proceeding from simpler to complex matters:" at the beginning of his commentary on theorem I 33, Proclus claims that it forms implicitly "a kind of boundary between the study of parallels and the study of parallelograms," as it "appears to state a property of equal and parallel lines but [. . .] also gives us, without openly doing so, the method of constructing a parallelogram," and as "the next theorem proceeds at once, as if the parallelogram had been constructed, to investigate the essential properties of such areas."[76] However, Proclus' concluding remark that "this is evident" (Ταῦτα μὲν οὖν σαφῆ) appears to refer to the obviousness of the meaning of theorem I 33 with regard to the following one rather than to the clarity produced by moving from simple matters (parallel lines) to complex ones (parallelograms).[77] As regards the second part of what, according to Proclus, ensures Euclid's clarity, his "laying the foundations of the science on the 'common notions,'" his remark that the theorem of I 30 "may also be evident to us from a common notion," is – counterintuitively – irrelevant here.[78] In the former case, τῶν κοινῶν ἐννοιῶν refers to "axioms;" in the latter, Proclus employs κοινῆς ἐννοίας in a more general sense of an "idea which is evident to everyone."[79] As I have argued above, the actual source of the clarity or evidence of the theorem in Proclus' account is the diagram. Three more of the passages we have surveyed refer to other kinds of first principles for Euclid's clarity, namely to a postulate and to two definitions. However, Proclus does not cite them as evidence for his claim that Euclid is clear by "proceeding from simple matters," i.e., from unproved principles. In one case, he praises the unambiguous specification of the subject of a definition;[80] another comment, too, is about the clarity *of* a definition (as part of the commentary on this definition), not about clarity

---

74 Cf. above, p. 73.

75 Cf. above, p. 82, n. 68.

76 *In Eucl.* 385.5–15 Friedlein (transl. Morrow).

77 Another passage (*In Eucl.* 418.19–23 Friedlein), which seems to recall the passage from the prologue by using the term "complexity," turns out to be irrelevant here: it is about the "complexity" of a set of definitions, i.e., of that which actually belongs to the "simple matters" in the axiomatic-deductive procedure "from simple to complex matters" (cf. above, p. 73). Moreover, Proclus argues that Euclid *dispenses with* such definitions or explanations of terms, that is, that he does not move from "simple matters" in this proposition (I 43), precisely (and ironically) because they would have been (too) complex in this case.

78 Cf. p. 76 with n. 32 for the full quotation and interpretation of Proclus' comment.

79 Cf. p. 73, n. 10 and 76, n. 30 for Proclus' use(s) of the term κοινὴ ἔννοια.

80 *In Eucl.* 156.11 f. Friedlein, cf. above p. 77 with nn. 40 f.

by means of a definition.[81] Finally, Proclus uses I, Post. 2 for defending proposition I 1 against criticism, but the deduction of I 1 itself is not based on this postulate.[82]

Apart from linking σαφήνεια with the 'logical' form, we have argued that the prologue associates σαφήνεια also with the *stylistic* form, as it mentions it together with two other rhetorical-stylistic terms: 'conciseness' (συντομία) and 'articulateness/distinctness' (τὸ διηρθρωμένον). In the passages which feature nonemphatic mentions of Euclid's clarity by means of σαφήνεια, clarity is mostly referred to a specific part of the text and presented as a quality produced by its author; however, it is not explicitly associated with style. Notably, none of the passages mentions again the other two stylistic terms from the prologue, although three of them could be understood as comments on the balance between clarity and conciseness: in two cases, Proclus commends disambiguating additions, in another, the eschewal of complex explanations of terms.[83]

On the whole, the nonemphatic occurrences of cognates of σαφήνεια neither provide strong evidence for the characterization of Euclid's σαφήνεια in the prologue nor contradict it, even though some of them imply a differing connotation of the word and other means for achieving it. Many of the passages use the terminology of σαφήνεια inconspicuously with regard to Euclid and his text without specifying the notion. This could imply that Proclus is not actually interested in the clarity of his author, but refers to it merely in order to underscore his own point. But as far as the meaning and means of Euclid's clarity are concerned, the semantic vagueness of the mentions of cognates of σαφήνεια leaves room for interpreting them in the way of the contextualizing emphatic appearance, i.e., as referring to form and style.

Two types of σαφές

The assumption that clarity, in Proclus' commentary, is of the text and produced by its author seems to be at stake in the only other passage (besides the initial metamathematical one) which addresses the foundations of clarity, namely, in the commentary to theorem I 20 ("In any triangle two sides taken together in any manner are greater than the remaining one.").[84] To the objection of the Epicureans that this theorem is "evident (δῆλον) even to an ass and not in need of constructive reasoning,"[85] Proclus replies that "the theorem is, to be sure, clear to sense-perception (σαφὲς κατὰ τὴν αἴσθησιν), but still not clear for scientific argument (σαφὲς κατὰ τὸν ἐπιστημονικὸν λόγον)."[86] This means that he distinguishes two kinds of σαφήνεια: on the one hand, a

81 *In Eucl.* 94.4 Friedlein, cf. above p. 77 with n. 39.
82 *In Eucl.* 215.24–26 Friedlein, cf. above p. 77 with n. 39.
83 *In Eucl.* 156.11 f. and 376.11–14; 418.19–23 Friedlein.
84 *In Eucl.* 322.1–323.3 Friedlein.
85 *In Eucl.* 322.4–6 Friedlein, my transl.
86 *In Eucl.* 322.15–17 Friedlein, my transl.

superficial one, and on the other, a deeper one which is based on knowledge. It is implied that Euclid's proof of the theorem is the means for acquiring the knowledge-based σαφήνεια of the proposition. In this way, the use of σαφές in the second type is consistent with most of the other instances of the term in Proclus' commentary because it relates it to the work of the author.[87]

By contrast, the phrasing σαφὲς κατὰ τὴν αἴσθησιν suggests that perception alone can be the basis of σαφήνεια. We have seen that this sense also occurs among the instances of the cognates of σαφήνεια, especially those of the first group, although the other sense prevails. In any case, it can hardly be the same kind of clarity which arises from perception or argument, which is exactly the point of Proclus' dichotomous distinction in reply to the Epicureans. It is presumably in order to take account of this situation that Morrow translates the two kinds differently "*evident* to sense-perception"[88] and "not clear for scientific thought" although the Greek has σαφές in both cases. "Evident" is also Morrow's suitable translation of δῆλον which Proclus uses in the paraphrase of the Epicurean critique instead of σαφές. Notably, Proclus goes on using different terms when giving examples for what is clear to perception but still in need of scientific explanation: it is τῇ αἰσθήσει σαφές ("clear to sense-perception") that fire warms, but τῇ αἰσθήσει δῆλον ("evident to sense-perception") that we move;[89] when returning to the theorem, Proclus uses yet another term: what it says is τῇ αἰσθήσει καταφανές ("manifest to sense-perception").[90] This means that σαφής appears to be applicable interchangeably for referring both to clarity by knowledge and proof and synonymously with other terms which indicate purely *evidential* clarity (as it would seem). I now proceed to explore the use of these other (and more) terms for 'clarity' in Proclus' commentary and their relation to cognates of σαφήνεια more systematically.

2.2.3 Alternative terminology of clarity

Ἐναργές/ἐνάργεια and related terms

Alongside σαφήνεια and its cognates, Proclus uses a variety of other terms in order to denote 'clarity:' δῆλος ('clear to the mind, manifest, evident,' 55 times), ἐνάργεια/ἐναργής

87 Although the subject of clarity is emphasized here by the distinction of the term, I nevertheless count this passage among the nonemphatic occurrences of cognates of σαφήνεια, as it does not foreground *Euclid's* clarity in the way of the passage from the prologue, but rather the necessity of proving this seemingly clear theorem.

88 My italics.

89 *In Eucl.* 322.18–24 Friedlein, my transl. Ironically, Morrow here translates each of the two terms (σαφές and δῆλον) as "clear."

90 *In Eucl.* 322.25–323.3 Friedlein, my transl. By translating the different terms (σαφές, δῆλον, καταφανές) differently in this passage, although all refer to 'evidential clarity,' I merely wish to make visible the rich terminology of clarity in Proclus which invites further investigation.

('manifest to the mind's eye, clear, distinct,' 25 times); φανερός ('visible, manifest,' 24 times), composites of φανερός (καταφανής, 'manifest, evident,' 14 times; ἐμφανής, 'visible, manifest, plain,' 9 times; περιφανής, 'conspicuous, evident,' προφανής, 'plain, clear,' and αὐτοφανής, 'self-evident,' each once), and τρανής ('clear, distinct,' 5 times).[91] Of these, we have just encountered δῆλος and καταφανής in the sense of 'evident' or 'manifest' in the passage distinguishing two types of σαφής. Semantically, these and almost all other terms of the above list point in the same direction, that of evidential clarity.[92] Let us consider these terms and their use in Proclus' commentary more closely.

The only technical rhetorical term on the above list is ἐναργής. The corresponding noun ἐνάργεια (Latin *evidentia*)[93] denotes above all 'pictorial vividness;' in Dionysius of Halicarnassus, it is "the stylistic effect in which appeal is made to the senses of the listener and attendant circumstances are described in such a way that the listener will be turned into an eyewitness [. . .] The sense uppermost in Dionysius' mind is plain sight."[94] The rhetorical term, thus, comprises both 'detailed representation' and the resulting 'visibility' and 'obviousness.' In the *Progymnasmata* (*School Exercises*), ἐνάργεια is, alongside σαφήνεια and συντομία ('conciseness'), one of the virtues of ἔκφρασις,[95] and it appears, accordingly, in the text as distinguished from σαφήνεια.

In Proclus' commentary, the noun ἐνάργεια is used almost exclusively (seven of eight times) of unprovable statements, especially in the explicit discussion of postulates (αἰτήματα) and axioms (ἀξιώματα).[96] According to Proclus, these principles differ from problems and theorems, respectively,[97] in that they have immediate clarity and do not need to be proved in order to be clear.[98] In fact, proving statements which are already absolutely evident (τὰ φανερώτατα) results in their losing some of their very ἐνάργεια ("evidential clarity") and is, thus, deemed counterproductive.[99] From these passages, therefore, ἐνάργεια emerges as 'immediate clarity,' much in accordance

---

91 Cf. LSJ s. vv.

92 That is, except for τρανής, on which see further below, p. 91.

93 Cf. Cic. *Luc.* 17.

94 Zanker 1981, 297 with reference to the *locus classicus Lys.* 7.

95 Webb & Weller 2012, 409; Kennedy 2003, 86.

96 *In Eucl.* 178.1–198.15 Friedlein, here 178.1 f. Cf. above, p. 73, n. 10 and p. 76, n. 30 for Proclus' use of the terms ἀξιώματα and κοιναὶ ἔννοιαι under which name Euclid's axioms have been transmitted.

97 *In Eucl.* 178.12–14 Friedlein.

98 To quote only one example, cf. *In Eucl.* 181.1–3 Friedlein on the ἐνάργεια of postulates: τὸ δὲ ὅλως ἀποδείξεως δεῖσθαι καὶ κατασκευῆς παρὰ τὴν ἰδιότητα τῶν ζητουμένων τῆς τῶν αἰτημάτων καὶ ἀξιωμάτων ἐναργείας ἀπολειπομένην ("but that a demonstration or a construction is needed at all results from the characteristic that conclusions lack the evidential clarity of postulates," transl. Morrow, modified; cf. also *In Eucl.* 185.6–8 Friedlein on *Elem.* I, Post. 1–3); on the ἐνάργεια of axioms/common notions cf. *In Eucl.* 194.4–8, 195.19–22, and 255.14–18 Friedlein, where the common notions are in fact identified with ἐνάργεια, on that of "the immediate premises" (τὰς ἀμέσους προτάσεις) in general cf. *In Eucl.* 194.15–18 Friedlein.

99 *In Eucl.* 195.19–22 Friedlein: ὁ γὰρ τοῖς φανερωτάτοις ἀπόδειξιν προσάγων οὐ βεβαιοῖ τὴν περὶ αὐτῶν ἀλήθειαν, ἀλλ' ἐλαττοῖ τὴν ἐνάργειαν, ἣν ἔχομεν ἐν ταῖς ἀδιδάκτοις προλήψεσιν ("He who adjoins a

with the rhetorical evidence, although the focus in Proclus is on 'obviousness' rather than 'vividness.' That which has ἐνάργεια, according to Proclus, is evident to a reader who simply looks at it and makes him immediately understand. This is also suggested by the tautological characterization of axioms or premises by the adjectives ἄμεσος ('immediate') and αὐτόπιστος ('self-evident') because of their having ἐνάργεια: "But some persons more accurately distinguish axioms from other premises and call 'axiom' a premise that is immediate and self-evident because of its clarity (τὴν ἄμεσον καὶ αὐτόπιστον δι' ἐνάργειαν πρότασιν), as Aristotle and the geometers say."[100] From the sources which Proclus names for this view, one might even conclude that this view is roughly contemporary with Euclid. As 'immediate obviousness' or 'evidential clarity,' ἐνάργεια is opposed, in most of the occurrences, to proof and provable statements which require a reader to go through a process of understanding before the statements become clear to them.

It is only in one single passage that Proclus says that the ἐνάργεια of a statement depends on its being proved, that is, in his commentary on what has been transmitted as Euclid's I, Post. 5. Proclus argues that it is a theorem rather than a postulate.[101] He claims that at the point where Euclid "uses it as obvious" (ὡς ἐναργεῖ προσχρώμενος), "it will be necessary to show that its obvious character (τὴν ἐνάργειαν) does not appear independently of demonstration (οὐκ ἀναποδείκτως προφαινομένην) but becomes intelligible (γνώριμον) by proof."[102] What Proclus means to say is that the theorem which Euclid treats "as obvious" (because he considers it a postulate instead)[103] is in fact not obvious but requires proof; however, he says that proof is necessary for understanding the *obviousness* of the theorem. This seems to be somewhat contradictory: if a theorem needs to be proved in order for its obviousness to appear, then it was not obvious to begin with, and it seems impossible that it can develop obviousness. What Proclus seems to mean is that the *theorem*, not its obviousness, becomes intelligible through proof; and that, thus proved, it has ἐνάργεια; but he opposes rhetorically effectively 'obviousness not requiring proof' and 'obviousness requiring proof.'

---

proof to things already abundantly evident does not confirm their truth but weakens the evidential clarity that they have when we accept them without instruction" (transl. Morrow, modified).

100 *In Eucl.* 194.4–8 Friedlein, cf. also 194.15–18: "and the sciences dealing with immediate premises that everywhere strike us because of their evidential clarity (τῶν τε περὶ τὰς ἀμέσους προτάσεις καὶ πάντη δι' ἐνάργειαν προσπιπτούσας) belong to a different class from the sciences that employ demonstrations;" and *In Eucl.* 255.14–18 Friedlein, where αὐτόπιστος is attached to ἐνάργεια itself: "Those that proceed from the starting-points are themselves of two kinds, as it happens, for they proceed either from common notions, that is, from self-evident clarity alone (τῆς ἐναργείας μόνης τῆς αὐτοπίστου), or from things previously demonstrated" (transl. Morrow, modified).

101 *In Eucl.* 191.20 f. Friedlein.

102 *In Eucl.* 193.5–9 Friedlein, transl. Morrow, modified.

103 On the originality of the lists of first principles and their designations which have been transmitted for the *Elements*, cf. above p. 22 f. As the titles heading the lists in the manuscripts are most probably non-Euclidean, Euclid, if at all, treated the principle transmitted under the name I, Post. 5 "as obvious."

The use of the adjective ἐναργής and the adverb ἐναργῶς by and large confirms what the use of the cognate noun has already suggested. It is used of the first principles[104] and specifically applied to axioms and postulates (in contradistinction to provable premises).[105] In one of the latter passages, in a quotation of Speusippus which supports Proclus' claim that "principles must always surpass what follows from them in simplicity (τῇ ἁπλότητι), indemonstrability (τῷ ἀναποδείκτῳ), and self-evidence (τῷ αὐτοπίστῳ),"[106] it is used of our "apprehension" (ἐπαφήν, literally: "touch") of these principles, which is "clearer than the sight of visible things."[107] In the characterization of axioms and postulates, it is coupled almost by way of hendiadys with words and phrases denoting 'non-necessity of proof.'[108] Because of the ἐνάργεια of the common notions, "propositions that contravene the common notions (τὰ ταῖς κοιναῖς ἐννοίαις μαχόμενα) are clearer (ἐναργέστερα) means of refutation than those that contradict theorems;" that is, if the converse of a proposition contradicts a common notion rather than the theorem, the proof (by contradiction) of this converse will have more evidence.[109] In the description of the congruence method, ἐναργής is used to characterize the "(grasp of) sense-perception."[110]

In view of these instances, it may surprise that many of the remaining occurrences of ἐναργής have the adjective and adverb applied to 'proof' and 'proving,' respectively. The most emphatic of these is a passage in which Proclus uses the superlative of ἐναργής of the proof of our sample proposition I 47 and – exceptionally – presents the ἐνάργεια of the proof as an accomplishment of its author: μειζόνως δὲ ἄγαμαι τὸν στοιχειωτήν, οὐ μόνον ὅτι δι' ἀποδείξεως ἐναργεστάτης τοῦτο κατεδήσατο ("I admire more the author of the *Elements*, not only for the very lucid proof which he made fast").[111] Of the meaning of ἐναργεστάτης, a similar statement toward the end of Proclus' comment

---

104 *In Eucl.* 255.24–26 Friedlein.

105 *In Eucl.* 196.4–6 Friedlein on the generality of *Elem.* I, Ax. 1–5; *In Eucl.* 193.5–7 Friedlein (of *Elem.* I, Post. 5): this is above-quoted passage (p. 87) which features also the adjective besides the noun. Proclus does not himself call I, Post. 5, as it is, "obvious," because he believes that it is a theorem and requires proof, but he is talking about Euclid's "using it as obvious;" what he doubts is not the obviousness of postulates, but that I, Post. 5 is a postulate.

106 *In Eucl.* 179.12–14 Friedlein.

107 ἐναργεστέραν ἐπαφὴν μᾶλλον ἢ τῶν ὁρατῶν ἡ ὄψις (*In Eucl.* 179.18 f. Friedlein, my translation).

108 *In Eucl.* 179.8 f. Friedlein: γνῶσις ἐναργὴς καὶ ἀναπόδεικτος ("clear knowledge without demonstration," of axioms); 188.3 f.: τοῦτο εἰ μὲν ὡς ἐναργὲς καὶ μὴ δεόμενον ἀποδείξεως συγχωροῦμεν ("If we admit that this statement is self-evident and does not require proof," of I, Post. 4, transl. Morrow; Proclus here reports the view of Geminus, while he himself proceeds to provide a proof).

109 *In Eucl.* 266.11 f. Friedlein: ἐναργέστερα γὰρ εἰς ἔλεγχον τὰ ταῖς κοιναῖς ἐννοίαις μαχόμενα τῶν τοῖς θεωρήμασιν ἀντιφασκομένων (transl. Morrow).

110 *In Eucl.* 243.6–8 Friedlein: καὶ γὰρ ἡ ἐφαρμογὴ καὶ ἡ ἀπὸ ταύτης ἰσότης δεικνυμένη παντάπασιν ἔχεται τῆς αἰσθητῆς καὶ ἐναργοῦς ὑπολήψεως ("for congruence, as well as the equality which is inferred from it, is completely dependent on the clear judgement of sense-perception," transl. Morrow). Proclus here quotes Carpus the Engineer (cf. p. 82, n. 70) on proposition I 4.

111 *In Eucl.* 426.10–12 Friedlein, transl. Morrow.

gives some indication; it uses the related term φανερός: τῆς δὲ τοῦ στοιχειωτοῦ ἀποδείξεως οὔσης φανερᾶς οὐδὲν ἡγοῦμαι δεῖν προσθεῖναι περιττόν ("But since the proof given by the author of the *Elements* is clear, I should not add anything superfluous.").[112] That is, it is not in need of explanation because it is completely evident. Similarly, Proclus describes the evidential clarity of I 21 by using the two terms: Τὸ μὲν δηλούμενον ὑπὸ τῆς προτάσεως φανερὸν καὶ ἡ ἀπόδειξις ἡ παρὰ τῷ στοιχειωτῇ τὸ ἐναργὲς ἔχει [. . .]. ("The meaning of the enunciation is obvious, the proof given by our geometer is evident [. . .]").[113] Proofs by contradiction (or "reductions to impossibility") in general "lead us always to something whose contradictory is evident."[114] Apart from Euclid's own proofs, also Geminus' demonstrating something[115] and Plato's proving I 47 by means of numbers are called ἐναργῶς ("evidently clear").[116] These uses of ἐναργής and ἐναργῶς of 'proofs' and 'proving' do not contradict the impression we have had from the use of the noun ἐνάργεια, that it characterizes, in most occurrences, statements which have 'evidence' without proof.[117] This is because there is nothing to prevent a proof itself from being characterized as 'evident' or 'immediately understandable' – whereas the respective statements to be proved are *not* 'evident.' They need to be proved in order to become clear (σαφής); it is only the resulting proof which is able to have ἐνάργεια. Apart from these dominant uses of ἐναργής/ἐναργῶς with respect to unprovable statements and proofs, they only occur in a handful of other passages in order to describe how natural or mathematical entities or situation appear or show something[118] or in order to state inconspicuously that "it is obvious that."[119]

From the evidence of ἐνάργεια and its cognate forms in Proclus' commentary it emerges that its (implied) objects are above all the 'elements' of deduction, that is, the unprovable principles on which deduction is based. Proclus does not provide information on the means by which their clarity is achieved. The evidence suggests that it is the mere content of these principles, or their ontological status, rather than the way in which Euclid phrases them, in virtue of which they have ἐνάργεια. For in many cases Proclus compares such principles with theorems and problems in terms of provability. From this we may conclude that even in those cases where ἐνάργεια is not said of those first principles, but of proof, it refers to the content, that is, to the evident nature of its logic and conclusiveness rather than to the form of the proof or how it is

112 *In Eucl.* 429.9–11 Friedlein, transl. Morrow.
113 *In Eucl.* 326.11 f. Friedlein, transl. Morrow, modified.
114 *In Eucl.* 254.22 f. Friedlein: αἱ δὲ εἰς ἀδύνατον ἀπαγωγαὶ πάντως μὲν εἰς ἀδύνατον τελευτῶσιν ἐναργές (transl. Morrow, modified). However, as we have seen, those proofs by contradiction which "contravene the common notions" have special ἐνάργεια (cf. above, p. 88).
115 *In Eucl.* 112.22–113.3 Friedlein.
116 *In Eucl.* 427.24–428.2 Friedlein with reference to Pl. *Resp.* 546c (Morrow 1992, 340).
117 Cf. pp. 86 f.
118 *In Eucl.* 89.4–7, 90.2–6, and 294.3–6 Friedlein.
119 *In Eucl.* 195.10 Friedlein.

done. This view is supported by the fact that, unlike σαφῶς, ἐναργῶς is rarely used with verbs of authorial action, that is, of verbs which refer to what Euclid 'says' or 'does,'[120] although the use of the adverb ἐναργῶς or the comparative adverb form ἐναργέστερον with verbs of saying is otherwise well attested.[121] Although some of the references to the ἐνάργεια of principles and proofs are quite emphatic[122] – above all the commendation of the proof of I 47 by means of the superlative form of ἐναργής – the impersonality of the references conveys the impression that the latter have 'evidential clarity' without the help of Euclid. In this way, the use of the term differs from how it is applied in the rhetorical treatises where it is, just like σαφήνεια, a quality of the text for which its author is responsible. As regards the meaning of the term, the emphasis of ἐνάργεια and its cognates in Proclus seems to be on 'immediate obviousness,' that is, on the fact that its objects can be seen effortlessly and right away, as opposed to something which is understandable only gradually and requires knowledge and proof.[123]

In a similar way, most of the nontechnical language of clarity, above all δῆλος and those terms which make use of the metaphor of visibility like φανερός and its composites, seems to be used in a similar way, that is, as having their usual connotations ('manifest,' 'evident,' 'obvious').[124] These terms are used both interchangeably and at times even coupled with ἐναργής, like in the above-quoted passage, which calls things which have ἐνάργεια "the most obvious ones" (τὰ φανερώτατα).[125] By contrast, they tend not to appear in the same passages as cognates of σαφής. An exception to this is the above-mentioned use of a number of these terms (but not ἐναργής) with σαφές in the passage which distinguishes two types of σαφές.[126] However, this passage is precisely where σαφές (κατὰ τὴν αἴσθησιν) is used synonymously with ἐναργές. In any case, ἐναργής, tellingly, is never applied in the same passage as σαφής.[127]

The reason for this seems to be the gist of σαφήνεια and its cognates in Proclus' commentary which we have noted: unlike ἐνάργεια, Proclus often ascribes σαφήνεια to the work of some author, primarily Euclid, and, by using the adjective and adverb with verbs denoting verbal action, marks it as a quality of the text; it appears to em-

---

120 The adverb ἐναργῶς is, as we have seen, used in few cases with verbs of authorial action (δηλοῦσιν, "show," 90.2 Friedlein, ἀποδείκνυσιν, "proves," 112.22 f. Friedlein, δείκνυται "has been shown," 427.25 Friedlein) but Euclid is not the subject of either of them.

121 Cf. Hdt. VIII 77 with λέγειν and Pl. *Ti.* 49a with εἰπεῖν (LSJ s.v.).

122 P. 88 f.

123 Cf. above, p. 87 f.

124 Cf. above, p. 85 f.

125 P. 86.

126 Cf. pp. 84 f.

127 In 195.10 Friedlein, ἀσαφέστερα and ἐναργές occur in the same sentence, but the latter is used inconspicuously in the sense "(it is) obvious (that)" (p. 89), so that it is not used in contradistinction to the former ("[these are] less clear," p. 82, n. 70), which is, in any case, not used with regard to Euclid's text, and thus not relevant for our purpose.

phasize the fact that a statement can be distinguished unmistakably from others and has to be understood in the respective way. The only alternative term for 'clarity' of our above list which is used along similar lines is the adjective τρανής. Derivated from τετραίνειν ('bore through, pierce, perforate'), it literally means 'piercing,' but is used metaphorically in the sense 'clear' or 'distinct' with verbs like εἰδέναι, μαθεῖν, ἀποδεῖξαι, and εἰπεῖν in classical Greek.[128] It occurs, however, but a few times and is only used once with respect to the text of the *Elements*, that is, to the clarifying potential of the construction, and here it seems to be synonymous indeed with how Proclus uses σαφής.[129]

To be sure, while the passages we have reviewed give some clues as for distinguishing Proclus' use of the rhetorical terms ἐναργής and σαφής in the way I have just described, they often appear indistinguishable – or distinguishable only in the light of the more conspicuous passages. Especially where these terms (and the other ones I have surveyed) are used nonemphatically, as in statements of the type 'it is clear/evident that,' they overlap almost completely – that is, σαφής is used synonymously with ἐναργής and other terms of 'evidential clarity' in these cases. Conversely, in the references to the clarity of proofs, ἐναργής also implies σαφήνεια (especially in the translation 'unmistakability'). For if a proof has 'evidential clarity,' it can hardly be misunderstood, and it may be 'obvious' precisely because the proof text is 'distinct' and 'unambiguous.' As a rule, the respective noun, used in each case only a few times, tends to be more telling and reliable, although this is not always the case.

Both the semantic overlap and difference are most conspicuous in the two passages which oppose types of ἐναργής and σαφής, respectively. They suggest and testify to the fact that either word *can* be used for either type of clarity, but the rhetorical effect requires that Proclus could have used more fittingly in each case different terms. The use of ἐνάργεια in the assertion that the obviousness of a theorem becomes intelligible only by proof is due to the foregoing assumption that Euclid treats it as "obvious" (ἐναργεῖ).[130] While the use of σαφήνεια would be more suitable, Proclus assimilates the term to ἐναργεῖ. Similarly, the use of σαφής in the first part of the passage quoted in the section on 'two types of σαφές,' where σαφὲς κατὰ τὴν αἴσθησιν is opposed to σαφὲς κατὰ τὸν ἐπιστημονικὸν λόγον,[131] seems to be rhetorically motivated: Proclus uses δῆλον in his quotation of the objection that the proof is "evident" and again links δῆλον and καταφανές with αἰσθήσει in the repetition of the formulation in the very same passage – terms which are used largely synonymously with ἐναργές in the commentary. But when distinguishing 'clarity based on perception' from 'clarity based on scientific argument,' a more real type, as it were, he uses σαφές even for this first in view of the second type for the sake of pointed parallelism.

---

128 LSJ s.v.

129 *In Eucl.* 375.25–376.2 Friedlein; otherwise, it is only used in philosophical passages of mathematical objects and the knowledge of them (*In Eucl.* 4.22, 11.18, 14.27, 89.2, and 376.1 Friedlein).

130 P. 87.

131 P. 84 f.

Terminology related to the notion of clarity

In addition to the various terms for expressing the notion of 'clarity,' Proclus employs other adjectives in conjunction with these terms, which further underline the gist of the notion(s) of clarity in his commentary. Of these, I have already mentioned ἄμεσος ('immediate'), αὐτόπιστος ('self-evident'), and ἀναπόδεικτος ('undemonstrated') which support the rhetoric of evidential clarity, and γνώριμος which means 'clear' in the general sense of 'understandable' or 'intelligible' and is often rendered as 'clear' in Morrow's translation of the commentary. The latter term may be used both in order to denote 'evidential clarity,' synonymously with ἐναργής,[132] or synonymously with σαφής in the sense of 'comprehensible' or 'clear on the basis of scientific knowledge.'[133]

Additionally, Proclus uses repeatedly πρόχειρος ('readily accessible,' 'easy'), εὐπόριστος ('easy to procure,' 'easily accessible'), εὔληπτος ('easily graspable'), εὐπετής ('easy'), εὐμήχανος ('well contrived'), ἀδίδακτος ('untaught'), and above all, ἁπλοῦς ('simple,' 'plain').[134] These words are often combined as hendiadys or tricolon with one another or with the aforementioned terms related to ἐναργής and support the notion of immediate, untaught, unproven, self-evident clarity. As is the case with ἐναργής alone, these often appear with respect to unprovable first principles.[135]

Interestingly, in one vital passage, several of these adjectives which express the evidential clarity of axioms and postulates appear diametrically opposed to ποικίλιος ('complex') which characterizes the construction not required by axioms and postulates.[136] It is noticeable that ποικιλία ('complexity') which emerges here as the opposite of ἐνάργεια and notions associated with it appears in one other passage as a

132 Cf., e.g., *In Eucl.* 181.8 f. Friedlein.

133 Cf., e.g., *In Eucl.* 418.23 Friedlein, where Morrow indeed translates γνώριμον as "clear."

134 The latter is not confined to the stylistic-cognitive sense, but is used also, among other things, in an ontological sense of the nature of mathematical objects.

135 Cf., e.g., *In Eucl.* 179.12–14 Friedlein; 181.4–9 Friedlein: ἄμφω μὲν οὖν τὸ ἁπλοῦν ἔχειν δεῖ καὶ εὔληπτον, τό τε αἴτημα λέγω καὶ τὸ ἀξίωμα, ἀλλὰ τὸ μὲν αἴτημα προστάττει ἡμῖν μηχανήσασθαι καὶ πορίσασθαί τινα ὕλην εἰς συμπτώματος ἀπόδοσιν ἁπλῆν ἔχουσαν καὶ εὐπετῆ τὴν λῆψιν, τὸ δὲ ἀξίωμα συμβεβηκός τι καθ' αὑτὸ λέγειν γνώριμον αὐτόθεν τοῖς ἀκούουσιν [. . .]. ("Both of these, postulate and axiom, must be simple and easy to grasp. But a postulate prescribes that we construct or provide some simple or easily grasped object for the exhibition of a character, while an axiom asserts some inherent attribute that is known at once to once's auditors [. . .]," transl. Morrow); 194.4–7 Friedlein; 195.19–22 Friedlein.

136 *In Eucl.* 179.2–8 Friedlein: [. . .] κατὰ ταῦτα δὴ καὶ ἐν μὲν τοῖς ἀξιώμασι ταῦτα λαμβάνεται, ὅσα αὐτόθεν εἰς γνῶσίν ἐστι καταφανῆ καὶ πρόχειρα ταῖς ἀδιδάκτοις ἡμῶν ἐννοίαις, ἐν δὲ τοῖς αἰτήμασι ταῦτα ζητοῦμεν λαβεῖν, ὅσα ἐστὶν εὐπόριστα καὶ εὐμήχανα, τῆς διανοίας οὐ καμνούσης περὶ τὴν λῆψιν αὐτῶν, οὐδὲ ποικιλίας δεόμενα οὐδὲ κατασκευῆς ("[. . .] so in the same way axioms take for granted things that are immediately evident to our knowledge and easily grasped by our untaught understandings, whereas in a postulate we ask leave to assume something that can easily be brought about or devised, not requiring any labor of thought for its acceptance nor any complex construction," transl. Morrow).

prerequisite of σαφήνεια.[137] This sets apart the two notions of clarity as conceived of by Proclus in the way we have suspected above: σαφήνεια, unlike ἐνάργεια, requires some (explanatory or stylistic) effort on the part of the author. In this way, σαφήνεια might be closely related to another quality of Euclid which Proclus commends, that is, ἀκρίβεια ('precision').

The relation of ἀκρίβεια ('precision') to σαφήνεια

'Ακρίβεια is another rhetorical term and closely connected to σαφήνεια in classical Greek rhetoric.[138] The noun and its cognate adjective ἀκριβής quite often occur in Proclus' commentary, though not only with respect to Euclid's and others' authorial practices. In the majority of cases, in fact, Proclus uses it with reference to (intelligible) mathematical and other intelligible (as opposed to sensible) objects, and of their understanding, concepts, dialectic, and sciences.[139] Here, Proclus regards ἀκρίβεια from an ontological and epistemological point of view, suggesting that the precision of the knowledge of an object (and, thus, of its exposition) depends on the precision of that very object.

Most of the remaining references have ἀκρίβεια as a quality of the mathematical account. With regard to Euclid's text, it occurs in the second part to the prologue of his commentary – in the praise of Euclid's other writings just after the 'catalogue of geometers'[140]

137 This is the passage quoted above in which Proclus notes that clarification would have required complexity (ποικιλίας γὰρ ἔδει πρὸς τὴν σαφήνειαν), that is, a complex set of definitions of terms (*In Eucl.* 418.19–23 Friedlein; cf. p. 83, n. 77). The notion of ποικιλία is quite prominent in Proclus' commentary (30 occurrences). Proclus uses it a few more times to refer to the kind of complexity which he thinks Euclid has avoided in view of his readership of beginners (see *In Eucl.* 269.24 f. Friedlein for the unsuitability of complexity in an elementary treatise, cf. p. 104). For example, he thinks that Euclid has avoided giving a "variety" of cases of construction instead of a single one (*In Eucl.* 212.9 and 227.10 Friedlein) and instead of using a particular complex method (321.9 Friedlein); and that he has focused on simple cases instead of more complex ones, leaving it to the reader to track down the complexity (414.11–14 Friedlein). While in these cases, ποικιλία seems to be opposed to simplicity, Proclus also has the notion of "complexity through simplicity" (ποικιλίαν αὐτῶν διὰ τῆς ἁπλότητος (44.17 Friedlein).

138 P. 36 with n. 187.

139 *In Eucl.* 4.10, 4.22, 7.5, 8.24, 12.14, 12.17, 13.3, 34.12, 34.16, 39.26, 43.12, 54.11, 59.10, 59.16, 65.5, 87.8, 139.26, 140.4, and 140.11 Friedlein.

140 πολλὰ μὲν οὖν καὶ ἄλλα τοῦ ἀνδρὸς τούτου μαθηματικὰ συγγράμματα θαυμαστῆς ἀκριβείας καὶ ἐπιστημονικῆς θεωρίας μεστά ("There are many other mathematical writings of Euclid, full of remarkable precision and scientific insight," *In Eucl.* 68.23–69.2 Friedlein, transl. Morrow). This remark, thus, does not refer to the *Elements*. This is because Proclus has just ended the 'catalogue of geometers' by praising Euclid's *Elements* (68.6–23 Friedlein) and now commends also Euclid's other works. While he does not attribute ἀκρίβεια to the *Elements* at the end of the 'catalogue,' he, after the remark about the other writings, goes on to talk about the merits of the *Elements*, and here he calls its proofs ἀκριβεῖς ("precise," 69.12 Friedlein).

and in a section on the aim of *Elements* I –[141] in the part on definitions,[142] and above all in the commentary of the propositions itself with respect to particular passages.[143] Moreover, Proclus uses the term prescriptively of the formulation of theorems as opposed to that of problems (and descriptively of a Euclidean theorem) in a paraphrase of Carpus of Antioch[144] and of authors other than Euclid.[145]

In all passages which refer to Euclid's text, Proclus uses ἀκρίβεια in order to characterize a theorem (that is, the enunciation of a theorem proposition) or a definition.[146] In all cases, he attributes their ἀκρίβεια to what Euclid "adds" (προστίθησι[ν]) or "added" (προσέθηκε[ν]).[147] The 'addition' always is a specification which eliminates a plurality of the interpretations of a statement or situation. Thus, 'precision' (ἀκρίβεια) seems to be equal to or to lead to 'disambiguation' and to be closely connected to the notion of σαφήνεια. Notably, the parts of the text to which Proclus refers as 'additions' are not marked as such in the Euclidean text. It is only just Proclus who singles out part of a statement and notes how the statement is ambiguous without this part.

For example, regarding theorem I 14, Proclus emphatically commends the "highest and unsurpassed degree of precision in scientific expression" (τὴν ἀκροτάτην ἀκρίβειαν καὶ ἀνυπέρβλητον τῆς ἐπιστημονικῆς ἑρμηνείας).[148] The subsequent commentary shows that Proclus ascribes the precision of (his text of) the enunciation to Euclid's 'additions,'[149] which I have italicized:

> If with any straight line, *and at a point on it* (καὶ τῷ πρὸς αὐτῇ σημείῳ), two straight lines *adjacent to one another* (ἐξῆς) and *not lying on the same side* (μὴ ἐπὶ τὰ αὐτὰ μέρη) make the adjacent angles equal to two right angles, the two straight lines will be in a straight line with one another.[150]

141 *In Eucl.* 83.5 Friedlein.

142 *In Eucl.* 113.9, 135.2, and 151.15 Friedlein.

143 *In Eucl.* 260.10, 292.13, 292.25, 295.7, 326.3, 385.16 Friedlein.

144 *In Eucl.* 242.10–14 Friedlein, cf. p. 82, n. 70, also on Carpus of Antioch.

145 That is, adhortatively of Proclus' own investigation (*In Eucl.* 178.5 Friedlein), of the distinction of axioms from other premises by "some" (τινες, 194.4 Friedlein), of a contribution by Porphyry (352.14 Friedlein), and of a proof by Ptolemy (367.4 Friedlein).

146 *In Eucl.* 134.27–135.4 Friedlein on I, Def. 10; 151.13–15 Friedlein on I, Def. 15; 242.10–14 Friedlein on theorem I 4, 260.10–13 Friedlein on theorem I 7, 292.12–25 Friedlein on theorem I 13, 295.7 f. Friedlein on theorem I 14, 326.27–327.8 Friedlein on theorem I 21, and 385.16 f. Friedlein on theorem I 33 (only here without the terminology of 'addition').

147 Cf., e.g., n. 149.

148 *In Eucl.* 295.7 f. Friedlein, my translation.

149 Literally, what Euclid "adds" (προστίθησιν, *In Eucl.* 295.10 Friedlein) or "added" (προσέθηκεν, *In Eucl.* 295.16 and 21 Friedlein).

150 Ἐὰν πρός τινι εὐθείᾳ καὶ τῷ πρὸς αὐτῇ σημείῳ δύο εὐθεῖαι ἐξῆς μὴ ἐπὶ τὰ αὐτὰ μέρη τὰς ἐφεξῆς γωνίας δύο ὀρθαῖς ἴσας ποιῶσιν, ἐπ' εὐθείας ἔσονται ἀλλήλαις αἱ εὐθεῖαι. *In Eucl.* 294.15–19 Friedlein, transl. Morrow. Interestingly, ἐξῆς is not part of Euclid's text and agreed to be mathematically not necessary. If this addition is indeed Proclus' own (rather than someone else's), which is implied by Morrow 1992, liii and 230, n. 120, Proclus would have 'improved' the text only to commend Euclid for the ἀκρίβεια of 'his' addition.

Proclus explains how each of the 'additions' rules out one of two possibilities of interpretation, that is, how they effectively lead to disambiguation. This means that these 'additions' in virtue of which theorem I 14 has ἀκρίβεια according to Proclus are of the same kind as the above-mentioned (italicized) 'additions' in definition I, Def. 17 ("the diameter *of the circle*") and problem I 12 ("from a given point *which is not on it*") which Proclus has mentioned as a source of σαφήνεια.[151] These examples suggest Proclus does not have formal or stylistic 'additions' in mind, but rather 'additional explanations.' Nevertheless, it is the "precision in scientific *expression* (ἑρμηνείας)" which Proclus applauds in view of Euclid's 'additions' to theorem I 14; that is, Proclus here refers ἀκρίβεια to 'style.'[152] Another passage, in which Proclus paraphrases Carpus of Antioch (and appears to approve of his thoughts), refers ἀκρίβεια to both style and mathematical explanation and suggests a close connection of ἀκρίβεια and σαφήνεια: after asking rhetorically, "What is obscure (ἀσαφές) and difficult about [problems]?" and in this way underlining their clarity, Carpus, according to Proclus, says:

> But the enunciation of a theorem [. . .] is a laborious matter and needs much precision (ἀκριβείας) and scientific acumen (ἐπιστημονικῆς κρίσεως) if it is not to appear redundant (πλεονάζουσα) or lacking in some element of truth, as is illustrated by this first of the theorems.[153]

The danger which Carpus seems to envisage is that of using either too many or too few words in formulating the enunciation. Whereas 'redundancy' refers to a stylistic deficiency, 'lack of truth' refers to a mathematical one. "Precision" (ἀκρίβεια), alongside "scientific acumen," is necessary for avoiding both. This is because form and content seem to be closely connected in the formulation of theorem enunciations: considering that they usually consist of no more than one sentence, using one word more, or one fewer, can make a difference in content (or "truth"). Ἀκρίβεια seems to be necessary in order to avoid pleonasm but also in order to counterbalance 'conciseness' (though the term συντομία is not explicitly mentioned here). Since the immediate context of Carpus' remark implies that theorem enunciations are obscure without ἀκρίβεια, the latter appears to be closely linked to σαφήνεια for which Proclus praises Euclid alongside with συντομία in the prologue. Similarly in another passage ἀκρίβεια appears to be in close proximity of σαφήνεια, and again, an opposition with συντομία seems to be implied: in commenting on the enunciation of I 7, Proclus states that it is "full of remarkable preci-

---

151 Cf. pp. 77 and 79.

152 ἑρμηνεία can also mean 'explanation,' but rather in the sense of 'interpretation,' which is not meant here. Proclus uses the word in the rhetorical sense of 'expression' or 'style' (LSJ s.v.).

153 τοῦ δὲ θεωρήματος ἐργώδη καὶ πολλῆς δεομένην ἀκριβείας καὶ ἐπιστημονικῆς κρίσεως, ἵνα μήτε πλεονάζουσα φαίνηται μήτ' ἐλλείπουσα τῆς ἀληθείας, οἷον δὴ καὶ τοῦτο πρώτιστον ὂν τῶν θεωρημάτων (*In Eucl.* 242.10–14 Friedlein, transl. Morrow, cf. p. 82, n. 70).

sion"[154] and that "the many added phrases (πάσαις [. . .] ταῖς προσθήκαις)" "make it irrefutable and unambiguous (ἀνέλεγκτος [. . .] καὶ ἀμφισβήτητος)."[155]

2.2.4 Summary of and conclusion from the review of the language of clarity

Let me sum up this investigation into the explicit evidence for notions of 'clarity' in the Euclidean commentators. Both Pappus and Proclus underscore the 'clarity' of the *Elements* emphatically in the place of the notorious 'obscurity,' that is, toward the beginning of and in the prologue to their commentaries, respectively. Proclus even terminologically alludes to the common reference to ἀσάφεια by using the positive σαφήνεια. The context of its use provides the term with a twofold connotation: on the one hand, the characteristics applauded in relation to σαφήνεια indicate that it pertains to the *logical* form; on the other hand, the terms used in conjunction with it denote stylistic qualities and indicate that it likewise refers to the *textual* form. In any case, σαφήνεια – in line with its technical rhetorical use – appears as a quality of the text for which Euclid, as author, is given credit. In the didactic context in which Proclus writes and which he assumes for Euclid as well, the term appears to refer to the fact that Euclid, by the 'unambiguousness' and 'distinctness' of *Elements* I, makes himself 'unmistakably understood,' i.e., to students of geometry.

In addition to the passage from the prologue, cognates of σαφήνεια occur many more times in a more or less inconspicuous way in order to characterize what are mostly mathematical statements or situations. The respective passages do not so much affirm Euclid's clarity but rather (the truth of) what Euclid (or Proclus himself) says. In this way, σαφής/σαφῶς are to be translated rather weakly as 'clear(ly)' or even as '(unmistakable, -y)' in most cases, even though the term is used then – as in the prologue – in the above way of its rhetorical provenance as 'unambiguousness,' 'distinctness,' or 'unmistakability;' sometimes it is employed differently in order to state that something is 'evident.' This distinction roughly overlaps with an explicit distinction by Proclus himself of (author-induced) 'clarity by proof' and '(evidential) clarity by perception.' Unlike in the prologue, Proclus does not explicitly relate σαφήνεια and its cognates to the logical or textual form, but, as I have argued, the emphatic praise of Euclid's σαφήνεια and its connotations seems to inform the unspecific uses

---

154 θαυμαστῆς ἀκριβείας πλήρης (*In Eucl.* 260.10 Friedlein); Morrow misleadingly translates "remarkably full and precise."

155 *In Eucl.* 260.11–13 Friedlein. In another passage, on theorem I 21, Proclus says that meaning of the enunciation is "evident" (φανερόν) and that its proof has "evidential clarity" (τὸ ἐναργές, 326.11 f.); shortly afterwards he mentions Euclid's "geometrical precision" (ἀκρίβεια) by referring to information which Euclid added to the enunciation (326.27–327.8). This means that Proclus attributes both 'evidential clarity' and 'precision' to the same theorem, though he does not state whether he thinks that the former is due to the latter.

of the word. Conversely, the great number of occurrences of the term underscores Euclid's emphatic characterization as 'clear.'

In addition, Proclus uses a range of other terms to denote 'clarity' which further enhance Euclid's presentation as clear. Most of these terms, however, refer to 'immediate, evidential clarity,' synonymously with Proclus' first type of σαφής. They are associated with the rhetorical term ἐνάργεια which – semantically relying on its rhetorical use – almost always refers to the 'evidential clarity' of the first principles and, in its cognate forms, also to the 'obviousness' of proofs. However, while in the rhetorical treatises, ἐνάργεια, just like σαφήνεια, is a quality of the text to be produced by the author, it here (unlike σαφήνεια) mostly appears as something which mathematical statements or proofs – independently from their author – either do or do not have. The counterintuitive use of the term with regard to proofs may be explained by the fact that a proof becomes evident if it uses σαφήνεια for a theorem to become clear (σαφής). Despite the differences in Proclus' use of σαφήνεια and ἐνάργεια, there is considerable overlap of the two terms and their cognates, especially wherever they are inconspicuously used to denote that something is 'clear' or 'clearly' said; in many of these cases, it is impossible to decide whether 'clear' means 'evident' or 'unmistakable/distinct/unambiguous.'

The terminology of 'clarity' in one sense or the other is flanked by terms which are related to it. A great number of terms – for notions such as 'simplicity,' 'self-evident clarity,' 'immediacy,' and 'accessibility' – characterize the first principles in support of the notion of ἐνάργεια. Fewer terms are related to the notion of σαφήνεια. The most important one is ἀκρίβεια ('precision'), another rhetorical term which is closely connected with σαφήνεια in the rhetorical treatises. In Proclus' commentary it is used with regard to theorems and definitions as a prerequisite to σαφήνεια and underlines the authorial effort necessary for this type of clarity. Proclus refers ἀκρίβεια to style and describes it as the result of disambiguating additions to a text which otherwise, as is implied, would be too concise. In this way, ἀκρίβεια is closely connected to the use of σαφήνεια in the prologue where it also appears in conjunction with συντομία ('conciseness').

In conclusion, both commentaries provide us with (late) ancient evidence regarding Euclid as 'clear;' the Proclean one, moreover, employs a rich terminology for the notion of 'clarity' as well as related notions. It uses the technical rhetorical terms much in accordance with the rhetorical treatises, as far as their connotations are concerned. All terms we have mentioned add to the impression of the clarity of the first book of the *Elements* which Proclus in one passage emphasizes explicitly. However, unlike in rhetoric, only σαφήνεια and, in connection with it, also ἀκρίβεια may refer to Euclid's share in this clarity. As earlier ascriptions of clarity to Euclid are not extant or known, the references to Euclid's clarity may not only reflect the rhetorical background of the terminology: considering Proclus' otherwise strong dependence on Aristotle for the interpretation of Greek mathematics,[156] his frequent use of the term ἀκρίβεια could indicate a reception of Aristotle's remarks on the ἀκριβολογία μαθηματική ("the precision of the mathematical manner of expression"),[157] even though Aristotle does not refer to Euclid's text directly nor, unlike Proclus, to particular parts of the kind of speech used when speaking of mathematics.[158] Although Proclus several times ascribes ἀκρίβεια (and sometimes also σαφήνεια) to disambiguating additions and gives some indication for the reasons for Euclid's clarity in the more pronounced passages, the text remains on the whole rather silent on the means and forms of σαφήνεια; unlike the general meaning of the term, these cannot be transferred from rhetoric which does not pertain to mathematical writing. The silence may have to do with the role which the notion of clarity plays in his own agenda; we have observed something similar regarding the use of obscurity in other commentators.[159] Indeed, I have hinted a number of times to the fact that Proclus appears to use the references to Euclid's clarity for emphasizing a point or for underscoring his own argument or ability to clarify.[160]

2.3 The function of the language of clarity in the commentaries of Euclid's *Elements*

In this final part of the chapter, I briefly attempt to interpret Proclus' references to (Euclid's) clarity with regard to the functions of his commentary. First, I consider whether the explicit evidence of the language of clarity in Euclid's commentators is at all matched by the commentary practice of Pappus and Proclus, i.e., beyond the immediate context of the passages which contain references to Euclid's clarity. I do so on the assumption that if Euclid really is as clear as both of his commentators profess, the commentaries should not consist in the clarification of obscure issues, or present themselves as doing so, but have rather different aims. Second, I conclude on the functions of the 'clarity' for Pappus and Proclus and reconsider, in light of this, (the commentators' references to) Euclid's own clarity.

---

156 Morrow 1992, lii and lxiv.

157 Cf. Section 1.3, 35 f.

158 In Aristotle's *Rhetoric*, ἀκρίβεια is not a prominent term. It occurs only once in the context of 'style' (λέξις) where it denotes the "finish" which is appropriate to written as opposed to oral style (*Rhet.* III 12.1414a14–19).

159 Cf. Section 1.4, 51.

160 Cf., e.g., pp. 77 and 84.

2.3.1 The commentary practice in the light of the language of clarity

Pappus, in the first volume of his commentary on *Elements* X, after having outlined the "aim, profit, and divisions" of Euclid's book (I 1–4),[161] offers a "thorough investigation" of "magnitudes" and of the relevant terms, comparing Euclid's views with Plato's and those of others (I 5–23).[162] Afterward (I 24–35), he describes the book according to parts by summarizing larger units and explaining the coherence and logic of these parts. Here, one could regard Pappus' activity as a clarification of the implicitness of Euclid's deductive structure,[163] but Pappus himself does not make this connection; nor does he seem to regard the implicitness as a source of obscurity. The second volume is not structurally dependent on Euclid's book. Rather, by and large it takes itself the form of propositions which seem to expand the thought of Euclid. Here, Pappus refers to Euclid only periodically and is not concerned, at least not primarily or systematically, with clarification of obscurity. In any case, he does not *present* what he does as 'clarification of obscurity.'

Proclus offers a rather detailed history of mathematics in his two-part prologue before dealing at length with the nature of definitions, axioms, and postulates. Here, he often debates and sometimes criticizes Euclid's classification of certain hypotheses, but he does not actually deal with Euclid's text, so clarification is not an issue. It is only halfway through his commentary that Proclus starts commenting on the propositions of *Elements* I one by one. It is here that Proclus paraphrases the condensed standardized Euclidean exposition in a more elaborate fashion and writes out Euclid's propositions in somewhat more familiar language, as it were. Moreover, he explains what the propositions mean, what Euclid does, and why he does it the way he does and not otherwise. Most importantly, he explicates that which remains implicit due to the logical form of the *Elements*, that is, the strict standardization of the deductive structure and style, and due to the according lack of a metamathematical introduction, reader address, structural remarks or debate on mathematical or terminological issues, method, and form.[164] For instance, Proclus systematizes the general proposition structure;[165] debates the choice of the propositions, and adds propositions which go beyond Euclid's book in their level of complexity; discusses Euclid's preference for a given proof, where others would have been possible but Euclid – due to the generic rule – has confined himself to one of them; supplies these alternative proofs; dis-

161 Here, *In Eucl. X* I 4 (p. 68 Thomson & Junge).
162 Here, *In Eucl. X* I 5 (p. 68 Thomson & Junge).
163 See, e.g., *In Eucl. X* I 25 (p. 89 Thomson & Junge), for his making explicit the structure of Euclid's book: "In this part, therefore, we have two series of six propositions, the first six putting together these six irrational lines, and the second six demonstrating the converse propositions."
164 Saito 2009, 804. Cf. above p. 17 for the implicitness of the *Elements* due to its deductive structure and p. 33 for the implicitness of the work because of its standardization in general.
165 Cf. Section 1.2, 23, n. 134.

cusses the sequence of the propositions and the place of particular propositions in the order of the book; explicates the axioms, postulates, and propositions which Euclid – according to the structural rules of the genre – actively presupposes by mentioning them earlier in the book instead of quoting them again explicitly in the place where they form part of a deduction;[166] explains technical mathematical terms; and discusses the metamathematical terminology of Euclid and others; compares Euclid's account with what others have said on the same subject; interprets the text from a philosophical perspective.[167] For the most part, he justifies Euclid against objections which have been raised in the past and against further potential criticism; it is only rarely that he himself criticizes Euclid, and in none of these cases is obscurity the object of his criticism.[168] Much more often, he favorably acknowledges Euclid's accomplishments, and "smooth[es] over" "lapse[s] from rigor," if there are any at all.[169]

This overview of Proclus' commentary practices shows that, like Pappus, he is not really concerned with the clarification of what is "genuinely obscure" according to Galen.[170] Most of his commentary practices can be subsumed under the terms systematization, explication, and unfolding. They are typical for "deuteronomic texts" in general,[171] but are especially enabled by the implicitness of Euclid's proof-centered and standardized style and the fact that it leaves many things "unspoken."[172] The practices in response to the implicitness of the *Elements* indicate a twofold readership: on the one hand, practices such as the discussion of the types of first principles and their terminology or the comparison of alternative proofs indicate a mathematically and philosophically trained expert audience; others such as the explanation of basic terms and methods or the schematization of the proof structure suggest that Proclus explains the Euclidean text to mathematical beginner students.[173] It is the latter type of commentary practices which could testify to a perception of obscurity among part of Proclus' readership.[174] Accordingly, one could regard these practices in terms of the clarification of obscurity. However, Proclus himself never openly calls Euclid's implicitness 'obscure' nor insinuates that it produces obscurity. Overall, he does not even present his own explications as 'clarification,' as he does not use the language of clar-

166 For Euclid's implicitness in this respect cf. Saito 2009, 809, n. 10; see 815 f. for the underlying deductive rule of never referring back to something but always proceeding forward in the deduction (see p. 10, n. 60).

167 Morrow 1992, xlviii–lii.

168 The few occasions on which Proclus criticizes Euclid are listed by Morrow 1992, liii.

169 Morrow 1992, liii.

170 Cf. above, Section 1.3, 45.

171 Netz 1998, 263–268 and 271–273. For Netz's term cf. Section 1.1, 9, n. 52.

172 Saito 2009, 804 on Euclid's implicitness and according "void" which has invited diverse interpretations and reconstructions.

173 Morrow 1992, lviii–l.

174 Such recipient-dependent obscurity corresponds to Galen's second type of obscurity (cf. Section 1.4, 45). On Euclid's implicitness as a possible source of obscurity cf. also below Section 5.2, 424.

ity most of the time when rendering things explicit. Conversely, when he does use it, he mostly does not refer it to his own clarification. But, as we have seen in our survey of 'clarity' in Proclus' commentary, Proclus sometimes presents himself as clarifier without however mentioning or insinuating Euclid's obscurity: rather, he implies that Euclid is already clear in some way and he himself makes him even clearer or clear in another way and for another readership, i.e., his student readership.[175]

In conclusion, Proclus' commentatory practices on Euclid, just like Pappus', do match the evidence of the terminology of clarity and clarification (and the omission of the topos of obscurity) in their commentaries: most of the time, Proclus is busy doing other things rather than clarifying the text. Where his practice might be understood as clarification, he mostly does not portray it in this way himself.

2.3.2 The function and perception of Euclid's clarity

The consistency of the explicit statements and implicit practices of the commentators should not let us take for granted their strong emphasis on clarity rather than on obscurity. From the fact alone that Pappus and Proclus dispense with the obscurity topos we cannot conclude straightforwardly that they always found Euclid clear, let alone that he was completely clear to (part of) their readers or even to contemporary readers of Euclid. It merely suggests that the two commentators did not find the topos useful for their own agenda as authors of commentaries. Seeing that the commentary tradition, which Pappus seems to address explicitly,[176] almost prescribes the use of this topos, they could have easily drawn on the tradition and insinuated that Euclid was unclear if they had wanted to explicate, unfold, and systematize the text under the pretext of the clarification of obscurity. However, they chose not to. On the one hand, they might have found it more difficult to present their practices in terms of 'clarification of obscurity' than a commentator on Aristotle or 'Hippocrates,' the main Greek authors of knowledge texts to have been charged with obscurity. On the other hand, the implicitness of the *Elements* provides its commentators with ample opportunity to comment on the text extensively according to their own purposes without having to insinuate its obscurity.

The decision to dispense with the topos of 'obscurity' does not yet explain the explicit occupation with *clarity* in Pappus and Proclus. In fact, the frequent reference to the notion in a Greek commentary to a scientific text is more striking than the one to obscurity. To be sure, clarity, unlike obscurity, does account for the canonicity of the source au-

175 Cf., above all, the passages *In Eucl.* 376.11–14 and 418.19–23 Friedlein from Proclus (quoted and discussed p. 79 with n. 52 and p. 80 with n. 55, respectively).
176 P. 72.

thor,[177] but it does not serve the commentator any *obvious* purpose. As we have seen in Section 1.4, it is only the identification of obscurity which enables the clarification of the commentary source author – which, for obvious reasons, is presented as the most important function of commentaries and clarification in the meta-commentarial reflections.[178] Compared with this, the high praise for the clarity of the source author in the commentaries of Euclid and the frequent references to Euclid's clarity are unexpected and in need of explanation.[179] The notion even seems to be counterproductive from the point of view of the commentator. For references to clarity such as 'this is clear'[180] or 'Euclid has stated clearly that'[181] give rise to the question why Pappus and Proclus would mention and explain these statements by Euclid at all, if indeed they are all that clear. In view of this situation, one must assume that clarity – no less than obscurity – is a rhetorical tool which serves particular purposes in relation to the function of the commentary.

In the case of Proclus, the rhetoric of clarity seems to fulfill several different functions in view of the different readerships to which the commentary is directed at the same time. With regard to either readership, Proclus uses the notion of 'clarity' both in order to underscore his aims in particular passages of his commentary where he diagnoses clarity as well as more generally in order to create a certain image of himself (and of Euclid). In the didactic context,[182] the notion of clarity serves the construction of Proclus' authority as a teacher who handles Euclid's text without difficulty and recognizes the structure of the book as well as on which principles and propositions another proposition is based. For instance, in his commentary on theorem I 4 he explains how this first theorem of the *Elements* is related to and based on the three preceding problem propositions and notes that with regard to triangles, equality refers to the "area" of the triangles;[183] afterwards, he refers to these and other explanations as that which should be anticipated "for the clarity (σαφήνειαν) of the theorem."[184] Moreover, the emphasis on the 'clarity' of Euclid's text in the context of teaching supports the (over)simplifying interpretation of Euclid who is made out even clearer than he actually is. For example, in his formalization of the general structure of the proposition,[185] he states what the part of the 'definition of the goal' "makes clear." As I have argued above, rather than drawing attention to how far the 'definition of the goal'

177 Cf. p. 51 with n. 232 on the paradox of obscurity in the commentaries on canonical (scientific) authors.

178 Cf. Section 1.4, 44 for Galen's definition of exegesis.

179 Cf. Section 1.4, 48 f. on the usefulness of the obscurity of the source authors from the commentators' point of view.

180 *In Eucl.* 385.16 Friedlein (discussed above, p. 75 with n. 24).

181 *In Eucl.* 156.11 f. Friedlein. (discussed above, p. 77).

182 Cf. pp. 74, 80, 82, 96 above.

183 *In Eucl.* 233.11–235.8 and 236.18 f. Friedlein.

184 *In Eucl.* 238.25 f. Friedlein.

185 This is argued by Netz in his 1999b; see below, Section 3.4 on the standardization of the structure of the proof.

adds to clarity of the Euclidean proposition, the rhetoric of clarity here serves Proclus' presentation of the 'definition of the goal' as part of his clearcut scheme of the proposition parts.[186] In another passage which we have discussed – Proclus' comment on definition I, Def. 17 that Euclid "makes it clear that he is not defining every diameter, but the diameter of the circle" – the rhetoric of clarity gives Proclus the opportunity to systematize different kinds of diameters and their terminology.[187] It is precisely the clarity (or unambiguousness) of Euclid's formulation of the subject of the definition ("diameter of the circle") which betrays Proclus' didactic aim because the formulation itself does not require any clarification. Seeing that the *Elements* itself in its original context presumably did *not* have the didactic function which Proclus insinuates,[188] his frequent ascription of σαφήνεια to Euclid seems above all to serve a particular function for himself as a commentator.

Apart from mathematical beginners, Proclus addresses, as we have seen, a mathematically more specialized and philosophical audience.[189] For his commentary goes beyond the requirements of students of elementary treatises. In this specialist context, the advertisement of clarity first of all serves the construction of Proclus' authority; it underscores the plausibility of Proclus' philosophical interpretation of Euclid in general and the truth of his explanations of Euclid's first principles and propositions in particular, often against other interpreters of Euclid. For instance, Proclus paraphrases what Euclid by his first definition "all but clearly says" in order to show that this definition is neither wrong nor imperfect; in so doing, he underscores the truth of the definition and enhances Euclid's authority; but the main aim of his emphatic reference to definition I, Def. 1 seems to be to underline the truth of his own preceding philosophical interpretation of the definition of 'point' (and of the objects of geometry in general).[190] Another example is Proclus' reference to the clarity of postulate I, Post. 2 for the sake of defending the construction of proposition I 1 against Zeno.[191] Morever, Proclus uses the notion of 'clarity' in order to justify introducing material or methods into his commentary which go beyond the aim and scope of the *Elements*. For example, in his commentary on I 47, Proclus uses the rhetoric of clarity in order to justify that he does not actually comment on the proposition since, he claims, it is clear as it is and he does not want to add superfluous explanations;[192] instead, he

186 *In Eucl.* 203.23–204.2 Friedlein, cf. p. 79.

187 *In Eucl.* 156.12–19 Friedlein. Cf. pp. 77.

188 For the original didactic function of the *Elements* according to Proclus cf. below, 103 and Section 2.2.1, 74.

189 On the two opposing tendencies in Euclid of addressing both very simple and very complex issues in the same text, cf. Morrow 1992, xlviii-l. For Proclus as the first "representative" of an influential, long-lasting tendency of commenting on the *Elements* "from a philosophical point of view" cf. Saito 2009, 804.

190 *In Eucl.* 93.6–94.7, and here 94.4 Friedlein, cf. above p. 77, n. 39.

191 *In Eucl.* 215.24–26 Friedlein, cf. above p. 77, n. 39.

192 Cf. p. 88 f.

presents Plato's and Pythagoras' numerical methods for finding scalene triangles which conform to the theorem of I 47.[193] Often Proclus includes additional, mostly more complex proofs and uses the notion of clarity to underline the reliability of his own proofs, e.g., when he refers to his proof of the falsity of the converse of I, Post. 4 as a "clearly demonstrable" statement.[194] With regard to this ample inclusion of additional material, 'clarity' also serves Proclus to justify the limitation of Euclid's scope and to (re)construct Euclid's own authority: He presents him as a (highly successful) didactic author who writes for beginner students; Euclid's "elementary treatise" surpasses all other "elementary treatises" by its didactic qualities, above all σαφήνεια, as we have seen Proclus claim in the passage of his emphatic praise of the latter. In his commentary on proposition I 8, Proclus defends Euclid's proof which is "without any complexity" (χωρὶς ἁπάσης ποικιλίας) against Philon's, as "its use of a variety (ποικιλίᾳ) of cases makes it unsuited for an elementary treatise;"[195] and many other times, Proclus points to Euclid's alleged aim to avoid complexity in order to justify Euclid's leaving out this or that additional proof or proceeding in a certain way. Although the passage I have quoted does not explicitly connect the avoidance of complexity and 'clarity,' it seems to imply such a connection. Against Philon's "elegant" omission of the theorem of I 7, Proclus defends Euclid's inclusion of it, as "this seventh is proved from the fifth and [. . .] provides an uncomplicated proof of the eighth."[196] In this way, the Euclidean method which Proclus commends for its avoidance of complexity in this passage consists precisely in "proceeding from simpler to more complex matters," which is how Euclid produces σαφήνεια according to Proclus' emphatic praise for it.[197]

In another passage the connection between (the avoidance of) complexity and clarity is not as straightforward. This passage which we have already quoted in our analysis of the language of clarity states that Euclid dispenses with explaining a number of terms which are necessary for defining the term "complements" (τὸ δὲ ὄνομα τῶν παραπληρωμάτων), as well as with the explanation of the term "complements" itself, because these explanations would have been too complex. The passage suggests, on the one hand, that Euclid avoids complexity for the sake of readability and, thus, clarity. On the other hand, Proclus here *associates* complexity with clarity, as he says that complexity would have been necessary for σαφήνεια, suggesting that the proof is *not* 'clear' as it is.[198] Indeed, according to what Proclus says in his prologue about

193 *In Eucl.* 427.24–429.8 Friedlein; cf. p. 89 with n. 116.
194 Cf. p. 78, n. 42.
195 *In Eucl.* 269.22–25 Friedlein, transl. Morrow, modified (cf. p. 93, n. 137).
196 τοῦτο τὸ ἕβδομον δείκνυται ἀπὸ τοῦ πέμπτου καὶ δείκνυσι χωρὶς ἁπάσης ποικιλίας τὸ ὄγδοον (*In Eucl.* 269.21–23 Friedlein, transl. Morrow).
197 Cf. pp. 73.
198 See pp. 80 and 93, n. 137.

what accounts for Euclid's clarity[199] – "its method of proceeding from simpler to more complex matters and its laying the foundations of the science on the 'common notions'" – the present proposition I 43 is not (entirely) clear; for it is not based on the first principles regarding its use of the term 'complements' (and the term parallelogram, which would have to be explained for the explanation of 'complements,' as Proclus states).[200] Here, Proclus' use of 'clarity' has a double function: first, with regard to his fellow mathematical experts, it serves to justify the limitation of Euclid's deduction and to construct Euclid as a simplifying writer of an elementary treatise for beginners; moreover, Proclus also underscores his own authority as a mathematical author and his ability to detect the gap in the deductive system. Second, with regard to his student readership, Proclus shows that he knows how to clarify Euclid and to provide (though not proper definitions) explanations of the terms which Euclid presupposes; as is indicated also by the didactic language of the passage,[201] the clarification itself must be directed to this student readership – for expert readers would know the terms. This implies a paradox, a 'paradox of clarity' similar to the 'paradox of obscurity,' as it were: if Proclus praises the clarity of the *Elements* as a didactic value, then why does the alleged student readership of Proclus require his clarification, and why is Proclus prepared to provide it – especially as Proclus suggests that Euclid decided to omit the complex definitions precisely for didactic reasons? In the present case, Proclus solves this paradox by suggesting that Euclid is clear in some way while he himself makes Euclid clear in another way (in which Euclid, we must infer, is *not* clear).[202] That is, Proclus indirectly refers to his own distinction of two ways in which something can be "clear" (σαφές):

> The term 'complements' was derived by the author of the *Elements* from the thing itself, since complements fill the whole area outside the two parallelograms. This is why he does not regard it as deserving of special mention in the Definitions. It would have required complexity for clarity (σαφήνειαν) in order for us[203] to understand what a parallelogram is and what the parallelo-

199 *In Eucl.* 74.13–16 Friedlein (transl. Morrow). Cf. p. 73 with n. 10 for the quotation.

200 To be sure, Proclus only wants the "foundations of the science" (ἡ [. . .] καταβολὴ τῆς θεωρίας), not every single proposition to be based on the "common notions" (κοιναὶ ἔννοιαι), not on the first principles, including the definitions, in general; but κοιναὶ ἔννοιαι here could stand for the first principles as *pars pro toto*. On the term cf. above p. 76, n. 30.

201 Cf. p. 80 for the connection of 'clarity' and 'understanding' in this passage; similarly, the passage from the prologue in which Proclus praises Euclid's σαφήνεια emphatically is full of references to understanding and learning (pp. 74 f. with n. 16).

202 Cf. above p. 100.

203 The first-person plural (ἵνα γνῶμεν) is to be understood as part of Proclus' didactic gesture (for he himself and his fellow experts do *not* require the complex clarification which he is about to deliver).

> grams that are constructed about the diameter are as a whole; for only after these had been explained would the meaning of 'complement' have become clear (γνώριμον).[204]

Without saying so explicitly, Proclus implies that while Euclid considered the meaning of τῶν παραπληρωμάτων ("complements") 'evident' in view of the diagrammatic construction, the proposition is still in need of explanation, i.e., not 'clear for scientific argument' in the sense of his earlier distinction of types of σαφές. This is the proper sense of σαφήνεια (as opposed to ἐνάργεια) in Proclus' commentary.[205] In this way, Proclus is able to present Euclid as clear and at the same time to provide his own clarification. Nevertheless, it is precisely because Proclus uses the term σαφήνεια consistently that a tension remains between Proclus' portrait of Euclid as σαφής and his present insinuation of Euclid's lack of σαφήνεια.

In a similar way as in the passage we have just discussed, Proclus' commentary often is relevant to both types of readers simultaneously, and the emphasis on clarity, accordingly, has a double function. This is mostly the case whenever Proclus commends the 'clarity' (σαφήνεια) or 'precision' (ἀκρίβεια) which problems or theorems have due to Euclid's disambiguating 'additions' to it, as we have seen him do numerous times.[206] For example, we have discussed his commentary on problem I 31 ("Through a given point to draw a straight line parallel to a given straight line"), in which Proclus refers to the clarifying addition to "point" in the analogous problem I 12 ("which is not on it [i.e., "a given infinite straight line"]"), which is missing from I 31 and must be "assume[d] in advance."[207] In the didactic context, the reference to the clarity of the formulation of problem I 12 provides Proclus with an opportunity to clarify problem I 31 accordingly; it underlines his interpretation of problem I 31 and his authority in doing so. In the expert context, the reference serves the implicit reflection about the necessary degree of explicitness in the formulation of theorems and problems and implies Proclus' stance in favor of the more explicit variant which makes the text unmistakable.

To be sure, Proclus could just as well have used the established topos of obscurity for all these purposes, just as other ancient Greek commentators have done. It is precisely the point that he did not; maybe because emphasizing clarity simply appeared more credible in the face of the Euclidean work. Proclus (as well as Pappus) seem to be able to count on Euclid's being relatively clear or at least not utterly unclear to (part of) their readers, in order for them to find the emphasis on clarity convincing. This is true especially regarding the fact that experts are among Proclus' intended

---

204 *In Eucl.* 418.15–23 Friedlein (Τὸ δὲ ὄνομα τῶν παραπληρωμάτων ἀπ' αὐτοῦ τοῦ πράγματος ἔλαβεν ὁ στοιχειωτὴς ὡς καὶ τούτων παρὰ τὰ δύο παραλληλόγραμμα συμπληρούντων τὸ ὅλον. διόπερ αὐτὸ καθ' αὐτὸ μνήμης ἐν τοῖς ὅροις οὐκ ἠξίωται. ποικιλίας γὰρ ἔδει πρὸς τὴν σαφήνειαν, ἵνα γνῶμεν, τί παραλληλόγραμμον, καὶ τίνα τὰ περὶ τὴν αὐτὴν διάμετρον τῷ ὅλῳ. τούτων γὰρ σαφηνισθέντων καὶ τὸ παραπλήρωμα μόνως ἂν ἐγένετο γνώριμον. Transl. Morrow, modified; cf. above p. 80).

205 Cf. pp. 82–84.

206 Pp. 79 f. and 93–95.

207 Pp. 79 f.

readership who in any case would see through Proclus' comments on the rather basic *Elements* I. Unlike commentators of other Greek scientific texts, Proclus can complement the Euclidean text extensively because of its implicitness but has to take into account the straightforwardness of the text. For the same reason, another interpretation of the rhetoric of clarity in Proclus seems unlikely: namely, that Proclus might even have invented the notion of the clarity of Euclid, while glossing over his obscurity and constructing his authority by pretending to find him clear. Again, he could have made perfect use of obscurity; instead, he seems to have used Euclid's clarity for his own purposes.

3 Textual standardization in the propositions of *Elements* I

3.1 Introduction

3.1.1 Method and corpus

By way of general introduction to the whole analytical part of this study, including Chapter 4, let me reflect on the method and corpus of the analysis and expand up on what I briefly said about it in Section 1.6.[1]

Methodically, I analyze the textual and diagrammatic standardization both quantitatively and qualitatively. The main method I use is the quantitative assessment of the textual objects and their degree of standardization. I complement the quantitative assessment with a qualitative analysis of the kind of textual objects which result from standardization regarding their standardizability and unambiguousness.[2] The quantitative analysis consists of making various counts of the textual features on all levels of the text: first in absolute terms, then with regard to their use in connection with other features and in a particular meaning, function, and part of the text. The textual features which I count comprise individual words, phrases, tenses, modes, verb forms, sections of text, letters, and diagrams. In evaluating these counts, I regard as 'standardization' (as defined above) any tendency to use the same formal elements in the same textual parts or function, that is, as a tendency of formal unification, toward formal uniformity or sameness. I regard this tendency as the result of a preferred application of certain formal elements in the place of alternatively available elements which the text applies less frequently, or in a less unified way, or not all. This means that I do not conceive of 'standardization' as one hundred percent uniformity, but only as a certain degree of sameness, without, however, predefining a minimum degree, i.e., a certain required percentage of unification which I count as 'standardization.' I merely consider in the case of each formal feature whether there is a tendency of formal unification, and how the degree of its standardization relates to one hundred percent standardization, which I regard as tantamount to the 'unambiguousness' of a feature.[3]

There are three reasons why I do not define a minimum degree, i.e., a certain required percentage of unification, for what counts as standardization: the first has to do with the transmission of the text. As the origin of the exact degree of the standardi-

---

1 Cf. above, pp. 64–66. The following remarks will also apply to the diagrammatic analysis in Chapter 4, while methodological issues which concern only the latter will be left for the introduction to Chapter 4 (cf. below, pp. 185–187).

2 This is the looser of the two senses in which I use the term 'unambiguousness' in the textual analysis (see p. 65); for the other one cf. the end of the paragraph.

3 On the different senses in which I use cf. above p. 65.

https://doi.org/10.1515/9783110670462-003

zation is uncertain,[4] it does not make sense to count as standardization only a certain exact degree of it because this implies that one must rely on the exact state of the text. The second reason is that a minimum degree would be of limited use because the exact degree of the standardization of the textual features which are analyzed is to some extent arbitrary: it depends on how they have been previously defined, since their predefinition already entails a certain predefinition of the degree of their standardization. I shall take this into account when choosing features for the analysis, and shall discuss the problem exemplarily. The third reason for my refraining from defining a certain percentage of uniformity which counts as standardization is that any such definition itself would be arbitrary. This is because the degree of unification has a different significance in the case of each formal feature. The significance of the unification of a formal feature depends on the frequency of its application in relation to alternative features which are available for fulfilling a certain textual function. Thus, one would have to determine for every single feature whether the degree of its unification is significant, that is, whether it indicates standardization in the sense of a preferred application of formal features. This could be proved for features which are used comparatively more often than other available features in the *Elements* compared with other texts.[5] However, hardly any texts are comparable for this purpose. Other elements texts cannot be used to show, as is my aim, the *generic* standardization of the *Elements*, but only the uniqueness of the standardization of the *Elements* compared with other elements texts. Texts from outside the genre differ so fundamentally in form that they do not use many of the main features of elements texts, and it is not certain whether their own features presented alternatives for elements authors at all. The generic form of the proofs is so unique that one can only be sure in the case of alternative features used in the text itself that they indeed represent alternatives. Therefore, I shall consider alternative textual features only as far as they are used within the *Elements* itself, and beyond it only more generally in terms of, e.g., alternative syntactical possibilities available in the Greek language or geometrical vocabulary attested prior to Euclid. This means that I might not be able to determine in each case whether the unification is indeed the result of a preferred application of the respective feature. However, this is not my purpose; it suffices to assess whether there is a at all tendency of using the same textual features in the same places or func-

4 Cf. above p. 10.

5 Cf. Netz's concept of "semantic markedness" which denotes "groups of words" (he coins it with respect to his notion of "formulae," see pp. 127 and 131, n. 107) which are "used more often than their semantic counterparts, relative to the distribution in other relevant texts" (Netz 1999a, 131; see below p. 155 f., n. 216). By contrast with "semantically marked" textual features, such that occur more often than others in the *Elements* but are the natural or only way of putting something (which are "regularly used" in Netz's terminology) do not conclusively indicate a preferred application, for instance the word καί (in the sense of "and"); the same applies to features which occur more often in the *Elements* than in other texts (which are "markedly repetitive," Netz, ibid.), for instance the terms for geometrical objects (if one simply compares their frequency in non-mathematical texts).

tions – which then is likely to be the result of a preference of these features over others. It is only on the basis of this analysis that I will consider in Chapter 5 whether the (preferred) application of the standardized features of the *Elements* is the result of an 'effort' to standardize, and to what end the work is formally standardized.[6]

For the statistical method, I rely on Netz's seminal 1999 study, who has applied it to Euclid and other mathematicians, although he started from different premises and drew different conclusions.[7] Nevertheless, since he has so comprehensively analyzed the generic features, and to a similar end as I intend to do, I not only use his method, but also start from the concepts he has developed and the terms he has coined. I discard some of them as unhelpful for my own purposes, but most of the time, I find them useful for making clear my own stance (while still acknowledging Netz's conceptual work), and redefine them to fit my approach.

As far as the text corpus of the analysis of the Euclidean standardization is concerned, I have mentioned the reasons for focusing on the first book of the *Elements* above: first, this decision has to do with the transmission and the ancient reception of the *Elements* and with the according importance of book I also for the present study. It is the best-known and most widely received part of the *Elements* for which most of the few diagrammatic fragments – discussed in Section 4.1 – survive and on which the only extant Greek commentary is based which we have surveyed in the preceding Chapter 2. We have seen that Proclus commends the *Elements* for its clarity and attributes its clarity (at least in part) to the form of the work. While I analyze Euclid's standardization in Chapters 3 and 4 independently of (the analysis of) Proclus' references to Euclid's clarity, it is nevertheless useful to base the chapter on the ancient reception of the form of the *Elements* in terms of clarity and the chapters on the form of the *Elements* on the same part of the work: on the one hand, because I use and discuss the most influential one of Proclus' conceptualizations of the form of the *Elements* for my analysis of the Euclidean text, that is, his scheme of the proposition parts; on the other hand, because I shall use the insights from the analysis of the commentators' reception of how clarity is related to form when I interpret the findings of the textual and diagrammatic analysis in Chapter 5.

Second, and most importantly, the decision to focus on *Elements* I is due to the fact that it is the most standardized book of the work. In this way, I assume, it is most apt to display the manifestations of the Euclidean standardization. Seeing that it is the least complex book at the same time, it prompts the question of the function of stan-

6 Cf. p. 2, n. 10. The term 'preferred application' includes the notion of an 'effort' of standardization only with respect to the respective feature which is preferred and thus used in a standardized way. It does not yet – in the manner of the term 'effort' (of standardization) – suggest a tendency toward standardization in general, which may only (but does not have to) be indicated by an overall tendency toward a preferred application of certain textual features. Netz 1999a, 134, e.g., denies such an "effort" regarding the standardization of the mathematical lexicon.

7 Netz 1999a. See above, pp. 57 f.

dardization. One might argue that book I is not representative in terms of its standardization, especially in terms of its structural standardization (both as regards the structure of its propositions and its macrostructure);[8] that the standardization of a sample naturally increases if one limits it to a small part of the work which contains only a certain type of propositions and at that the easiest specimen of this type. However, on the one hand, the differences between the books regarding standardization are only gradual; all other books of the work are also extremely standardized; the arithmetical and geometrical books of the work are standardized to a similar degree and in the same way despite their difference in content, except for very few qualitative differences, e.g., as regards the way in which the letters are (standardly) used and what they (standardly) represent in the diagram.[9] On the other hand, I do not aim at offering a representative picture of the standardization of the *Elements* to begin with, but rather at showing how the standardized form relates to clarity; this can be done best by using the part of the work which displays the workings of standardization most visibly. In doing so, I shall bear in mind possible reasons for the special standardization of book I outside of the general motives for and mechanisms of standardization in the *Elements*: that the strong standardization of its first book has to do precisely with the easy contents as the text lends itself to standardization most easily and that book I was especially affected by additional standardization in the later course of transmission, perhaps because of its greater diffusion.

As regards book I itself, I do not always take all its propositions into account; sometimes, especially in Section 4.4, even smaller samples will prove sufficient for the statistical analysis.[10] Sometimes I do not need to conduct statistical research, either because a single proposition can stand for all or because I can rely on previous statistical evaluations and merely wish to demonstrate how the formal features previously recognized relate to unambiguousness and clarity by way of example; in these cases, I shall use I 47, arguably the most famous proposition of the *Elements*. This is the proposition of what has come to be known as the 'Pythagorean theorem,' which proves that in a right triangle the square on the hypotenuse is equal to the sum of the squares on the two other sides. Within book I, this is one of the longest propositions, and one of the most complex ones as far as the number of geometrical objects involved is concerned. For reasons which will become clear in Section 4.4.4, the symmetry of the diagrammatic figuration of I 47 makes this proposition particularly apt for the analysis of the diagrammatic letters.[11]

---

8 Cf. Netz 1999b, 294 on Proclus' scheme of the standard structure of the proposition, which is not as widely applicable as Proclus pretends, but adequately describes only the propositions of the first book of Euclid's *Elements* (cf. pp. 23, n. 134 and 178 f.).

9 Cf. p. 283, n. 175.

10 Cf. especially Section 4.4.4 on the order of the letters in multi-lettered 'letter words.'

11 Cf. below, p. 387.

While the analysis of the text and diagrams in Chapters 3 and 4 is focused on book I, I shall make occasional reference to other books wherever necessary (mostly in Chapter 4). This is the case if certain features of textual and diagrammatic standardization which appear exclusively in some of the subsequent books add to understanding the reasons for standardization; in Section 4.2, for instance, I consider the diagrammatic representation of some of the solid figures of books XII and XIII in the manuscripts of the *Elements*, as *Elements* I does not deal with solid figures.[12] Other reasons for referring to books beyond the first one have to do with the transmission of the *Elements*: in Section 4.1, I discuss the ancient papyrological evidence, most of which survives for *Elements* I, but one of the most important testimonies is a fragment that includes a diagram for proposition II 5.

I shall refrain from taking into account other elements texts or scientific or 'knowledge texts' in general.[13] I have already noted my hesitation regarding a comparative approach when I explained why I dispense with comparing other texts for the sake of determining the exact degree of the standardization of the textual features of the *Elements*:[14] among texts of the theoretical tradition of Greek mathematics, I focus on the example of the *Elements* because I am interested in the meaning and function of the *generic* standardization of elements texts, which seems to be at its greatest and most clearly visible in this work.[15] That is, I do not aim to prove the uniqueness of the exact degree of the standardization of the *Elements within* the genre.[16] As a next step, to be sure, one could compare other, often somewhat less strictly standardized exponents of Greek theoretical mathematics in order to distinguish degrees and kinds of standardization and to determine the *particular* function of the standardization of the *Elements* of all works of the genre. Suitable comparanda for this purpose could be the Euclidean *Data* which refer to the contents of *Elements* I–VI;[17] the first four books of Apollonius' *Conics*, which are comparable to the *Elements* as regards their aim;[18] some of those works which contain, according to Netz, 'ludic' elements, often resulting in structural non-standardization, for instance, the books by Archimedes which mostly present new discoveries, Apollonius' *Conics* III, Hypsicles' 'book XIV' of the *Elements*, or even *Elements* XIII which differs from the rest of the work by the playful

---

12 I only make brief references to the role of letters in the arithmetical books VII–IX (p. 283, n. 175), and to the absence of the συμπέρασμα from them (pp. 23 f., n. 125).

13 I use this term interchangeably with 'scientific texts,' for which cf. p. 3, n. 17.

14 Cf. above p. 108 f.; see also p. 66 of Section 1.6.

15 Cf. pp. 10 f. and 66.

16 The exact degree of Euclid's own contribution to the original standardization of the *Elements* is also impossible to find out (cf. p. 10).

17 Asper 2014, 464.

18 Cf. above pp. 7 f. and 12.

character of its form;[19] and 'book XV'[20] of the *Elements*.[21] Moreover, the works of later writers of theoretical mathematics, e.g., Nicomachus, Pappus, or Proclus, could be compared. These works make use (to some extent) of many of the characteristic features of the language and style of elements texts, but within a different text form, e.g., in the case of Pappus, the 'collection' (συναγωγή) of the same name. In this way, their mathematical prose differs from that of (pre-)Hellenistic elements texts[22] and is interesting to compare regarding its standardization. However, I have decided to analyze thoroughly only (the first book of) the *Elements*, since an analogously detailed analysis would go beyond the scope and purpose of this project.

Outside Greek theoretical mathematics, one should hardly be able to find any scientific text which compares to the *Elements* (or other elements texts) in terms of its overall standardization. The main reason, however, why I decided against an inter-generic comparison is that the textual form of elements texts or, more broadly, Greek theoretical mathematics, differs fundamentally from that of other genres, because many of its characteristic features do not occur elsewhere, or only in terms of a reception of the Euclidean form: wherever mathematics or mathematical passages occur in other scientific texts, such as philosophical, astronomical, or geographical ones, these are either direct testimonies of Euclid's predecessors or are informed by the Euclidean style in general. Thus, comparison with them in any case amounts to an intra-generic one, which, as I have explained, is not my aim here. Besides the works of theoretical mathematics, few texts of the large field of practical mathematics have survived (and hybrids of the two traditions: Hero and Diophantus), although the elements tradition has dominated the transmission of Greek mathematics. Texts extant from the practical tradition also have a standardized form, but it is a completely different genre, which does not use the features of theoretical mathematics.[23] Quantitative comparisons regarding the degree of standardization with samples from other, not only mathematical 'knowledge texts,' say, by Herodotus or Aristotle, could be useful, but the choice of such texts would be somewhat arbitrary regarding their genre, epoch, and context. Moreover, regarding the comparison of other elements texts, comparably extensive quantitative analyses would also exceed the scope of this study. Therefore, I have decided to examine thoroughly one single text only.

---

19 Cf. above p. 7.

20 Cf. above, p. 1, n. 7.

21 Netz's 1999a analysis, which uses large samples from Hellenistic or pre-Hellenistic authors of elements texts, including the Euclidean *Data*, Hypsicles' book XIV of the *Elements*, and all major authors of the second type, does not show fundamental differences regarding their application of the main textual features, but as it is not concerned with standardization *per se* (pp. 57 f.), it is not aimed at comparing how much the generic features are standardized in each of them.

22 Asper 2001, 78.

23 Asper 2007, 197.

3.1.2 Levels of the analysis of textual standardization

In the study of the textual form in Chapter 3, I consider each of the three textual levels separately (lexicon, grammar, and structure). I advance from the smallest unit of these from which the text is built (lexicon) via that which modifies and connects these units (grammar) to that which comprises those larger units within the form of the proposition (structure). In doing so, I start from questions conventionally connected to the realms of lexicon, grammar, and structure, in order to determine the degree and kind of standardization of each of them.

Questions on the lexical standardization concern the nature and scope of the vocabulary: which and how many kinds of words do occur? How many words do occur in relation to words that are contemporarily available and in relation to the number of concepts which are to be expressed? With regard to the standardization of syntax, I ask how many and which sentence constructions, modes, tenses, and particles the text employs, and for what purpose. In order to determine if and how the structure of the proof is standardized, I look for a recurring order of parts that are distinguished and recognizable by their contents or formal characteristics, that is, above all, by their lexicon and grammatical features.

Thus, while I look at each textual level separately, the three levels may overlap in the sense that it is difficult to answer the questions specified for each without referring to the other two levels. This overlap has stirred different ways of conceptualizing the form of elements texts with the help of innovative categories, which I discuss in the course of the analysis. Nevertheless, I decided to assess the three levels separately according to traditional linguistic and literary criteria for analyzing the form of a text. The reason for this is that I do not want to anticipate the actual analysis by using categories which (only) appear to describe the form of the *Elements* adequately. Rather, I aim at analyzing the text without bias, if at all possible, because I hope to capture the specificity and extremeness of the standardization of the *Elements*, and its relation to unambiguousness, precisely by using general, non-generic criteria. Therefore, I will separate in the analysis what is closely bound up in the text while noting where lexicon, grammar, and structure interlock, and what this reveals about the standardization of the text.

3.2 Lexicon

In this section, I first focus on the standardization of single words, and then consider Netz's concept of 'formulae' as a means of systematizing the phenomenon of the standardized occurrence of multi-word phrases in the *Elements*.

3.2.1 Words

Restriction of the lexicon

As far as its scope is concerned, the *Elements*' lexicon turns out to be extremely limited and highly invariant.[24] It is limited largely to ordinary words,[25] using few to no rare words. I 47, e.g., is free of *hapax legomena.* Also, the number of words that only occur in the Euclidean *Elements* or in the elements genre in general is quite small, since most of the words referring to mathematical objects have been in use before, and the verbs denoting geometrical activities are mostly common verbs that assume a specific technical but easily explicable meaning in the mathematical texts. The only group of words that are specific to mathematical texts are the 'letter words,' i.e., combinations of letters referring to lettered diagrams. As this group accounts for 74 of the 379 words of I 47, e.g., it is quantitatively significant; the meaning of these conglomerations of letters, however, can be reduced to a single one ('the geometrical figure denoted by the respective letters in the diagram adjunct to the text and the kind of nominalizing article preceding these same letters in the text').[26] As a side note, this most 'mathematical' part of the *Elements*' lexicon only occurs in the main part of the proposition that is concerned with one specific case and, thus, refers to the letters of one specific diagram. By contrast, letter words are neither part of the beginning of the proposition which sets out what is going to be proven in general terms nor (in the case of theorem propositions) of its ending which basically repeats the text from the beginning.[27] Since the lexicon of these parts of the proposition does not differ otherwise and since the general and concrete parts are equally standardized, I do not consider them separately in my analysis of the choice and use of words.

What is more telling than the limitation *to* ordinary words is the extreme limitation of the ordinary words themselves.[28] Two word classes, the article and prepositions, make up comparatively large portions of Euclid's text:[29] the article (84 times) and the prepositions (33 times) together account for almost one third of the text of I

24 Asper 2007, 116 f.; Netz 1999a, 104–126.

25 Netz 1999a, 89.

26 Cf. Netz 1999a, 104, n. 35: "Approached from the outside, the whole *A,B* phenomenon is characteristic of Greek mathematics, best understood as a single word, highly declinable. Approached from the inside, each *A,B* word has its own individuality." I will not consider the 'letter words' in the present chapter, but only in Section 4.4 in the analysis of the diagrammatic standardization.

27 Cf. Section 1.2, 23 f. and Netz 1999a, 148–150, who in this respect distinguishes only *two* parts of the proposition, namely the "letterless, general enunciation" from "the rest of the proposition, structured around the proof." In Section 4.3.2 on the 'constructing section,' I shall propose another two-part distinction of the Euclidean proposition.

28 Netz 1999a, 120.

29 Cf. Netz 1999a, 105, who chooses Aristotle for comparison. Since Aristotle makes ample use of articles and prepositions himself, the contrast between elements works and other Greek texts (both scientific and more literary) would be even more apparent.

47. The dominance of these two word classes is due to their standardized use in the formation of composite words denoting concrete objects of a diagram.[30] Considering that the article consists of only one and the class of prepositions of very few word types to begin with,[31] the lexicon is not only limited to certain kinds of ordinary words but indeed to very few different words at all.

Distinctness of the lexicon

The mathematical lexicon is not only limited by its limitation to ordinary words and to certain word classes, but also by what Reviel Netz has called "the one-concept-one-word principle," which is to say both that each word has one and only one meaning, and that each concept only has one verbal correspondent.[32] The practice of using one term per concept restricts the vocabulary insofar as it is responsible for the fact that Euclid's work is largely clear of synonyms, i.e., different terms for the same thing, and also of near-synonyms.[33] In addition, this practice also accounts for the opposite effect, that is, the elimination of homonyms, i.e., terms having different independent meanings.[34] As a result, the lexicon is limited up to the point where virtually every term is unambiguous, since there is in each case only one term which refers distinctly to a specific object.

Here is an alphabetical list of all the words which occur in proposition I 47 (with the exception of letter words). I provide, as far as possible, the undeclined form of the words, the respective number of occurrences, and the respective meaning(s).[35] In the case of particles, I list their respective function(s) in the proposition apart from their overall meaning. The list is to show by way of example that the "fixation of reference is distinct:"[36]

ἄγειν	1, 'to draw,' sc. a straight line
ἀναγράφειν	1, 'to erect,' sc. a square
ἀπό	12, 'from,' introducing the line of a triangle 'on' which a square is or should be erected

30 See below Section 3.2.2, 124 f. for what Netz calls "object formulae."

31 Cf. Netz 1999a, 97, n. 18 for his distinction of "word-tokens" and "word-types."

32 Netz 1999a, 108–113, here 108.

33 Netz 1999a, 111.

34 Cf. Netz 1999a, 112 f.

35 The translations of the terms are based on the classic translation by Heath (1926) which I use throughout in this study unless otherwise stated. However, unlike Heath, I always use Greek diagrammatic letters in the translation so that it fits the Greek text and the lettered diagrams that I supply in Chapter 4.

36 Netz 1999a, 125. Cf. below, Section 4.4.3, 304 for Netz's use of the term "fixation of reference" in the sense of the "specification" of the diagrammatic letters (with reference to his 1999a, 19 f.).

ἄρα	7, 'therefore,' introducing conclusions in the proof part *or* introducing the conclusion (part of the proposition) as part of the phrase ἐὰν ἄρα
αὐτός	7 'it' *or*, in attributive position, 'same'
βάσις	4, 'base'
γάρ	4, 'for,' introducing the construction part *or* introducing an explication in the proof part[37]
γωνία	9, 'angle'
δέ	4, 'and,' used as correlate of μέν for forming complex assertions from simple ones[38]
δεικνύναι	2, 'to show'
δεῖν	1, 'to have to,' in the form ἔδει within the phrase ὅπερ ἔδει δεῖξαι, 'the very thing which had to be demonstrated'
δή	remains untranslated and expresses the obviousness of mathematical statements in the proof part, either by itself *or* as part of the phrase διὰ τὰ αὐτὰ δή ('for the same reason') for introducing an "analogical demonstration"[39] *or* as part of the phrase ὁμοίως δή ('in the same way') for introducing a demonstration which will have to be done analogically with the preceding one without being done explicitly in the text[40]
διά	2, with accusative: 'because of' (as part of the phrase διὰ τὰ αὐτὰ δή, 'for the same reason') *or* with genitive: 'through'
διπλάσιον	2, 'double'
δύο	5, 'two'
εἶναι	18, 'to be,' in the form ἔστω ('let there be') as a main verb *or* in the forms ἐστί(ν)/εἰσί(ν) ('is'/'are') as copula
ἑκάτερος	4, 'each of two' *or*, if "repeated in ref. to each of two parties,"[41] 'respectively'
ἐν	4, 'in'
ἐπεί	3, 'since'
ἐπί	3, 'on'
ἐπιζευγνύναι	2, 'to join (sc. two intersecting lines)'
εὐθεῖα	4, 'straight line'
ἐφεξῆς	1, 'adjacent,' of an angle
ἔχειν	3, 'to have'
ἴσος	15, 'same'

37 "γάρ strutturante" and "γάρ esplicativo" in the terminology of Acerbi 2012, 173.

38 Acerbi 2012, 174.

39 Acerbi 2012, 176 ("dimostrazioni 'analogiche'").

40 Acerbi 2012, 176 ("dimostrazioni 'potenziali'").

41 LSJ s.v.

καί	15, 'and,' 'also,' used in a connective function for combining either construction imperatives or assertions (also as part of the conjunction καὶ ἐπεί) *or* for introducing co-assumptions[42]
κεῖσθαι	1, 'to lie,' that is, 'to be situated'
κοινός	1, 'common,' of an angle to be added 'to each' of two other angles
λέγειν	1, 'to say,' only in the phrase λέγω ὅτι, 'I say that'
μέν	4, untranslated antithetical correlate of δέ
μέρος	1, 'side,' of a line that has been intersected by another line into two parts
μή	1, 'not'
ὁ	84, 'the,' as the definite article of nouns denoting geometrical objects that are often left out in the description of specific objects, so that the noun has to be reconstructed with regard to the rest of the phrase, that is, the preposition that may follow the article and a number of letters
ὅλος	3, 'whole'
ὁμοίως	1, 'similarly'
ὁπότερος	1, 'either of two'
ὀρθογώνιος	3, 'rectangular'
ὀρθός	8, 'right,' of an angle
ὅσπερ	1, only in the phrase ὅπερ ἔδει δεῖξαι, 'the very thing which had to be demonstrated'
ὅτι	1, 'that,' only in the phrase λέγω ὅτι, 'I say that'
πάλιν	1, 'again'
παραλληλόγραμμον	3, 'parallelogram'
παράλληλος	3, 'parallel'
περιέχειν	2, 'to contain,' of the sides or lines of a right triangle forming the right angle, that is, the legs
πλευρά	6, 'side,' of a triangle
ποιεῖν	1, 'to form (sc. a geometrical figure)'
πρός	2, 'at'
προσκεῖσθαι	1, 'to be added'
σημεῖον	1, 'point'
τε	2, untranslated correlate of καί ('and'), used for conjoining two objects
τετράγωνον	15, 'square'
τις	1, 'a'
τρίγωνον	7, 'triangle'

42 Acerbi 2012, 173 f.

ὑποτείνειν	2, 'to subtend,' of the line or side of a right triangle extending under the right angle, that is, the hypotenuse
ὑπό	9, 'by,' introducing the points 'by' which an angle is contained

What we can see here is the distinctness of the relation of the words and the concepts to which they refer. Although the majority of the words (37 of 55) occur more than once – note the small number of *hapax legomena* "in a relative sense" (i.e., words which only occur in this text) even in such a small sample[43] – virtually every word occurring in proposition I 47 has only one meaning. Wherever I list more than one meaning, these amount to different standardized functions of a term which nevertheless can be regarded as one and the same in terms of meaning. Conversely, the frequent occurrences of most terms are due to the fact that almost no meaning occurs in different entries on this list because no meaning is covered by two different words – except in the case of two of the most basic of Greek words, namely καί and δέ. They are synonymous only with regard to their common basic meaning; however, within the Euclidean proposition, the two words fulfill different (only partly overlapping) functions and are not always used interchangeably.[44] What adds to the distinctness is the text's avoidance of semantic nuances: the realm of the words of this list and their corresponding concepts in relation to the other word-concept entities is clear-cut.[45]

The related tendencies of the fixed connection of words and concepts and of the repetition of the lexicon (in recurring meanings) neither are in any way necessitated by the mathematical content nor have come about for want of linguistic alternatives. Rather, they are the result of an authorial-generic choice. The Greek language prior to the emergence of the elements genre, on the one hand, would have known alternative expressions for words that have become standardized in elements texts, and, on the other hand, many of the words available were such as to cover not only one, but a whole range of meanings. This means that the *Elements* displays a strong tendency to minimize linguistic variability and to exclude alternative words and meanings. The above list, therefore, is a result of the avoidance of lexical variety and indeterminacy. To give an example: the adjective ὀρθός, occurring altogether eight times in proposition I 47, has only one standardized meaning, even though this adjective, starting from its basic meaning 'straight,' has assumed various meanings in classical Greek: 'upright, standing' when referring to people's posture, buildings, or a crop; 'straight' as opposed to σκολιός ('crooked') and πλάγιος ('aslant'); 'safe, prosperous,' e.g., of prophecies; 'right, correct,' e.g., referring to messages, accounts, and speech; 'true, real, genuine;' 'just,' mostly of νόμος and λόγος; 'straightforward' and 'excited' of a person's

---

43 For this term cf. Netz 1999a, 104 f., who arrives at similar results regarding his mathematical sample texts (Apollonius' *Conics* I 15, Archimedes' *On the Sphere and the Cylinder* I 32, Euclid's *Elements* II 10, and Euclid's *Optics* 34) compared with his sample from Aristotle.

44 Cf. Acerbi 2012, 173.

45 Cf. Netz 1999a, 108 and 126.

character and state of mind; 'right' of an angle.[46] Of these meanings, only the last one occurs in I 47, where ὀρθός in every single case refers to γωνία ('angle') denoting an angle that is equal to 90 degrees. In addition to this, one should note another three instances of the composite adjective ὀρθογώνιος ('rectangular'). To refer to the concept 'angle,' in turn, Euclid (along with the other elements authors) uses only one word, namely γωνία, occurring nine times in proposition I 47, where apparently at least one other, that is, γλωχίς, would have been available by his time.[47] This means that in the case of ὀρθός readers are not going to be confused by the variant application of different meanings of the word, as to which would be the one Euclid currently has in mind, while in the case of the concept 'angle' they are not going to puzzle over the use of different words for referring to it.

If one considers the lexicon of I 47 in the context of *Elements* I, one realizes that the distinctness of the words and meanings I have listed does not only pertain to the immediate context of our sample proposition: even in the whole work there are virtually no synonyms for the words used in I 47; accordingly, there is no word on our list of the lexicon of I 47 which does not recur in the *Elements* – they are repeated over and over. Again, most of the words figuring in I 47 have the very same meaning they have here throughout the *Elements*. It is true that in some cases, it is only possible to grasp the generality of a meaning of a term if one takes a larger sample into account.[48] For example, in I 47, the preposition ἀπό is used with the verb ἀναγράφειν ('to erect') and introduces the line of a triangle 'from' which a square is drawn, i.e., 'on' which it is erected. Beyond I 47, ἀπό is used in the *Elements* not only to depend on ἀναγράφειν but also on other verbs;[49] the general sense (which also applies in I 47) always is to mark a geometrical object as the point of origin starting 'from' which another object is drawn. It is mainly some of the particles of I 47 (δή, δέ) which have additional functions in other (types of) propositions.[50] The lack of synonyms and homonyms and the according repetition of the highly limited lexicon become even more apparent in view of the larger sample of *Elements* I. And even with regard to the whole work, the words of I 47 retain their meanings, and the concepts expressed by the lexicon of I 47 keep their verbal correspondences. The lexicon only becomes larger as more concepts and according words occur due to the different subjects of the books of the *Elements*.

---

46 LSJ s.v.

47 Cf. Netz 1999a, 108 with reference to Hero *Def.* 15. Netz lists another 14 pairs or trios of mathematical terms of which Euclid uses only one in each case.

48 Since the lexicon of *Elements* I is limited because of the limited number of geometrical objects with which the book deals, I will make reference also to other books here.

49 E.g., ἐπιζευγνύναι ἀπό τινος ἐπί τι ('to join something to something'), cf. *Elem.* III 2 (vol. 1, p. 96.2 f. Heiberg & Stamatis).

50 The particle δή is regularly used in problem proposition to introduce the definition of the goal (δεῖ δή); δέ is used in the proof part for introducing "co-assumptions" (Acerbi 2012, 173).

Most of the remaining apparent homonyms in the *Elements* are words which recur in different fixed conjunctions of words. For example, the verb ἀναγράφειν is used in I 47 and beyond with τετράγωνον ('to erect a square'), but it is also used regularly with a couple of other nouns, e.g., εὐθύγραμμον ('rectilinear figure') or παραλληλόγραμμον ('parallelogram'). Rather than being homonymous, these uses of ἀναγράφειν can be subsumed under the general sense 'to erect (a figure).'[51] Another case to consider is that of verbs of which specific forms occur in different phrases. The verb δεῖν ('to have to'), for instance, as we have seen in the above list of words of I 47, occurs only at the end of the proposition within the phrase ὅπερ ἔδει δεῖξαι ("the very thing which had to be demonstrated") in I 47. Apart from this use, the form δεῖ ('one has to,' 'it is necessary to') is the only other to be used in the *Elements*: it is standardly employed as part of the phrase δεῖ δή ('Thus it is required to') for introducing the 'definition of the goal' of problem propositions (as such it is absent from the theorem proposition I 47). Taken by itself, the meaning of the verb is the same in either of its two uses. However, if one considers the respective recurring context and function of the two forms within the proposition, ἔδει and δεῖ appear as two distinct uses; and considering that these two uses are connected to two distinct forms of the verb beyond which it is not used, the two verb forms almost appear as two different words. Nevertheless, one certainly would not regard the two ways in which the verb δεῖν is used as homonymous.[52] The only words of the above list of I 47 for which I list several 'meanings' each are the particles, but here 'meaning' refers to the 'function' or 'use' in the respective part of the proposition, while these uses in each case are based on the respective well-known sense of the particle. The only real homonymy within a treatise or book of the elements genre which is focused on a single subject area appears to be between what Netz calls "first-" and "second-order discourse," that is, between the mathematical language proper and the 'logical' language of the proof proposition.[53] Since the *Elements* (unlike most extant mathematical treatises) comprises different subject areas, it contains another type of instance of homonymy: terms which are used differently in the context of different mathematical fields. However, not all examples noted by Netz may present proper instances of homonymy.[54] For example, the term τετράγωνος does not only occur in the geometrical books of the *Elements* as a neuter noun to mean 'square' (as it does in I 47), but also in the arithmetical books where it is used as an adjective with ἀριθμός to denote a 'square number.' However, rather than being homonymous, the term seems to be employed metaphorically in its arithmetical use.

---

51 Cf. Netz 1999a, 112 on the "question whether 'the center of a circle' and 'the center of a sphere,' for instance, use the word 'center' homonymously."

52 Cf. Netz 1999a, 112 on cases in which indeed "special senses become attached to certain conjugations of the verb and no others" but which still do not qualify as homonyms.

53 Netz, 1999a, 112. Cf. Mugler 1958 s.v. ἀνάγειν. For the distinction between 'first- and second-order language,' 'terms,' or 'discourse' cf. Netz 1999a *passim*.

54 Netz 1999a, 112.

As in the case of homonymy, most of what seem cases of synonymy can be reduced to "differences in use," as Netz puts it, that is, with regard to their immediate verbal context and their function within the proposition.[55] For example, ἄγειν, γράφειν, ἀνάγειν, and ἀναγράφειν all mean 'to draw,' but each of these verbs are used primarily for drawing a specific type of object. The first two, ἄγειν and γράφειν, refer to the drawing of lines; while ἄγειν means 'to draw (a straight line),' γράφειν is used 'to draw (a curved line).' The latter two, ἀνάγειν and ἀναγράφειν, refer to drawing perpendicular lines: thus, ἀνάγειν means 'to erect (a straight line)' and ἀναγράφειν 'to erect (a figure).'[56] While there used to be a number of real synonyms in the early days of the genre, e.g., στιγμή ('point') besides σημεῖον, which occurs in I 47, or γλωχίς next to γωνία, the word for 'angle' used in I 47 (the example which I have discussed above), only few synonyms have survived in Euclid. These are mostly "second-order" terms,[57] such as ἀποδεικνύναι and δεικνύναι (of I 47), which are used interchangeably in the sense 'to show/prove' in the proof part.[58] Other examples from I 47 include διὰ τὰ αὐτὰ δή ('for the same reason'), which can be substituted by ὡσαύτως δή ('in like manner') to introduce an analogous case in the proof part;[59] and ἄρα ('therefore') which introduces conclusions in the proof part and can be substituted by ὥστε in order to avoid two consecutive ἄρα.[60]

To sum up: there are only a few synonyms, which are, at that, restricted to second-order language, and even fewer homonyms, which only occur between first- and second-order language. Thus, the lexicon borders on one hundred percent standardization and is extremely, almost completely unambiguous.

Extent of the lexicon

I have tried to show that in order to get hold of the extreme degree and specific kind of unambiguousness of the lexicon of the *Elements*, one mostly cannot define the meaning of single words individually. For doing so, it is necessary to include the immediate verbal surrounding of a word as well as its larger context, such as the place within the proposition and the kind of the proposition. This is because the meaning of many single words is inextricably linked with the sequence of words in which they occur; it amounts to their use(s) as part of one or several sequence(s). For this reason, it may prove difficult, paradoxically, given the unambiguousness of the lexicon, to define the meaning of a word straight away. This can be seen in the above list in which I tried to provide meanings for all the words occurring in I 47. In order to define the

55 Netz 1999a, 111.
56 Netz 1999a, 110 and 114. For the second-order meaning of ἀνάγειν cf. above p. 121, n. 53.
57 Netz 1999a, 108–111.
58 Netz 1999a, 111.
59 Acerbi 2012, 176.
60 Acerbi 2012, 173 f.

meaning of a word, one has to refer to one or more other words to which it is connected; e.g., for the meaning of an adjective to the noun it refers to, for the meaning of a preposition to the verb or (implied) noun from which it depends, for the meaning of one particle to another with which it forms a compound particle, and for the meaning of a verb form to the proposition part it which it appears. Sometimes, the reference to the verbal context is particularly vital, e.g., in the case of an article used in place of the full terms for geometrical objects to nominalize the letters which refer to a diagrammatic figure in the elliptic designation of particular diagrammatic objects; compare their most simple type τὸ Α (sc. σημεῖον, 'the [sc. point] A'). Even though in most cases their meaning does not change fundamentally compared to what one might guess they mean individually in a mathematical context, their distinctness and technicality come about just by their being tied to a standardized sequence of words, so the respective superordinate phrase needs to be referred to in the definition of its components.

The strict standardization of the words of the *Elements* in relation to other words implies that certain sequences of words come to be standardized in the same way as single words. The most obvious of these are conglomerations of words which are used standardly in order to express more complex concepts, for example, terms for geometrical entities like ὀρθογώνιον τρίγωνον, the standard term for the concept 'right triangle,' or ἡ ἐφεξῆς γωνία for 'adjacent angle,' and standardized conjunctions of particles and other little words, e.g., διὰ τὰ ταὐτὰ δὴ καί ('for the same reason'), which regularly introduces the transfer of one result of deduction to a comparable case. These standardized phrases are formed from words which reappear as part of other standardized phrases (e.g., τρίγωνον, γωνία, or δή). In this way, the lexicon remains unambiguous and limited at the same time. Instead of inventing new words or attaching various meanings to known ones, the text economically establishes new terms and concepts by reusing and combining single words.

The standardization of phrases is not restricted to compound nouns or adverbial phrases such as I have mentioned, but even extends to larger sequences of words which need not be restricted to the same word class and syntactical unit, but may comprise several of them, and even contain main clauses and their dependent constructions; cf., e.g., the standardized introduction of the definition of the goal λέγω ὅτι ('I say that') and the regular proposition ending ὅπερ ἔδει δεῖξαι ('the very thing which one had to prove'). These phrases have a standard meaning in elements texts just like certain single words do. One could continue the list of words and corresponding meanings occurring in I 47 which I have provided above by adding groups of words occurring in I 47 that have a standard meaning. Thus, in order to make Netz's "one-concept-one-word principle" fit the Euclidean practice, one needs to qualify in it either "concept" (in the sense of 'meaning') or "word" (in the sense of 'denotation' or 'term').[61]

61 On the possibility of a 'one-concept-one-formula principle' see the subsequent section, pp. 127 and 171, n. 283.

3.2.2 Phrases

Netz's concept of 'formulae'

Reviel Netz has argued that basically all words in the *Elements* form part of smaller or larger standardized phrases for which he uses a highly evocative term: "a web of formulae is cast over the corpus."[62] The term 'formulae,' which in the context of Classical scholarship points to Homer and oral poetry theory, also suggests, especially when applied in a mathematical context, the mathematical non-textual origins of the term: except for the most simple types of word conglomerations, some of which I have quoted at the end of the preceding section, many of the standardized mathematical phrases which Netz detects are "not verbatim repetitive"[63], but contain one or more variables.[64]

Thus, in Netz's taxonomy, the formulaic phrases do not accumulate additively but rather form formulaic hierarchies. This means that complex formulae may govern subclauses or even whole sentences by forming overarching structures into which smaller subordinate formulaic "building-blocks"[65] are inserted which may again contain subordinate formulaic phrases.[66] The hierarchical workings of the formulaic language according to Netz will become clear from a short review of his five types of formulae which he distinguishes by subject matter,[67] for the different types are not combined randomly, but each implies a different syntactical value and inner complexity. I present them in order of increasing complexity, not abiding by Netz's order.

Types of formulae according to Netz

First, there are "object formulae" denoting "diagrammatic 'things.'"[68] These are, first of all, the generic terms for geometrical objects. But mostly, Netz seems to have in mind the characteristic, often elliptic terms for specific geometrical objects which are

62 Cf. Netz 1999a, 161. For his thorough analysis of the phenomenon in his sample *Elements* II 1, see 1999a, 127–167.

63 Netz 1999a, 162. For other reasons for the variability of formulae discussed by Netz see below pp. 130 f.

64 Netz 1999a, 145 f. and 155.

65 Netz 1999a, 158.

66 See, e.g., the "phrase-structure tree" of *Elements* II 1 in Netz 1999a, 146, fig. 4.1.

67 While Netz states that only some of his types of formulae are defined by their content, and others by their form (1999a, 137), in fact all of his five types are unambiguously characterized by their subject matter.

68 Netz 1999a, 133.

identifiable by their letters in a diagram. I will call these terms 'letter phrases.' Commonly, they only consist of the articles of the respective noun terms for geometrical objects, the diagrammatic letters of the object, and, in the case of more complex geometrical terms involving several noun terms, prepositions. The noun to which the article refers may be added, but is often left out. The variable element of these 'letter phrases' is either what I call 'letter word,' that is, the unit of one or several letters which refer to the concrete geometrical object, or yet another 'object formula,' which itself contains a 'letter word' as its variable element.[69] For example, Netz's formula 8 for 'square' "τὸ ἀπὸ τῆς AB – 'the (neut.) on the (gen. fem.),'" which governs his formula 2 for 'line' "ἡ AB – 'the (fem.) {2 letters or more},'"[70] is realized in I 47 in the non-elliptical form τὸ ἀπὸ {τῆς {ΒΓ} [sc. γραμμῆς]} τετράγωνον ("the square on {[sc. the line] {ΒΓ}} – τετράγωνον ("square") could have been left out like γραμμῆς ("line").[71]

Second in line are what Netz calls "predicate formulae," which denote among other things relations between diagrammatic objects, and, thus, may incorporate "object formulae."[72] Cf. Netz's formula 78 "{object} ἴσον ἐστὶ {object},[73] which can be found many times in I 47, e.g., ἴση ἐστὶν ἡ [. . .] ΔΒ τῇ ΒΓ ("{the [sc. line] [. . .] {ΔΒ}} is equal to {the [sc. line] {ΒΓ}}").[74]

The third type are the so-called "construction formulae," which may govern the former two types.[75] At their core, they have a verb form postulating or acknowledging the construction of diagrammatic objects; cf., e.g., Netz's formula 29 "ἀναγεγράφθω ἀπὸ {formula 2: line, in the genitive} τετράγωνον {formula 6: area}," appearing in I 47 as Ἀναγεγράφθω [. . .] ἀπὸ {τῆς {ΒΓ} [sc. γραμμῆς]} τετράγωνον {τὸ {ΒΔΕΓ} [sc. χωρίον]} [. . .].[76] This is already a self-contained sentence: "[. . .] let there be described on {the [sc. line] {ΒΓ}} {[sc. the area] {ΒΔΕΓ}} as a square [. . .]").[77]

---

69 Cf. Netz 1999a, 133–136.

70 Netz 1999a, 133 f.

71 *Elem.* I 47 (vol. 1, p. 63.9 Heiberg & Stamatis). Curly brackets indicate the variable element(s) within the formula (Netz does not put them consistently, especially in the Greek).

72 Cf. Netz 1999a, 142–144.

73 Netz 1999a, 143.

74 *Elem.* I 47 (vol. 1, p. 64.1 f. Heiberg & Stamatis).

75 Cf. Netz 1999a, 136 f.

76 *Elem.* I 47 (vol. 1, p. 63.11 f. Heiberg & Stamatis). For his formula 29 cf. Netz 1999a, 137.

77 My translation according to Netz's formalization by means of his formula 29, which interprets τὸ ΒΔΕΓ in the present passage as short for τὸ ΒΔΕΓ χωρίον ("the area ΒΔΕΓ," cf. 1999a, 134, n. 16) rather than as τὸ ΒΔΕΓ τετράγωνον ("the square ΒΔΕΓ"). One might question this on the grounds that τετράγωνον is the most obvious supplement to the neuter article as it is explicitly related to τὸ ΒΔΕΓ, being mentioned immediately before it. In Netz's interpretation, τετράγωνον seems to be used predicatively ("the area ΒΔΕΓ as a square"); if one, however, supplements τὸ ΒΔΕΓ with τετράγωνον, then τὸ ΒΔΕΓ (sc. τετράγωνον) is an apposition to the preceding τετράγωνον, which is, then, not a predicative noun but the subject of the imperative: "let there be described a square (sc. namely) ΒΔΕΓ" (cf. Heath's

The two remaining types are superordinate to all other types of formulae and govern complex sentences. One of them is “argumentation formulae” which are understood as “expressions validating an argument.”[78] Some of them are what can be translated into our modern neat mathematical formulae using symbolic ‘language,’ e.g., Netz’s formula 63 “ἐπεὶ {formula 35: cutting a line into equal and unequal segments, expressed in the indicative instead of the imperative} ἄρα {formula 9: rectangle, its two lines being the two unequal segments} μετὰ τοῦ {formula 8: square, its line being the difference between the unequal and the equal cut; in the genitive} ἴσον ἐστὶ {formula 8: square, its line being the equal section; in the dative},” which is essentially $(a + b)(a - b) = a^2 - b^2$.[79] None of the types of “argumentation formulae” which Netz finds in *Elements* II recurs even remotely similarly, let alone in the exact same form in I 47.

The other type of formulae that govern complex structures are “second-order formulae” which are not intrinsically mathematical, but which structure the proposition and work as “signposts within arguments.”[80] As such, they often introduce sentences, like the already mentioned λέγω ὅτι (“I say that”), Netz’s formula 44.[81] The variable element in this formula is whatever forms the rest of the sentence, that is, the goal of the respective proposition, as in I 47: λέγω, ὅτι {τὸ ἀπὸ τῆς ΒΓ τετράγωνον ἴσον ἐστὶ τοῖς ἀπὸ τῶν ΒΑ, ΑΓ τετραγώνοις} (“I say that {the square on ΒΓ is equal to the squares on ΒΑ, ΑΓ}.”).[82] Note that Netz does not mention it as part of the formula, presumably because it cannot be pinned down in general for all individual cases in terms of a combination of specific ones of his formulae.

The usefulness of the concept

Netz’s systematic approach to the mathematical language is indeed intriguing and impressive because it offers a comprehensive theory of how the lexicon is employed in context. However, it remains to ask how convincing and helpful this theory is, above all as to whether it adequately explains the specifics of the *Elements*’ extreme formal standardization, which, as I have argued above, is their main formal feature. Extreme standardization does seem to be central to the formulaic system since formulae are

more concise translation “[. . .] let there be described on BC the square BDEC,” which, by using the definite article with “square,” somewhat obscures the syntactical structure, but nevertheless appears to supplement τὸ ΒΔΕΓ τετράγωνον (1926, 349). This would mean that there is another, more general formula for ‘square’ (τὸ ΑΒΓΔ [τετράγωνον] or τὸ ΑΒ [τετράγωνον] to indicate the square by means of its “two opposite vertices,” Netz 1999a, 134, n. 16) besides Netz’s formula 8 τὸ ἀπὸ τῆς ΑΒ (τετράγωνον) which only works if a square is to be located with respect to a certain line.

78 Netz 1999a, 139–141, here 139.

79 Netz 1999a, 141.

80 Netz 1999a, 137–139, here 138.

81 Netz 1999a, 138.

82 *Elem.* I 47 (vol. 1, p. 63.9 f. Heiberg & Stamatis).

units of words that are standardized through repetition.[83] Rather surprisingly, Netz, though he admits that the system is "quite economical," denies that its purpose was complete disambiguation. While we have said that the text on the level of the lexicon follows a "one-concept-one-word principle," it does not, according to Netz, abide by an analogous "one-formula-per-concept" on the superordinate level, for it contains a few synonyms (though no homonyms).[84] Even beyond the question of the semantic standardization of formulae, the formulaic system as a whole does not seem to be fully cogent and unambiguous, as it could be conceptualized differently. The possibility of a different conceptualization, however, would entail that the formulae are not so greatly standardized after all.

The number of formulae

It becomes most apparent that the system of mathematical formulae is to some extent arbitrary when one attempts to determine the exact number of formulae. Netz non-exhaustively lists 105 different formulae, 71 of which represent all formulae occurring in his sample *Elements* II. The absolute number of different formulae in the *Elements* he estimates at "a few hundred."[85] He seems unsure about the portion of the text constituted by formulae in relation to the amount of non-formulaic words. On the one hand, "all the text is formulaic,"[86] and indeed, all the words occurring in I 47, e.g., seem to be covered by Netz's formulae. On the other hand, Netz wonders which of his formulae are in fact perceived as formulae and which are, therefore, more formulaic than others and which part of the formulae he lists is in fact formulaic, and which is not.[87] In other words, the number of formulae in the *Elements* hinges on the underlying definition of 'formula.' In Netz's definition, the degree of relative repetition is the only criterion for 'formularity:' on the one hand, the frequency of a phrase in relation to its frequency in other texts (he calls such a term "markedly repetitive"); on the other hand, in relation to "equivalent expressions," "relative to the distribution in other relevant texts" (such a term is "semantically marked").[88] Here, one would like to

---

83 Netz shows convincingly that mere repetition is not a sufficient criterion, but only certain kinds of relative frequency (1999a, 130–132, cf. above p. 109, n. 5).

84 Netz 1999a, 134. For the economy of the formulae (and the greater economy of the lexicon) cf. also 155 and 161.

85 Netz 1999a, 160. To be sure, Netz conceives of the mathematical formulae as an "open-ended system" (1999a, 155) and rejects the idea of a "set, once-and-for-all list of formulae," and it is true that the formulaic system (if one admits to its validity at all, see below Section 3.3.3, 174 f. for an alternative conceptualization of the mathematical language) is in general open-ended because of the "generative nature of formulae" (158); however, the text of Euclid's *Elements* as it is must be expected to display an exact number of formulae.

86 Netz 1999a, 161.

87 Netz 1999a, 159.

88 Netz 1999a, 131 (cf. p. 109, n. 5 and p. 131, n. 107).

know exactly how many times a phrase needs to occur with regard to other phrases and texts for it to be considered formulaic. For example, some of Netz's large "argumentation formulae" naturally occur a lot less frequently than most of his more concise formulae, if at all they are repeated recognizably.

Other important criteria for the definition and reckoning of the *Elements*' formulae concern their *form*. Netz does not mention them in his definition of 'formulae' but brings them up in the course of his discussion of the concept.

First, size: with a measure of common sense, Netz defines formulae as "groups of words," but admits that there are also "one-word-long formulae."[89] But then, what is the difference between any standardized word and a one-word formula? For if the relative frequency is the only criterion, then standardized one-word names for geometrical objects like γωνία or τετράγωνον should be considered formulae, even though not mentioned as formulae in Netz's survey, for they may be inserted into Netz's tripartite formula 23 as its second element: "ἔστω {any unlettered object formula} {a lettered equivalent}, 'let there be {some object} < namely > {some lettered object}.'" With regard to this formula, one may further ask up to which point formulae can be divided into smaller formulae. That is, is ἔστω only a part of the tripartite formula, or is it itself a formula? I have the feeling that single words like γωνία or ἔστω are formulaic if they are standardized within certain formulae, but that they are hardly formulae themselves.

At the other end of the scale, is there an upper limit to the size of a formula? How large can it be in order to stay recognizable as a formula? In Netz's phrase-structure tree which depicts the dependency of the formulae in II 2, the biggest formulae amount to complex sentences, and, therefore, even to whole parts of the proof, because – except for the largest part of the proposition, which is the demonstration– standardized parts of the proof consist of just one standardized sentence. Here is where the standardization of the lexicon would not only interlock with that of syntax, but also with that of the structure of the proof. Ultimately, one could even consider the existence of a 'proposition formula' or different 'proposition formulae' according to different types of propositions. The reason why Netz does not conceptualize such structural formulae even though he mentions the prospect of doing so[90] may be that they would contain too many variable elements for them to form a recognizable pattern. In his taxonomy, as we have seen above, the two words λέγω ὅτι alone represent a formula, although they are regularly superordinated to some x, that is, a dependent clause of the statement which is the definition of the goal of theorem propositions.[91]

---

89 Netz 1999a, 132.

90 "On a higher level still, it is possible to see the entire proposition structure as such a matrix [. . .]. It is through formulae that the structure of the proposition should be approached." (Netz 1999a, 138 and 166 for his "phrase-structure tree" of *Elem.* III 2).

91 Cf. Netz 1999a, 138, conceiving of the formula even more broadly in structural terms: "[. . .] the formula is constituted not only by its two words, but also by what precedes it (setting-out) and what follows it (definition of goal)."

Second, internal structure, to which the question of size is very much connected. Although the former, like the latter, is not included in Netz's definition of 'formula,' it emerges as an essential condition from his discussion of the phenomenon. He imagines formulae as "molecules" as opposed to the "atomic" lexicon and regards their "internal structure" as a prerequisite of their perceived "unity."[92] This view seems to preclude one-word formulae altogether, even though Netz, as we have seen, explicitly allows them. To be sure, the simplest type of an internal structure would be a compound of two or more words.

Above all, however, Netz seems to have in mind hierarchical internal structures which include variables.[93] From this perspective, the formulaic nature of what Netz calls "lettered object-formula" is at least questionable.[94] Strictly speaking, they do not include any variable element x, for it is only because of the letters denoting a concrete object that the respective noun which would refer to the object in a generic way is omitted. In any case, these phrases do not contain a self-contained element, for the letters following the respective article do not mean anything without the preceding article.[95]

Besides size and structure, the third criterion for determining the number of formulae is the degree of formal variability allowed for them to be still considered one and the same formula. On the one hand, Netz counts some phrases as one single formula which could as well be deemed several distinct formulae. Netz himself shows awareness of the two possibilities in specific cases.[96] On the other hand, many formulae listed by Netz resemble each other so closely that one wonders why they do not go by one single formula, but are counted as different ones[97] or are even held to belong to different types of formulae.

For the latter, let us have a look at the (limits of the) formal variability of Netz's 'construction formulae' 21–39. These are so named because they demand the construction of a geometrical object. At their core, they have the third-person perfect passive imperative of a construction verb and the term for a geometrical object which forms the subject of the imperative.[98] Take, for instance, the above-quoted implementation of Netz's 'construction formula' 29 in I 47: ἀναγεγράφθω γὰρ ἀπὸ {τῆς {ΒΓ} [sc. γραμμῆς]}

92 Netz 1999a, 156 and 163.

93 Netz 1999a, 145.

94 Cf. Asper 2003b, 8. For Netz's term, see, for example, 1999a, 161.

95 Cf. Asper 2003b, 8, who proposes to understand what Netz calls 'formulae' with Bruno Latour's term "immutable mobiles" as "(schriftliche) Dekontextualisierungsprodukte, die der Wissenschaftssprache durch Normierung intersubjektive Akzeptanz sichern wollen." This definition, however, does not say anything about the form which the "Normierung," that is, standardization, takes, whether that of formulae or of something else.

96 Cf., e.g., 1999a, 140 and 151.

97 Cf. Netz's formulae 75–77 ("{object} ἴσον ἐστὶ {object} μετὰ {object}," "{object} ἴσον ἐστὶ {object} καὶ {object}," "{object} ἴσον ἐστὶ {object} τε καὶ {object}"), which, as Netz himself notes, "are three ways of representing [sc. the modern mathematical formula] a + b = c" (1999a, 143).

98 Netz 1999a, 137. E.g., formula 29 mentioned above. I use the term 'construction verb' for verbs which are used to denote the diagrammatic construction of geometrical objects.

τετράγωνον {τὸ {ΒΔΕΓ}} [. . .] ("for let there be described on {the [sc. line] {ΒΓ}} a square {[namely] {ΒΔΕΓ}}).[99] The verb ἀναγράφειν reappears toward the end of the proposition – in the indicative mode: καί ἐστι τὸ μὲν ΒΔΕΓ τετράγωνον ἀπὸ τῆς ΒΓ ἀναγραφέν [. . .] ("And the square ΒΔΕΓ is described on ΒΓ [. . .]").[100] Netz seems to conceive of this as a different formula altogether, that is, as a "transformation" of a "construction formula" into a "predicate formula," "one describing a property."[101] This interpretation is debatable, as the second passage still contains the same construction verb, only that it is now used to *describe* rather than to demand the construction. Apparently, Netz regards the imperative mode of the verb as a decisive part of the construction formula, so that a change of mode also entails a change of formula. To this, one could object from a lexical point of view that it is the conjunction of a specific construction verb and a specific geometrical-object noun which constitutes a standardized phrase and which should be regarded as one and the same (though formally variable) formula, whether the verb occurs in the perfect passive indicative or imperative, in the aorist infinitive, or in participle form, and whether the geometrical-object noun is the subject or object of that verb.[102] This view would imply that the *Elements* contains much fewer formulae than are listed by Netz.

Reasons for the formal variability of formulae

Only the 'second-order formulae' of Netz's system remain invariant throughout the *Elements*. All others appear formally variable. There are three main reasons for this formal variability of formulae.[103]

First, ellipsis, that is, part of the "full version" of the formula is omitted. For example, we have seen that 'object formulae' sometimes mention the geometrical-object noun, and sometimes leave it out. In Netz's view, such ellipsis does not change the "identity" of the formula.[104]

Second, a change of syntactical position resulting in a change of grammatical form. It depends on the word class of the core element of the formulae whether and how (much) they are affected by the syntax. On the one hand, geometrical objects may be mentioned in different cases while indisputably conforming to one and the

---

99 *Elem.* I 47 (vol. 1, p. 63.11 f. Heiberg & Stamatis); for the syntactical structure of this formula cf. above p. 125 f., n. 77).

100 *Elem.* I 47 (vol. 1, p. 65.2 f. Heiberg & Stamatis).

101 Netz 1999a, 150 on an analogous example which contains the verb προσκεῖσθαι ('to be added'), for which cf. below p. 132, n. 113.

102 For a more general formalization of the mathematical language in the place of Netz's "formulae," one which joins the lexical and the grammatical side of the Euclidean standardized phrases cf. Section 3.3.3, 174 f.

103 Netz 1999a, 158, n. 47.

104 Netz 1999a, 148 f. and 152 on how "the process of ellipsis may also be the source of the formulaic status itself."

same formulae. Formulae constituted by verbs, on the other hand, transform into different (types of) formulae according to Netz if the mode of the verb changes, as we have just seen in the example of 'construction formulae.'[105]

The third reason for variation, not considered by Netz, is the change in the order of the words or elements within formulae. For instance, in statements about the equality of two geometrical objects, the three elements often appear in a different order than in Netz's formula 78 "{object} ἴσον ἐστὶ {object}; in I 47, e.g., the adjective and copula appear in the middle position as in the formula but also as the first of the three elements and as the last one. Either way, they would most probably be recognized by Netz as manifestations of his formula.[106] However, one could question the formulaic status of phrases involving the words ἴσον ἐστί precisely on the grounds that they do not display a fixed form but are the formally non-standardized recurrence of the most common and non-technical way to state the equality of two things.[107] This is especially worth considering as the phrase ἴσον ἐστί does not only occur at differing places within the supposed phrase, but also in different forms depending on the gender and number of the noun term to which it refers.[108] Another example is Netz's above-quoted formula 23 "ἔστω {any unlettered object formula} {a lettered equivalent}."[109] Its three elements appear in this very order in a great number of propositions, but the core verb form ἔστω is also used as part of a variation of the phrase.[110] The question which arises from these examples is whether a set order of the elements is constitutive for formulae or whether it just helps their immediate recognition.

Conceptualizing formal variability

The supposed formal variability of the mathematical formulae can be conceptualized in different ways which originate from different views on the formal stability of formulaic phrases and result in different numbers of formulae in a text. Netz, e.g., assumes a "canonical form," of which there are "variations," that is, deviations which do not change the "canonical form," above all due to ellipsis,[111] and "transformations"

105 Netz 1999a, 150 f.

106 E.g., *Elem.* I 47 (vol. 1, pp. 63.4–7 and 64.3–65.16 f. Heiberg & Stamatis).

107 Netz's responds to this objection by admitting that the formula is not semantically marked, but markedly repetitive, and therefore fulfills one of his two prerequisites for 'formulae' (1999a, 143). However, he ignores the problem of the changeability of the order of the elements.

108 Καὶ ἐπεὶ ἴση ἐστὶν ἡ μὲν ΔΒ τῇ ΒΓ, ἡ δὲ ΖΒ τῇ ΒΑ, δύο δὴ αἱ ΔΒ, ΒΑ δύο ταῖς ΖΒ, ΒΓ ἴσαι εἰσὶν ἑκατέρα ἑκατέρᾳ· καὶ γωνία ἡ ὑπὸ ΔΒΑ γωνίᾳ τῇ ὑπὸ ΖΒΓ ἴση· (*Elem.* I 47 [vol. 1, p. 64.1–5 Heiberg & Stamatis], "And, since ΔΒ is equal to ΒΓ, and ΖΒ to ΒΑ, the two sides ΔΒ, ΒΑ are equal to the two sides ΖΒ, ΒΓ, respectively").

109 Cf. p. 128.

110 The discussion precisely of the uses of ἔστω at the end of the subsequent Section 3.3.3 will be central for my evaluation of the concept of formulae regarding the question of how the language of the *Elements* is standardized.

111 Netz 1999a, 148–150.

which generate new formulae having their own canonical form.[112] Netz accounts for the close dependency of formulae which have been differentiated through transformation by means of his numbering.[113]

To "variation" and "transformation," one might add 'loss of formulaic status' which applies if a phrase deviates from the original (or canonical form) beyond recognition. It is, of course, in many instances debatable which of the three is the case, the main issue being what constitutes the supposed canonical form.[114] Consider again, for instance, the implementation of Netz's construction formula 29 in I 47: ἀναγεγράφθω γὰρ ἀπὸ τῆς ΒΓ τετράγωνον τὸ ΒΔΕΓ" ("for let there be described on ΒΓ the square ΒΔΕΓ.")[115] What is the canonical form of this formula? Here, the text appears pretty much in accordance with Netz's formalization, that is, ἀναγεγράφθω ἀπὸ {formula: line, in the genitive} τετράγωνον {formula: area}.[116] As we have seen, Netz regards as crucial parts of the formula both its lexical elements as well as the grammatical form of the verb, so that both the use of a different verb, e.g., συνεστάτω ("let there be constructed"), and a change of the mode of the verb, e.g., to the indicative, entail that a different formula is being used. In the first case, this would be a different formula altogether,[117] the second case would count as a 'transformation' of the original formula into a new formula according to Netz's view.[118]

As I have suggested above, one could instead, from a lexical point of view, regard the conjunction of the term for the geometrical object (τετράγωνον (τὸ ΑΒΓΔ)), the verb denoting its diagrammatic construction (ἀναγράψαι) and the oblique syntax of the verb (ἀπὸ τῆς ΑΒ) as the core of a more general formula, that is, ἀπὸ τῆς ΑΒ τετράγωνον τὸ ΑΒΓΔ ἀναγράψαι ('to describe on the line AB a square ΑΒΓΔ'). This formula recurs in different standardized syntactic positions and, accordingly, in different 'variations' of its form, with the noun term τετράγωνον (τὸ ΑΒΓΔ) taking the position of either subject or object and with the verb ἀναγράψαι occurring in different modes and tenses.[119] I should like to call my formalization of it 'core' rather than 'canonical form' of the for-

112 Netz 1999a, 150–154.

113 Cf. his formulae 31 "προσκείσθω αὐτῇ εὐθεῖα ἐπ' εὐθείας, 'and let a line be added to it in a straight line'" and 31a "πρόσκειται αὐτῇ {formula 2: line}, 'a line has been added to it'" (1999a, 150).

114 Netz does acknowledge a number of cases in which a different conceptualization would be possible (e.g., 1999a, 140) and puts the definition of "canonical form" (or: "identity") up for debate (1999a, 149).

115 *Elem.* I 47 (vol. 1, p. 63.11 f. Heiberg & Stamatis).

116 On the exact syntax of the phrase and the formula to be inserted after τετράγωνον cf. p. 125 f., n. 77.

117 Formula 27 "συνεστάτω {any figure}" according to Netz 1999a, 137. Euclid uses συνεστάτω with τετράγωνον, e.g., *Elem.* X 96 (vol. 3, p. 171.9 f. Heiberg & Stamatis), but not, as ἀναγεγράφθω, with ἀπό to indicate the line "on" which the square is to be constructed.

118 See Netz 1999a, 150–154.

119 Τετράγωνον either works as the object or subject, taking τὸ ΑΒΓΔ as an apposition to it in certain variations of the phrase. Even more generally, one could substitute τετράγωνον in the formula with

mula as the phrase does not itself feature in exactly this form in the *Elements*, nor is the implementation of the phrase which contains the aorist infinitive any more frequent than nor formally superior to other variations of it and in this way 'canonical.'[120]

From a grammatical point of view, one could think of a different formalization of phrases such as Netz's formula 29 "ἀναγεγράφθω ἀπὸ {formula 2: line, in the genitive} τετράγωνον {formula 6: area}." The 'core' of the 16 different construction formulae which Netz lists only for his sample book *Elements* II is "{construction verb in the third-person perfect passive imperative} {geometrical object-term in the nominative}."[121] There are numerous direct implementations of this 'core' in the *Elements*, e.g., ἐπεζεύχθω ἡ ΓΔ ("let [sc. the line] ΓΔ be joined").[122] The core of verb form and noun term can be extended by one or more elements. These may supply, as in our example from I 47, e.g., the place of the construction (ἀναγεγράφθω γὰρ *ἀπὸ τῆς ΒΓ* τετράγωνον τὸ ΒΔΕΓ," "for let there be described *on ΒΓ* the square ΒΔΕΓ").[123] The following example adds the result (and again the place) of the demanded construction to the formulaic 'core:' τετμήσθω ἡ ΑΓ *δίχα κατὰ τὸ Ε* ("let ΑΓ be cut *in two at Ε*").[124] This represents a combination of Netz's formulae 34 ("τετμήσθω {formula 2: line} δίχα," "let {formula 2: line} be cut in two") and 33 ("τετμήσθω {formula 2: line} κατὰ {formula 1: point}," "let {formula 2: line} be cut at {formula 1: point}"), which he notes

---

certain other, hardly generalizable terms which Euclid uses with the phrase ἀπὸ τῆς ΑΒ [. . .] ἀναγράψαι: εὐθύγραμμον ('rectilinear figure,' e.g., *Elem.* VI 18 [vol. 2, p. 69.2–5 Heiberg & Stamatis]), παραλληλόγραμμον ('parallelogram,' e.g., *Elem.* VI 29 [vol. 2, p. 93.16–18 Heiberg & Stamatis]), and παραλληλεπίπεδον ('body with parallel surfaces,' 'parallelepipedal solid,' e.g., *Elem.* XI 27 [vol. 4, p. 45.9–12 Heiberg & Stamatis], LSJ s.vv.).

120 The above phrase which contains the aorist infinitive ἀναγράψαι occurs, first, in the enunciation of problem propositions; here it appears in its most general form without containing any lettered object terms (cf. *Elem.* I 46 (vol. 1, p. 62.2 Heiberg & Stamatis): Ἀπὸ τῆς δοθείσης εὐθείας τετράγωνον ἀναγράψαι. / "On a given straight line to describe a square;" on the jussive function of this infinitive cf. below pp. 155–157). Second, it occurs in the definition of the goal of problem propositions; here it contains one lettered 'object formula' (cf. *Elem.* I 46 [vol. 1, p. 62.3 f. Heiberg & Stamatis]: [δεῖ δὴ] ἀπὸ τῆς ΑΒ εὐθείας τετράγωνον ἀναγράψαι. / "[thus it is required] to describe a square on the straight line AB."]. This means that the aorist infinitive never occurs with the object formula τὸ ΑΒΓΔ. I have made it part of the above formulaic 'core' ἀπὸ τῆς ΑΒ τετράγωνον τὸ ΑΒΓΔ ἀναγράψαι precisely because I have modelled the latter not only on the kinds of implementations of it which contain the infinitive but more generally on all grammatical variations of the phrase. The reason why I have chosen as 'core' the form containing the infinitive is that I find the uninflected form suitable for collecting this and analogous standardized phrases containing construction verbs in an alphabetical list (in continuation of my list of the standardized terms of I 47 pp. 116–119), without, however, meaning to indicate that the form of the phrase containing the infinitive is the original one or basis of the other variations of the phrase. I have chosen the aorist infinitive because the infinitive of construction verbs only ever occurs in the aorist for reasons of aspect.

121 Netz 1999a, 137.

122 *Elem.* I 6 (vol. 1, p. 15.9 f. Heiberg & Stamatis), corresponding to Netz's formula 26 (1999a, 137).

123 *Elem.* I 47 (vol. 1, p. 63.11 f. Heiberg & Stamatis), my italics.

124 *Elem.* I 16 (vol. 1, p. 25.6 Heiberg & Stamatis), my translation and italics.

separately although in his sample *Elements* II, the phrase τετμήσθω δίχα is never mentioned without the intersection being given by means of κατά, while conversely δίχα can be left out of the phrase 'τετμήσθω {line} κατὰ {point}.' It seems that the phrase 'τετμήσθω {line}' ('let the line be cut'), which is an implementation of the 'core' we suggested for construction formulae, can be extended by several words and phrases, rather than there being a number of distinct formulae which contain the imperative τετμήσθω. Finally, the core of verb form and noun term is often extended by phrases which specify the relation to other objects, as one can see it happen twice in the following example: Ἤχθω γὰρ ἀπὸ τοῦ Α σημείου τῇ ΑΓ *εὐθείᾳ πρὸς ὀρθὰς* ἡ ΑΔ καὶ κείσθω *τῇ ΒΑ ἴσῃ* ἡ ΑΔ ("For let ΑΔ be drawn from the point A *at right angles to the straight line ΑΓ* and let AD be made *equal to BA*").[125] In Netz's system, phrases as '{line, in the dative} πρὸς ὀρθάς' and '{object, in the dative} ἴσος' are 'predicate formulae' which may form part of 'construction formulae,' e.g., of formula 38 "ἤχθω {formula 2: line} πρὸς ὀρθάς {formula 2: line, in the dative}" which comprises formula 72 "πρὸς ὀρθάς."[126] If one wanted to extract a consistent 'formula' from the Euclidean text which contains the elements of formula 38, one would have to add to it 'ἀπὸ {formula 1: point, in the genitive}' ('from {point}'), which is how the phrase always appears in Netz's sample book II as well as beyond. But as πρὸς ὀρθάς is used also as part of other phrases, e.g., with the respective forms of τέμνειν from book II, it might be more fittingly described as the fixed term for 'at right angles' which is an adverbial either used absolutely or with the dative of (different types of) 'line' and which is regularly combined with other fixed terms so as to form larger phrases. Accordingly, 'ἤχθω {line}' is the 'core' phrase which can be extended in different ways and which could itself be described as an implementation of the above general 'core' of construction formulae '{construction verb in the third-person perfect passive imperative} {geometrical object-term in the nominative}.'

Such a focus on the recurrence of verb forms such as the imperatives of construction verbs would, on the one hand, conflate all of Netz's 'construction formulae' and reduce the number of formulae in a useful way. On the other hand, our 'core' of construction formulae would not be able to take into account the respective syntactical changes of lexical phrases such as 'ἀπὸ τῆς ΑΒ τετράγωνον τὸ ΑΒΓΔ ἀναγράψαι' which we had identified as the 'core' of the formulae with construction verbs from a lexical perspective. That is, while the formula '{construction verb in the third-person perfect passive imperative} {geometrical-object term in the nominative}' would cover the imperative phrase ἀναγεγράφθω γὰρ ἀπὸ τῆς ΒΓ τετράγωνον τὸ ΒΔΕΓ from I 47, it would not pertain to the same lexical phrase if the verb changes to the indicative as in the same proposition (καί ἐστι τὸ μὲν ΒΔΕΓ τετράγωνον ἀπὸ τῆς ΒΓ ἀναγραφέν) or to the infinitive, as in I 46 (ἀπὸ τῆς ΑΒ εὐθείας τετράγωνον

125 *Elem.* I 48 (vol. 1, p. 65.18 f. Heiberg & Stamatis), my italics.

126 Netz mentions them first as part of his "construction formulae" without noting their formulaic status (1999a, 138) and only retrospectively, in the discussion of "predicate formulae" notes that they are themselves formulae, that is, his formulae 72 and 78 (143).

ἀναγράψαι).[127] This means that depending on one's focus one may differently formalize the standardized phrases of the *Elements* and that a shift of focus, e.g., from a lexical to a grammatical one, cuts across the foregoing systematization.

Synonymity of formulae

As I have briefly mentioned above, Netz's formulae are not only formally variable, but also, to some extent, semantically.[128] If one compares the formulae of Netz's list semantically as if they were one-word terms, one is faced with a number of (near-) synonyms. There will not be any synonyms in the strict sense as each word of the formula, its syntax, and, as we have seen with Netz's construction formulae, even the grammatical forms of the words add to the meaning of a formula which can hardly be exactly the same twice. However, there are a number of phrases which have roughly the same meaning and perform the same function in the text and which are therefore 'near-synonyms.' Compare, for instance, Netz's formulae 3 ("ἡ ὑπὸ τῶν ΑΒΓ [γωνία]," "the [sc. angle formed] by ΑΒΓ") and 4 ("ἡ πρὸς τὸ Α [γωνία]," "the [sc. angle] next to the point A"), which he cites as synonyms on the grounds that both of them mean 'the (particular) angle.'[129] However, while the two terms have the same overall meaning, either of them has a different nuance which is why they are only 'near-synonyms:' while 3 explains the angle with respect to the point at which it is situated, 4 locates it with regard to the two lines (in turn, defined by three points) which form the angle. Depending on the context of the proposition, one of the two terms may be more suitable than the other. Their subtle difference is enabled precisely by the extremely standardized nature of the lexicon: either term is composed of several standardized words and is therefore itself highly standardized, whereas their appearance in Netz's list of formulae and uniform, synonymous translation as "angle" suggests otherwise.

The probability of variability and of (partial) semantic overlap increases with the length of the formulae. Netz's formulae 75–77 ("{object} ἴσον ἐστί {object} μετὰ {object}," "{object} ἴσον ἐστί {object} καὶ {object}," "{object} ἴσον ἐστί {object} τε καὶ {object}"), are a case in point: they are listed as "three ways of representing [sc. the modern mathematical formula] a + b = c," that is, effectively, as synonyms.[130] The only difference between them is that the operation of 'adding' one object to the other is expressed either by μετά (literally, 'along with,' 'together with,' as in formula 75), or by καί ('and,' as in 76), or by τε καί ('both . . . and,' as in 77). The Euclidean lexicon allows for this variation in accordance with the various possibilities of the Greek language to express the basic, non-technical concept of joining to objects in the sense of 'and' or

127 Cf. above pp. 133, n. 120.
128 Cf. above, p. 127.
129 1999a, 134; my additions in square brackets.
130 1999a, 143; cf. above p. 129, n. 97.

'with.'[131] Such variation, however, hardly accounts for the difference and therefore synonymity of the alleged larger, superordinate phrases. The standardized 'core' term seems to be again, as in the case of the above-discussed formula 78, merely 'ἴσον (ἐστί)' together with its oblique syntax, that is, the dative of the *comparatum*. It can be used to express the equality of two objects, or of the sum of two objects and another object, of the sum of two objects and the sum of another two objects and so forth.

The non-standardization of 'formulae'

I have been meaning to show to things: first that it turns out impossible to distinguish formulae in the *Elements* without there being at the same time several alternative ways to do so. They do not appear as distinct unities. This might be because the text "does not lend itself naturally to such divisions, because parts do not come in such neat bits."[132] If one tries to conceptualize the Euclidean language in terms of 'formulae,' that is, in terms of a system of standardized phrases, in Netz's way or differently, the language appears to be to some extent formally and semantically variable and non-standardized. This effect of non-standardization and ambiguity does not fit the extremely standardized nature of the mathematical language as we have encountered it when analyzing the mathematical lexicon. My impression is that the synonymity among the formulae and their lack of formal standardization are only a result of the formulaic system rather than characteristic of the mathematical language, which is just the opposite, that is, extremely standardized. In this way, thinking of the mathematical language in terms of formulae creates problems which would not have come up otherwise.[133]

The second thing which I wanted to show by analyzing Netz's concept of formulae is that it does not seem to focus sufficiently on the nature of the grammatical standardization. He does sometimes consider grammatical aspects such as the mode and tense as key elements of construction formulae;[134] but he does not conceptualize the language systematically in terms of grammatical standardization but in terms of standardized phrases which are ultimately larger standardized lexical units (even though containing standardized grammatical elements). There is no doubt, however, that the language of the *Elements* is standardized beyond the level of words and phrases. Therefore, it seems to be necessary to investigate the nature of the grammatical standardization more closely in its own right and eventually to merge the lexical and grammatical per-

---

131 Cf. above p. 131.

132 This is part of a quotation by Netz on Proclus' division of the Euclidean propositions (cf. p. 23 with n. 134 and Section 3.4.1 below); I quote it in full in the following note.

133 Cf., again, Netz (1999b, 291): "The truth is that the question, 'What part is missing from it?' is only meaningful starting from Proclus' scheme. Starting from the text itself, it does not lend itself naturally to such divisions, because parts do not come in such neat bits."

134 Netz 1999a, 136: "Objects are constructed, and they are constructed in specific ways, with specific verbs, in specific tenses and modes: in a word, constructions have their formulae."

spective in order to come to terms with the standardization of the mathematical language. Reconciling the lexical and grammatical side of the mathematical language as we tried in an attempt to adjust the concept of 'formulae' has not led to convincing results; the grammatical side could only be highlighted at the expense of the lexical and vice versa. This is why I presume that the result of conflating both perspectives certainly will not be 'formulae' in the sense which Reviel Netz had intended.

3.3 Grammar

First, I focus on the standardization of two related aspects: sentence constructions and verb forms (in particular, modes and tenses).[135] Second, I return to the subject of 'formulae' and choose one example for investigating the standardization of smaller or larger sequences of words, that is, Netz's formula 23 "ἔστω {any unlettered object formula} {a lettered equivalent}."

3.3.1 Sentence constructions

Number and types of sentence constructions

The construction of sentences is extremely standardized in the *Elements*. Compared to the diversity of sentence constructions available in Greek and their varied use in Greek prose, the number of constructions used in the *Elements* is extremely limited.

The dominant type of sentence construction is the simple paratactic main clause.[136] In I 47, one notes 26 occurrences compared to only three cases of hypotaxis. Only two of these include genuine subordination (each time of a causal clause, introduced by καὶ ἐπεί, to a subsequent main clause), while the third is only a seeming one: in the definition of the goal of the proposition, an object clause beginning with ὅτι is grammatically subordinated to the short main clause λέγω; however, since all other theorems in the *Elements* are likewise introduced by λέγω ὅτι,[137] the phrase technically works as a stan-

135 For a comprehensive analysis of particles in elements texts, cf. Acerbi 2012, 172–174.

136 Here and throughout, I use the term parataxis in the broad Greek sense of "coordination in general" (cf. Smyth §2168a), as referring both to asyndetic arrangements of main clauses, and to the coordination of main clauses (or simple sentences) and of compound sentences (that is, several simple sentences coordinated by particles) by means of connectives, as opposed to Smyth's use of it for "the arrangement of two independent sentences side by side, though one is *in thought* subordinate to the other" (§2168). Since interpunction is a matter of post-antique convention (see Acerbi's suggestions based on his analysis of the mathematical syntax in 2012, 213), I count as a main clause (or simple sentence) every occurrence of '(implied) subject + predicate.' On the prevalence of parataxis in the mathematical demonstrations of the elements genre cf. Acerbi 2012, 212 f.

137 An only slight exception is presented by λέγω δὴ ὅτι in I 46.

dardized particle which announces a direct, independent statement.[138] Netz's conceptualization of the λέγω ὅτι as a formulaic phrase preceding the variable element of the formula takes account precisely of this standardized use of the two words. The impression that the λέγω ὅτι is a standardized phrase is the result of the standardization of both the lexicon and the syntax. There are numerous alternative ways in the Greek language to express the concept 'to say, claim,' but Euclid invariantly uses the verb λέγειν; this verb can be constructed either with an object clause introduced by ὅτι or with the infinitive, but Euclid always uses the first option,[139] and he always uses the first person singular active where he could alternate between the singular and the plural as well as the active and the passive voice, as he does in other parts of the proposition.[140] Because of the strict standardization of the two words which belong to different syntactic levels – λέγω is the main clause, ὅτι the first word of the subordinated object clause – the actual syntactic relations of the sentence become inverted and the lines between what is lexicon and syntax get blurred:[141] the subordination of ὅτι is no longer perceived as it has come to form an inseparable unit with λέγω; tellingly, Netz's formula is "λέγω ὅτι" without a comma, while Heiberg & Stamatis always print "λέγω, ὅτι" in accordance with the conventional comma placement in the modern editions of Greek texts. As Netz's conceptualization of the formula for the definition of the goal shows, the sentence seems to have the structure 'λέγω ὅτι + the rest' rather than 'λέγω + ὅτι . . .,' and the goal, which

---

138 Similarly, the definition of the goal in problem propositions is regularly introduced by δεῖ δή ('thus it is required'); it is only once in *Elements* I that δεῖ is used with a different particle (δέ, cf. *Elem.* I 22 [vol. 1, p. 30.24 Heiberg & Stamatis], which is one of only two instances in the *Elements*), but here it is used with a different function. The main clause δεῖ δή is always followed by an infinitive construction which states what one has 'to do' and repeats in concrete terms the goal as it is stated in the enunciation in general terms. E.g., in the opening proposition I 1, the definition of the goal δεῖ δὴ ἐπὶ τῆς ΑΒ εὐθείας τρίγωνον ἰσόπλευρον συστήσασθαι ("Thus it is required to construct an equilateral triangle on the straight line AB") essentially repeats after the standard introductory phrase δεῖ δή the enunciation ἐπὶ τῆς δοθείσης εὐθείας πεπερασμένης τρίγωνον ἰσόπλευρον συστήσασθαι ("On a given finite straight line to construct an equilateral triangle," *Elem.* I 1 [vol. 1, p. 7.1–6 Heiberg & Stamatis]). Because of its standardized appearance, the introductory δεῖ δή, just like the λέγω ὅτι, is not really perceived as a superordinate main clause, but rather as an introductory signal for the mentioning of the goal by means of the subsequent infinitive construction. In this way, the infinitive is similarly independent as in the enunciation, where it has a jussive function and is used like finite verb in the imperative (cf. p. 155–157). In fact, the syntax could be explained in the same way in both cases, if the infinitive of the enunciation is interpreted "as being governed by an understood verb form with directive connotation, like an imperative or δεῖ (alternatively, it could be interpreted as having itself a "directive" or, as I have called it, jussive meaning, see Acerbi 2021, 113).
139 See pp. 153 f. for a possible reason for the choice of syntax after λέγειν and Section 5.1 for a discussion of the 'logical' necessity of the linguistic standardization in this case.
140 Cf. the recurring synonymous phrases ὁμοίως δὴ δείξομεν, ὅτι ('similarly we can prove that'), which uses the first person plural active, and ὁμοίως δὴ δειχθήσεται, ὅτι ('similarly it can be proved that'), which contains the third person singular passive.
141 Cf. the introduction to this chapter for the characteristic overlap of the different textual levels in the *Elements*.

is grammatically part of the subordinated part of the sentence, becomes the seeming main clause while the actual one λέγω along with ὅτι steps back syntactically – it is only a phrase, repeated over and over. At the same time, the level of lexicon seems to dominate over the level of syntax. While the syntax of the sentence is just as standardized as its lexicon, it is precisely because of this standardization that it is not recognizable anymore and that λέγω, ὅτι becomes λέγω ὅτι, a lexical phrase.

A similar example of the seeming subordination of statements is that of clauses which follow upon the phrases ὁμοίως δὴ δείξομεν, ὅτι ("similarly we can prove that") and ὁμοίως δὴ δειχθήσεται, ὅτι ("similarly it can be proved that").[142] They are employed to introduce analogous cases in the proof part. Because of the repetition of these lexical phrases which, like the λέγω ὅτι, extend over different parts of the sentence, the respective clauses after ὅτι are only formally subordinated and appear as independent, main clause-like statements. While the standardization of the λέγω-ὅτι phrase is one hundred percent in theorem propositions, the two phrases which introduce analogous cases are used synonymously; either one or the other occurs in those proofs which introduce analogous cases.[143] Nevertheless, the standardization in this case, though it is not complete, is still very strong, as it reduces various lexical and grammatical possibilities to two standardized lexical phrases. Most obviously, the construction after δεικνύναι is variable; the verb can be constructed not only with an object clause introduced by ὅτι (or, alternatively, ὡς), but also with a participle construction, either with an 'accusative + participle,' or with a simple participle if its subject coincides with that of the participle, or it can be used in the personal passive.[144] Because of the invariant use of the object clause with the same conjunction, ὅτι becomes part of the standardized phrase,[145] which, in turn, becomes the set introductory signal for a quasi-direct statement.

Genuine hypotaxis is restricted to very few occurrences and to very few types in the *Elements*, which ultimately can be subsumed under the broad category of conditional clauses.[146] The only other type of subordinate clause which appears in the *Elements* is numerically insignificant: ὥστε + infinitive, which occurs only three times. To

---

142 Cf. p. 138, n. 140.

143 ὁμοίως δὴ δείξομεν, ὅτι ("similarly we can prove that") occurs five times in *Elements* I, ὁμοίως δὴ δειχθήσεται, ὅτι ("similarly it can be proved that") twice.

144 Cf. the other of the two standardized functions of δεικνύναι in the *Elements*, that is, ἐδείχθη δέ ("but it was proved"), which is used in the proof part to refer back to a foregoing result of the proof and is either constructed impersonally with a ὅτι clause (twice in *Elements* I) or, more often, personally (10 times), e.g., ἐδείχθη δὲ καὶ ΓΑ τῇ ΑΒ ἴση· ("But ΓΑ was also proved equal to AB;," *Elem.* I 1 [vol. 1, p. 8.3 f. Heiberg & Stamatis]).

145 For this reason, I take the standardized phrases to be ὁμοίως δὴ δείξομεν, ὅτι and ὁμοίως δὴ δειχθήσεται, ὅτι, that is, including the conjunction, rather than merely ὁμοίως δὴ δείξομεν and ὁμοίως δὴ δειχθήσεται (cf. Netz's formula 41; 1999a, 138).

146 I list the propositions in which the respective types of subordinate clauses occur in the following section where I discuss the types of subordinate clauses in more detail regarding their standardized places in the proposition.

this, one could add nine occurrences of ὥστε + indicative, but these clauses are "often so loosely connected with the leading verb as to be practically independent and coordinate."[147] The type of conditional clause which appears most regularly in *Elements* I[148] is the causal-conditional 'paraconditional'[149] sentence which is standardly introduced by the causal conjunction (καὶ) ἐπεί (71 times in 34 propositions of *Elements* I). The second most frequent are conditional periods introduced by ἐάν (twice each in 14 propositions), while conditional clauses beginning with εἰ appear less often (altogether 16 times in 13 propositions). Two clauses introduced by the temporal conjunction ὅταν (in I 11 and 12) likewise fall under the category of conditionals. Finally, 12 propositions have altogether 20 relative clauses, most of which have a conditional connotation.[150] It is important to note that the number of subordinate clauses is not

---

147 Smyth §2255. Heiberg & Stamatis invariantly mark ὥστε + indicative as (somewhat) independent clauses by punctuating the preceding clause with a point above the line, whereas Heath always translates "so that" after a comma, thus marking the consecutive clause as a formally subordinate one.

148 66 times in 36 of 47 propositions.

149 The term originates from Stoic logical theory: παρασυνημμένον δέ ἐστιν, ὡς ὁ Κρῖνίς φησιν ἐν τῇ Διαλεκτικῇ τέχνῃ, ἀξίωμα ὃ ὑπὸ τοῦ "ἐπεί" συνδέσμου παρασυνῆπται ἀρχόμενον ἀπ' ἀξιώματος καὶ λῆγον εἰς ἀξίωμα, οἷον "ἐπεὶ ἡμέρα ἐστί, φῶς ἐστιν." "A paraconditional [inference] is, as Krinis says in his *Art of Dialectic*, a proposition which is bound together by the conjunction 'since' and which begins with a proposition and ends with a proposition, for example, 'Since it is day, it is light.'" (Diogenes Laertius VII 71, translated by Inwood/Gerson 2008, 19; cf. Acerbi 2012, 169, who formalizes the structure as "sistema di subordinata causale + propozione principale introdotto da ἐπεί").

150 Most of the relative clauses in *Elements* I have the present indicative and, thus, qualify both for "ordinary relative clauses" which "define more exactly a definite antecedent" (Smyth §2553) and for "simple present relative clauses" which have an "indefinite" antecedent and "often have the value of *general* conditions" (§2562). As the relative clauses of *Elements* I always refer to the geometrical objects with which a proposition deals, the question of the definiteness or generality of the relative clause and its antecedent brings up the notorious mathematical-philosophical question of the ontological status of the geometrical objects in elements texts (cf., e.g., Section 4.3.2, 246). The matter is further complicated by the internal structure of the proposition: the general proposition at the beginning is proved by means of a particular case of it which is (in theorem propositions) generalized again at the end (cf. Section 1.2, 23). This means that the objects of a proposition appear at its beginning and (in theorem propositions) at its end in a general way, while they become concretized in the main part of the proposition by the letters by which they can be identified in a lettered diagram. I believe that the relative clauses have a conditional force in any case. It is generally agreed that the propositions do not deal with real objects, but with generic ones – even in the lettered main part of the propositions where the objects are described in the same way as at the beginning and end except for their lettering. Throughout, the relative clauses ascribe to the objects certain conditions under or by means of which the proof must be performed. They are conditional as their ascriptions present conditions that are the case as opposed to other conditions that are not. The ultimate proof for the conditionality of the relative clauses is that the negation they take wherever they are negative in *Elements* I, both in the enunciation and conclusion as well as in the concrete main part, is invariantly μή, which always is the negation of conditional clauses. For example, the (concrete) conclusion of problem I 12 has μή as the negation of its relative clause just like the other three negated relative clauses of this proposition: Ἐπὶ τὴν δοθεῖσαν ἄρα εὐθεῖαν ἄπειρον τὴν ΑΒ ἀπὸ τοῦ δοθέντος σημείου τοῦ Γ, ὃ μή ἐστιν ἐπ' αὐτῆς, κάθετος ἦκται ἡ ΓΘ· ("To a given straight

only limited within the unit of the proposition, but also within each sentence. One does not record more than one level of subordination by adverbial or relative clauses per sentence. Wherever there is more than one subordinate clause per sentence, the hypotactic sentences are themselves paratactically coordinated. Cases of double subordination, if any, are realized by combinations of subordinate clause and participle construction. The reason for this could be that the hierarchical relation of different types of subordination is more readily comprehensible and unambiguous than that of two adverbial clauses.

However, the limited use of subordinate clauses cannot be explained by the presence of participle constructions. Non-finite clauses which in theory may substitute subordinate clauses and could account for the comparatively rare appearance of the latter are just as infrequent as finite subordinate clauses: in the sample proof I 47, there are no AcI, NcI, articular infinitives, and supplementary participle, and only one genitive absolute and non-absolute circumstantial participle each. A similar picture emerges from a survey of all propositions of *Elements* I: of the constructions mentioned, only circumstantial participles including genitive absolute are of some significance, while the articular infinitive appears only twice in total,[151] the others not at all.[152] The genitive absolute occurs in only six propositions (altogether 16 times). Non-absolute circumstantial participles can be found more regularly (45 times in 25 propositions); however, in 12, that is, nearly half of these 25 propositions, there is only one form of ἔχων, -ουσα, -ον, which in connection with a noun loses its own verbal force and simply mean "with."[153] In most cases, both the genitive absolutes and the *participia coniuncta* denote a condition, in a similar manner as the relative clauses.

As we have seen, the choice of sentence constructions and their quantitative application is highly limited. The propositions get by with very few types of sentence constructions. Of this extremely limited selection, it is again one single type, the paratactic main clause, which dominates large portions of the text, while the text avoids any accumulation of subordination.

line AB from a given point Γ, which does not lie on it, the perpendicular ΓΘ has been dropped.," *Elem.* I 16 [vol. 1, p. 20.25–27 Heiberg & Stamatis]). On the implications of the qualification of geometrical objects as "given" ones in different types of propositions and different parts of the proposition cf. Acerbi 2011a). Here, the condition of Γ is that it is not on the straight line AB, as opposed to another condition that is not the case, namely that it does lie on the straight line AB.

151 Both times in the accusative after διά (*Elem.* I 4 and 29 [vol. 1, pp. 11.9 and 42.4 Heiberg & Stamatis]).

152 There are of course the regular occurrences of jussive infinitives in the enunciations of problem propositions and of the infinitives after δεῖ in the definition of the goal of problem propositions, but as I have pointed out above (see p. 138, n. 138), both cases do not represent types of real syntactical subordination: the first, because it equals the use of imperative in a main clause, the second because the superordinate δεῖ because of its standardization merely works as an introductory signal for an independent statement.

153 LSJ s.v. ἔχειν A 6.

Places of sentence constructions

The appearance of most syntactical constructions other than parataxis is confined to specific small parts of the proposition. In this way, they interrupt the paratactic scheme only scarcely and expectably. It is especially the prominent types of subordinated clauses which are assigned fixed places within the course of the proposition. Here is where the levels of structure and syntax (and lexicon) overlap.

Among subordinate clauses, the use of conditional clauses is paradigmatic; since they account for the greatest part of what I have described as 'genuine hypotaxis,' I will focus on describing their preferred places in the proposition *exempli gratia*.[154] The different types of conditional clauses have standardized functions within the propositions and always appear at the same places. Iterative conditionals occur, if at all, twice per proof; they feature exclusively in the very first sentence of theorem propositions (the enunciation), which announces the claim in the form "*if* x is the case, *then* so is y,"[155] and in the conclusive part of the respective propositions, which repeats the initial claim.[156] This conditional form of theorem propositions alternates with a non-conditional form, a simple "declarative statement."[157] The non-conditional form is preferred for reasons of conciseness: either when the subject of both parts of the conditional sentence would be the same – in this case, the content of the antecedent is expressed by "an attributive participle of the grammatical subject" of the declarative statement – or when the antecedent can be replaced by a single geometrical term which can be integrated into the declarative statement.[158] The distribution of the conditional and the non-conditional form of theorem propositions in Euclid's *Elements* is fairly equal.[159] A "bridge" between the two forms is the rare use of genitive absolute in the place of the conditional clause; it presents a way to accommodate a

---

154 I have already described above the standardized places of ὅτι clauses, which, in my view, do not present cases of real subordination. Further, I will leave out here consecutive clauses which are rare as well as relative clauses.

155 Netz 2004b, 6.

156 I 4, 6, 8, 13, 14, 15, 21, 24, 25, 26, 27, 28, 41, 48. Cf. Acerbi 2012, 172.

157 Acerbi 2021, 113.

158 Cf. Acerbi 2021, 113 with n. 4, who, for an example of the second case, refers to the enunciation of I 5 ("The angles at the base of isosceles triangles are equal to one another."), in which the term "isosceles triangles" substitutes the antecedent of a conditional version of the statement ("If a triangle have two equal sides, the angles at its base are equal to one another.").

159 The conditional form appears 14 times in *Elements* I (I 4, 6, 8, 13–15, 21, 24–28, 41, 48) and 224 times in the whole work, the non-conditional one 20 times in the first book (I 5, 7, 16–19, 29, 30, 32–40, 43, 47) and 146 times in the whole work (Acerbi 2021, 114). In the *Data*, the conditional form predominates (in 91 of 94 propositions; ibid., 114, n. 6); similarly, in Apollonius, 176 theorem enunciations take the conditional form, 32 the non-conditional one (ibid., 117, n. 17). In Archimedes, the ratio of the two forms is more balanced: here, the non-conditional form (98 times) even slightly prevails over the conditional one (72 times; ibid., 117). In view of this, it seems doubtful to explain the "pervasive[ness]" of the conditional form in later authors by the "multipl[ication]" of "the specific assumptions on which a

condition in the non-conditional form of the enunciation.[160] While iterative conditionals only appear at the beginning and end of propositions, they also are the only kind of subordinated clause (with the exception of relative clauses) to appear at these places. Since the conditional clause of these compound sentences always takes the subjunctive + ἄν, it is always introduced by ἐάν. In this way, the standardization of the syntax implies a standardization of the lexicon within the proposition: the word ἐάν is the only conjunction to occur at the beginning and ending of the propositions, and it is only there where the conjunction, if at all, occurs.[161] Exceptionally, ἐάν is replaced by ὅταν ("whenever") in the enunciation and conclusion of III 20.[162] The use of the iterative conditional itself (and even more so its strict standardization) is noteworthy. For the "canonical format of the conditional statements (συνημμένα litt. 'connected') recommended in Stoic logic" is the 'simple' conditional with εἰ + indicative in the protasis.[163] The point of using ἐάν (with subjunctive) instead of the traditional εἰ (with indicative) seems to lie in the "requirement of attaining a maximum of generality."[164]

The 'paraconditional' clauses only occur at the beginning of the demonstrative part of the proposition, which undertakes to prove the claim.[165] One could think of these clauses – there may be up to four per proof – as a subpart of the demonstrative section. Since this "initial segment" of the proof refers back to what has been established in the foregoing parts of the proposition, that is, exposition and construction, it has been

theorem rests" "in less elementary treatises" (ibid., 117). A different explanation could be the on-going process of formal standardization (cf. Section 5.3, 432 f.).

160 Acerbi 2021, 49 and 266. In *Elements* I, genitive absolutes replace ἐάν clauses in I 5, 16 and 32 (see below p. 148).

161 Outside of *Elements* I, clauses introduced by ἐάν also occur in other parts of the proposition, mostly in the proof part: these are either "non-instantiated" or "instantiated" citations of previous enunciations, that is, citations of enunciations in their original general or in concrete form; the "non-instantiated" ones "are all of doubtful authenticity" (Acerbi 2021, 267).

162 Acerbi 2021, 266, n. 227. The usual place for ὅταν to occur is in the definitions where it means "exactly when," "if and only if" (ibid. 266), and in the citations of these definitions in the proof part (*Elem.* I 11 [vol. 1, p. 19.14–16 Heiberg & Stamatis] and I 12 [vol. 1, p. 20.21–24 Heiberg & Stamatis].

163 Acerbi 2021, 265 with n. 225 with reference to Diogenes Laertius VII 71 (cf. p. 140, n. 149 above).

164 Acerbi 2021, 266, see also 116 f.

165 I 1, 2, 3, 5, 6, 7, 9, 10, 11, 12, 13, 14, 15, 16, 17, 18, 20, 21, 22, 24, 26, 28, 30, 31, 32, 33, 34, 35, 36, 42, 43, 44, 47, 48. For the term, cf. above p. 140 with n. 149.

named "anaphora."[166] The logical syntax of the 'paraconditional' "anaphora" is adequately described by Acerbi.[167]

> [T]he 'anaphora' is a system comprising a causal subordinate clause, introduced by ἐπεί 'since' and with a verb form in the present or perfect indicative (the perfect stem is used whenever a direct reference is made to a constructive act), followed by a principal clause with a verb form in the indicative or in the imperative, possibly introduced by ἄρα 'therefore' [. . .]. The causal subordinate, which I shall call the 'antecedent,' contains the reference, the principal clause (the 'consequent') contains its consequence.

While this type of subordinate clause exclusively appears at the head of the proof part, it is, conversely, also the only type of subordinate clause to appear at this place. Linguistically, the "causal subordinate clauses" differ from conditional clauses only in the kind of conjunction by which they are headed (ἐπεί).[168] "Semantically," as Acerbi notes, "they are stronger than a conditional," as "[t]he connector announces both that the second is a consequence of the first and that the first subsists."[169] The standardization of this conjunction is one hundred percent, and in this case, it is not, as in the case of ἐάν, implied by the standardization of the syntax, but due to a lexical standardization. To be sure, by the time that the Stoics reflected on the 'paraconditional' as a logical structure, ἐπεί had come to be a standard part of it, but before, a range of other causal conjunctions would have been available. While the 'paraconditionals' always include ἐπεί, this conjunction is used in different ways: either (rarely) by itself or together with other conjunctions and particles (καὶ ἐπεί, ἐπεὶ οὖν, ἐπεὶ καί, ἐπεὶ γάρ, πάλιν ἐπεί, διὰ, τὰ αὐτὰ δὴ ἐπεί); for this reason, it may not appear that strictly

166 The term has been coined by Federspiel, who originally used it for referring to a function of the article ("fonction analogique de l'article," 1995, 250, n. 3). While he recognized a part of the proposition amidst ("au sein de") the demonstration which presents a "transform[ation]" of the exposition or the construction into the indicative (1995, 253), he did not use the term "anaphore" or "partie anaphorique" for this part until his 1999 paper on Apollonius' *Conics* II (411), in which he first explicates the function of referring back to the exposition and construction. Subsequently, Acerbi used the term for conceptualizing the initial part of the proof section in more detail (2007, 273–275; see also ibid. 2012, 169). Afterward, both scholars provided detailed analyses of the "anaphora" (Federspiel 2015, not acknowledged by Acerbi 2021, 163–165, and, on the "paraconditional," 269–275). Whereas Federspiel 1995, 253 seems to conceive of the "anaphora" as a part of the proposition of its own, Acerbi sticks with Proclus who does not acknowledge the "anaphora" as a proposition part, emphasizing that the "anaphora" is only the "initial segment" of the proof (while itself devoting to it a subchapter of its own as he does with the conventional proposition parts, i.e., chapter 4.4, 163–165).

167 2021, 163.

168 Acerbi 2021, 169.

169 Diogenes Laertius VII 71 (ἐπαγγέλλεται δ'ὁ σύνδεσμος ἀκολουθεῖν τε τὸ δεύτερον τῷ πρώτῳ καὶ τὸ πρῶτον ὑφεστάναι, transl. Acerbi 2021, 269).

standardized after all.[170] But all these joint uses are again standardized as they all have different logical-rhetorical functions which Acerbi thoroughly discusses.[171]

The appearance of conditional clauses whose *protasis* starts with εἰ is confined to those propositions that contain one or more indirect proofs of two kinds: either a ‘proof by contradiction,’ which proves a claim by proving the impossibility of its opposite, or a ‘proof by contrapositive,’ which infers a conditional statement ‘if A, then B’ from its contrapositive ‘if not B, then not A.’ Proofs by contradiction can be divided into three parts: in Acerbi’s terminology, the “initializing clause,” the “reference to the impossible,” and the “negation of the reduction assumption.”[172] Often, but not always, the “initializing clause” is formed by means of a conditional clause, with the εἰ-clause stating “the negation of what must be proved” and “coincid[ing] with the negation of the statement that immediately precedes the initializing conditional itself.”[173] The εἰ clause may appear in a standardized abbreviated form: the two most frequent phrases in the *Elements* (and the only ones to appear in *Elements* I) are εἰ γὰρ μή (“for if not”)[174] and εἰ (γὰρ) δυνατόν (“(for) if possible”), the latter “mainly when negative statements are reduced to the impossible (cf. I 7). Proofs by contrapositive are rarer in the *Elements* (23 occurrences, five in book I). The most frequent form includes a conditional clause introduced by εἰ γάρ (“for if,” cf. I 8 and 29); differently, in I 19 and 25, the proof by contrapositive is introduced by the elliptic εἰ γὰρ μή, which negates the omitted antecedent of the following unreal conditional.[175] As a matter of fact, proofs by contrapositive may always be transformed into the proofs by contradiction.[176] With regard to the use of proofs by contrapositives in the *Elements*, this indicates a choice not only of the logical but also of the linguistic form and implies the standardized use of the two types of indirect proof: proofs by contrapositive seem to have been preferred for shorter indirect arguments.[177]

There may be several conditional clauses introduced by εἰ per proof, up to three in the propositions of *Elements* I. Most of the time, the first of them appears just after the definition of the goal in almost all propositions which contain indirect proofs.[178] This is the place which normally undertakes the ‘construction’ of the geometrical objects. However, the methods of indirect proof change the order and function of the

170 On the varying modes (indicative or imperative) and tenses (present or perfect) of the ‘paraconditionals’, cf. Section 3.3.2 on the standardization of modes and tenses.

171 Acerbi 2021, 273–275.

172 Acerbi 2021, 252 f.

173 Acerbi 2021, 253. Of *Elements* I, cf. proofs 4, 6, 14, and 26.

174 I 27, 39, 40. Cf. Acerbi 2021, 253 for “elliptical” and “non-elliptical antecedents” and for the remaining types of the former to occur in the *Elements*.

175 Acerbi 2021, 261 f.

176 Acerbi 2021, 260 f.

177 Acerbi 2021, 261.

178 Of the propositions of *Elements* I which contain indirect proofs, only I 7 presents an exception to the above rule: here, the proof by contradiction follows directly upon the enunciation.

proposition parts, as it were. In indirect proofs, the construction, if there is one at all, is part of the proof, because it is assigned on hypothetical grounds.[179] For this reason, the beginning of the proof must be announced prior to the construction. But the proof, in these cases, does not begin in the usual way, that is, by means of a 'paraconditional', but in a way which makes clear that an indirect proof ensues: the usual way for an indirect proof to start is by means of an εἰ clause, although, as we have seen, it may be missing in some cases. The syntactical standardization in the case of these conditional clauses comes down to a lexical one: although the antecedent is usually in the present indicative, its mode and tense are often not recognizable as the verb is omitted; one may only conclude on it from the form of the subsequent main clause.[180] In any case, the conjunction is always εἰ. In this way, it is already the use of this conjunction at its standardized place – immediately after the definition of the goal – which predicts the special function of the ensuing construction and proof. Even though the irrealis mode is very rare in Euclidean indirect proofs, the standardization of εἰ indicates that what follows will result in something which is 'impossible' or 'not the case' or which is different from the defined goal.[181] The standardized place of the first occurrence of these εἰ clauses is after the definition of the goal, but further εἰ clauses may appear

---

179 Even in propositions which do not contain indirect proofs, the construction part is not always present. Of the propositions of book I which do contain indirect proofs, I 14 and 19 do not feature any construction part, I 25 (which contains two εἰ clauses) features a construction only after the second εἰ clause, and I 4 and 8 contain constructions of a special kind (that is, involving the technique of 'applying' (ἐφαρμόζειν) an object to another one, on which see Section 4.3.3, 260 f.); these are provided not in the usual syntax of the construction but in the form of a genitive absolute, which occurs *before* the εἰ clause of the respective proposition but has itself a conditional meaning (see below p. 159). In several propositions with indirect proofs, the construction is not assigned directly after the εἰ clause, but after another exposition by means of a sentence containing ἔστω. Thus, this exposition is part of the indirect proof like the construction, whereas the actual exposition occurs at the usual place after the enunciation in these propositions. The indirect proof of I 14 only contains an exposition, but not a construction. Apart from the construction which is part of the indirect proof, I 39 (as well as the supposedly spurious I 40) contains a construction assignment which precedes the definition of the goal. It follows directly upon the exposition – if one agrees with Heiberg that the first definition of goal which is attested after the exposition (*Elem.* I 39 [vol. 1, p. 53.2–3 Heiberg & Stamatis]) is spurious.

180 The tense of the verb of the consequent is always present or future, the mode indicative or imperative (Acerbi 2021, 267).

181 Some proofs by contrapositive contain unreal statements (*Elements* I 19, 25), but in these cases, the "antecedent of the conditional is omitted because its negation is anticipated as a separate statement which opens" the proof by contrapositive (Acerbi 2021, 262). Unlike "in present-day mathematical discourse," Euclidean proofs by contradiction never use the irrealis. According to Acerbi, the reason is that "present-day mathematicians have a much less developed mathematical sense of generality than Greek mathematicians had" (ibid. 267 with n. 233). However, the absence of irrealis in Euclidean proofs by contradiction is remarkable because the ancient Greek tradition of indirect proofs beyond Greek mathematics both in pre- and in post-Euclidean literature regularly uses counterfactual statements (that is, in the irrealis), which might be the very reason why modern mathematicians use unreal conditional clauses.

along the indirect proof. As they are contextualized by the respective first εἰ clause of the proposition, they are recognizable as instances of the standardized type of conditional clauses in indirect proofs.

There is one other function which εἰ clauses can have: rarely, they are used to differentiate cases within a proof.[182] The differentiation often but not always takes the form εἰ μὲν οὖν . . . εἰ δὲ οὔ (in *Elements* I, only in proposition 13).[183] The common (and unique) linguistic feature of this use of εἰ clauses is the negative οὐ in the second element which is unusual in conditional clauses and can be explained by "the rhetorical function of marking the division into cases;" perhaps, it may even be used to distinguish these εἰ clauses from the indirect proof type of εἰ clauses which often start off with the phrase εἰ γὰρ μή (although the former are already formally distinguished from the latter by their dichotomic or even multiple structure indicated by μέν . . . δέ (. . . δέ etc.).[184] These εἰ clauses may feature after the definition of the goal (like those of indirect proofs), but also after the construction or toward the end of the proof in the part which introduces a demonstration that will have to be done analogically with the one described without being done explicitly in the text.

Apart from these rare εἰ clauses which distinguish different cases of the same proposition, the εἰ clauses of indirect proofs are also the only type of subordinate clause to occur at the place where they commonly appear, that is, after the definition of the goal. The regular 'paraconditional' introduction of the proof normally (though not always) occurs at a different place, namely *after* the construction. Moreover, the methods of indirect proof change the conventional form of the proof: in this way, the appearance of the εἰ clauses of indirect proofs very often coincides with the absence of paraconditional ἐπεί clauses, as in *Elements* I, 8 propositions out of 12 with εἰ do not have ἐπεί;[185] conversely, 8 propositions out of those 11 without ἐπεί have εἰ.[186]

---

182 "The preferred form is an instantial imperative that introduces each case of the disjunction." (Acerbi 2021, 267, n. 234).

183 For all instances in the *Elements* cf. Acerbi 2021, 267.

184 I quote Acerbi 2021, 267. To be sure, a differentiation of cases often accounts precisely for the conditional μή in relative clauses or with participles (cf. above p. 140 f., n. 150), but here, the choice of οὐ rather than μή has to do with putting an emphasis on distinguishing cases of the same proof as opposed to presenting condition which leads to a different outcome of the proof. In any case, the "simple mode" of the conditional is not a sufficient reason for the conditional to take οὐ. Cf. 2021, 266 f. on the change of the mode and negation of the six definitions between *Elements* X 47 and 48 (ἐάν + subjunctive) upon their quotation in proposition X 66 (my italics): "For this reason, in these propositions the conditional is set in the simple mode (that is, it is in the form of εἰ 'if' + indicative), *so that* the kind of negation changes from μηδέτερον to οὐδετέρα." Rather, the reason for switching to οὐ could be (as suggested above) that it is the standardized negation of conditionals which differentiate cases.

185 Of the following propositions which contain indirect demonstrations with εἰ, those which lack ἐπεί clauses are printed in italics: I *4*, 6, 7, *8*, 14, *19*, *25*, 26, *27*, *29*, *39*, *40*. Cf. Acerbi 2012, 172.

186 Of the following propositions which lack ἐπεί clauses, those which contain indirect demonstrations with εἰ are printed in italics: I *4*, *8*, *19*, *25*, *27*, *29*, 37, 38, *39*, *40*, 41. Thus, 33 of 37 with ἐπεί do not have εἰ; and 33 of 36 without εἰ have ἐπεί.

The standardized recurrence of the main three types of conditional clauses is striking and quickly recognizable. As each type of clause is inseparably connected to a particular part of the proof, they function as structuring devices, while the conjunctions of the conditional clauses by their standardized and unambiguous use within those structuring devices work as 'signposts' of the argument structure.[187] This effect is even more striking since the conditional clauses, naturally headed by their conjunction, mostly start off the respective part of the proposition. Here is where the standardization of lexicon, syntax, and structure interconnect tightly and produce disambiguation.

The use of participle constructions with respect to their position within the proof is not as strictly regulated as that of subordinate clauses but it is still standardized to some extent. Genitive absolutes, for example, do not regularly occur in *Elements* I, but only in six out of 48 propositions. But in these few propositions they regularly recur at a few standardized places (altogether 16 times). The most frequent use is the above-mentioned one in the enunciation and conclusion of theorem propositions where the genitive absolute substitutes the ἐάν clause (in I 5, 16, and 32). The reason for choosing the participle construction over the subordinate clause in these propositions seems to be that it allows the enunciation to be formulated as a simple statement rather than as a conditional.[188] There may be more specific reasons for formulating an enunciation of a theorem as a simple statement: as Acerbi points out, the simple form of I 32 including a genitive absolute printed in italics (παντὸς τριγώνου *μιᾶς τῶν πλευρῶν προσεκβληθείσης* ἡ ἐκτὸς γωνία δυσὶ ταῖς ἐντὸς καὶ ἀπεναντίον ἴση ἐστίν [. . .] / "An jedem Dreieck ist der *bei Verlängerung einer Seite* entstehende Außenwinkel den beiden gegenüberliegenden Innenwinkeln zusammen gleich [. . .]") could be due to the analogy with the style of the enunciations of I 16–20 which all begin by the words παντὸς τριγώνου ("an jedem Dreieck").[189] For if one rephrases the enunciation of I 32 in terms of a conditional, the genitive παντὸς τριγώνου becomes unnecessary.[190]

Compared with *participia coniuncta* which, just like relative clauses, invariantly occur in all parts of the proposition, genitive absolutes are easier to locate; they are restricted to the beginning and ending, and to the demonstrative part. While genitive absolutes most of the time replace either ἐάν or ἐπεί clauses, the subordinated clauses which are typical for the parts in which they appear, *participia coniuncta* only in some cases substitute a standardized subordinated clause, most prominently the ἐάν clause of the beginning and ending; in other cases, they add another condition to the one given by a subordinate conditional clause. In any case, even if they perform the function of

---

187 Netz uses the term "signposts" differently for a group of phrases he calls "argumentation formulae."

188 Acerbi 2021, 49 f.

189 The genitive absolute is printed in italics. Cf. *Elem.* I 16–20 and 32 (vol. 1, pp. 24.15–17, 26.2 f. and 20 f., 27.14 f., 28.9 f., 44.7 f. Heiberg & Stamatis).

190 Acerbi 2021, 266. Beyond *Elements* I, genitives absolutes also occur together with ἐάν clauses to formulate "constraints" in the enunciation of theorem propositions, either as part of the ἐάν clause (*Elem.* X 2 [vol. 3, 6.12–15]) or introducing the conditional sentence as a whole (*Elem.* X 1 [vol. 3, 4.5–10]).

clauses standardly introduced by ἐάν, ἐπεί, or εἰ, they are not unambiguously identifiable in the way subordinate clauses are, since they lack the conjunctions that hold apart the different types of conditional clauses and point to their specific nuances.[191] Consequently, they do not have a structuring function in the proposition like subordinate clauses do. This observation points us to the question why certain constructions are preferred over others.

Preference of paratactic main clauses over finite subordinate clauses

The strong preference for paratactic main clauses over subordinated clauses might issue from the desire to avoid ambiguity. From a formal point of view, parataxis is, as a matter of fact, more straightforward and less complex than hypotaxis.[192] Since the reader does not need to reconstruct different hierarchical levels of each sentence, they can read the text in a linear way.

To be sure, ancient readers of Greek are well accommodated to a multi-variant syntax rich in hypotaxis. However, it remains a fact that parataxis minimizes potential sources for misunderstanding. Readers can focus solely on the relation between each paratactic clause, while those of more hypotactic texts would need to busy themselves both with the structure of more complex sentences *and* with their relations to one another.

By contrast with the "coordinative parataxis" of other mathematical texts, the *Elements*' parataxis is "argumentative" in nature,[193] making abundant use of particles, which establish the logical relation of a sentence to its preceding one right at the outset (that is, at the first or second position of the sentence). That is to say, not only does the *Elements* show a preference for paratactic main clauses in general; but it also shows a preference for a particular type of paratactic main clause which happens to be more capable of unambiguousness than the alternative type.

If one compares paratactic main clauses with subordinate adverbial and relative clauses specifically, it is with respect to tenses and modes that main clauses are rather unambiguous, whereas subordinate adverbial and relative clauses display a considerable degree of indeterminacy. It is true that even in independent clauses, the tenses do

191 Cf. Acerbi 2012, 170 for his classification of the beginning of I 16, which uses a genitive absolute instead of the regular conditional clause with ἐάν to express a condition, as "enunciat[o] di teorimi in forma assertoria," as though it did not have the 'if . . . then' structure that, according to himself, is the regular structure of the beginnings of theorem propositions.

192 Cf. Smyth §2160 for the distinction of "compound sentences" (unification of simple sentences) and "complex sentences" (subordination).

193 Cf. Acerbi 2012, 212, who describes the dominant syntax of the demonstrations as "paratassi argomentale" as opposed to the "paratassi coordinative" of the "procedures" and "algorithms," two types of mathematical discourse to which Acerbi allocates separate stylistic codes that, according to him, have been partly suppressed by the dominant style of the geometrical propositions. On the argumentative nature of the propositions cf. also 172.

not express absolute time except in the indicative,[194] while in other modes they only express the stage of action.[195] In dependent clauses, however, the tenses never express absolute time, but always refer to the stage of action. If time is at all expressed, like in clauses in the indicative or in indirect discourse when the optative "represents the indicative after a past tense of a verb of saying or thinking,"[196] then only relatively to the time of the superordinate verb.[197] That is, as far as its time is concerned, a subordinate clause cannot be interpreted except by taking the superordinate sentence into account, for it does not have any time itself. What is more, unlike subordinate clauses in Latin, whose time is determined according to the standardized *consecutio temporum*, Greek subordinate clauses do not give any formal hints as to their temporal relation to the respective main clause.[198] The time of subordinate clauses, thus, is underdetermined and needs to be worked out by the reader. It is, therefore, with respect to their time that subordinate clauses as such are non-standardized and ambiguous.

The greater standardization and unambiguousness of the expression of time in independent in comparison with dependent clauses is enforced by the standardized use of the modes in independent clauses. For the modes except for the indicative have a temporal dimension themselves. Subjunctive, optative, and imperative all refer to future time. While the application of modes in independent sentences unambiguously follows a definite scheme, the one in dependent sentences is prone to variation (within certain defined limits).

The use of the modes in subordinate clauses was formed in analogy to that in main clauses;[199] however, because of the relatedness of dependent clauses, their designated mode may change according to the mode and tense of the respective superordinate clause. The mode of the subordinate clause either may be affected by a secondary tense in the main clause and switch to optative from indicative or subjunctive,[200] or may be assimilated to the mode of the main clause, if the connection to the thought of the main clause is particularly close.[201] In both cases, it mostly remains for the writer to choose between two alternative modes. Thus, under certain conditions, finite subordinate clauses can be constructed in several synonymic ways.[202] Conversely, there are homonymic types of subordinate clauses which partly arise from the possibility of synonymous clause construction, which adds more functions to the existent modes. In these

194 Smyth §1851.
195 Smyth §1859.
196 Smyth §1862.
197 Smyth §2177.
198 Smyth §1850a.
199 Smyth §1769.
200 Smyth §2176.
201 Smyth §§2183–88.
202 E.g., in purpose clauses, it is possible to use the future indicative or the present or aorist subjunctive, or even the oblique optative, which can generally substitute the respective regular mode after secondary tenses.

homonymous types, the mode of the predicate can only be interpreted unambiguously with regard to the conjunction of the clause or even to the kind of superordinate main clause, if the combinations of conjunction and mode (and tense) of the predicate are homonymous.[203] Thus, the use of modes in subordinate clauses is to a certain extent non-standardized and prone to variation, whereas main clauses require a deliberate, definite, and unambiguous decision in this respect.

Since, then, in a highly paratactic text, the predicate of almost each clause is determinate and self-explanatory regarding its mode and tense, such a text contains a considerable degree of unambiguousness by contrast with a work dominated by hypotaxis. One might now ask why Euclid used subordination at all if he really was interested in being unambiguous above all. It appears that he uses hypotaxis wherever necessary, that is, where one thought is subordinated to another and where parataxis would obscure the strong dependency of the two thoughts. In this case, hypotaxis is more unambiguous than parataxis.[204]

Preference of finite subordinate clauses over other types of subordination

I have tried to show that the formal features of paratactic main clauses and of finite subordinate clauses might explain why the former appear a lot more often than the latter in *Elements* I. Similarly, finite subordinate clauses appear more frequently than the remaining kinds of subordination of larger syntactical units, that is, non-finite sentence constructions. I suggest that this is, again, because finite subordinate clauses are more apt to be unambiguous than the analogous participle or infinitive constructions. Not only do they occur a lot more frequently in my sample, but one can also see how they are directly preferred over the other two types.

In the case of participle constructions, this is visible from the fact that, wherever they appear in *Elements* I, they are used unexceptionally as circumstantial participles in a conditional or causal-conditional sense and perform the functions of one of the main types of subordinate clauses employed in the sample. The only type of subordinate clause

203 E.g., the optative in the *protasis* of a conditional clause can refer either to a condition in the present, if a potential optative is superordinated, or to a repeated or general condition in the past, if the main clause has an imperfect. The latter kind of optative has been called "iterative" (against this view cf. Smyth §2340a), but could more reasonably be compared to the oblique optative, for it occurs after a secondary tense and seems to substitute an indicative that is only rarely, but still used (§2342). This optative can only be interpreted by taking the superordinate clause into account. The oblique optative is the most conspicuous case of homonymous clause construction, since it occurs in many different types of subordinate clauses after secondary tenses in the main clause and is only disambiguated by the respective conjunction of the subordinate clause.

204 It is, of course to some extent, possible also for paratactic clauses to express a dependency of thoughts by means of connective particles, but the connection between main and subordinate clause is much stronger.

that is never substituted by a participle construction is the infrequent object clause introduced by ὅτι depending on forms of δεικνύναι, which could have made way for supplementary participles. This means that Euclid does not use circumstantial participles only in some senses, while covering all other senses by the respective subordinate clause. Rather, he employs both types primarily to convey a conditional sense. In view of this, it is significant that he uses one of these types almost twice as much as the other.[205]

This preference could be explained by the greater unambiguousness that finite subordinate clauses are able to produce compared with participle clauses because of their verbalized nature. They explicitly spell out all information that is presented, if at all, in a much more condensed form in the equivalent participle constructions. Most importantly, they are plain about the part of the sentence which they stand for – e.g., adverb or object – and if for an adverb, about the kind. All finite subordinate clauses (except relative clauses) are introduced by a conjunction that immediately gives away the overall sense of the subordinate clause. By contrast, the sense of a circumstantial participle must be reconstructed mostly from the context. While some circumstantial participles may give some clues about their adverbial sense by means of their tense, negation, or little words added to them,[206] the text does not even make use of these possibilities.[207] Furthermore, the mode of the predicate betrays how what is said in the finite subordinate clause is supposed to relate to reality according to the speaker. Again, in the case of participle clauses, this has to be figured out from the context. Additionally, clauses in the indicative specify the tense of the predicate, whereas the tense of participles, just like that of non-indicative predicates in subordinate clauses only specifies the aspect of the verbal action, not its temporal dimension. And finally, the subject of subordinate clauses is either mentioned explicitly in the nominative case or is directly derivable from the ending of the predicate, while the indirect subject of a circumstantial participle is not a part of that participle but must be reconstructed from its antecedent. As a result, participle constructions do not only condense information which requires unraveling, but also leave out some particulars. From a formal point of view, thus, they are underdetermined, even though an ancient reader who is used to their abundant application in Greek might not have perceived them as such. Their lack of unambiguousness might be the reason why their use in the mathematical texts is limited. In the face of the advantages of subordinate clauses, the question arises as to why participles are used at all despite their relative

205 Compare 112 finite subordinate clauses with 62 circumstantial participles.

206 E.g., the future tense indicates that the participle expresses a purpose, the negation μή suggest that the participle is used in a conditional sense, a concessive use is signaled by the little word καίπερ, while ἅτε and ὡς point to different causal senses. In the case of supplementary participles, it is not the participle but the predicate verb that indicates the supplementary use of the participle that may be substituted by a ὅτι clause.

207 This is partly because it rarely makes use of the kinds of adverbial uses of the participle that allow for such disambiguation.

ambiguity. To this, one could respond that their use is after all extremely limited both with regard to the number of participles and to the number of semantic uses in the Euclidean text, and that a Greek text which completely dispenses with participles would be very disturbing in the eyes of a Greek reader.[208]

It seems noteworthy that the text does not show a general aversion to the grammatical category of participle. In their other main function, that is, as attributes of nouns (or as nominalized adjectives, which is essentially the same thing), they feature abundantly in the *Elements*.[209] In this function, they share all features of attributive adjectives and are just as unambiguous. Interestingly, *Elements* I displays a preference for using either adverbial or attributive participle in a proof, while avoiding using both simultaneously or none at all.[210] One could interpret this situation either as a matter of stylistic variation or of stylistic disambiguation. The latter seems to be the case. Rather than simply avoiding too many participles per proof, Euclid appears to prevent participles with different syntactical functions from coming together, just as he generally seems to cultivate the repetition of items with identical function and meaning even within the same propositions while avoiding grammatical homonyms.

A similar preference for finite subordinated clauses is suggested by the rare appearance of infinitive constructions in *Elements* I. While causal(-conditional) relations are most often expressed with the help of ἐπεί clauses that are characterized by the explicitness of subordinate clauses that I have sketched out above, the rare occurrence of the causal διὰ τὸ + infinitive phrase might be due to its implicitness and indeterminacy; the infinitive lacks all the specifications of a finite predicate. The more striking case of ruling out the infinitive as a valid syntactical alternative is the above-mentioned case of a ὅτι clause after λέγω, which grammatically can be, but in Euclid is never replaced by AcI. One of the reasons presumably is again the greater explicitness and, thus, unambiguousness of the finite clause. What is more, the statement

---

208 Cf. Smyth's remark on an eventual absence of particles in a Greek text: "The absence of connectives in a language so rich in means of coordination as is Greek is more disturbing than in other languages." (§2165a). Besides, some participles appear to have their own specific functions: Acerbi 2012, 181–183 points to their use in a small number of proofs (though not in *Elements* I) within the so-called 'method of exhaustion' which is operated through the repeated subtraction of parts of a geometrical figure (182). Here, a row of circumstantial (conditional) participles attached to the subject is used to depict an infinitely recursive process. Acerbi explains this construction as a contamination of the 'language of the demonstrations,' which he defines as the dominant stylistic code of Greek mathematics, by one of the two remaining stylistic codes which he recognizes, that is, 'the language of the procedures' (181–183); cf. above, p. 149, n. 193.

209 125 times in 32 proofs. A great many of these attributive participles are aorist passive forms of διδόναι, which assign to geometrical object the status of "Givens" (on this feature of the Greek mathematical language cf. Acerbi 2011a).

210 While there are only two proofs which have neither attributive nor adverbial participles, both types feature only in 13 of the remaining 46 proofs. Conversely, those proofs that lack one of the two types, will most probably have the other one (out of 16 propositions without attributive participles, 12 have adverbial ones, and out of 22 propositions lacking adverbial participles, even 20 have the attributive type).

grammatically subordinated by ὅτι is more of a direct statement following the standardized phrase λέγω ὅτι. The seemingly dependent statement is, thus, part of the dominant paratactic scheme marked by directness, whereas an AcI after λέγω would emphasize the dependent nature of the statement.

Conclusion

To sum up briefly my analysis of the standardization of syntax in terms of the application of sentence constructions: first, the number of types of sentence constructions used is fairly limited. Second, the types of sentence constructions chosen are used in a standardized way, according to specifically reserved places within the proposition, which suggests an extremely close, even constitutive, relation of syntax and structure. The standardization of the types of sentence constructions is again strong but not perfect and complete. And last, it is the particular choice of types of sentences constructions that contributes to the impression of unambiguousness achieved by extreme standardization. The few types that are preferably used are more unambiguous to begin with than those that are – for their very ambiguity, I believe – avoided.

3.3.2 Verb forms

Now I shall focus on the standardization of verb forms. I shall start from the standardization of modes and tenses, respectively, within the proposition, then consider the standardization of mode-tense combinations, and the standardization thereof according to diathesis as well as grammatical person and number. Finally, I shall analyze the relation of the standardized verb forms to the limited number of verbs employed.

Modes and tenses

The use of the modes and tenses has been considered in detail by Fabio Acerbi.[211] However, his review of those highly standardized grammatical features is very descriptive, carving out only the forms resulting from standardization, rather than problematizing the standardization itself and explaining it. Drawing on his results, I shall survey the specifics of the modes and tenses in Greek mathematics with a focus on their standardization and avoidance of ambiguity.

In order to appreciate the extreme standardization of modes and tenses in the *Elements*, it is necessary, like in the case of sentence constructions, to start by comparing the ones used in the *Elements* to the full range which the Greek language offers.

211 2012, 170–172.

From my analysis of the 26 verb forms of my sample proposition I 47 and the 972 of my larger sample *Elements* I, it appears that the number of modes and tenses used regularly and even more so of those used frequently is very limited.

Of the four Greek modes, the one by far most frequently employed is the indicative without ἄν: 732 of 97 verb forms in *Elements* I and, accordingly, 21 of 26 in I 47 are in the indicative. Its predominance may be linked to the paratactic paradigm, since independent statements are more likely to be in the indicative than finite subordinate clauses which often require or simply tend to have the subjunctive or optative. The prevalence of the indicative strengthens the tendency of the mathematical syntax to be unambiguous. For, as we have seen above, a writer of paratactic main clauses needs to make an unambiguous decision about their mode, and if they choose indicative, they will by definition have to be explicit and unambiguous about tense, too, since the tenses have a temporal significance in indicative independent statements. The second most frequent mode is the imperative, employed 187 times in *Elements* I and five times in I 47. Although it appears much less often than the indicative, its relatively strong presence should rather strike non-mathematical readers, for it is far less common in most other genres. The remaining modes are much less present in the propositions: the subjunctive still occurs 40 times in *Elements* I, the indicative with ἄν only once, and the optative not at all.[212] Additionally, I count the infinitive as a mode where it is applied with a jussive connotation in the place of a finite verb. This is the case altogether 13 times.[213] This is to say that on average, these four modes appear extremely rarely, and that there are many Euclidean propositions in which they cannot be found at all, e.g., in my sample proof I 47.

Similarly, the number of tenses used, especially those used to a significant degree in the *Elements* is very much limited. Again, the text presents us with the absolute dominance of one type (the present tense) and the relative dominance of another type which is a lot more prominent in the *Elements* than in many non-mathematical texts (the perfect tense).[214] Even though the present tense is the most frequent, it is the perfect tense which seems to lend the text its distinctive character.[215] This impression, for one thing, is due to the unusualness of the perfect's appearing in such large numbers; for another thing, it has to do with the use of these tenses in combination with specific verbs, which produces rare and remarkable verb forms.[216] I shall go into this a little

212 Acerbi 2012, 170–172, who does not distinguish between the use of the modes with and without ἄν. See 177 on the stronger presence of the optative in pre-Euclidean demonstrations.

213 Acerbi 2012, 171.

214 Of 972 verb forms in *Elements* I, 654 are in the present, and still 157 in the perfect tense; of 26 in I 47, 19 are in the present and 4 in the perfect tense.

215 Cf. Acerbi 2012, 171: "Dà la propria impronta allo stile dimostrativo."

216 Cf. Netz's notions of the "marked repetitiveness" and "semantic markedness" of linguistic phenomena, which he uses in addition to the notion of mere repetitiveness in order to evaluate the language of Euclid in a comparative way (cf. p. 109, n. 5). The perfect forms in Euclid are 'markedly repetitive' because they occur more often than in other comparable (or simply: compared) texts, and they are se-

further below when talking about the use of the verb forms. The other tenses available in the Greek language are much less prominent (if they appear at all) than the two that I have mentioned. In I 47, there is no pluperfect, while imperfect, aorist, and future each occur only once. Regarding the larger sample, however, the future stands out from the remaining tenses: it is used 80 times in *Elements* I,[217] compared to 48 appearances of imperfect, 33 of aorist, and still none of pluperfect.[218]

Notably, the text is dominated by primary tenses, that is, present, perfect, and future tense. Considering that the propositions follow a paratactic paradigm whose prevailing mode is indicative, most of the verb forms have a temporal value, with the dominant temporal dimension being that of the primary tenses. The past tenses do appear in the main clauses, too. However, their significance in terms of their temporal value is even smaller than the number of their occurrences suggests. First, the imperfect is confined to the standardized concluding formula of the Euclidean proposition, and its presence within the overall temporal scheme is, thus, only marginal. Second, of the 33 aorist forms in *Elements* I, more than one-third (13) occur with the jussive infinitive and, thus, do not carry any temporal meaning. While one more of the 33 aorist forms occurs with the subjunctive, there are only 19 aorist forms remaining which, together with the marginalized application of the imperfect make up all occurrences of secondary tenses. Thus, the verbs of the *Elements* are highly standardized toward a primary tense paradigm.

The application of modes and tenses in Euclid, thus, is standardized with regard to the limitation of their types. Accordingly, also the number of combinations of modes and tenses is necessarily limited. Let us now consider the limitations and standardization of these combinations.[219]

mantically marked through their occurring within specific forms which have a specific, extraordinary meaning (for which see below, Section 4.3.2, 245–248).

217 Acerbi's remark: "A volte, l'indicativo futuro sostituisce il presente [. . .] ma si tratta di una variante priva di sistematicità e irrilevante dal punto di vista stilistico." (2012, 170), relates to the role of the future indicative in the *apodoseis* of the conditional clauses presenting the theorem at the very beginning of the proposition. Acerbi himself highlights the significance of the rare substitution of the present by the future indicative in this context by pointing to the semantics of the future tense in the main clause of conditional sentences at the beginning (and ending) of the proposition in comparison with the present tense (2012, 170, n. 4): "Ma si ricordi che il future rende ancora più espicita la necessità della relazione di conseguenza espressa in un enunciato [. . .]." He does not, however, mention anywhere in his analysis of the modes and tenses of the geometrical propositions that the presence of the future tense is otherwise quite significant in the language of the *Elements*. For the appearance of the future tense in the standardized course of the proposition is not confined to those conditionals, but is also prominent in the demonstrative part, both with certain verbs that standardly occur in the future tense, and with others whose appearance is not restricted to that one tense (see below, 159 with n. 233).

218 Acerbi 2012, 170–172.

219 See the survey by Acerbi 2012, 170–172.

Mode-tense combinations

Which mode-tense combinations are used in the *Elements*, and how frequently are they used? The by far most frequent hybrid is the present indicative, which accounts for 547 of 972 verb forms in *Elements* I and 18 of 26 in I 47. Since this is only a count of the actual verb forms appearing in the text, the actual number of occurrences of present indicative is actually much higher. For a large portion of the instances of present indicative in the *Elements* is in the form of the copula ἐστίν/εἰσίν, which may regularly be left out by ellipsis. In I 47, e.g., one would need to add another four instances of elliptic use of the present indicative to the 18 actual appearances of present indicative verb forms. The second most frequent combination, which, however, trails considerably behind the dominating present indicative, is the perfect imperative with 119 occurrences in *Elements* I and 4 in I 47. These two prevailing mode-tense combinations are followed by the future indicative (80 times in *Elements* I), the present imperative (68 times), the imperfect indicative (48 times), the present subjunctive (38 times), the perfect indicative (38 times), the aorist indicative (19 times), the aorist infinitive with a jussive function (13 times), the imperfect indicative with ἄν (4 times), and the aorist subjunctive (twice). To illustrate the low numerical significance of these nine mode-tense combinations within each single proposition, one may consider their occurrences in I 47: only four of them (present imperative, future indicative, imperfect indicative, and aorist indicative) appear once each, the others not at all. This means that even of the few types of mode-tense combinations that are possible because of the limited use of modes and tenses, only some occur, and of these, again only a few types dominate the text. Thus, the modes and tenses are not only standardized themselves, but also with regard to one another. As a result, the application of mode-tense combinations is standardized by the limitation of their types, and the number by which they are used.

The limitation of types of mode-tense combinations is partly implied by the limitation of types of sentence constructions. For considering the dominance of paratactic main clauses, the occurrence of combinations such as aorist/present subjunctive with ἄν or aorist/present optative without ἄν, whose use is limited to subordinate clauses, cannot be very frequent to begin with. The fact that the oblique optative is, in fact, completely absent, can be explained by the predominance of primary tenses in the superordinate main clauses. Further, because of the limitation of types of subordinate clauses, some combinations like those with subjunctive without ἄν which are restricted to purpose clauses do not occur (because the *Elements* does not make use of purpose clauses).

Let us now explore the use of the nine mode-tense combinations of the *Elements* regarding their places and functions within the propositions,[220] starting with the type that occurs most frequently. The overall predominance of present indicative is mainly due to the fact that it is the dominant mode-tense combination of the demonstration

220 See Acerbi 2012, 170.

part of the proposition, which, in turn, is the core part of each proposition which also dominates it in quantitative terms. Acerbi summarizes its main function in the demonstrative part as follows: "esprime la semplice presa d'atto di una successione, ordinate logicamente, di stati di cose relative agli oggetti matematici introdotti nell' esposizione e nella costruzione."[221] Another function it can have in the demonstrative part is within the *protasis* and *apodosis* of conditionals which introduce demonstrations by contradiction.[222] The second part of the proof in which the present indicative regularly figures is the beginning and ending of theorem propositions.[223] Here, it has the same function as in the demonstrative part, if the theorem is phrased as an assertion;[224] if it has the form of a conditional clause, the present indicative forms part of the *apodosis* where it expresses the regular (that is, iterative) implication of a given condition. Thirdly, the present indicative is regularly present in the metamathematical[225] first person λέγω and in the third person δεῖ which standardly introduce the definition of the goal in theorems and problems of construction, respectively, as part of the phrases λέγω ὅτι ("I say that") and δεῖ δή ("Thus it is required to"). Finally, in dependent clauses, the present indicative figures after ὅτι and in relative clauses that mostly set out more specific conditions of the objects posited.

The use of the second most frequent mode-tense combination (perfect imperative) is restricted to directions on how the geometrical objects 'should be constructed.'[226] These construction-related imperatives appear mostly in a section of the proposition exclusively dedicated to the construction of the objects. But they may also feature in the demonstrative part following upon the section of the construction, in case further objects are introduced which 'should be constructed' in the course of the demonstration.[227] The perfect imperatives, thus, though they are concentrated in one part of the proposition, may be located at other stages of the proof, too. However, as we shall see below, their function standardly relates to the concept of the construction of geometrical objects.[228] This could indicate that the mode-tense combinations are not necessarily standardized

221 Acerbi 2012, 170.

222 Acerbi 2012, 170. In some of those propositions which contain a demonstration by contradiction, the conditionals in question may appear not only in the demonstrative part, but already in the preceding construction part, but, as I have pointed out above [. . .], in proofs working by contradiction, construction and demonstration are hardly distinguishable, since the construction (which is to turn out impossible) is already part of the demonstration.

223 The beginning and ending of a theorem proposition can be viewed as one part, since the ending mostly repeats verbatim the beginning (before concluding with a standard three-word formula); see above, Section 1.2, 23 f.

224 Acerbi 2012, 170.

225 See Netz 1999a, 137–139 for his "second-order formulae" and above, p. 126.

226 I deal with these imperatives and the meaning of the perfect more comprehensively below in the context of 'diagrammatization' (Section 4.3.2, 245–248).

227 Acerbi 2012, 171.

228 P. 165 f.

with regard to proposition parts but with regard to functions within the proposition that may extend across proposition parts, but this again raises the question of how exactly we define 'proposition part' which I will discuss below.[229]

The same strict standardization characterizes the use of the less frequent mode-tense combinations. The greatest number of occurrences of the future indicative can be found in the demonstrative part of the propositions where it appears altogether 46 times and assumes two main functions. Most of the time, it substitutes the present indicative in statements about the objects that *will be* true under certain conditions. In fact, these future indicative forms make their most prominent appearance in propositions I 4 and 8 (12 and 10 occurrences, respectively), that is, in two of the fewer demonstrations which are not structured by ἐπεί, but introduced by genitive absolutes, to which the clauses with the future indicative form direct or, in the course of the argument, indirect *apodoseis*.[230] From the frequent appearance of the future indicative in I 4 and 8 (22 times, which leaves only 15 more occurrences in *Elements* I), one can see that this first function of future indicative within the demonstrative part is less prominent in the remainder of *Elements* I, since the overall number of occurrences of this type is not so large. The second function of future indicative in the demonstrations is as part of the two standardized metamathematical statements which indicate toward the end of the demonstration that one more deduction which is necessary for accomplishing the proof can be made in analogy to the previous one.[231] The third more regularly appearing type of occurrence of the future indicative is at the beginning and ending of roughly one third (14) of the propositions of *Elements* I, for the most part theorem propositions,[232] of which the book has 33 in total. It is the most usual mode-tense combination in theorem enunciations that take a conditional form, appearing in 12 of 20 of them, and the standard mode-tense combination in the main clause of those conditional enunciations which start with an ἐάν clause, occurring in all but three of 14.[233] Finally, there remain four occurrences in the ὅτι clause after λέγω, where the future substitutes the usual present tense.

---

229 See Section 3.4.1, 179–181 on Netz's distinction of understanding proposition parts as "stylistic" or "conceptual units" and on the related question of the standardization of the proposition structure.

230 Notably, both demonstrations include proofs by contradiction. In two other demonstrations (of I 27 and 29), two or one main clause, respectively, in the future tense follows upon *participia coniuncta* rather than genitive absolutes.

231 ὁμοίως δὴ δείξομεν ("similarly we can prove," cf. *Elem.* I 14, 17, 20, 39, 40 [vol. 1, pp. 23.4, 26.14, 29.4, 53.14, 54.14 Heiberg & Stamatis]) and ὁμοίως δὴ δειχθήσεται ("similarly it can be proved that," cf. *Elem.* I 15, 16, 27, 47 [vol. 1, pp. 24.6, 25.20 f., 39.10, 64.18]).

232 It also appears in I 7, which is a special case falling out of the dichotomy 'problem vs. theorem proposition,' for it demonstrates the impossibility of a problem of construction in the standard form of a theorem proposition, stating the impossibility of the construction problem as the theorem to be proven.

233 This means that the occurrence of future indicative in the *apodosis* of conditional clauses in the enunciation and conclusion of the proposition is not an insignificant stylistic variation of the standard application of the present indicative, as Acerbi has it (see p. 156, n. 217); to the contrary, it is – at least

The appearance of present imperatives is almost exclusively restricted to the exposition part of the proof which sets out the objects relevant for the proof. Here it is by far the dominant mode-tense combination; only occasional relative clauses (in the indicative) that are attached to the imperative ones present an exception to this rule. The present imperative, which, by contrast with the perfect imperative, is always active in *Elements* I, appears with εἶναι in most cases (51 times in 42 proofs). The resulting forms ἔστω and ἔστωσαν ("Let there be") mostly appear at the place which follows the enunciation of the theorem and precedes the definition of the goal by means of λέγω or δεῖ, but at times also only after the definition of the goal in the middle of the construction. Either way, they have the function of setting out the objects or certain characteristics of them. For the present imperatives of εἶναι are so closely linked with the concept of setting out the objects that they indicate the *concept* of the exposition of the objects wherever they appear.[234]

Apart from these imperatives of εἶναι, there are 12 occurrences of present imperative with other verbs. All of them are verbs which say something about the relation of the geometrical objects or about what they do to each other (ἔχειν, 'to have;' ποιεῖν, 'to make;' ἐμπίπτειν, 'to fall on;' συμπίπτειν, 'to meet;' τέμνειν, 'to cut'). Nine out of these twelve imperatives, like the majority of the ἔστω/σαν instances, appear in the part before the definition of the goal where they substitute or complement the exposition by means of ἔστω/σαν.[235] These verbs do not merely posit geometrical objects, but posit them by giving additional information about them.

Of the three remaining present imperatives which appear *after* the definition of the goal, that is, at the usual place of the construction, one is again expositional,[236] occurring in a sentence with γάρ (which, as I will show later, usually marks expositional sentences without ἔστω/σαν)[237] and nowhere near perfect imperatives denoting construction. The other two occur together with perfect imperatives denoting the con-

---

as far as *Elements* I is concerned – just the other way around: the use of the present tense is marginal in these cases.

234 See below, Section 3.4.1, 179–181 on Netz's distinction of "conceptual" and "stylistic" parts and on the standardization of the structure of the proof.

235 The verb forms are: ἐχέτω ("let have," *Elem.* I 8, 26, 41 [vol. 1, pp. 16.7, 36.14, 55.2 Heiberg & Stamatis], of two triangles that should have equal bases [I 8] or one side equal to one side [I 26], and of a parallelogram that should have the same base with a triangle [I 41]), ποιείτω ("let make," *Elem.* I 13, 27, 28 [vol. 1, pp. 21.5, 39.3, 40.8 Heiberg & Stamatis], of a straight line that should make two angles [I 13, 27], of a straight line that should make an angle equal to another one [I 28]), ποιείτωσαν ("let make," *Elem.* I 14 [vol. 1, p. 22.14 f. Heiberg & Stamatis], of two straight lines that should make two angles), τεμνέτωσαν ("let cut," *Elem.* I 15 [vol. 1, p. 23.13 Heiberg & Stamatis], of two straight lines), ἐμπιπτέτω ("let fall on," *Elem.* I 29 [vol. 1, p. 41.11 Heiberg & Stamatis], of a straight line that should fall on two other straight lines).

236 Ἐμπιπτέτω ("let fall on," *Elem.* I 30 [vol. 1, pp. 42.21 Heiberg & Stamatis], of a straight line that should fall on three other straight lines).

237 See below, Section 3.3.3, 172–174.

struction of geometrical objects.[238] Even so, they do not seem to be used *pro perfectis* in a constructional sense, but, like the instances of the cognate ἐμπίπτειν, to add an expositional aspect, only at a later stage of the proof. Acerbi, by contrast, regards all present imperatives in the *Elements* as exceptions from the perfect-imperative paradigm of the exposition and construction, which he seems to consider as a unit.[239] He argues that within this "ambiente di perfetto," the present imperatives ἔστω/σαν are an "exception," to which one needs to assign a "'presenziale,'"[240] while the remaining present imperatives must be understood as "praesen[tia] pro perfect[is]."[241] By contrast, I argue that it is precisely the tense which distinguishes the mere postulation of the objects and of their relation to each other from their full construction – apart from diathesis: all present imperatives are active, all perfect imperatives passive.[242] This hypothesis is supported by the fact that there is almost no lexical overlap between the two groups of present and perfect imperatives. The only two verbs to appear in both forms are ἐπιζευγνύναι ('to join to') and τέμνειν ('to cut'). Generally, the verbs used for perfect imperatives describe activities to have been done by the mathematician drawing the diagram, while those applied in the present tense are more descriptive in character, even though they, ironically, are active, demanding what features the objects should have or what they should 'do' to one another.[243] Thus, the specific function of all *present* imperatives is to accomplish the exposition of the geometrical objects,[244] whether they appear before or after the definition of the goal:[245] they are 'exposition imperatives' as opposed to 'construction imperatives.'

The occurrence of the imperfect indicative without ἄν is confined to the very last sentence of the propositions, where it appears without exception, that is, 48 times in *Elements* I. It standardly confirms that the result of the demonstration is "what *was to*

238 Both times συμπιπτέτωσαν ("let meet," of two straight lines, *Elem.* I 27, 44 [vol. 1, pp. 39.6, 59.8 Heiberg & Stamatis]).

239 2012, 171. I, too, shall argue that the function of the exposition and the construction is essentially the same as far as the drawing of the diagram and the accomplishment of the proof on basis of it are concerned, but that it is precisely for this reason that the linguistically standardized differentiation between the two (which I am about to describe) is remarkable and in need of explanation. In Section 5.1, I shall explain it as an instance of unnecessary formal standardization (p. 418).

240 2012, 170.

241 2012, 171.

242 Generally on diathesis as one means of the standardization of the verb forms in the *Elements*, see below, pp. 163 f.

243 On the impersonality of the construction imperatives cf. below Section 4.3.2, 245–248.

244 For this reason, it seems wrong either to classify the most frequent of present imperatives in the *Elements* – ἔστω – as part of a 'construction formula,' as Netz does (1999a, 137), or to regard all uses of present imperative as a violation of the perfect rule, as Acerbi does (see above). For the present tense has its own function when used with the imperative, namely to perform the exposition of the objects.

245 See below Section 3.4.1, 178–182 on the problem of distinguishing 'parts' of the proposition.

be proven" (ὅπερ ἔδει δεῖξαι). The imperfect with ἄν only occurs four times in two proofs, each time within a demonstration by contradiction.[246]

The present subjunctive with ἄν regularly appears in propositions which begin and end with a conditional with ἐάν (36 times in 13 proofs); only in proposition I 21, the aorist subjunctive is used instead of it.[247] Besides, the present subjunctive occurs only scarcely, that is, twice in ὅταν clauses in the demonstrative part of propositions I 11 and 12.

The perfect indicative has two main functions in the Euclidean proofs: first, it standardly occurs at the end of all problem propositions which state that the respective problem of construction 'has been' solved.[248] Here it corresponds with the aorist infinitive at the starting sentence of the respective proof, which has the standardized function to announce at the head of the problem proposition what one 'should construct.' Second, it appears 15 times in the demonstrative part, where it serves to denote what can be said of the geometrical objects as a result of their foregoing construction. Besides, it is the most common mode-tense combination in those rare instances of ὅτι clauses after λέγω which appear in problem propositions. In the latter, the definition of the goal is regularly introduced by δεῖ δή instead of λέγω ὅτι, but in six of them, a λέγω ὅτι sentence is added to the δεῖ δή sentence, and in five of these six cases, the perfect indicative is used.[249]

The aorist indicative exclusively occurs in the demonstration part. Its standard function, appearing 14 times in *Elements* I, is to introduce a metamathematical remark, which calls to mind a fact that "has been demonstrated" (ἐδείχθη/σαν δέ) at an earlier stage of the demonstration or book.[250] There remain only four other instances of aorist indicative within demonstrations, three of them with the construction verb ἐμπίπτειν and one in the composite form ἐστιν ἀναγραφέν ("is described [sc. on a straight line]," of a square), which seem to be unsystematic uses.[251]

246 *Elem.* I 19, 25 (vol. 1, pp. 27.22, 28.2, 35.15, 18).

247 *Pace* Acerbi (2012, 170), who claims that the present subjunctive is used in the ἐάν clauses of the beginning and ending only with the verb εἶναι, while the aorist subjunctive is "obbligatorio per verbi diversi da 'essere,'" the present tense is indeed the standard form in our sample of *Elements* I, occurring with all kinds of verbs; the forms that appear are: 16 times ἔχῃ ("has"), twice ὦσιν ("are"), six times ποιῇ ("makes"), twice ποιῶσιν ("make"), twice τέμνωσιν ("cut"), and four times (ᾖ), while the aorist subjunctive does not appear once with εἶναι, but in total only twice in the form συσταθῶσιν (*Elem.* I 21 [vol.1, pp. 29.10, 30.19]).

248 That is, 14 times in propositions I 1, 2, 3, 9, 10, 11, 12, 22, 23, 31, 42, 44, 45, 46. Again, I 7, the 'problem proposition in disguise' (see p. 159, n. 232) falls out of alignment: it has the future indicative that is regularly used in theorem propositions.

249 *Elem.* I 9 (τέτμηται, "has been bisected"), 10 (τέτμηται, "has been bisected"), 11 (ἦκται, "has been drawn"), 12 (ἦκται, "has been drawn"), 22 (συνέσταται, "has been constructed"); vol. 1, pp. 17.21, 18.15 f., 19.20, 20.27, 31.16 Heiberg & Stamatis.

250 Acerbi 2012 does not mention this prominent function of aorist indicative.

251 Ἐνέπεσεν ("falls upon," of a straight line: *Elem.* I 44, 45, 46 [vol. 1, pp. 59.3, 61.2, 62.16 Heiberg & Stamatis]; ἐστιν [. . .] ἀναγραφέν: *Elem.* I 47 (vol. 1, p. 65.2 f. Heiberg & Stamatis). Acerbi 2012, 171 points to these instances, but does not offer any explanation for them.

Thus, the nine mode-tense combinations that we have identified in *Elements* I are extremely standardized with respect to their use within the proposition. They perform specific functions at specific stages in the presentation of the proof and, thus, occur in set parts and sentences within the proposition. Even the seemingly omnipresent present indicative does not provide an exception to this rule. The use of the mode-tense combinations according to these functions is indeed one hundred percent standardized in some cases, in others it is not completely standardized but nevertheless extremely fixed so that it is expectable and recognizable from a reader's point of view. Because of their standardized recurrence at specific stages of the proof, the mode-tense combinations acquire a firm meaning which is an important and meaningful part of the respective statement. That is, the meaning of a particular predicate is not only defined by the verb that has been used, but to a considerable degree also by the mode-tense combination in which it appears. While this may seem to be true in general of verb forms in Greek, in which the system of modes and especially tenses with their range of temporal and aspectual connotations is extremely nuanced, it is especially noteworthy with regard to the mathematical language because of the standardized use of mode-tense combinations. They are never used randomly, idiosyncratically, or spontaneously, but always carry the meaning of the respective part of the proof.

Diathesis

The standardization of the verb forms extends even further since it includes also the standardized use of diathesis. For most of the mode-tense combinations appear only either in the active/medium[252] *or* in the passive voice. Thus, the present indicative, the present imperative, the imperfect indicative (with and without ἄν), and the aorist infinitive are used in the active voice only,[253] whereas the perfect imperative and the aorist subjunctive are used exclusively in the passive voice. The remaining mode-tense combinations are indeed used both in the active *and* in the passive voice, but the functions of each mode-tense combination that I have distinguished above, again, are confined to either active or passive. Of the perfect indicative, for example, the instances in the conclusion of problem propositions and in the ὅτι clauses of the definition of the goal occur exclusively in the passive, while the occurrences within the demonstrative part are almost all active (with the exception of two instances of ὑπόκειται, "is assumed as hypothesis"). It is only the future indicative which is used less systematically in relation to diathesis: of the two future forms of the metamathematical δεικνύναι, one is active (δείξομεν, "we can prove"), the other one passive (δειχθήσεται, "it can be proved"), and the remaining forms of future indicative in the demonstrations and those at the

252 Most verb forms which are not passive are active; only of some verbs, the medium is used because of the special meaning of the medium form.

253 Only the aorist infinitive is used five out of 14 times with the medium συστήσασθαι ("to construct").

beginning and ending of the propositions are active with the exception of three occurrences of the form (οὐ) συσταθήσονται ("they cannot be constructed"). One can see that the application of diathesis can be related to the use of the modes and tenses and is, in this regard, fairly standardized, while exceptions to the rule are confined to specific verbs (especially συνίστασθαι, "to construct").

Grammatical person and number

Apart from mode, tense, and diathesis, also grammatical person and number serve as means of the standardization of verb forms in the *Elements*. The text uses only third and first person. The predicates of all *mathematical* statements in the *Elements*, whether expositional, constructional, or deductive, are in the third person. Whether they are singular or plural depends on the implied number of geometrical objects. Metamathematical statements appear both in the first and in the third person,[254] but the first person occurs only with two of several standardized metamathematical remarks: one of them is plural (δείξομεν, "we can prove"), the other one singular (λέγω, "I say"). Thus, the use of diathesis on the one hand and of grammatical person and number on the other complements the standardized application of the modes and tenses within the different parts of the proof. The standardization is not complete in the sense that the overall third-person paradigm is interrupted by a few first-person forms and that not all metamathematical statements are made in the first or third person. Similarly, the use of first-person forms varies between singular and plural. However, the individual uses of first and third person as well as of first-person singular and plural are completely standardized. It is especially the standardization of the first-person forms which stands out because they are so rare compared with third-person forms in the *Elements*. In this way, they contribute even more to the general impression of standardization. Their standardization is especially remarkable because it raises the question as to the meaning of the first person: in fact, it is because of their standardization that the first-person forms are just as impersonal as the third-person forms.

Grammatical verb forms and specific (classes of) verbs

As we have seen, the application of the mode, tense, diathesis, person, and number of verbs in the *Elements* is highly, though not completely standardized. In order to be able to estimate the significance of this grammatical standardization for generating meaning in the mathematical proposition, one needs to examine the use of the stan-

254 Examples for third-person forms are: δεῖ ("it is required"), ἔδει ("it was required"), ἐδείχθη ("it was proved"), δειχθήσεται ("it can be proved").

dardized grammatical verb forms in relation to the use of the verbs as lexical units. The restriction of certain verb forms to specific verbs and vice versa has already been hinted at a couple of times and deserves further consideration.

On the one hand, the standardized forms of the verbs each occur only with a limited, fairly small number of verbs. In *Elements* I, for example, the first-person singular present indicative active is used exclusively of λέγειν (λέγω, "I say"); the third-person aorist indicative passive only of δεικνύναι (ἐδείχθη/σαν, "it was proved"); the third-person aorist indicative active only of ἐμπίπτειν (ἐνέπεσεν, "falls upon"); the third-person present imperative active of εἶναι (ἔστω/σαν, "let there be") and of few verbs that are used to describe how geometrical objects are related to one another or what they do to each other, that is, ἔχειν (ἐχέτω, "let have"), ποιεῖν (ποιείτω/σαν, "let make"), συμπίπτειν (συμπιπτέτωσαν, "let meet"), ἐμπίπτειν (ἐμπιπτέτω, "let fall on"), ἐπιζευγνύναι (ἐπιζευγνύτωσαν, "let join to"), and τέμνειν (τεμνέτωσαν, "let cut");[255] and the third-person perfect imperative passive solely of construction verbs. The number of the verbs with which one verb form is used is low for two reasons: first, because of the "one-concept-one-word principle,"[256] which results in a small vocabulary to begin with, and second, because of the restriction of verb forms – if not to one single verb – to semantically defined *groups* of verbs like, for instance, construction verbs.

On the other hand, as a result of the lexical limitation, the verbs or groups of verbs to which certain grammatical forms are limited, are mostly not themselves limited to these verb forms, but appear with a number of mode-tense combinations. For example, the present indicative appears most often with the verb εἶναι, that is, in the third-person forms ἐστίν and εἰσίν. The use of εἶναι, however, is not confined to the present *indicative*, but it occurs prominently in the third-person present imperative ἔστω or ἔστωσαν. Admittedly, εἶναι is a special case, since its two uses in the *Elements* coincide with its two main syntactical functions, that is, as copula and main verb. What is important, however, is that its two appearances relate to two very distinctive kinds of forms, which a reader can easily identify with the respective function of εἶναι in the proposition. Because of the restriction of the verb to these two kinds of forms and their standardized use regarding a specific function in the course of the proof, the two forms of the verb εἶναι – ἐστίν/εἰσίν and ἔστω/σαν – do not just appear as different forms, but even as different *words*. Here is where the level of syntax overlaps with that of lexicon, because the grammatical form is just as decisive for the meaning of the verb form as is the actual lexical unit.

Something similar can be observed by comparing the different uses of construction verbs within the proposition. Their usual place is, of course, in the construction part, where they appear as third-person perfect passive imperatives. In most cases,

---

255 On the ambivalence of ἐπιζευγνύναι and τέμνειν which are used both as construction verbs and as verbs describing the relation of geometrical objects, and with the concomitant verb forms, cf. above p. 161.
256 Cf. above, Section 3.2.1, 116.

the construction part follows the definition of the goal, but, as we have argued above, the construction is expressed by construction verbs in the form of third-person perfect passive imperatives wherever they occur, which may be the case also at other places of the proposition.[257] While the regular appearance of construction verbs in theorem propositions is indeed confined to the imperatives demanding the step-by-step construction of the geometrical objects, they regularly appear in two more forms in problem propositions which are concerned with a specific problem of geometrical construction and make up roughly one third of all propositions of *Elements* I. Here, apart from appearing as third-person perfect passive imperatives, they occur, first, as aorist active infinitives at the beginning of the proposition, where the infinitive has an imperative force, but is never replaced by an imperative; second, they take the form of third-person perfect passive indicatives, by which the request of the beginning is rephrased at the end in terms of an accomplishment. Apart from these more common uses, construction verbs appear only a few times as perfect passive indicatives in the unusual instances of λέγω-ὅτι clauses in problem propositions. This means that the regular syntactical uses of construction verbs are threefold. Again, these different uses are realized by means of only three distinct forms.[258]

Let us take as an example the common verb ἄγειν, which in Euclid's *Elements* always has the technical sense 'to draw (a line/perpendicular).' It occurs 25 times as a finite verb (or in the place of a finite verb) in *Elements* I. These 25 occurrences are confined to only three different forms, that is, to the very forms corresponding to the three main functions of construction verbs which I have just mentioned: the aorist active infinitive ἀγαγεῖν ("to draw [a line]," three times in the starting sentence), the perfect passive indicative ἧκται ("has been drawn," three instances in the final section, two in λέγω-ὅτι sentences of problem propositions), and, most prominently, the perfect passive imperative ἤχθω ("let [a line] be drawn," 17 times in the construction part). Like in the case of εἶναι, the restriction to very few forms and the repeated use of these few forms does not make one perceive so much the presence of the lemma word ἄγειν, but rather of the specific forms ἀγαγεῖν, ἧκται, and above all ἤχθω.[259]

---

257 See Section 3.4.1, 179–181 Netz's distinction of "conceptual" and "stylistic" parts.

258 Or five, if one takes the singular-plural difference of the finite forms into account, but this difference is not essential to my present argument.

259 Admittedly, in the rest of the *Elements*, one encounters a few more finite forms of ἄγειν, first of all the plural imperative ἤχθωσαν ("let [several lines] be drawn," e.g., *Elem.* II 1 [vol. 1, p. 68.6 Heiberg & Stamatis]), and the rare subjunctive forms ἀγάγωμεν ("[if] we draw," *Elem.* IV 15 [vol. 1, p. 177.7 f.], IV 16 [vol. 1, p. 179.2 f.], XII 2 [vol. 4, p. 81.4, 14], XII 10 [vol. 4, p. 106.12 f. Heiberg & Stamatis), ἀχθῇ ("[if a line/perpendicular] is drawn," *Elem.* III 19 [vol. 1, pp. 121.23 f., 122.17 f.], VI 2 [vol. 2, pp. 43.13 f., 44.9 f.], VI 8 [vol. 2, pp. 54.22 f., 56.15 f.], X 32 [vol. 3, p. 54.17 f. Heiberg & Stamatis]), and ἀχθῶσιν ("[if perpendiculars] are drawn," *Elem.* XI 35 [vol. 4, p. 66.5–10 Heiberg & Stamatis]). Here is, thus, another use of construction verbs: they may occur in the protasis of conditionals at the beginning of propositions or at different places in the proof part, e.g., in the references to possible analogous proofs, whose func-

Here, again, is where syntax and lexicon overlap. For if one wanted to memorize the lexicon of *Elements* I, it would neither be useful to remember solely the lemma word ἄγειν, nor to know the whole paradigm of this verb, but it would be effective to learn precisely three forms. This may seem to be spoken too much from the point of view of the modern, non-native learner, but in fact, at least the two perfect passive forms of construction verbs will not have been very familiar to an ancient reader either. The form ἦκται, for instance, occurs only 180 times in the whole corpus of extant Greek literature. More than one quarter of all occurrences (54) are in Euclid, and another 80 in other mathematical writers, most of whom are later than Euclid,[260] while it appears only twice in contemporary mathematicians.[261] The remaining occurrences for the most part either stem from late antique linguistic (20)[262] or Christian (13)[263] texts. This leaves us with just eleven appearances in other authors, of which only four date from before Euclid.[264] This is to say that the use of the form ἦκται is confined to later technical, above all mathematical texts and that it should have been extremely uncommon to the lay reader and maybe even to the mathematical reader in Euclid's own time.[265]

Since each of these peculiar forms is connected with a distinctive act in the course of the description of the proof, they do not only stand out as distinct lexical units from a formal point of view, but also because each of them assumes a specific meaning which is not restricted to the lemmatical one. This becomes particularly apparent where several forms of the same construction verb occur within one and the same proof. In I 10, for instance, τέμνειν ("to cut [in half a straight line]") occurs in three different shapes according to the three different functions it assumes:[266] in the form τεμεῖν in the introductory

tion it is to lay down a condition for the hypothesis to be proven. This condition may be a construction that is necessary to be undertaken before the actual proposition starts.

260 16 occurrences in Apollonius of Perge, ten in Theodosius, two in Hero, three in Theon, 17 in Pappus, five in Serenus, one in Proclus, eleven in Eutocius, two in Olympiodorus, and 13 in the scholia on Euclid.

261 In Ptolemy and Autolycus.

262 One in Pseudo-Zonaras, one in Photius, one in Eusthatius, two in Choeroboscus, three in the *Etymologicum Geminum*, three in the *Etymologicum Gudianum*, four in the *Etymologicum Magnum*, and six in the *Etymologicum Symeonis*.

263 One in the Septuagint (Isaias), three in authors referring to the Septuagint passage (Eusebius, Procopius, Cyrillus), one in Epiphanius, one in the *Chronicum Paschale*, one in Ioannes Damascenus, and two in Nicophorus I.

264 Two each in Herodotus and Plato.

265 Here we enter the field of pre-Euclidean mathematical literature that is no longer extant. The question is how far Euclid has not only codified the mathematical knowledge of his time, but also the mathematical language.

266 While the use of construction verbs proper in the *Elements* is indeed confined to the three verb forms mentioned above, τέμνειν is (like ἐπιζευγνύναι) an exceptional case, as it can work both as a verb denoting the construction of geometrical objects by the geometer, and as a verb describing what geometrical objects do or should do to each other. In the first case, it takes one of the three standardized forms, in the second, it appears predominantly in the present tense (see above, p. 161).

sentence of the proof ("to cut [in half a straight line]!"),[267] as τετμήσθω in the construction part ("[a straight line] should be cut [in half]"),[268] and in the shape τέτμηται ("[a straight line] has been cut [in half]") both in the conclusive section and in what is one of the rare instances of a λέγω-ὅτι clause in a problem proposition.[269]

'Forms of verbs'

We have seen that the use of the verb forms is extremely restricted and standardized to the point that even the individual forms of specific verbs themselves, not only general grammatical features of the verb are repeated. It is not primarily verbs or modes that are standardized, but rather individual *words* which I would like to call 'forms of verbs.'[270] This perception is enhanced by the uncommonness of many of the verb forms in discourses other than Greek theoretical mathematics. Especially the perfect forms are striking in the eyes of the lay reader and recognizable as standardized words.[271] For example, compare a form like ἐπεζεύχθω/σαν, appearing 30 times in *Elements* I, or the omnipresent present imperative ἔστω/σαν. As we have seen, both the verb ἐπιζευγνύναι and εἶναι do not only occur in the form of third-person perfect passive imperatives or third-person present active imperatives, respectively; nor are the former restricted to the verb ἐπιζευγνύναι or the latter to εἶναι. However, because of their standardized repetition, these unusual verb forms are striking, recognizable, and memorable as lexical items of their own.

3.3.3 'Formulae' reconsidered: the use(s) of ἔστω/σαν

Now, having analyzed the standardization of the mathematical syntax, I shall come back to the question whether the *Elements* has been written by generating and using formulae or whether its formulaic nature is a corollary of something else. Netz's 'construction formula' 23[272] "ἔστω {any unlettered object formula} {a lettered equiva-

267 *Elem.* I 10 (vol. 1, p. 18.1 Heiberg & Stamatis).

268 *Elem.* I 10 (vol. 1, p. 18.6 Heiberg & Stamatis). On the meaning of the perfect tense cf. below Section 4.3.2, 245–248.

269 *Elem.* I 10 (vol. 1, p. 18.8, 15 f. Heiberg & Stamatis). It can take the same form here as at the end of the proposition, because it expresses the same idea in both cases, that is, the result of the construction as a fulfilment of the goal of the proposition; only in the λέγω-ὅτι clause it is part of an hypothesis imagining the goal as fulfilled, whereas at the end, it confirms the fulfilment of the goal right after the demonstration has proven that it has indeed been fulfilled.

270 These standardized 'forms of verbs' seem to be 'formulaic;' in the following section, in which I reconsider Netz's concept of "formulae," I shall argue that the standardization of 'forms of verbs' is precisely the effect of the strict standardization of lexicon and syntax.

271 Cf. Netz's notion of "marked repetitiveness" (p. 109, n. 5).

272 Netz 1999a, 137.

lent},” which I have mentioned when I first introduced Netz’s concept,[273] shall serve again as an example.

Netz’s formula “ἔστω {any unlettered object formula} {a lettered equivalent}”

It can be found in this very form in 37 of the 48 proofs of *Elements* I, with the introductory third-person imperative of εἶναι (ἔστω/ἔστωσαν)[274] governing two noun phrases in the nominative case, first the lettered and second the unlettered variant of the same object formula. This fixed order makes the phrase conspicuous and recognizable in spite of the fact that only its first element is verbatim the same, whereas the other are variables which are fixed only in kind.

The tripartite phrase may be extended by the single or double repetition of the sequence of the second and third element,[275] or, as in the case of I 47, by what Netz calls ‘predicate formulae:’ Ἔστω τρίγωνον ὀρθογώνιον τὸ ΑΒΓ ὀρθὴν ἔχον τὴν ὑπὸ ΒΑΓ γωνίαν· (“Let there be a right triangle ΑΒΓ having right the angle ΒΑΓ.”).[276] Here, the ‘predicate formula’ ὀρθὴν ἔχον (“having right”)[277] depends on the second ‘object formula’ and is in turn superordinated to another ‘object formula’ (τὴν ὑπὸ ΒΑΓ γωνίαν, “the angle ΒΑΓ”).

Wherever this phrase occurs in the *Elements*, it has the same meaning, that is, it is not homonymous: it posits a certain kind of geometrical object x by the concrete name y. Having this standardized meaning, it performs a standard function within the proof, that is, to name both the kind and the diagrammatic denotation of the main geometrical objects with which the proof is going to deal. Consequently, the phrase standardly occurs at the same place in each proof which is just after the general enunciation of the claim, introducing the part of the proof which sets out its concrete scenario.[278]

No doubt this tripartite ‘standard phrase,’ as I shall call it in view of Netz’s formula, is standardized both in form and content in the sense that it occurs like this *wherever it occurs*. This is, of course, a circular argument since its standardization can only be judged by taking into account also its ‘non-occurrence.’ The cases of ‘non-

---

273 That is, in Section 3.1, pp. 128 and 131 on the lexicon.

274 In the following, I shall mean the phenomenon that is represented both by the singular and by the plural form of the imperative, even when mentioning only the singular.

275 E.g., in six proofs of *Elements* I (2, 11, 12, 31, 42, 45 [vol. 1, pp. 8.13 f., 18.20 f., 20.1 f., 43.13 f., 56.1 f., 60.4–6 Heiberg & Stamatis]), the imperative ἔστω/σαν announces two different objects that are each referred to by the sequence ‘abstract object + concrete object’ and juxtaposed by μὲν . . . δέ; in two proofs (23, 44 [vol. 1, pp. 32.15–17, 58.13–15 Heiberg & Stamatis]), even three different objects organized by μὲν . . . δὲ . . . δέ follow upon the imperative; in I 43 (vol. 1, p. 57.7–10 Heiberg & Stamatis), ἔστω is repeated before the second repetition of the original second and third element.

276 *Elem.* I 47 (vol. 1, p. 63.8 f. Heiberg & Stamatis).

277 Netz does not mention this formula explicitly; but cf. his (predicate) formula 86 ὀρθή (1999a, 144, n. 131).

278 Netz 2004b, 6.

occurrence,' however, are difficult to count for obvious reasons. I shall first look at all those instances in which only part of the 'standard phrase' phrase occurs, namely its definitive element ἔστω (that is, even at those that occur outside the standard place of the 'standard phrase'), and second at the location where the whole phrase otherwise standardly occurs (the beginning of the proof proper just after the general enunciation), but seems to be replaced completely by something else.

Alternative use(s) of ἔστω

An analysis of the alternative uses of the constitutive invariant of the phrase, that is, the imperative ἔστω, may shed a different light on the supposed standardization of the phrase. Indeed, the use of ἔστω is by no means confined to the Netz's formula. It occurs altogether 12 times: either roughly in its usual place after the general enunciation of the proposition's claim, but not within the form of the supposed 'standard phrase' (I 30, 41, and 48); or in the exposition just after the main objects of the proposition have been introduced by means of the standard phrase, in a relative clause or in an independent main clause characterizing (part of) the concrete object mentioned in the standard phrase (I 3, 22, and 24–25); or in the construction part of the proof which introduces further (parts of) objects (I 6, 14, 26 [2x], and 29).[279] As far as their mathematical content is concerned, all of these are occurrences within clauses which posit particular qualities of certain parts of objects or relations between objects. Syntactically, the difference between the use of ἔστω as part of the 'standard phrase' and in the alternative constellations comes down to the difference between the two main uses of εἶναι, i.e., as a main verb (in the formula) and as copula (in all other occurrences), and is a minor one.[280]

Following Netz's formalization, however, one would need to invent a whole new formula to cover the core of the 12 clauses. It would be a combination of ἔστω, 'object,' 'predicate formula,' with the order and number of the elements being much less fixed than in the supposed formula we have been looking at. Cf., e.g., I 30: Ἔστω ἑκατέρα τῶν AB, ΓΔ τῇ EZ παράλληλος ("Let both AB and ΓΔ be parallel to EZ").[281] By contrast, the supposed 'standard phrase' simply posits objects, while information about the parts and relation of the objects is supplied only afterward and grammatically linked

279 Cf. my above distinction of uses of ἔστω in different parts of the proposition within my discussion of present imperatives (Section 3.3.2, 160); this distinction does not fully overlap with the present one, as ἔστω turns out to be used in different functions even at the usual place of the exposition.

280 Whereas Heath's translation of the phrase ἔστω εὐθεῖα ἡ AB "Let AB be a straight line" suggests that it is a "specification per se" and "assert[s] a relation between a symbol and an object," it would not make sense when it is written out in full: "let the straight line AB be a straight line" (Netz 1999a, 43 f. with reference to Heath's translation). Rather, therefore, it "assert[s] an action," translating as "Let there be a straight line, [viz.] AB." (Netz 1999a, 44), with AB working as an index (according to Peirce "signifying by virtue of some deictic relation with their object," Netz 1999a, 47) rather than as a symbol.

281 *Elem.* I 30 (vol. 1, p. 42.19 Heiberg & Stamatis).

to the third element either by means of a preposition, participle, or relative clause, leaving the tripartite phrase unaltered.

Implications of the different use(s) of ἔστω

Both form and content of the 'alternative' use(s) of ἔστω, thus, seem to differ from that of the supposed 'standard phrase.' However, both types contain ἔστω and at least one lettered object phrase. Thus, there are at least three possible conclusions from the phrasal use of ἔστω, if one wishes to conceptualize it in terms of formulae (and to retain Netz's formula alongside): either (1) one assumes several types of standardized phrases containing ἔστω – in this case, one would have to formalize the 'alternative use,' which is hardly possible following the rules of Netz's formulaic system. It is presumably precisely for this reason that Netz does not undertake the formalization. Or (2) one understands the 'alternative use(s)' as variations of the 'canonical form,' which is Netz's formula 23.[282] In this case, the formalization according to Netz would ignore the actual variability of the phrasal language. Or, finally, one recognizes only Netz's formula while regarding all other uses of ἔστω as non-formulaic and, thus, unstandardized. In either case, the conceptualization in terms of formulae does not seem to be able to account for the standardization of the contextualized use of the lexicon. From Netz's point of view, at least, one would need to account for the presence of highly variable, non-formulaic parts of text.

In terms of form, thus, the conceptualization in terms of formulae which in the manner of lexical units correspond to certain more complex lexical concepts[283] suggests variability, in terms of meaning, it suggests lots of near-synonyms. For instance, the ἔστω phrase formalized by Netz and the 'alternative use' seem semantically quite distinct. Nevertheless, the latter may be used in the place of the former. For while the phrases that I have summarized as 'alternative use' posit certain features or relations of geometrical objects, they imply completely the content of the 'standard phrase' which is the exposition of a geometrical object pure and simple. The sentence that I have quoted from I 30 could be rephrased as Ἔστωσαν δύο εὐθεῖαι αἱ ΑΒ, ΓΔ. ἔστω ἑκατέρα τῇ ΕΖ παράλληλος ('Let there be two straight lines ΑΒ, ΓΔ. Let each of them be parallel to ΕΖ.'). This shows that the use of ἔστω in I 30, although it deviates from Netz's formula, may adequately substitute the 'standard phrase' in the place where it quite regularly occurs: for it fulfills the standard function of the exposition within the

282 However, Netz only recognizes ellipsis and change of grammatical form due to change of syntactical position as legitimate variations, whereas he conceptualizes other changes as transformation (see above, pp. 131 f.).

283 Even though Netz denies the existence of a one-concept-one-formula principle ("one-formula-per-concept," 1999a, 134) on the grounds that there are a couple of synonyms, his taxonomy conveys the idea that each formula has a specific semantical point of reference.

proof. Therefore, if one chooses to conceptualize both uses by means of formulae and considers their use within the proposition, these formulae are partly synonymous.

Alternatives for ἔστω phrases

After having covered all uses of ἔστω in *Elements* I in order to determine whether Netz's formula is at all valid or by what other formalization the use(s) of ἔστω might be conceptualized, I shall now question the standardization of ἔστω itself. Since ἔστω is the only fixed element in Netz's formula, the place to look for verbal alternatives would be where it most regularly occurs, that is, in the exposition part (45 times), where it most often governs the first main clause of the exposition. In all cases in which ἔστω is not the first verb form of the exposition part, other third-person imperatives occur instead.[284] As I have pointed out above in my analysis of the present imperative, these are above all present imperatives of verbs describing what relations geometrical objects have or what they do to each other;[285] there is only one perfect imperative of a construction verb which usually forms the predicate(s) of the construction part.[286] The core of the syntactical constructions that use these verb forms is, again, a third-person present (or, exceptionally, perfect) imperative as predicate and a geometrical object as subject, but, since most of these verbs (by contrast with εἶναι) require an object, the core construction is usually extended by an accusative object or else by a prepositional expression.

Again, one might want to determine the semantical relation of these other imperatives with Netz's ἔστω-formula. To be sure, taken by themselves, the imperatives have a different meaning than ἔστω; however, all but one of them shares with it tense and diathesis and, thus, a significant part of the meaning of the verb form. As we have seen in the case of the alternative uses of ἔστω, the sentences that appear in the usual place of the 'ἔστω phrase' and contain other imperatives can be rephrased so as to contain the 'standard phrase': Δύο γὰρ εὐθεῖαι αἱ ΑΒ, ΓΔ τεμνέτωσαν ἀλλήλας κατὰ τὸ Ε σημεῖον· ("Let two straight lines AB, ΓΔ cut one another in point E.")[287] is, thus, a valid way to say Ἔστωσαν δύο εὐθεῖαι αἱ ΑΒ, ΓΔ τέμνουσαι ἀλλήλας κατὰ τὸ Ε σημεῖον· ('Let there be

284 Cf. I 13–15, 21, 27–29.

285 Cf. (γωνίας) ποιείτω/σαν ("let make [angles]," I 13, 14 [vol. 1, pp. 21.5, 22.13–15 Heiberg & Stamatis]), (γωνίαν ἴσην / γωνίας ἴσας) ποιείτω ("let make an angle/angles equal," I 27, 28 [vol. 1, pp. 39.2 f., 40.7 f. Heiberg & Stamatis]), τεμνέτωσαν ἀλλήλας ("let cut one another," I 15 [vol. 1, p. 23.13 Heiberg & Stamatis]), ἐμπιπτέτω ("let fall," I 29 [vol. 1, p. 41.11 Heiberg & Stamatis]), the subject being "a straight line," or "two straight lines," respectively, and βάσιν ἐχέτω ("let have as baseline," I 41 [vol. 1, p. 55.1 f. Heiberg & Stamatis]), which uses ἐχέτω (instead of ἔστω) for describing a property of a geometrical object, namely a parallelogram.

286 Συνεστάτωσαν is in the perfect tense ("let be brought together," I 21 [vol. 1, p. 29.14 f. Heiberg & Stamatis]). See above, p. 161 for the important semantical difference of present and perfect imperatives in the *Elements*.

287 *Elem.* I 15 (vol. 1, p. 23.13 f. Heiberg & Stamatis).

two straight lines AB, ΓΔ, which cut one another in point E.').[288] Thus, in terms of semantic functions, especially the present imperatives in the *Elements* are synonymous, even though the meanings of the respective verbs differ. If, however, the text consisted of (even only partly) synonymous formulae, it would be highly ambiguous. From what we learned about the standardization of the *Elements*' lexicon and syntax, this is not the case, on the contrary, its features are extremely distinct and unambiguous.

Let us examine the relationship of the uses of ἔστω and other present imperatives more closely. It is remarkable that ἔστω in almost all cases in which it governs the first main clause of the beginning proper of the proposition occurs right at the head of that clause, whereas the imperatives substituting it regularly do *not* appear at the beginning of the clause. In this particular position, thus, ἔστω seems to work as a signal which announces the expositional part or even the beginning of the actual proposition after its introductory 'title sentence.' The different imperative forms that occur wherever ἔστω is missing, by contrast, do not.[289] This may be because they involve several geometrical objects which are assembled around the predicate more naturally regarding the Greek rules for word order. In any case, the absence of the signaling force of ἔστω seems to be felt, for the particle γάρ is inserted after the first word of every exposition (except I 30) in which the first word is not ἔστω/σαν, apparently for the sole reason of compensation, whereas it almost never occurs *if* the first word is ἔστω/σαν (except in I 18 and 20). A formalization which accounts for the linguistic and stylistic standardization of the beginning of the concrete main part of each proposition in *Elements* I would be: "ἔστω/σαν at the beginning of the sentence {geometrical object as subject} OR {} γάρ {geometrical object as subject} {third-person imperative present of a verb describing the relations of geometrical objects or perfect of a verb of construction}."[290] The complementary use of ἔστω/σαν and γάρ + a different third-person imperative is the defining feature of this sentence of the proof. It forms what I would like to call its 'set frame,' a mixture of standardized lexicon and syntax. It now turns out that one single present imperative – ἔστω – performs a special function as opposed to other present as well as perfect imperatives, namely when introducing the proof proper. The remainder of this part of the proof is standardized according to the rules of the standardization of the lexicon and syntax; for instance, if there are more main clauses to follow, they, again, will have third-person imperatives as their predi-

288 In this manner, one can extract the 'ἔστω phrase' even from the only one sentence which contains a perfect imperative and thereby conveys a sense of construction rather than of exposition.

289 For this reason, the ἔστω phrase might be more suitably termed 'second-order formula' than (with Netz) 'construction formula,' according to Netz's terminology.

290 Cf. Acerbi's summary of the linguistic designators of the exposition: "introdotta da una particella di primo livello *oppure* da verbo 'essere' incipitario con valore 'presenziale;' verbi all' imperative; comparsa delle lettere denotative" (2012, 174). There is only one proof in *Elements* I whose introduction is not either ἔστω (in the first position) or γάρ (in the second position), which is I 7, while there are two propositions which feature both components (I 18 and 20).

cates, that is, ἔστω or other ones.[291] In a similar manner, other 'set frames' consisting of standardized lexicon and syntax can be formalized for other parts of the proposition.[292] They consist of verbatim recurring 'markers' like conjunctions, particles, or verbs + conjunction on the one hand and combinations of modes and tenses on the other hand. It is only these 'set frames' that I would regard as 'formulaic.' In a way, all such 'set frames' work as what Netz calls 'second-order formulae,' as they organize the structure of the text on a meta-level, working as "signposts within arguments."[293]

'Set frames' and the nature of the standardization of the mathematical language

The language of the *Elements*, I argue, is standardized not in terms of 'formulae,' but in terms of (a) lexicon and (b) grammar, and, as it turns out, (c) structure. I argue that the particularly strong standardization of larger units of texts such as sentences and parts of sentences in the *Elements* can be grasped more suitably as a coincidence of the standardization of lexicon and syntax *in particular parts of the proof.* For the syntax and part of the lexicon are inextricably linked with particular stages of the argumentation: especially the standardization of syntax cannot be assessed without making reference to the structure of the proposition. This does not mean that our 'set frames' are locally tied to specific places within the proposition; as we have seen, construction imperatives may also occur in the proof part and indicate a part of the construction there.[294]

The impression of a formulaic organization of the language is due precisely to the interaction of the different linguistic dimensions. For if an extremely standardized use of a very limited number of words coincides with the standardized application of very few types of constructions in certain parts of the proposition, certain combinations of words in a certain grammatical form must necessarily recur which one may regard as formulaic. However, the origin of this nature of the text does not seem to be the "formulae" themselves, as Netz has it. For him, formulae are a cognitive tool fulfilling deductive functions, above all mirroring logical relations and making them more transparent through their "internal structures."[295] In my view, however, the text does not "mirror" logical relations but establishes them in the first place. In doing so, it uses a strictly standardized language. The text only appears to be put together from standardized building blocks because of the overall limited number of geometrical objects involved and operations (both mathematical and argumentative) performed. Whether the phrasal nature of the text has originated in the application of

291 In *Elements* I, cf. propositions 5, 8, 12, 16 for a construction imperative and 3, 22, 24–25 for another ἔστω in the exposition after the introductory ἔστω.

292 Cf. Acerbi's table which systematizes "i marcatori che permettono di identificare le varie parti specifiche" (2012, bottom of 174 and top of 175).

293 Netz 1999a, 47.

294 On how we have to understand 'proposition part' cf. the subsequent Section 3.4.1, 178–182.

295 Netz 1999a, 158 and 165.

formulae in Netz's sense or a combination of lexicon and syntax makes a huge difference regarding the question of standardization. While formulae, according to Netz, have just come about as the most useful way of thinking the proof and talking about it, the extreme combined standardization of lexicon and grammar which, in my view, has led to the impression of the text's being formulaically organized, must have been explicitly practiced. The standardization, thus, would be differently motivated.

By embracing the origin of the formulaic nature, our concept of 'set frames' is able to account for the degree of the *Elements*' standardization more adequately. Netz's 'formulae,' to be sure, meet the requirements of our own loose definition of 'standardization;' 37 of 48 possible occurrences in the case of his formula 23 which we have examined indeed indicate a certain standardization, but the conceptualization does not do sufficient justice to the *extreme* standardization of the *Elements*. The new focus on the standardization of the different levels of the mathematical language in different proposition parts solves both problems of Netz's concept which we had identified above: the inadequacy regarding standardization and the missing focus on grammar. It is by focusing on lexicon *and* grammar in terms of 'set frames' that one could more fittingly grasp the specific nature and scope of the Euclidean standardization. Nevertheless, singling out items for investigation of standardization, in any case, remains a task prone to a certain danger of circularity, as explained already above, and the exact degree of standardization can hardly be determined.

3.4 Structure

3.4.1 Proposition structure (microstructure)

In this third part of my analysis of textual standardization in the *Elements*, I focus on the microstructure, that is, that of the unit of the proposition, before turning briefly to its consequences for the macrostructure, that is, the setup of each of the *Elements*' books.[296] It will be the shortest part, because what remains to say about the standardization of the proposition structure mostly follows from what has been said about the standardization of lexicon and especially syntax. Conversely, it has turned out impossible to talk about the standardization of syntax without referring to the parts of the proposition that are constituted by this very syntax using the terms that go back to the Neoplatonist Proclus.[297]

---

296 Netz 1999b, 282, n. 3 uses the terms "small-scale" and "large-scale style" instead of 'micro-' and 'macrostructure.'

297 *In Eucl.* 203.1–207.25 Friedlein. We do not have any earlier explicit conceptualization of the standardized structure. Apollonius' Καθόλου πραγματεία, mentioned in Marinus' commentary of Euclid's *Data*, might have been the first metadiscussion or theory of the typical structure of elements texts, but is no longer extant (Asper 2014, 479).

The Proclean scheme

Proclus' systematization of the parts of the mathematical proposition is still held to be an adequate description of the Greek deductive practice and forms the basis of practically all modern research on Greek mathematics. Even Netz, who questions its validity,[298] takes it as a basis for the first volume of his edition of *The Work of Archimedes* and sums it up conveniently for our purposes; I shall give the corresponding Greek section of our sample I 47 in the middle column of the following table next to its respective definition by Netz on the left-hand side and Heath's translation on the right-hand side:[299]

Proclus' parts of the proposition according to Netz	The parts of I 47	Translation by Heath[300]
"Enunciation, in which the claim of the proposition is made, in general terms, without reference to the diagram. It is important to note that, generally speaking, the enunciation is equivalent to a conditional statement that if x is the case, so is y."	Ἐν τοῖς ὀρθογωνίοις τριγώνοις τὸ ἀπὸ τῆς τὴν ὀρθὴν γωνίαν ὑποτεινούσης πλευρᾶς τετράγωνον ἴσον ἐστὶ τοῖς ἀπὸ τῶν τὴν ὀρθὴν γωνίαν περιεχουσῶν πλευρῶν τετραγώνοις.	In right triangles the square on the side subtending the right angle is equal to the squares on the sides containing the right angle.
"Setting-out, in which the antecedent of the claim is restated, in particular terms referring to the diagram (with the example above, x is re-stated in particular reference to the diagram)."	Ἔστω τρίγωνον ὀρθογώνιον τὸ ΑΒΓ ὀρθὴν ἔχον τὴν ὑπὸ ΒΑΓ γωνίαν·	Let ΑΒΓ be a right triangle having the angle ΒΑΓ right;
"Definition of the goal, in which the consequent of the claim is re-stated, as an exhortation by the author addressed to himself: 'I say that . . .,' 'It is required to prove that . . .,' again in the particular terms of the diagram (with the same example we can say that y is re-stated in particular reference to the diagram)."	λέγω, ὅτι τὸ ἀπὸ τῆς ΒΓ τετράγωνον ἴσον ἐστὶ τοῖς ἀπὸ τῶν ΒΑ, ΑΓ τετραγώνοις.	I say that the square on ΒΓ is equal to the squares on ΒΑ, ΑΓ.

298 The origin of the scheme and its relation to the Euclidean practice is the topic of his 1999b.
299 2004b, 6.
300 1925, 349 f. Here as always in the present study, I have modified Heath's translation so as to contain *Greek* diagrammatic letters (cf. p. 116, n. 35).

(continued)

Proclus' parts of the proposition according to Netz	The parts of I 47	Translation by Heath
"Construction, in which added mathematical objects (beyond those required by the setting-out) may be introduced."	Ἀναγεγράφθω γὰρ ἀπὸ μὲν τῆς ΒΓ τετράγωνον τὸ ΒΔΕΓ, ἀπὸ δὲ τῶν ΒΑ, ΑΓ τὰ ΗΒ, ΘΓ, καὶ διὰ τοῦ Α ὁποτέρᾳ τῶν ΒΔ, ΓΕ παράλληλος ἤχθω ἡ ΑΛ· καὶ ἐπεζεύχθωσαν αἱ ΑΔ, ΖΓ.	For let there be described on ΒΓ the square ΒΔΕΓ, and on ΒΑ, ΑΓ the squares ΗΒ, ΘΓ; [I. 46] through Α let ΑΛ be drawn parallel to either ΒΔ or ΓΕ, and let ΑΔ, ΖΓ be joined.
"Proof, in which the particular claim is proved."	καὶ ἐπεὶ ὀρθή ἐστιν ἑκατέρα τῶν ὑπὸ ΒΑΓ, ΒΑΗ γωνιῶν, πρὸς δή τινι εὐθείᾳ τῇ ΒΑ καὶ τῷ πρὸς αὐτῇ σημείῳ τῷ Α δύο εὐθεῖαι αἱ ΑΓ, ΑΗ μὴ ἐπὶ τὰ αὐτὰ μέρη κείμεναι τὰς ἐφεξῆς γωνίας δυσὶν ὀρθαῖς ἴσας ποιοῦσιν· ἐπ' εὐθείας ἄρα ἐστὶν ἡ ΓΑ τῇ ΑΗ. διὰ τὰ αὐτὰ δὴ καὶ ἡ ΒΑ τῇ ΑΘ ἐστιν ἐπ' εὐθείας. καὶ ἐπεὶ ἴση ἐστὶν ἡ ὑπὸ ΔΒΓ γωνία τῇ ὑπὸ ΖΒΑ· ὀρθὴ γὰρ ἑκατέρα· κοινὴ προσκείσθω ἡ ὑπὸ ΑΒΓ· ὅλη ἄρα ἡ ὑπὸ ΔΒΑ ὅλῃ τῇ ὑπὸ ΖΒΓ ἐστιν ἴση. καὶ ἐπεὶ ἴση ἐστὶν ἡ μὲν ΔΒ τῇ ΒΓ, ἡ δὲ ΖΒ τῇ ΒΑ, δύο δὴ αἱ ΔΒ, ΒΑ δύο ταῖς ΖΒ, ΒΓ ἴσαι εἰσὶν ἑκατέρα ἑκατέρᾳ· καὶ γωνία ἡ ὑπὸ ΔΒΑ γωνίᾳ τῇ ὑπὸ ΖΒΓ ἴση· βάσις ἄρα ἡ ΑΔ βάσει τῇ ΖΓ ἴση, καὶ τὸ ΑΒΔ τρίγωνον τῷ ΖΒΓ τριγώνῳ ἐστὶν ἴσον· καὶ τοῦ μὲν ΑΒΔ τριγώνου διπλάσιον τὸ ΒΛ παραλληλόγραμμον· βάσιν τε γὰρ τὴν αὐτὴν ἔχουσι τὴν ΒΔ καὶ ἐν ταῖς αὐταῖς εἰσι παραλλήλοις ταῖς ΒΔ, ΑΛ· τοῦ δὲ ΖΒΓ τριγώνου διπλάσιον τὸ ΗΒ τετράγωνον· βάσιν τε γὰρ πάλιν τὴν αὐτὴν ἔχουσι τὴν ΖΒ καὶ ἐν ταῖς αὐταῖς εἰσι παραλλήλοις ταῖς ΖΒ, ΗΓ. ἴσον ἄρα ἐστὶ καὶ τὸ ΒΛ παραλληλόγραμμον τῷ ΗΒ τετραγώνῳ. ὁμοίως δὴ ἐπιζευγνυμένων τῶν ΑΕ, ΒΚ δειχθήσεται καὶ τὸ ΓΛ παραλληλόγραμμον ἴσον τῷ ΘΓ ετραγώνῳ· ὅλον ἄρα	Then, since each of the angles ΒΑΓ, ΒΑΗ is right, it follows that with a straight line ΒΑ, and at the point Α on it, the two straight lines ΑΓ, ΑΗ not lying on the same side make the adjacent angles equal to two right angles; therefore ΓΑ is in a straight line with ΑΗ. For the same reason ΒΑ is also in a straight line with ΑΘ. And, since the angle ΔΒΓ is equal to the angle ΖΒΑ: for each is right: let the angle ΑΒΓ be added to each; therefore the whole angle ΔΒΑ is equal to the whole angle ΖΒΓ. And, since ΔΒ is equal to ΒΓ, and ΖΒ to ΒΑ, the two sides ΒΑ, ΔΒ are equal to the two sides ΖΒ, ΒΓ respectively, and the angle ΔΒΑ is equal to the angle ΖΒΓ; therefore the base ΑΔ is equal to the base ΖΓ, and the triangle ΑΒΔ is equal to the triangle ΖΒΓ. Now the parallelogram ΒΛ is double of the triangle ΑΒΔ, for they have the same base ΒΔ and are in the same parallels ΒΔ, ΑΛ. And the square ΗΒ is double of the triangle ΖΒΓ, for they again have the same base ΖΒ and are in the same parallels ΖΒ, ΗΓ. Therefore the parallelogram ΒΛ is also equal to the square ΗΒ. Similarly, if ΑΕ, ΒΚ be joined, the parallelogram ΓΛ can also be proved equal to the square ΘΓ; therefore the whole square ΒΔΕΓ is equal to the two squares ΗΒ, ΘΓ.

(continued)

Proclus' parts of the proposition according to Netz	The parts of I 47	Translation by Heath
	τὸ ΒΔΕΓ τετράγωνον δυσὶ τοῖς ΗΒ, ΘΓ τετραγώνοις ἴσον ἐστίν. καί ἐστι τὸ μὲν ΒΔΕΓ τετράγωνον ἀπὸ τῆς ΒΓ ἀναγραφέν, τὰ δὲ ΗΒ, ΘΓ ἀπὸ τῶν ΒΑ, ΑΓ. τὸ ἄρα ἀπὸ τῆς ΒΓ πλευρᾶς τετράγωνον ἴσον ἐστὶ τοῖς ἀπὸ τῶν ΒΑ, ΑΓ πλευρῶν τετραγώνοις.	And the square ΒΔΕΓ is described on ΒΓ, and the squares ΗΒ, ΘΓ on ΒΑ, ΑΓ. Therefore the square on the side ΒΓ is equal to the squares on the sides ΒΑ, ΑΓ. Therefore etc.[301]
"Conclusion, in which the conclusion is reiterated for the general claim of the enunciation."	Ἐν ἄρα τοῖς ὀρθογωνίοις τριγώνοις τὸ ἀπὸ τῆς τὴν ὀρθὴν γωνίαν ὑποτεινούσης πλευρᾶς τετράγωνον ἴσον ἐστὶ τοῖς ἀπὸ τῶν τὴν ὀρθὴν περιεχουσῶν πλευρῶν τετρα γώνοις· ὅπερ ἔδει δεῖξαι.	Q.E.D.

The nature of the structural units of the Euclidean proposition

As it is, this scheme suggests an enormous standardization of the proposition in terms of structure and therefore supports our argument in favor of the *Elements*' extreme unambiguousness. However, while Proclus claims that it "holds for 'every problem and theorem,'"[302] Netz argues that the Euclidean text is by far not as standardized as the scheme would suggest, because the supposed parts were often (a) "missing"[303] to begin with or (b) "intermingling"[304] and so no longer recognizable as distinct parts. The parts which, according to Netz, may be missing are the definition of the goal, the setting-out, the construction (from problem propositions), the conclusion (again, only from problem propositions), and at times even the enunciation. This would leave the proof as the only non-negotiable part. As for parts intermingling with one another, the construction may appear not only at its supposed place, but also within the setting-out and the proof.[305]

301 Tellingly, Heath does not bother to spell out the end of the proof, because its ending is a foreseeable repetition of standardized elements. Similarly, he does not repeat the enunciation in the terms of the general conclusion in the next and final section, but only gives the abbreviation of the concluding formula (1926, 350).

302 Netz 1999b, 282 on Proclus, *In Eucl.* 203.1–2 Friedlein.

303 Netz 1999b, 288–290.

304 Netz 1999b, 290–294.

305 Proclus, too, admits to the missing of some parts (Netz 1999b, 287), but his rules seem to have been formulated tacitly with regard to theorem propositions (288–290).

From this angle, the scheme would suit, if at all, only the theorem propositions of the *Elements*, and, at that, only those of the more standardized book I, but not the whole work, let alone the Greek mathematical practice in general. In view of this, the structure of the *Elements* was not so fixed after all. Netz concludes that the Proclean parts can be found in the *Elements*, if at all, only as "conceptual parts," that is, as parts of the content of the proof which do not correspond to a definite portion of the text, but not as "stylistic parts," that is, as "specific chunk[s] of text."[306]

But the picture is more complicated and does not lend itself to either definition of proposition 'part.' For some parts listed by Proclus are indeed 'parts' in the sense of Netz's stylistic units which regularly appear at their designated place, such as the enunciation, which always appears at the beginning and is missing only very rarely; the proof, standardly occurring just before the conclusion; the setting-out; and the conclusion.

The view that the latter two are often missing from the proposition is based on a particular conceptualization of these parts in stylistic terms: the 'conclusion,' if defined with Netz as a repetition of the enunciation in the same general terms (that is, without containing any letters), to which a standardized tripartite sentence is added, is missing from most problem propositions.[307] However, many problem propositions do have some kind of conclusion mirroring the enunciation, however, it is formulated in particular rather than in general terms (cf. I 2), for a construction as is the goal of a problem proposition can, as Netz notes, only be accomplished of a concrete object.[308] While this sentence which refers back to the beginning is missing from some propositions (e.g., from I 1 to which it is added by Proclus in order for his sample proposition to fit the scheme),[309] problem propositions do feature a concluding sentence – only ὅπερ ἔδει ποιῆσαι as opposed to ὅπερ ἔδει δεῖξαι in theorems. This means there are two main types of conclusions, one of which is not covered by Proclus, but his general intuition that each proposition ends with a standardized conclusion in terms of a 'stylistic part' turns out correct.

As for the setting-out, if it is defined by the occurrence of ἔστω or even of one particular phrase containing ἔστω, it is indeed absent from many propositions of *Elements* I. But as I have shown above, the setting-out as the "stylistic part" following upon the enunciation is more suitably defined by the alternative use of either ἔστω or of γάρ with another (mostly present) imperative.[310] To be sure, the setting-out may be taken up again in the construction by further occurrences of ἔστω and other present imperatives, but where it occurs for sure is after the first sentence. Similarly, the definition of the goal regularly appears after the exposition (understood as above).[311]

---

306 Netz 1999b, 291.

307 Netz finds it especially compelling that it is missing from I 1, Proclus' example (1999b, 289).

308 Netz 1999b, 289.

309 Netz 1999b, 289.

310 See above, p. 172.

311 P. 32.

Since the remaining part construction often "intermingle[es]" with a number of other parts, it does not count as a "stylistic part" in the sense of Netz's definition, and it further causes those parts in which it appears to be no longer "stylistic parts," since they become indistinguishable from the intruding part. But the distinction between "stylistic" and "conceptual parts" is misleading, for it is precisely because of its language or style that the "conceptual part" of construction may be recognized wherever it appears: its 'set frame,' as I wish to call to it, consists of construction verbs in the perfect imperative. Similarly, and even more obviously, the definition of the goal catches one's eye precisely as the definition of the goal because of the verbatim recurrence of δεῖ δή or λέγω ὅτι, respectively, even if it appears a second time away from its usual place. It is by standardized stylistic features that a structural unit is constituted and remains recognizable.

If, conversely, a stretch of text does not adhere to the standardized style, it cannot be considered the structural part of the stylistic features which it lacks. For instance, while Netz argues that our sample I 47 does not stick to the Proclean scheme as it has a second construction *after* a period of proof, my stance is that the sentence in question does not present a 'construction.' This is because the mere occurrence of construction verbs does not result in the act of construction as it is standardly addressed at some point of the proposition; what is crucial is the combination of such a verb with a particular verb form, but the present case has a participle form instead of the perfect imperative: ὁμοίως δὴ ἐπιζευγνυμένων τῶν ΑΕ, ΒΚ δειχθήσεται καὶ τὸ ΓΛ παραλληλόγραμμον ἴσον τῷ ΘΓ τετραγώνῳ· ("So, joining the < lines > AE, BK, it shall be similarly proved that the parallelogram ΓΛ, too, is equal to the square ΘΓ").[312] It seems to be for a particular reason that the perfect imperative is not used: ὁμοίως δειχθήσεται regularly introduces a deduction analogical to one already performed. It is, therefore, not necessary to go through the whole construction and proof again, but it suffices to mention the analogy so as to make the proof complete. If, by contrast, one line of argument requires another construction, it should appear in the respective style, that is, have construction verbs in the perfect imperative as predicates.

The Euclidean propositions are indeed organized by structural units which are, to be sure, defined and arranged by their logical content and their relation to one another,[313] but constituted by formulaic 'set frames' of standardized syntax and vocabulary. The logical order of these structural units in the simplest type of proposition is the one of the Proclean scheme. Enunciation and conclusion, obviously, have their place at the beginning and ending, respectively, and the proof proper in most cases presupposes an exposition of the objects, a definition of the goal, and a construction, appearing in this very order. A deviation from this sequence or the number of parts is

---

312 *Elem.* I 47 (vol. 1, p. 64.18 f. Heiberg & Stamatis). Construction verbs appearing in the perfect indicative do not count as "construction formulae" in Netz's systematization (see above, p. 130).
313 Cf. Netz 1999b, 288.

mostly necessitated by the content, but it is precisely by their standardized style that the (conceptual) parts remain recognizable to a reader who is used to the linguistic practice of the elements genre. The most subtle difference between parts is the one between setting-out and construction; both are constituted primarily by third-person imperatives differing only in tense and diathesis, but in most cases, at least in *Elements* I, the text makes this however slight difference.[314]

Thus, Proclus' conception of the structure of the proofs is not as inadequate as Netz has it. The general idea that the proofs are composed of stylistically well-defined parts of text which recur in a fairly standardized sequence is generally sustained by the text; indeed, it is the strong structural standardization of the text which only just allows for such schematization and even invites it. Nevertheless, the scheme which Proclus sets out is, with regard to Euclid's *Elements*, nothing but an "analytical tool,"[315] since it has come into being long after Euclid and his contemporaries, and has not served them as a "style manual."[316] Rather, it seems to have served Proclus, whether or not he has invented it, for a simplifying harmonization of Euclid for didactical purposes.[317]

Apparently, Proclus has not randomly chosen his sample for teaching the regular structure of the propositions of the classical elements text and for regularizing contemporary mathematics, but because of its very standardization. He has chosen Euclid over other authors and, within Euclid's *Elements*, the first book of the *Elements* over the remaining books.[318] In doing so, he testifies implicitly to the unparalleled strict structural standardization of Euclid's *Elements* as standing out even from the extant

---

314 Cf., somewhat differently, but also with a wider focus on the whole elements literature Netz 1999b, 287, n. 9: "It is very often possible to state the construction within the setting-out itself. [. . .] Then the definition of the goal would have referred not to virtual objects, [. . .] but to actual rectangles and squares, and no construction would have been required. Very often, then, this is a purely stylistic decision, how to distribute the burden of construction between the setting-out and the construction (in other cases of course, the added construction really is very clearly auxiliary to the construction demanded to set out the proposition itself. Pythagoras' theorem, Elements I 47, is an obvious case)."

315 Netz 2004b, 6.

316 Cf. Netz 1999b. The two main arguments in favor of this view are first that the system in terms of terminology is nowhere attested before Proclus, and second that mathematical authors after Proclus such as Eutocius write exactly according to the scheme (or have been subject to subsequent regularization; 1999b, 295) whereas Euclid, as we have seen, does not stick militantly to it.

317 Netz 1999b, 303.

318 A number of times, Proclus "hints" to extending his commentary beyond the first book, but later on, he also seems to doubt his being able to do so, and in any case, there is no external evidence that he actually commented on books other than the first (Morrow 1992, lv f.).

corpus of Greek mathematical elements texts, which are already characterized by much standardization and systematization compared with other contemporary scientific traditions. The detail and perseverance of the standardization of the Euclidean proof, however, which enables to deal with the simplest and most complex hypotheses of diverse geometrical content alike, is indeed unparalleled. The propositions of book I, because of their comparative compactness and straightforwardness, especially conform to a foreseeable structural scheme. It is, of course, not certain, whether the degree of the textual standardization of *Elements* I encountered by Proclus was originally Euclidean.

3.4.2 Book structure (macrostructure)

Textual vs. logical order: the implications of the list form of the "Satzsystem"

The textually highly standardized propositions of the *Elements* are arranged according to subject matter as part of 13 books in total. From a formal-stylistic point of view, the "large-scale structure" of the individual books follows directly from the microstructure of the propositions, which are simply lined up one after the other without any transitional devices, explicit grouping or headings, except for being numbered.

The textually linear arrangement, to be sure, conceals that the propositions are logically arranged,[319] with the logic of the arrangement following the axiomatic-deductive method: starting from certain unprovable principles, geometrical theorems are proven and problems of construction solved by means of deduction. The deductive text proper begins with simple theorems and problems and moves on to ever more complex ones, with the results of prior deductions themselves becoming premises on which further deduction is based. As a rule, thus, later propositions tend to presuppose earlier ones, although the exact choice and sequence of propositions in many cases is questionable and has been debated at least since Proclus.[320] Remarkably, this logical, deductive macrostructure is only implicit and not formally indicated, except by the increasing length and elaborateness of the propositions, even across books, since the topics of the books tend to become more complex. In modern editions, it is made explicit by tables indicating which axioms, postulates, definitions, and preceding propositions form the basis of which proofs.[321] However, even the introductory lists of principles preceding the propositions may be the re-

319 For the ability of lists to conceal their own internal structure cf. Mainberger 2008.
320 See above, p. 99 f.
321 Cf., e.g., Thaer 1970.

sult of wishing to make the logical structure of the proposition chains explicit in the first place.

For one thing, the list of axioms (just like the definitions) proves the strong linguistic and structural standardization that has been the subject of this chapter, and the standardizability in terms of explicit standardization by the very possibility of their being set up. Concluding on the unprovable principles on which the proofs are based requires an unambiguous use of the geometrical terms and the unambiguousness of all proofs. For another thing, the setting-up of introductory lists is telling as regards the perception of the standardized deduction, as it seems to indicate a certain dissatisfaction with the deductive part as it is, which leaves its own premises implicit. I shall return to this observation in Section 5.2 on standardization and communication.

Overlap between textual levels

I have opted for an analysis according to textual levels precisely because it is suited best to capture the high degree and specificity of standardization, and the relation of standardization to unambiguousness. Indeed, it is the very limitation of the analysis to individual textual levels that has revealed the specific overlap, mutual relatedness, and even indistinguishability of the features of lexicon, syntax, and structure, because this overlap is the direct result of individual standardization. The link between the standardization of lexicon and grammar is particularly strong. For it turns out that the lexical standardization of terms in the *Elements* is not confined to the unit of the single word, but may pertain to specific grammatical 'forms of words' on the one end of the scale and to combinations of words on the other. Such compound structures which standardly refer to a single concept may consist of only a few words or even amount to clauses, with the lines between words and sentences becoming conceptually blurred in the analysis. The standardized meaning of a single word may be limited to only one or a small number of grammatical forms regularly used, or it may be bound inextricably to the exact phrase or to the kind of sentence in which it standardly occurs. This is where the levels of lexicon and grammar merge.

Similarly, the standardization of structure is closely bound up with that of lexicon and syntax. For the individual parts of the proposition are each not only characterized but also constituted by a specific choice of vocabulary as well as by a distinctive combination of constructions, tenses, and modes. Therefore, it is not possible to talk about structure without talking about lexicon and syntax. Conversely, the meaning of particular verb forms, words, groups of words, and constructions which function as structuring devices or which perform a particular function within a specific part of the proposition can only be understood by explaining the very structure within or toward which they are working.[322] This is true even more of those words whose function differs depending on the part of the proposition in which they appear and whose meaning cannot be fully grasped without making reference to that part.

322 Asper 2007, 117.

4 The standardization of lettered diagram and diagrammatization

This chapter deals with the standardization of the diagram in the *Elements*. It is concerned with two main questions: whether diagrams were standardly applied and, if so, whether they were applied in standardized form. Unfortunately, original diagrams of Euclid are no longer extant. Nevertheless, it is generally assumed that the diagrams were present in the Euclidean text. However, regarding my subject of the formal standardization of the *Elements*, it is vital to know whether diagrams were indeed part of the text and what they looked like. Therefore, I will examine the issue from scratch.

Let us consider first whether there is any explicit evidence of them. Since the Euclidean *Elements* completely lacks metamathematical remarks and consists only of first principles and proofs, and since methodological reflections are scarce even in those elements texts which include metamathematical paratexts, the texts do not feature a methodology of the diagram. However, diagrams are sometimes explicitly mentioned in both the texts of Euclid and his contemporaries as well as in the later commentaries with reference to their source texts. Remarkably, a few propositions in the works of each of the surviving Greek mathematicians explicitly acknowledge that they rely on a diagram: either by stating that they are accomplished διὰ γραμμῶν ('by letters') as opposed to δι' ἀριθμῶν ('by numbers'),[1] or by referring to a 'drawing' (καταγραφή) or 'figure' (σχῆμα). Often, they specify to which drawing among several they refer: for example, to 'the first.'[2] The mathematical commentators, too, refer to the diagrams of their source texts by means of cognates of γραφή ('drawing') and σχῆμα ('figure'),[3] while using the term διάγραμμα metonymically for the whole '(diagrammatic) mathematical proposition' or 'mathematics' in general.[4] While the term, by its etymology, strongly hints to

1 Netz 1999a, 35 with passages from Hero, Ptolemy, Pappus, and Proclus.

2 For καταγραφή cf. *Elem.* III 33 (vol. 1, p. 140.19 f. Heiberg & Stamatis): ὡς ἐπὶ τῆς πρώτης καταγραφῆς ("as in the first drawing"), *Elem.* III 33 (vol. 1, p. 142.6 f. Heiberg & Stamatis): ὡς ἔχει ἐπὶ τῆς δευτέρας καταγραφῆς ("as it is in the second drawing"), *Elem.* III 33 (vol. 1, p. 142.24 Heiberg & Stamatis): ὡς ἔχει ἐπὶ τῆς τρίτης καταγραφῆς ("as it is in the third drawing"), *Elem.* IV 5 (vol. 1, p. 158.6 Heiberg & Stamatis): ὡς ἔχει ἐπὶ τῆς δευτέρας καταγραφῆς ("as it is in the second drawing"), *Elem.* IV 5 (vol. 1, p. 158.11 f. Heiberg & Stamatis): ὡς ἔχει ἐπὶ τῆς τρίτης καταγραφῆς ("as it is in the third drawing"), *Elem.* X 110 (vol. 3, p. 197.12 Heiberg & Stamatis): ὡς ἐπὶ τῶν προκειμένων καταγραφῶν ("as in the preceding drawings"), *Elem.* XII 4 (vol. 4, p. 91.27 Heiberg & Stamatis): Ἐπὶ γὰρ τῆς αὐτῆς καταγραφῆς ("In the second drawing"), and various passages in Apollonius, while Archimedes uses σχῆμα (Netz 1999a, 35–37).

3 Cf. Pappus, Proclus, and Eutocius for καταγραφή ('drawing,' 'delineation'), διαγραφή ('delineation,' 'outline,' 'diagram'), ὑπογραφή ('outline,' 'sketch,' 'diagram'), and ὑπογεγραμμένη (as ὑπογραφή), cf. Netz 1999a, 35–37.

4 Netz 1999a, 35–38 argues that this use of the term in fact mirrors the prominent role of the diagram as being a "predominant" (14) and "necessary" (12) feature of Greek mathematics. Plato was the first to use the term in this way, and was probably followed by Xenophon and Aristotle in doing so, although we cannot be sure about the meaning of the term in Aristotle (Netz 1999a, 35).

https://doi.org/10.1515/9783110670462-004

the fact that diagrams were somehow involved, it does not indicate unequivocally that they were physically part of the text. In any case, the few explicit mentions of drawings could well indicate an occasional use of diagrams, but they do not necessarily imply a standard application of diagrams. Moreover, it is not sure to what kind of diagrams the above terms refer: they do not give any indication of the form of the diagrams and whether the form was standardized in case the diagrams were standardly applied.

This means that we have to approach the question of the standardization of the diagrams differently and from different angles: rather than relying on explicit evidence for their application, we have to look for implicit indications of the use of diagrams. For this purpose, I analyze, on the one hand, the diagrams that have been transmitted for Euclid's *Elements* from ancient and Byzantine times. On the other hand, I examine the *text* of the *Elements* for its diagrammatization, that is, for ways in which diagrams have shaped it and for traces which diagrams have left in it. This chapter will be by far the longest of the study because it will prove complex to determine the relevance, indications, and conclusiveness in each case. I analyze the transmitted diagrams and text separately, precisely because it is uncertain whether the transmitted diagrams have anything to do with the ones which might have been originally included in the text. I begin with the transmitted diagrams before turning to the analysis of the text but will reevaluate the form of the transmitted diagrams in the light of the implications of the textual construction for the existence and form of the diagrams. In the absence of explicit evidence of and about Euclidean diagrams, we will not be able to find out whether diagrams were indeed materially present as a drawing, but only, if at all, whether they were logically presupposed by the text, that is, whether the proposition texts were written based on ready diagrams. However, I will not distinguish in the analysis between the 'logical' presupposition and the 'material' presence, that is, the presence of the diagrams as lettered drawings. This is because if the former is proved, the latter is merely a question of transmission.

The diagrammatic standardization shall be analyzed analogously with the textual standardization, that is, both with regard to the diagram as a literal part of the proposition, and with regard to how the text is shaped by the diagram. However, in the absence of directly transmitted diagrams, the diagrammatic standardization cannot be assessed straightforwardly as in the case of the textual standardization, but only indirectly. This implies that it cannot simply be measured quantitatively. Rather, the quantitative method will only be applied at certain stages of the analysis. For example, in the case of the ancient fragments which transmit diagrams, it is first of all necessary to determine on the basis of the text and its presentation whether they stem from book manuscripts of Euclid's *Elements* before comparing the form of the diagrams of the diverse fragments and assessing it quantitatively. In the case of the well-transmitted diagrams of Byzantine manuscripts, a quantitative analysis can be accomplished quite easily, but since these manuscripts are so much later than Euclid, the quantitative method alone does not help to decide whether the Byzantine diagrams indicate the use of diagrams in Euclid's own text. Here, it will prove fruitful to analyze the methods of presentation in relation to the text of the propositions and to the layout of the manuscripts, and to compare the diagrams of different manuscripts in these regards. Finally, also the diagrammatization of the *text* cannot always be analyzed quantitatively. This is because the text is necessarily more standardized through the standardized application of a diagram. I look for traces of the diagram, on the one hand, in terms of a (standardized) non-standardization and underdetermination of the text. This is on the assumption that a standardized application of *both* text and diagram entails a double standardization of the 'logical' form and could result regularly in lesser standardization on the part of the literary and logical form of the text. On the other hand, I look for standardized features of the text which do not have any 'logical' function on the hypothesis that the standardization of such features which are logically not explicable could betray the underlying presence of a diagram by reflecting that a diagram is being looked at.

The chapter has four parts: the first two will be concerned with the extant diagrams, the latter two with the diagrammatization of the text. I begin by analyzing the diagrams of the ancient papyrus and ostraca fragments (Section 4.1) before turning to the diagrams of the Byzantine manuscripts (Section 4.2). Under the heading 'diagrammatization of the text,' I will consider first the different ways in which the proposition text can be diagrammatized before focusing on the implications of the parts of the text which construct the diagrams (Section 4.3). In the second part on the diagrammatization of the text, I examine the ways in which the diagrammatic letters are used in the text (Section 4.4).

4.1 Diagrams I: the ancient evidence of diagrams for Euclid's *Elements* transmitted on papyri and ostraca

The original text of the *Elements* is no longer extant; neither are ancient manuscripts which transmit larger parts of elements texts from which we could tell whether the works originally included diagrams. However, some very scarce ancient evidence has survived on papyri and ostraca. I shall consider here only the evidence which survives for what appear to be either propositions or content closely related to the propositions of Euclid's *Elements*.[5] It is potentially important because it presents the only ancient evidence of diagrams which we have for the *Elements*. Some of the fragments are quite early and could contain indications of Euclidean diagrammatic conventions. Here, as in the analysis of the Byzantine manuscripts in Section 4.2, my aim is not to regard and judge the diagrams by modern standards, but rather to view them from the perspective of the Euclidean *text*.[6]

5 In this section, my remarks are based, on the one hand, on the respective editions of the papyri, and, on the other hand, on the synoptic overview of the ancient textual and diagrammatic evidence of Euclid's *Elements* provided by Fowler 1999, 209–217, including Plates 1–4 between pp. 6 and 7.

6 In this way, my approach differs fundamentally from Saito & Sidoli 2012, who simply emphasize the "overprecision" and lack of "visual accuracy" (*passim*) of the medieval diagrams by contrast with certain ancient diagrams (however, not the ones of the Euclidean papyri).

4.1.1 Overview of the ancient evidence and problematization

In total, four papyrus fragments (one from Herculaneum, the rest from Egypt) and one set of six ostraca (from Elephantine) survive,[7] and each of these five testimonies contains at least one diagrammatic figure. Altogether, they feature nine figures, that is, for nine different propositions. More proofs than diagrams are extant, but where no diagram is extant, this seems to have to do with the fragmentary transmission rather than with the non-inclusion of a diagram. Just like the proposition texts, the diagrams, too, are in some cases fragmentary. Nevertheless, all of them reveal what characterizes their form such as whether they have been drawn with the help of a ruler or not, or whether they are lettered or not.

The fragmented text of the ostraca bears only very loose connection to the treatment of the respective topics by the *Elements* – two of the six fragments whose text can be related to two specific *Elements'* propositions do not contain any drawing, while on the ostraca which contain traces of a diagram almost no text is legible.[8] By contrast, the seven texts containing the remaining seven diagrams which survive on the four papyri fragments most certainly contain parts of propositions from the *Elements*; however, much of their own text deviates from them or truncates them.[9]

7 There is one more testimony (P. Mich. III 143), which contains only definitions (the first ten of *Elements* I) that are not accompanied by diagrams. It has been edited and discussed in detail by Turner *et al.* 1985 and more briefly by Fowler 1999, 215. The editors date it to the third century AD (14) and consider it a "private extract from the *Elements*" (13); they presume it to have been written by a schoolmaster in preparation for his next lesson on the grounds of the similarity of the handwriting in P. Oxy. i 29 (15) – ascribed, in turn, to a "grammaticus" (14). While diagrams are not included in the definition lists in the modern editions, they do appear in some of the Byzantine manuscripts – however, only *in margine* – by contrast with the diagrams which are included in the propositions in our papyrus evidence and in the Byzantine manuscripts (for which see below). This could indicate that they have been added by another hand later. Still, their original non-inclusion in the definitions is, despite the evidence of the Michigan papyrus, not definite. The reason why I will not take the Michigan fragment into account is that I do not consider the definitions at all in the present study.

8 The whole set (P. 11999; 12002; 12007; 12008; 12609; 12611), published by Mau & Müller 1982, seems to revolve around the material of XIII 10 and 16 (Fowler 1999, 209); but only the better-preserved P. 12002 and P. 12609 can be allocated to XIII 10 and XIII 16, respectively. The fragments that mainly consist of a diagram are P. 11999 and P. 12611 (the latter presumably contains the second of the two diagrams of XIII 16; cf. Mau & Müller 1982, 5).

9 The papyrological evidence of diagrams for Euclid's *Elements* includes: P. Herc. 1061 (diagrams for I 9 and I 10), published by Heiberg 1900 and republished by Angeli & Dorandi 1987; P. Oxy. i 29 (diagram for II 5), published by Grenfell & Hunt 1898; P. Fay. 9 (diagram for I 39), published by Grenfell *et al.* 1900; and P. [inv.] Berol. 17469 (diagrams for I 8, I 9, and I 10), published by Brashear 1994. Besides these fragments, there is one more fragment from Oxyrhynchus, which is as yet unpublished (Turner *et al.* 1985, 17, n. 9, with reference to "E.G. Turner in a letter [confirmed by P. Parsons];" Brashear 1994, 29 and Fowler 1999, 216, n. 44 also mention the fragment). Although Fowler (1999, 216) concludes that "[. . .] neither the earliest evidence in the Elephantine ostraca, nor that in the fragment that most resembles a professional scribe's copy, P. Fay. 9, follows the received text," the dependence from the Euclidean text

Although the textual fragments on ancient papyri which present direct evidence of propositions of the *Elements* all contain diagrams, it is first of all the scarcity of the evidence which prevents one from generalizing the results. The diagrams which survive are so few compared with the potential number of diagrams which might have been drawn for the *Elements* that it could be that we just happen to have a wholly unrepresentative sample. What is more, the more significant part of the evidence stems from the first centuries AD (the dating of one of the fragments is not entirely secure, but it is in any case much later than Euclid).[10] Evidence from BC is limited to the indirect evidence of P. Herc. 1061 from around 100 BC, which preserves an essay by Demetrius Lacon,[11] and to the ostraca from Elephantine dating around 300 BC and presenting by far the oldest testimony of Euclidean geometry.[12] Remarkably, among this small body of evidence, there are two propositions including diagrams which feature on two different papyri, that is, I 9 and I 10 on P. Herc. 1061 and the most recently published Berlin papyrus P. [inv] Berol. 17469. This overlap allows for a comparison of the characteristics of the diagrams which could enable a more objective assessment of what diagrams in the *Elements* would have looked like.

In order to evaluate the significance of the fragments for our present question of the contemporary standardization of diagrams, it is vital to know how close the fragments were to the Euclidean original, who its authors or copyists were, and what functions the texts had. For answering these questions, it is necessary to start from the *text* of the fragments, that is, its proximity to the Euclidean text and the form of its presentation. By contrast, previous research – e.g., the most comprehensive discussion of the papyrological evidence of the *Elements* to date by Fowler – relied (also) on the *drawings* for determining the relevance of the papyri. It reflexively inferred the proximity of the fragments to the Euclidean text not only from the quality of the text but also from the qualities of the diagrams transmitted by them.[13] In so doing, it does

---

is indisputable in the case of the papyri. This is why those of the paypri which were already available to Heiberg have been used by him for establishing the text and even for excluding I 40 as spurious for its not being transmitted on P. Fay. 9 between what are I 39 and I 41 according to the manuscripts.

10 P. Oxy. i 29 is dated to the third or fourth century AD by its editors B. P. Grenfell and A. S. Hunt, and to 75–125 AD by Sir E. Turner (Fowler 1999, 211); P. Fay. 9 is dated by its editors B. P. Grenfell, A. S. Hunt, and D. G. Hogarth to the latter half of the second century AD (Fowler 1999, 213); and P. [inv.] Berol. 17469, published by W. Brashear, is assigned to the second century AD (Fowler 1999, 215).

11 Fowler 1999, 210.

12 These fragments are believed to bear some significance for the dating of Euclid himself, i.e., to 300 BC, but Fowler points out that "one complication in this proposal is that we cannot be sure of their relationship to the received text of the *Elements*. It is not inconceivable that the fragments represent a preliminary version of the final text, not the other way around, though they provide, in themselves, slender evidence for such a proposal" (1999, 209 f.). Fowler himself assumes that the text is the result of someone's referring to the *Elements*' text by "working through" the material instead of simply copying it (209) but does not exclude the above-quoted alternative. See below on the original purposes of the texts transmitted on papyri (pp. 197–199).

13 Cf. Fowler 99, 209–216 and the debate which he relates.

not only take for granted the preexistence of diagrams in Euclid and in the earliest ancient manuscripts of the *Elements*, but also presupposes knowledge about what diagrams in them would have looked like. The treatment of the fragments by Fowler and by the scholars generally seems to assume that diagrams were indeed part of the earliest manuscripts of the *Elements* on which the fragments which have come down to us are based, and that these diagrams were technically drawn, along the lines of: 'the more orderly or technical the diagram looks, the more professional the scribe, and the more reliable the fragment.' However, the papyrus fragments present our earliest evidence of the diagrams of the *Elements*, and it is from these fragments, if at all, that we can make inferences about whether the diagrams were regularly included in the texts in the first place, and about what they might have looked like in case they were. This means that we cannot draw any conclusions about the existence and look of the diagrams in Euclid directly from the papyrus diagrams themselves. Rather, we first have to determine the relation between the respective fragment and Euclid's *Elements* on the basis of the text of the fragment. Then, we can draw tentative conclusions about the function of the text of which the fragment forms a part. Subsequently, we are going to analyze the characteristics of the diagrams systematically, according to the criteria 'materials' (how the ground and means of drawing predetermine the form of the diagrams), 'drawing techniques' (whether they have been drawn by means of ruler and compass or not), and 'lettering' (whether or not there is any, and in what form). Finally, we attempt a synoptic evaluation of the textual and diagrammatic sides of the fragments.

4.1.2 The relation of the text of the papyrus fragments to Euclid's *Elements*

In none of the cases is the text of the fragments identical with that of the modern editions which is established (mainly) based on the Byzantine manuscript transmission;[14] it is often badly damaged, but where it is readable, it deviates from the Euclidean text in terms of wording, structure, completeness, or the placement of the diagram.[15] Nevertheless, all of the papyrus fragments suggest that they refer directly to Euclid's *Elements*. By contrast, as we have seen, the text of the ostraca only indicates thematical relatedness to some Euclidean propositions. This means that only the papyrus fragments qualify at all for possibly attesting the use of diagrams in Euclid's *Ele-*

14 As has been pointed out above, Heiberg, in his edition, has considered the papyrological evidence as far as it was known to him; however, for the most part, this evidence deviates so much from the congruent text transmitted by the different Byzantine manuscripts that it has not proved helpful in correcting it.
15 E.g., the text of P. Fay. 9 structurally comes closest to that of the standard edition, but throughout deviates a lot from it in detail. Heiberg (quoted by Fowler 1999, 214 f., n. 43) notes that the papyrus deviates strongly from the generally accepted text of the manuscripts; Fowler attributes the "difficulties of this fragment" to "the fact that no single complete line of text survives" and "that the lines of the fragment seem to contain a very irregular number of letters" (213).

ments. Accordingly, I do not consider the ostraca in the present analysis of the text of the ancient evidence of the *Elements*. However, in the subsequent analysis of the diagrams of the ancient fragments, I do consider the diagrams of the ostraca more generally with regard to drawing conventions which they show, because they are effectively the only contemporary testimony from Euclid's own time.

In the present review of the relation of the text of the papyrus fragments to the *Elements*, I focus on two points: (1) 'macrostructure,' that is, the order of the propositions, and (2) 'microstructure,' that is, the structure of the propositions themselves, including the position of the diagram within the text. I do not consider deviations in the exact wording from the manuscript-based Euclidean text, as these are not relevant for my present purpose of determining the relation of the fragment to the Euclidean text.

The number and order of the propositions

Let us first look at the macrostructure of the texts of the fragments in relation to the respective parts of the *Elements*. Larger changes of the thematical framework of the individual books can hardly be told from them, as each of them only contains the remains of very few, but, in any case, no more than three propositions. Three fragments contain larger or smaller parts of two propositions (P. Herc. 1061, mainly of I 9 and I 10;[16] P. Oxy. i 29, of II 4 and II 5; P. Fay. 9, of I 39 and I 41), one contains parts of three propositions (P. [inv.] Berol. 17469, of I 8, I 9, and I 10).

These small sequences appear to follow the proposition order of the *Elements* and, in so doing, indicate that they refer to this work. Only the proposition order of P. Fay. 9 deviates from the manuscript transmission: propositions I 39 and I 41 of the manuscripts are given in direct sequence on the papyrus. In this case, the intermittent proposition (I 40) has been deemed spurious on the basis of the evidence of the papyrus, so the order of the propositions of the fragment has been taken to be the originally Euclidean one.

P. Herc. 1061 is a special case because it quotes from the sequence I 9–10 within the context of another work and forms a piece of indirect evidence. The reference to Euclid is focused on I 10, and I 9 appears to be related for the sake of proving it,[17] including a quotation from I 3 on which, in turn, its own proof is based. In quoting the earlier proposition explicitly, the text decidedly deviates from the Euclidean practice which never explicitly includes underlying proofs or axioms. The deviation from this practice is due to the fact that the text takes a passage from the *Elements* out of its original context, that is, out of the deductive structure: I 3 can no longer be presupposed as already proven and for this reason must be inserted into the text.[18] For the same reason, the text quotes definition I, Def. 15 before it cites from the sequence I

16 The fragment also quotes *Elem.* I, Def. 15 and a small part from *Elem.* I 3.

17 Heiberg 1900, 17.

18 Cf. Turner *et al.* 1985, 17 with reference to Heiberg 1900, 15–17 and Heiberg 1903a, 47.

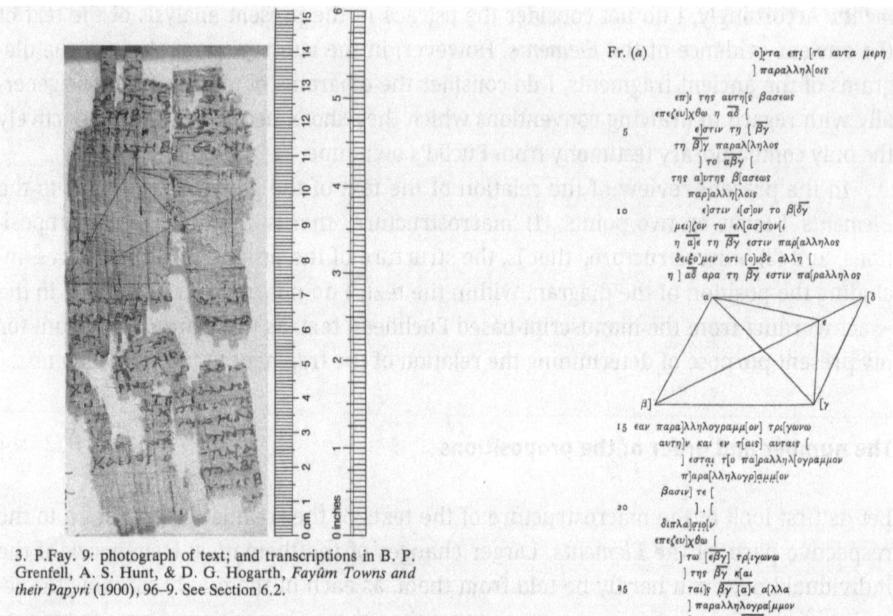

(a)

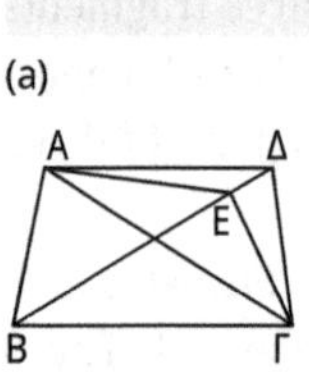

(b)

Fig. 1: (a) P. Fay. 9 (Fowler 1999, Plate 3); (b) I 39 according to Heiberg & Stamatis.

9–10. Although the fragment focuses on two successive Euclidean proofs, the inclusion of different parts from *Elements* I shows how taking a sequence out of its context disrupts the original deductive proposition order.

With its main focus being on I 10, the Euclidean reference of P. Herc. 1061 is probably limited to the propositions mentioned. The three other fragments, which seem to be pieces of direct evidence, give no clear indication that the text is necessarily restricted only to the surviving propositions: P. Fay. 9 starts after the beginning of I 39 and breaks off after I 41, and the fragmentary state of P. [inv.] Berol. 17469 does not show what precedes I 8 and what follows after I 10. The text of P. Oxy. i 29 starts after the damaged beginning of the sheet with the enunciation or conclusion of II 4; thus, the text may have contained the first propositions of book II, and even book I, beforehand. However, the text ends somewhat before the end of the sheet with the enunciation and diagram of II 5, so it could be that this copy from Euclid ended here, but it could well be that it continued on another sheet; a closer inspection of the text quoted from II 4 and II 5 might give an indication as to this – I shall come back to it in the following section.

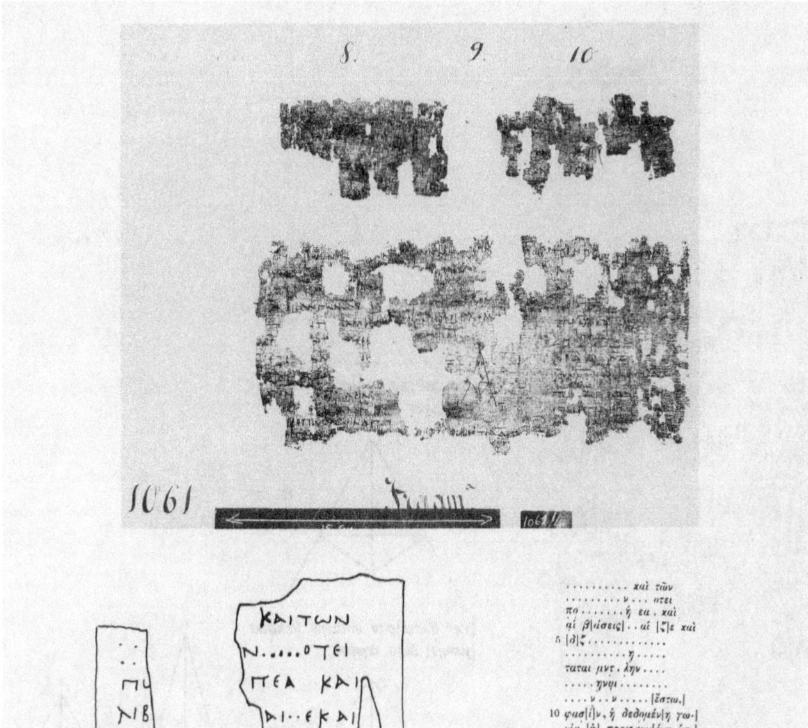

1. P. Herc. 1061: photograph of columns 8, 9, and 10; and drawings and transcription of column 9 in J. L. Heiberg, Quelques papyrus traitant de mathématiques, *Oversigt over det Kgl. Danske Videnskabernes Selskabs Forhandlinger* 2 (1900), 164–5. See Section 6.2.

Fig. 2: P. Herc. 1061 (Fowler 1999, Plate 1).

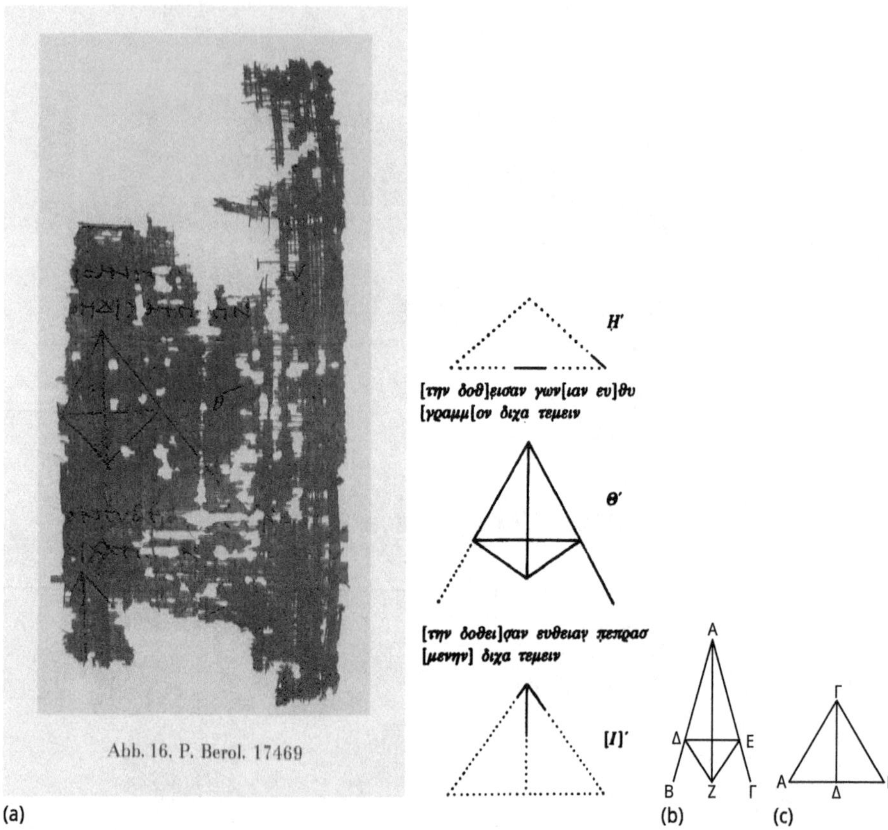

Fig. 3: (a) P. Berol. [inv.] 17469 (Brashear 1994, 30 and Plate 16); (b) I 9 according to Heiberg & Stamatis; (c) I 10 according to Heiberg & Stamatis.

The scope and structure of the proposition texts

Second, let us consider the structure of the individual proposition texts of the papyri in relation to the propositions of the *Elements*. Only the proposition text of one of the fragments (P. Fay. 9) clearly sticks to the internal structure of its respective propositions (I 39 and I 41). The diagram extant for I 39 follows after the conclusion of I 39 and precedes the enunciation of I 41; it is drawn into the space left between the two proposition texts in the middle of the column. Thus, this fragment appears to present both the structure of the proposition texts and their respective diagrams in the way which we encounter in the later manuscript transmission.[19]

19 The order of the text and diagram corresponds precisely to the one we find later in P, which always features the diagram at the end of the proposition text. Differently, in B, it tends to appear at the end of the *page* (see Section 4.2, 218 f.).

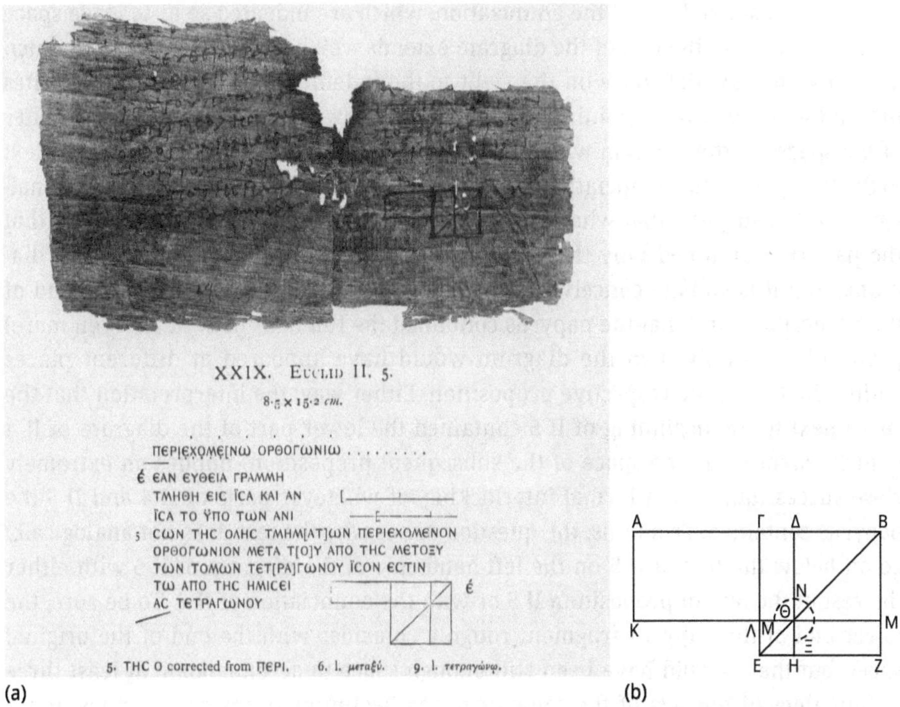

Fig. 4: (a) P. Oxy. i 29 (Fowler 1999, Plate 2); (b) II 5 according to Heiberg & Stamatis.

It is possible that the same applies also to another one of the fragments (P. Oxy. i 29), which contains parts of the text of propositions II 4 and II 5 and the diagram of II 5: it could either be that the papyrus originally contained both the full proposition texts of both II 4 and II 5, or that it only contained the enunciations of both II 4 and II 5.[20] The fragment recognizably transmits the enunciation of II 5, and above it, almost unrecognizably, either the enunciation or the conclusion of II 4. The possibility that it is the enunciation is indicated by what could be traces of the diagram of II 4 below the remains of II 4; the diagram would then appear in analogical position with that of II 5, that is, after the enunciation. More precisely, the diagram of II 5 appears on the right-hand side of the column; its upper part is located on the same

20 The first interpretation is favored by Turner (in personal communication with Fowler, quoted 1999, 211) who holds that the text comes from a "longer (? complete) manuscript of Euclid," thus assuming that the enunciation of II 5 in it is following naturally upon the conclusion of II 4; for the second one, see Fowler 1999, 213. A third possibility is that the papyrus contained the full text of II 4, but only the enunciation of II 5, but such a mixture seems to be rather unlikely, and the suggestion, accordingly, has as yet not been put forward by anyone.

level as the last two lines of the enunciation, which are indented so as to leave space for the figure, and the rest of the diagram extends well below the end of the enunciation. By analogy, the space on the right to the indented lines 1–4 of II 5 indicates that in the place of the adjoining hole in the papyrus could be the lower remainder of the space for the diagram which belongs to the preceding II 4.[21] If the space next to the first part of the enunciation of II 5 indeed contained the II 4 diagram, the analogy with II 5 suggests that what is left of the text of II 4 is its enunciation, and that the papyrus contained only the theorems (and problems) and the respective diagrams. But it is equally conceivable that the diagram of II 4 appears at the end of the proposition, and that the papyrus contained the full texts of both (or even more) propositions – only then the diagram would have appeared at different places within the text of the respective proposition. Either way, the interpretation that the space next to the beginning of II 5 contained the lower part of the diagram of II 4 (which extends into the space of the subsequent proposition) implies an extremely close succession, even a formal interlocking, of whatever parts of II 4 and II 5 the papyrus contained. From this, the question arises why the text does not analogically go on below the text of II 5 on the left-hand side of the diagram of II 5 with either the rest of the text of proposition II 5 or with the enunciation of II 6. To be sure, the lower end of the papyrus fragment roughly coincides with the end of the original sheet, but there would have been still enough space to accommodate at least three to four lines of the rest of the same or of the beginning of the next proposition.[22] This could provide an argument for the interpretation that the papyrus contained only the enunciations and diagrams of II 4 and II 5; in this case, the text could have continued on the next sheet with the enunciation and diagram of II 6 because the combination of text and diagram would not have fit anymore below the enunciation and diagram of II 5. By contrast, if the text were to continue with the rest of proposition II 5, there would be no reason why the text would not go on in the empty space below the enunciation. Therefore, it appears to be somewhat more likely that the papyrus contained of only the enunciations and diagrams of II 4 and II 5 (rather than the proposition texts as a whole), and maybe analogously also all other enunciations and diagrams of book II or even of the complete *Elements*.

---

21 This is the interpretation by Turner, communicated and supported by Fowler 1999, 212 f., which convincingly corrects the suggestion of the editors Grenfell & Hunt. They hold that the rest of a straight line visible next to the indented (textual) line 4 is part of a figure consisting of two parallel "horizontal strokes" which unprecedentedly illustrate the phrase εἰς ἴσα καὶ ἄνισα (cf. Fig. 4[a] = Fowler 1999, Plate 2). To this, Turner rightly objects that line 2 (which marks the beginning of II 5) is indented too, although this line could well have extended until the end of the page if it were not for (part of) a larger figure taking up the space next to the indented lines (see Fowler 1999, 212).

22 This problem is not addressed by Fowler (nor by Turner nor by Grenfell & Hunt), even though he does acknowledge "the generous space immediately below the last line and the second paragraphus [*sic*] at the end of the fragment" (1999, 213).

This proposition structure is also recognizable on P. Berol [inv.] 17469, though in reverse order: the papyrus has of I 8–10 only the diagrams and the enunciations of the problems;[23] each time, the diagram comes first.[24]

P. Herc. 1061, as has already been pointed out, weaves other parts of book I into its account of I 9–10. Fowler describes its structure as follows:

> Column 8 lines 9–17 cites *Elements* I Definition 15, the definition of a circle; column 9 cites *Elements* I 9, the proposition on bisecting an angle, and its figure and a summary of the proof are given in column 10; column 9, lines 12–15 cites *Elements* I 3, a proposition that is used in the proof of I 9; and finally column 11 gives the enunciation and figure of I 10.[25]

Thus, the reference to I 9 and I 10 is utterly non-uniform and deviates from the structure of the Euclidean propositions: the account of I 9 is longer than in Euclid, including an explicit reference to another proposition and a summary of itself, with the diagram following at the very end upon the summary; I 10 is reduced to enunciation and diagram. The comparison with the presentation of I 9 might suggest that the diagram of I 10 marks the end of the reference to the Euclidean passage.[26]

Preliminary conclusions on the functions of the texts of the ancient papyrus fragments

Two of the papyrus fragments cannot have been part of a larger or even complete manuscript copy of the *Elements* which reproduces the text for the sake of preserving or communicating it. In the case of P. Herc. 1061, this is already clear from its being an indirect testimony to the Euclidean work within another treatise, but even the part which it quotes, although it centers around two successive Euclidean propositions, does not adhere to the structure of the *Elements* (neither in terms of proposition order nor in terms of the structure of the individual texts). P. Berol [inv.] 17469 does adhere to the order of the propositions and could be part of a larger text that does so, but it only gives the diagrams and enunciations instead of copying the full proposition texts. It seems to be part of a collection of only the enunciations and diagrams of the respective propositions.[27] The possible functions of such selective reception of the *Elements* are manifold; it could be a 'thematic index,' or it could originate in the context

23 Brashear 1994, 29, emphasizes that neither can there be seen traces of the body and end of the Euclidean propositions anywhere on the fragment, nor is there any space left or recognizable on the fragment where they could have been found.

24 Brashear 1994 does not discuss the unusualness of the diagrams preceding the text in the face of the Byzantine manuscript transmission in which diagrams tend to appear at or toward the end of the proposition texts or at the end of the page (see Section 4.2, 218 f.).

25 Fowler 1999, 210.

26 Fowler 1999, 210.

27 Brashear 1994, 29.

of mathematical, either autodidactic learning or institutionalized education.[28] In any case, the aim of it cannot have been to copy the *Elements* for the sake of preserving it, for the most crucial part, the proof, is missing.

Therefore, only the remaining two of the fragments – P. Fay. 9 and P. Oxy. i 29 – could have been part of a text which presented the propositions in the way of the full text of the *Elements*, both as far as the proposition order and the structure of the propositions themselves are concerned. P. Fay. 9 presents the two (presumably) successive propositions I 39 and I 41 successively and does not give any indication of being confined to only the two visible propositions. Since the text, moreover, seems to adhere to the structure of both Euclidean propositions, it indicates that it could have been part of a larger manuscript copy of the *Elements*. This is also suggested by the formal qualities of the text, that is, the correct spelling, the 'Normalzeile,' and the 'handwriting,'[29] which are in turn matched by the form of the diagram (for which see the following section). What does not seem to fit the form of the fragment is that the text of the propositions deviates strongly from the now accepted text of the editions. As Fowler puts it, "the fragment remains as a tantalizing unresolved puzzle: that the surviving papyrus fragment that most closely resembles a professional scribe's copy differs so much from the received text."[30] The deviations from the Euclidean text could be explicable either by the quality of the underlying copy or by the purpose of improving the text, or by the writer's own mistakes, but, in any case, they suggest some distance from the Euclidean text. Regarding our current purpose, this implies

28 Cf. Fowler 1999, 215: "There are several possibilities for its origin: it could be a teacher's or student's notes on the text, an *aide-mémoire* [. . .] It could be a thematic index to some or all of the *Elements* [. . .] this is often done for volumes of music, which give the first two or three bars of each piece, and modern catalogues of manuscripts, which give the incipits of each item." The editor of the text, W. Brashear, deems it an "Auszug, der wohl zu Privatzwecken, vielleicht von einem Schullehrer angefertigt wurde," but in any case not a "originalgetreue Abschrift des euklidischen Werkes" (1994, 29). Netz 1999a, 19, n. 31 argues that it "is fair to assume that the original papyrus had more propositions, treated in the same way. My guess is that this was a memorandum, or an abridgement, covering the first book of Euclid's Elements. Had someone been interested in carrying out the proof, the lettering would have occurred on a copy on, e.g., a wax-tablet." He goes on to claim that the theory he shall put forward later in his book on the time and manner of the lettering – "prior to the mathematical reasoning" (19), "a semi-oral dress-rehearsal during which letters were assigned to point" "is in agreement with the evidence from the papyri." I think, however, that it is impossible, as Netz attempts, to pin down the exact order in which certain parts of the text/diagram entity came into being, and much less to say whether the letterless diagrams are in accordance with such a theory or not.

29 Fowler 1999, 214, with reference to Heiberg and the editors Grenfell, Hunt, and Hogarth; Fowler himself disagrees on Heiberg's recognition of a "Normalzeile," finding the lines differing considerably in length (1999, 213).

30 Fowler 1999, 214 f. Because of its ambivalent implications, the fragment has been considered varyingly "eine verwilderte Überlieferung" or "an imperfectly remembered exercise" and "part of a regular book" ("ein wirkliches Buch," Heiberg and the editors Grenfell, Hunt, and Hogarth quoted by Fowler 1999, 214 f., n. 43, and 214, respectively).

that one should be cautious to draw conclusions from the fragment about the status and look of the diagrams in Euclid.

P. Oxy. i 29 less clearly indicates whether it could have been part of a larger or full copy of the *Elements*: of both of its successive propositions II 4 and II 5, it only shows one of the abstract parts, respectively, and so could be an excerpt from the *Elements* in the way of P. Berol. [inv.] 17469. As such, it could have originated either as part of a collection of enunciations and respective diagrams, or it "could have been part of a manuscript of notes by someone working through book II of the *Elements*," as Fowler suggests with regard to the quality of the writing and writing material ('its quickly written hand on poor quality papyrus').[31] Although he thus concludes that the "fragment does not appear to come from a scribe's copy of the work," this possibility cannot be excluded in view of the fragmentary state of the papyrus: Since the papyrus seems to break off soon after the II 5 enunciation, and since it is not clear which part of II 4 precedes, one cannot rule out that the papyrus presented the two propositions in full and maybe even a larger part of the *Elements* or the work as a whole. Turner claims precisely that it was part of "a longer (? complete) manuscript of Euclid."[32] In advocating this view, he mentions the existence of a *kollesis* and the "empty space at the left," which indicated "that this was a generously laid out manuscript;" moreover, he professes a somewhat more favorable view of the quality of the writing[33] – by contrast with that of the drawing (on which see the following section).

4.1.3 A review of the diagrams of the ancient fragments

I now review the transmitted diagrams as regards their form and classify them according to their common features. Subsequently, I attempt to align my findings with that of the previous section on the text of the fragments. In doing so, I consider again

31 Fowler points out that the quality of the drawing points in the same direction (1999, 213); on the diagram of the fragment see the following section. Netz 1999a, 19, presumes the same function ("memorandum," "abridgement") for this fragment as for the Berlin papyrus.

32 Turner, quoted by Fowler 1999, 211. That Turner by "manuscript" indeed means a book manuscript and not a "manuscript of notes," as Fowler calls it, is clear from this opposition and from his mentioning of a *kollesis* and the margin on the left-hand side as an argument in favor of his view, neither of which would probably have characterized a collection of notes.

33 "The handwriting is a medium to large rounded capital, roughly bilinear, fairly quickly written, clear to read but with no claim to be regarded as calligraphic" (Turner, quoted by Fowler 1999, 211). Besides these arguments in favor of his manuscript hypothesis, Turner adduces "the fact that this is a II 5." It is not evident why this would be an argument for the text's coming from a longer manuscript – is it because it is a proposition from somewhere in the middle of the work, or because it is a proposition from book II rather than from the much better known and more often quoted book I, or because II 5 specifically for its rather complicated content and long proof would not have been dealt with except by a professional copying the whole book?

all papyrus fragments (not only the two which might come from manuscript copies) as well as the ostraca to see whether there is some congruence as to the form of diagrams on a larger scale.

The materials on which the ancient diagrams are transmitted

Ostraca[34] pose special difficulties to the drawing of diagrams. Originally parts of pottery for daily use such as vessels, they would be curved to a greater or lesser degree and thus uneven, with the drawing surface being rather hard and originally fairly smooth, but likely to have a more or less damaged surface with holes and split-offs from the shattering of the object of which they were a part and from subsequent storage. For these reasons, it would be difficult or even impossible to properly employ a ruler or compass on such a surface. The surface, moreover, cannot be prepared for mathematical use; rather, a piece of pottery suitable not only in terms of its surface but also of size has to be chosen to begin with. Ostraca provide mostly very limited space, and often no larger or defined areas for a geometrical drawing. For these reasons, ostraca do not qualify materially for containing whole book manuscripts, on the one hand, or for displaying technical and precise drawings on the other hand. As far as our specific set of ostraca is concerned, the editors neither comment on the size of the two fragments which contain diagrams in relation to the size of the diagrams, nor on the place of the diagrams, nor on the surface state. But by the general characteristics of ostraca, the use of common technical means of geometry can be excluded from the start for the two diagrams.

As for papyrus, size is not an issue since long roles may be produced by means of the *kollesis* of papyrus sheets, notwithstanding the potential cost of the material. Thus, papyrus is, generally, apt for accommodating large amounts of text, even a whole book or a whole work like the *Elements*, and drawings of some size. A *kollesis* is only recognizable in the Oxyrhynchus papyrus, but the state of transmission does not allow for excluding a larger context for the other fragments. The size and shape of the writing and drawing space is usually dictated by the chosen form of the column, which is in any case straight, bounded by an imaginary or actually drawn upright rectangle. The diagrams are fitted into the space defined by the column, or the space between columns (but must not, at least not in case of the primary use of the

34 Cf. Netz 1999a, 14–17, who discusses the media which are attested for having been available to certain Greek mathematicians (and in general to authors engaging in mathematical writing) apart from papyrus, and their advantages and disadvantages regarding the drawing of diagrams: ("wetted and tamped") sand/"dusted surface," "ostrakon or wax-tablet," pinakes, "leukoma" (wood painted white; suggested by Jackson (1920, 193, cf. Netz 1999a, 15, n. 15) as a medium used by Aristotle, and mathematical columns out of stone or marble. The essential difference of all of them, including papyrus, compared with the media now available ("pencil/paper, chalk/blackboard, and [. . .] computer/printer"), is that "none had ease of erasing and rewriting" (Netz 1999a, 16).

papyrus, be adjusted to the size and form of a fragment, which is always the case with drawings on ostraca, which are, by definition, secondarily used). In practice, of course, the space and place available for diagrams is often confined by the overall size and amount of material available or determined by aesthetic norms. Accordingly, the diagrams may be generously drawn in the middle of a column between two passages of text, as appears to be the case with P. Fay. 9 and P. [inv.] Berol. 17469 or squeezed inside the textual space in the lower right-hand corner of a column, as might be the case in P. Oxy. i 29. In no case do the diagrams fit into the narrow space between columns.[35]

Concerning the surface, the material comes in very different qualities, ranging from low-quality rough and thick to high-quality smooth and thin papyrus. Generally, rulers and compasses may be applied. Usually, the recto-side of it, i.e., the one with the fibers running horizontally along the page, parallel with the lines of writing, is the one on the inside of the papyrus which is used for the original writing purpose (while the backside may be written on by subsequent users of the material). This is because the writing is facilitated by this orientation of the fibers. Conversely, for geometrical drawings, this means that they only some of the time smoothly follow the fibers, since they do not only run horizontally, but may extend vertically over more than one strip. In this case, the copyist might have encountered some difficulties in drawing straight lines against the grain, especially if the fibers were rough, as well as vertically to the strips without sliding. The pen even might have got stuck where the papyrus strips are glued together. In general, the material allows for the application of the common technical means for producing geometrical drawings, but their quality and look may be impeded by the quality of the surface, with the surface taking in the ink unevenly, so that it either runs or leaves no traces at all. As a result, certain lines appear thicker than others, and lines may deviate slightly from their original course despite the application of a ruler.

The papyrus fragments preserved for the *Elements* come in very different qualities, ranging from the "[t]hick and poor quality papyrus" of P. Oxy. i 29, the obstructiveness of which might have somehow contributed to the extreme thickness of some of the drawing's lines[36] to the higher quality of P. Fay. 9, which appears to have allowed for using a pen that produced very delicate and evenly drawn lines.[37] Neither

35 See for such a practice the manuscripts for Ptolemy's *Geography*.

36 Quotation by Turner in Fowler 1999, 211. Neither of them, however, ascribes the quality of the drawing to the quality of the paper. Rather, Fowler attributes the thickness to the lines' being redrawn in the cause of working through the proof (see below, p. 204). However, it cannot be excluded that the thickness might have come about in the event that the copyist felt that the lines initially drawn were too thinly or not properly reproduced on the papyrus and caused the exact opposite by his efforts to change the situation.

37 Turner in Fowler 1999, 214, presumes the use of "different, sharper pen" for the letters and lines of the diagrams (sc. than for the text), but does not elaborate on the quality of the papyrus.

do the likewise rather thin lines of the figures of P. [inv.] Berol. 17469 seem to be impeded by the quality of the underlying material,[38] however, the crossing of several lines at the top of the diagram of I 9 results in a thick and therefore imprecise point.

The editors do not comment on whether the surfaces of the papyri have been prepared or smoothed in advance to facilitate the application of a diagram, either as a whole or in the very place of the diagram. However, it is clear from the photographic reproductions that the diagrammatic space has not, in any case, been organized anyhow by means of ruled lines which could provide orientation for drawing the diagram in terms of direction or in terms of length and proportion. One can recognize neither a framing in the form of a square or rectangle which predefines the space into which the drawing must be fit and the lines of which may be used themselves as part of the drawing, or for the alignment of lines inside the frame which are parallel or perpendicular to them.[39] Nor can one recognize any internal organization of the diagrammatic space by means of vertical and horizontal lines in the manner of a grid, as in modern quadrille paper or scale paper, or of the preparation of maps by drawing meridians and parallels of latitude that we find in Ptolemy.

Drawing techniques

As one would expect, the diagrammatic traces transmitted for the material of the *Elements* on ostraca indicate free-hand drawing; the straight lines have been drawn without a ruler, the circular arc without a compass.[40] In this way, they do not qualitatively differ from the whole sample of evidence transmitted on papyri; for although the material generally enables the use of such means, these are not automatically applied. Of the four papyrus fragments, two feature ruler-drawn constructions[41] (P.

38 Its editor Brashear, though, does not comment on the quality of the papyrus. The Herculaneum papyrus P. Herc. 1061 is a special case, as it is, after having been unrolled in a special procedure, in a very bad state, so that "the photographs show of little use" (Fowler 1999, 210) and do not inform about the original state of the papyrus (Fowler 1999, 210).

39 For such framing rectangles, cf. Ptolemy's *Geography* which even explicitly assigns the drawing of such a frame. In this as in other cases mentioned further below (Section 4.5), the comparison with Ptolemy proves productive, as unlike Euclid, he makes methodological remarks on the construction of maps and diagrams. The silence of Euclid on such issues has, on the one hand, generic reasons which have to do with the rules of the axiomatic-deductive method: he never explains why or how he does something, but simply does it (see Section 1.2, 17, on the general lack of metamathematical remarks in Euclidean mathematics). On the other hand, the contrast with Ptolemy regarding the description of the construction may be due to a different take on the importance of (diagrammatic) construction in their texts (see below, p. 413).

40 Accordingly, the editors characterize the drawing of P. 12611 as "Zeichnung eines grob gezogenen Kreisbogens mit Sehne" and of P. 11999 as "grobe[] Handzeichnung" (Mau & Müller 1982, 4 and 6).

41 On why they are not really 'constructions,' but only pseudo-constructions despite being drawn with the help of rulers, see below, pp. 205–211.

[inv.] Berol. 17469 and P. Fay. 9), while another two contain roughly, that is, free-hand drawn sketches (P. Herc. 1061 and P. Oxy. i 29).[42] A compass is not employed in any of them as the respective propositions do not revolve around circular figures; neither do the fragments necessitate or display the use of another means of drawing, because they only feature objects which consist of straight lines. In the present section, I examine to what effect rulers are used or not used in the ancient papyrus fragments, mostly regarding what impression they convey in terms of straightness, precision, and correctness. In so doing, I do not mean to presume that ancient diagrams were necessarily aimed at these qualities.[43] However, the editors of the ancient diagrammatic fragments have speculated about their function not only based on the quality of the text and writing (as we have seen above), but also based on the appearance and, particularly, the 'precision' of the diagrams.[44] For our question of whether diagrams were part of the ancient manuscripts and if so, in what form, it is indeed vital to determine the function of the extant papyri. However, rather than drawing conclusions on this function directly from the degree of precision of the diagrams, we must first explore what diagrams in ancient manuscript were expected to look like. To this end, I examine in the present section to what aesthetic effect or mathematical purpose technical instruments were employed, in what ways (if any) free-hand and ruler-drawn diagrams differed, and whether the fragments could be grouped together based on the quality of the diagrams.

In all three roughly drawn diagrams of P. Herc. 1061 and P. Oxy. i 29, straightness is only achieved to a certain extent so as to appear to characterize the supposedly straight lines as straight rather than circular or curved. The diagrams of the Herculaneum papyrus seem to have been drawn fairly quickly and without much care. Ac-

42 I use the term 'roughly drawn' throughout in this sense, that is, 'free-hand drawn' as opposed to 'ruler-drawn,' referring to Fowler's use of the phrase with regard to the diagram of P. Oxy. i 29, although he seems to use the phrase to describe the *specific* look of this free-hand drawn diagram, not necessarily the general quality of its being drawn without a ruler.

43 In Section 4.3, I will explore what expectations are raised by the textual construction as to what a diagram should look like.

44 In fact, only Brashear, the editor of the Berlin papyrus, refers explicitly to the "precision" ("Präzision") of the diagrams (1994, 29), and he does so in a neutral manner, without drawing conclusions about the function of the text of the papyrus. However, the notion of '(im)precision' seems to be a criterion for determining the function in other editors and commentators of the ancient papyri: For instance, Fowler thinks that the Berlin papyrus – as it contains enunciations and unlettered figures – supports his interpretation of the P. Oxy. i 29 "as a similar text of enunciations and unlettered figures, *though* that text was much more casually written and its figure more crudely drawn" (1999, 216, my italics). Turner assumes a different function for the latter papyrus ("a longer [? complete] manuscript of Euclid") but argues similarly that "this was a generously laid out manuscript. But the figure is drawn without rule, free-hand, and carelessly [. . .];" that is, he seems to think that the style of the figure somehow contradicts his manuscript hypothesis (quoted by Fowler 1999, 211, my italics). Regarding the writer of P. Fay. 9 he notes that "he was taking some trouble (using a rule), and perhaps these indications are enough to suggest that we have part of a real manuscript for the *Elements* here" (quoted by Fowler 1999, 214).

cordingly, they do not attest to much effort being spent on achieving straightness. In the diagram for I 9, the straight line AB appears as slightly curved and bent to the right, and the straight line AΓ does not seem to have been drawn in one stroke: rather, the copyist appears to have lifted the pen once before he continued drawing it. In the diagram for I 10, again, the two recognizable sides of the triangle are bent to the inside of the figure.[45]

In P. Oxy. i 29, the thickness of some of the lines seems to indicate that the copyist has gone over the sketch at least one more time.[46] The overall impression conveyed by the diagram from Oxyrhynchus is that the figure appears somewhat crooked but that it accomplishes the straightness of the lines more successfully (if that was the aim) than the ones from the Herculaneum papyrus. Only lines AΓ and EZ appear to be slightly curved (in order to be able to refer to the lines of the unlettered diagram of the papyrus, I mention the letters as they can be found in the diagrams of the manuscript tradition in accordance with the text of proposition II 5).[47] As we have suggested above, one way of interpreting the multiple tracing of some of the straight lines, especially of ΓE (or rather of its compartments ΓΛ and ΛE, for they seem to have been redrawn independently from one another), is to assume that they are the result of trying to clarify the figure in terms of the straightness of its lines. An observation in favor of this interpretation could be that in the case of EZ, the tracing toward the side of Z does not overlap with the first stroke of the line: the upper version of the line appears to be the correct one in terms of straightness, drawn arguably in correction of the first crooked one.[48] Fowler has suggested another interpretation: he argues that the text "could have been written by somebody who wrote down the enunciation, drew the figure, then worked through the proof doubling the lines on the figure as he did so, but without writing down any of the details of the proof or labelling the figure;" according to him, this explains why only "the diagonal line is undoubled: the diagonal enters only in the construction of the figure and does not play any part in the proof," so the diagonal line does not get redrawn while the writer works through the proof part.[49] I will discuss the usefulness of an unlettered sketch below. Here, I would only like to note that Fowler's argument is weakened by the fact that the lines that are referred to most frequently in the course of the proof (the segments of AB: AΓ, ΓB, AΔ, ΔB, ΓΔ) are not the ones which appear to be the thickest, while, conversely,

45 Cf. Fig. 2 p. 193 above.
46 Cf. Netz 1999a, 15, who points out that the impossibility of "erasing and rewriting" is a central property of all ancient media available to mathematicians.
47 Cf. Fig. 4[b] p. 195 above.
48 "Arguably," because it could also be the other way around: the copyist who traces the sketched-out lines while reading through the text might have failed to have the doubling line match the first, correctly drawn line, as might have been also the case in the top left-hand corner of the figure, where point A appears to have been drawn at the correct place at first, before the crooked doubling of line AΓ shifted its place.
49 Fowler 1999, 213.

the thickest ones (ΓΕ or its segments ΓΛ and ΛΕ) are not mentioned as often.[50] The diagram's sketch-like quality becomes apparent not only from the non-straightness of the straight lines. Some of the lines and points became so thick and bulky that they no longer match the definition of 'line' and 'point,' respectively. This does not seem to be an effect of the ink's running on the papyrus but will likely have to do with the process of redrawing the diagram. For whatever reason, some of the lines were drawn multiple times.

By contrast with the free-hand drawn diagrams, all extant ruler-drawn diagrams are precise regarding the straightness of the lines. However, beyond the question of straightness, each of the diagrams seems to display a different degree of precision, judging by the look of them: the drawing of I 39 on the Fayum papyrus seems to be extremely fine, and it is only in the upper-left point A that the convergence of the four lines AB, AΓ, AE, and AΔ is a little inexact and leaves a somewhat prominent mark.[51] Compared with this diagram, the two better preserved of the three diagrams of the Berlin papyrus convey a general sense of imprecision:[52] in the diagram for I 9, the lines ΔE and ZE do not even meet AΓ so as to produce point E on the latter line;[53] the straight line AZ runs beyond Z, and point A, just like point Γ in the diagram for I 10 (and there also the uppermost part of the line ΓB) are rather bulky. The bulkiness (but not the gap) may be explained by the tendency of the papyrus to absorb the ink very quickly where the pen stopped for too long.

Thus, from the four papyri fragments, it turns out that the use of a ruler, which, on the face of it, implies that the diagram in general is precise and has been properly constructed, guarantees precision only regarding the straightness of straight lines. It is only in this respect that the fragments which contain diagrams that are drawn without a ruler and those which contain diagrams that are drawn with a ruler may form the distinct groups. Otherwise, the dichotomy 'roughly drawn/ruler-drawn' does not seem to hold: first, because all four fragments differ considerably regarding the precision of their drawings, and second, because one of the ruler-drawn diagrams appears rather 'roughly drawn' despite the use of a ruler.

The significance of the application of rulers regarding the precision of the drawing may be even more doubtful if one does not only consider the general appearance of the drawing but examines the drawings as to whether they really depict the objects

50 Fowler's argument is supported, however, by the fact that the thinness of KM and especially its compartment ΛM which looks just as thin as the diagonal can also be explained by his hypothesis regarding the diagonal: KM, too, is only mentioned in the construction part and does not recur in the proof.

51 I refer to the capital letters as they can be found in the reproductions of the diagram in the modern editions (cf. Fig. 1(b) above p. 191).

52 Again, the letters used for referring to the unlettered diagrams stem from the diagram as it is shown in the modern editions (cf. Fig. 3(c) p. 194 above).

53 To be sure, the gap between the triangle and AΓ could also be due to abrasion of the writing, but if one extended the lines of the triangle, there a gap would still remain.

and object relations described by the text. It is clear that those three diagrams which have been drawn without a ruler cannot but imitate the precisely defined objects and relations, just as they have been shown to only imitate straightness. Of the ruler-drawn diagrams which are concerned, P. Fay. 9 is so badly transmitted that one can hardly judge its mathematical precision. Let us, therefore, have a closer look at the diagram on the Berlin papyrus for I 9, the proposition which assigns the bisection of a given rectilinear angle. One can see even without measuring the lines of the triangle ΔEZ that they are not equally long, and therefore do *not* form an equilateral triangle (τρίγωνον ἰσόπλευρον) as the text assigns, but, if at all, an isosceles one (ἰσοσκελές), with the line ΔE being, in any case, much longer than ΔZ and EZ. Thus, the diagram does not feature an object which is constructed as a basis for the proof which shows that the construction problem has been solved.[54] Likewise, it is obvious from merely looking at the drawing that the line ΔE is not bisected into two equal halves by the line AZ. This affects the outcome of the proof itself: for if the two sides of the line ΔE on either side of the line AZ are not the same, then the angle BAΓ would not have been bisected either – which was precisely the construction task, but which has not been accomplished, as far as one can tell from the diagram. Apparently, ΔZ and EZ are not the same – thus, ΔEZ turns out to be not even an isosceles triangle – which is why the angles ΔAZ and EAZ are not the same, and BAΓ is not bisected by AZ.

The depiction is not only imprecise in terms of occasionally thicker points, but also obviously wrong, in that it has not been technically constructed, notwithstanding the fact that it was drawn with the help of a ruler. More precisely, I 9 presupposes (besides I, Def. 20 and the theorem of I 8) the constructions of I 1 and I 3.[55] First, I 3 is necessary to "cut off from the greater a straight line equal to the less,"[56] in order to cut off from the line AE the line AΓ equal to AΔ (Fig. 5). The construction of I 3 requires drawing a circle and implies the use of a compass. I 3 itself depends on the construction of I 2 (Fig. 6).[57] Second, I 1 is necessary "(. . .) to construct an equilateral triangle on a given finite straight line"[58] for constructing the equilateral triangle ΔEZ on the base ΔE (Fig. 7). However, the diagram of I 9 on the Berlin papyrus does not

54 The point is not that the proof would actually work even on the basis of an isosceles triangle with the base ΔE, but that there are separate terms for the two kinds of triangles, which are employed unambiguously in the *Elements*, and that the triangle depicted on the Berlin papyrus is clearly not an equilateral one.

55 Acerbi 2007, 2543.

56 *Elem.* I 3 (vol. 1, p. 9.16 f. Heiberg & Stamatis): δύο δοθεισῶν εὐθειῶν ἀνίσων ἀπὸ τῆς μείζονος τῇ ἐλάσσονι ἴσην εὐθεῖαν ἀφελεῖν.

57 *Elem.* I 2 (vol. 1, p. 8.11 f. Heiberg & Stamatis): "To place at a given point (as an extremity) a straight line equal to a given straight lines." πρὸς τῷ δοθέντι σημείῳ τῇ δοθείσῃ εὐθείᾳ ἴσην εὐθεῖαν θέσθαι.

58 *Elem.* I 1 (vol. 1, p. 7.1 f. Heiberg & Stamatis): ἐπὶ τῆς δοθείσης εὐθείας πεπερασμένης τρίγωνον ἰσόπλευρον συστήσασθαι.

show any sign of one of these auxiliary constructions.[59] Without constructing the figure properly in the way suggested by the deductive system of the *Elements*, in which more complex propositions follow upon proofs that they build on for their own constructions and proofs,[60] the diagram must turn out to be non-metrical: i.e., the "quantitative relations inside the diagram" do not "correspond to such relations between the objects depicted." This means that the "schematic diagram" can do no more than convey a general impression of the "qualitative relations of the geometrical configuration" and the evolving look of the figure, but cannot show exact proportions.[61]

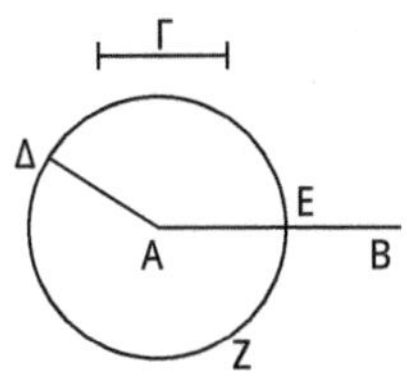

Fig. 5: I 3 according to Heiberg & Stamatis.

If one compares the ruler-drawn diagram for I 9 of the Berlin papyrus with the 'roughly drawn' diagram for I 9 of the Herculaneum papyrus, it turns out that the former is, in fact, no more qualitatively precise and no more technically constructed than the latter. To be sure, in it, the straight lines are indeed straight, but 'straight' (εὐθεῖα) is only one adjective that characterizes objects in the *Elements* among many

59 Interestingly, such necessary auxiliary constructions will turn out to be absent from the manuscripts, too (and from the modern editions which are largely oriented toward the Byzantine depictions), and I will debate the repercussions of this situation below, Section 4.2.2, 223.

60 Other ways of construction besides those performed in foregoing propositions are theoretically possible, as the *Elements*, as a rule, only presents one single solution to a certain construction problem or proof of a certain theorem, while many times others would be available, many of which the commentary reception has felt the need to supply (cf. Saito 2009, *passim*). A means which the *Elements* generally does not consider is the use of measurements: it works entirely without indicating them, which suggests that they are neither to be employed for the construction nor for tracking the deductive assertions in the diagram (more on this in Section 4.3.3, 251 f., where I examine the textual construction regarding how it helps constructing the diagram). That is, merely measuring the length of line AΔ and then cutting that length off AΓ so as to produce AE on it would not be considered an option, as it is not a proper construction according to the construction method of the *Elements*.

61 Netz 1999a, 18, who argues that, in general, the diagrams transmitted for the *Elements*, i.e., above all by the manuscript tradition, are not "metrical" in the above sense and asks whether they were even meant to be in the first place. I shall pursue this question on the basis of a close analysis of the textual construction in *Elements* I below. The opposition "quantitative" and "qualitative" may be misleading: what seems to be meant is that the object relations mentioned in the text are not shown quantitatively correctly in the diagram, but that their qualities are "schematical[ly]" represented by the diagram. The inexactness of one of the ancient diagrams regarding proportions is explicitly acknowledged by the editors of the ostraca with regard to the drawing of P. 11999 ("Proportionen nach der groben Handzeichnung auf P. 11999 geschätzt," Mau & Müller 1982, 6).

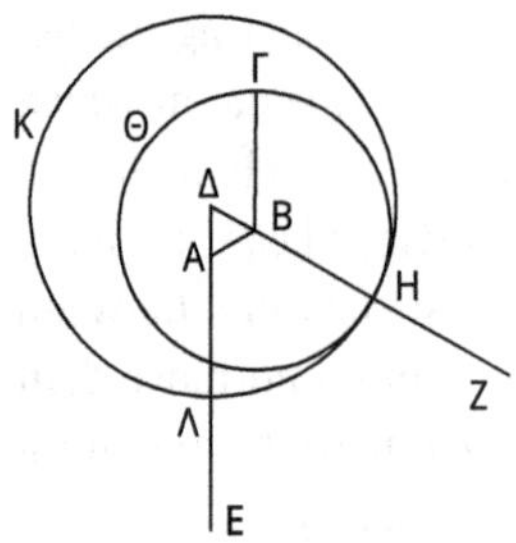

Fig. 6: I 2 according to Heiberg & Stamatis.

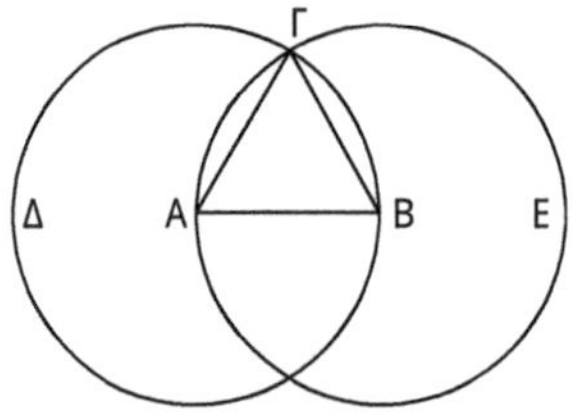

Fig. 7: I 1 according to Heiberg & Stamatis.

which do so explicitly (e.g., ἰσόπλευρον, 'equilateral') or implicitly as part of the definition of an object. Apart from straightness, virtually none of the qualities that either characterize objects or relations between objects are realized in the diagram for I 9 (and in that for I 10) on the Berlin papyrus, nor can they possibly be unless the objects are constructed rather than approximately drawn.

From this perspective, it appears unjustified to attach too much importance to the use of a ruler when one wishes to assess the technicality and professionality of the manuscript copy in which it appears. It is true that, from an emic point of view, the use of a ruler might nevertheless indicate a serious occupation with a text, e.g., of a professional scribe. However, starting from the text only, the use of a ruler makes only a small difference in view of the overall imprecision of the diagram in relation to the precise textual assignments. In fact, the supposedly roughly drawn diagram for I 9 of the Herculaneum papyrus turns out to display the quantitative relations of the figures described possibly more precisely than the equivalent ruler-drawn one of the Berlin papyrus, because it is more easily recognizable from the former that the triangle is supposed to be equilateral and that AZ is supposed to bisect ΔE, and also the angle BAΓ.[62] Thus, although it is technically no more precise than the other, it gives a better (i.e., more precise) impression of the objects and object relations described.

62 In Heiberg's drawing of the diagram of the papyrus (1900, 164), point Δ is wrongly called Λ, for a Δ is feebly readable in the papyrus, as Heiberg himself acknowledges in his printing of the diagram. Another mistake of Heiberg's drawing is that it does not extend the line AZ from ΔE toward A: even though the part of the line inside the triangle is much clearer, traces of the extension of this line toward A are faintly visible in the original (cf. Heiberg's note in the apparatus, 1900, 165). A mistake of

It may be suspected with some certainty that the figure transmitted for I 39 on the Fayum papyrus is just as imprecise with respect to the objects and object relations mentioned by the text, although only a few parts of its lines and none of the described object relations are recognizable because of its fragmentary and somewhat distorted state. As far as one can see, no auxiliary lines that indicate a technical construction can be recognized in the diagram. This is remarkable because proposition I 39 ("Equal triangles which are on the same base and on the same side are also in the same parallels")[63] – like I 9 – presupposes a deductive chain for the construction of the proof by contradiction:[64] first, the construction of I 31 (Fig. 8), that is, of "a straight line parallel to a given straight line [. . .],"[65] which, again, presupposes for its construction proposition I 23 (Fig. 9), which solves the problem of "constructing a rectilineal angle equal to a given rectilineal angle [. . .],"[66] which presupposes, in terms of construction, I 22 (Fig. 10), i.e., "to construct a triangle [. . .],"[67] which presupposes the above quoted chain of dependent constructions of I 1–3.[68] This means that the diagram, if properly

---

the original diagram itself is that it calls the two parts of the angle ΒΑΓ by the letters Η and Θ (cf. Heiberg's note ibid.).

63 *Elem.* I 39 (vol. 1, p. 52.18 f. Heiberg & Stamatis): τὰ ἴσα τρίγωνα τὰ ἐπὶ τῆς αὐτῆς βάσεως ὄντα καὶ ἐπὶ τὰ αὐτὰ μέρη καὶ ἐν ταῖς αὐταῖς παραλλήλοις ἐστίν.

64 Subsequently to the construction, the proof presupposes the proof of I 37.

65 *Elem.* I 31 (vol. 1, p. 43.11 f. Heiberg & Stamatis): [. . .] τῇ δοθείσῃ εὐθείᾳ παράλληλον εὐθεῖαν γραμμὴν ἀγαγεῖν.

66 *Elem.* I 23 (vol. 1, p. 32.12–14 Heiberg & Stamatis): τῇ δοθείσῃ γωνίᾳ εὐθυγράμμῳ ἴσην γωνίαν εὐθύγραμμον συστήσασθαι.

67 *Elem.* I 22 (vol. 1, p. 30.23 f. Heiberg & Stamatis): [. . .] τρίγωνον συστήσασθαι.

68 Acerbi 2007, 2544. In his "Tabelle deduttive," Acerbi lists (besides the underlying definitions, postulates and axioms) also all of the proofs underlying the above quoted constructions, which, in turn, themselves presuppose underlying constructions; I have concentrated on delineating the chain of foregoing propositions which is necessary for doing the construction of the proof by contradiction, without having to prove all underlying theorems (again by means of construction). Acerbi's table, just like the specification of the numbers of the underlying proposition directly in the text by Heath, only gives the numbers from the construction part onward, but does not specify the constructional basis for the exposition. This concerns, on the one hand, the relationship of the textual parts of exposition and construction, and, on the other, the construction of the diagram: how was it meant to come into being, if the most important objects were not to be constructed – it could only be by means of a sketch – and what was, then, the significance of the subsequent construction part, for even though it is thought to be based on previously solved construction problems and, thus, doable, it cannot really be done in a half-finished but only sketchy diagram. The problem may not seem vital with regard to such propositions which, at the beginning, posit very simple objects and relations which do not require a proper construction anyway, as, e.g., I 23, which simply requires a line, a point on it, and an angle, but it does come up in propositions like I 39 which posits two equal triangles on the same base. This means that either the understanding of the deductive system in the main modern editions is wrong in not taking into account the exposition part, and the objects of the exposition relied on previous constructions for their own constructions, just like those of the subsequent construction part; or the nature of the exposition part betrays that the diagram as a whole was not meant to be constructed in the first place, not even by the construction part. Either way, it would be difficult to account for the

constructed, would have to bear evidence of the underlying constructions in the form of auxiliary lines which reproduce, as it were, the diagrams resulting from the underlying constructions: that is, it would have to contain an auxiliary line (corresponding to AΔ in I 31) connecting A to BΓ, which is necessary for constructing AE through point A parallel to BΓ. The construction of AE, then, with the help of this line requires, according to I 31, the construction of the angle in A on one side of this line which is equal to the angle formed by BΓ and this auxiliary line on the other side of the line (these angles, thus, would form the alternate interior angles at the auxiliary line when it cuts the parallel lines AE and BΓ).[69] The construction of an angle equal to another at a straight line, in turn, requires, with I 23, the construction of a triangle, of which the angle to be constructed, formed by the future AE and the auxiliary line, would be a part and which would thus be visible where the auxiliary line cuts AE. The construction of a triangle from three lines equal to three given straight lines (i.e., of the given angle, that is, the triangle constructed at the given angle) would require, however, I 22. This proposition accomplishes the triangle construction with the help of a straight line and two circles, which would have left traces in the diagram of I 39. More auxiliary circles would have to be expected from the constructions of I 1–3 which are necessary for accomplishing the construction of I 22. However, the fragments of the diagram for I 39 do not bear evidence of any of the underlying constructions. This means that the discrepancy between its precise appearance due to the use of ruler and fine writing material and, at a closer look, its emerging non-technicality is even greater than in the case of the diagrams of the Berlin papyrus, and its use of a ruler all the more deceiving.

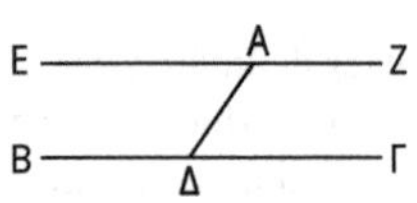

Fig. 8: I 31 according to Heiberg & Stamatis.

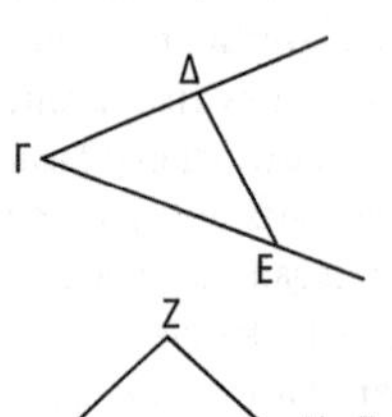

Fig. 9: I 23 according to Heiberg & Stamatis.

conceptual difference implied by the exposition which apodictically posits objects, and the construction which at least purports that the objects be brought into being geometrically. I shall return to this in the part specifically devoted to the ramifications of the nature of the textual construction (Section 4.3.3, 272).

69 Proofs by contradiction present a cognitive paradox in the sense that their constructions on the basis of formerly solved construction problems are, by the end of the proof, said to lead to impossible situations.

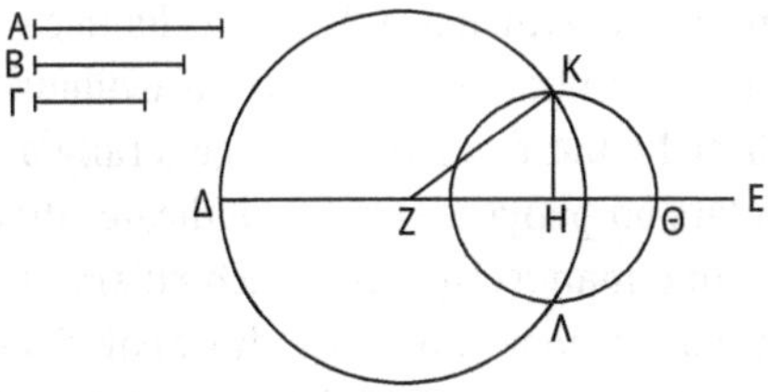

Fig. 10: I 22 according to Heiberg & Stamatis.

It emerges that none of the ancient ruler-drawn diagrams is properly, that is, technically or geometrically, constructed, let alone metrical, and any more precise than the roughly drawn ones, except regarding the straightness of the straight lines. Rather, beyond the 'ruler-drawn/roughly drawn' dichotomy, each of them displays an individual drawing style and conveys a different impression of precision.

While all of the objects contained by the diagrams transmitted for the *Elements* from antiquity theoretically can be drawn by means of ruler and compass only, as the small ancient diagrammatic evidence revolves around only one specific type of objects, i.e., two-dimensional rectilinear ones,[70] the *Elements* is not limited to these objects.[71] Some of the objects dealt with in the further course of the work cannot be drawn in any case solely by using these two most basic instruments. This concerns the depiction of perspective in the case of three-dimensional objects that involve circular lines.[72] I shall consider these objects both in my review of the diagrams of the manuscripts and in my analysis of the textual construction, although they do not occur in *Elements* I, which only contains rectilinear and circular two-dimensional objects and is, therefore, not representative in this respect.

To conclude from the review of the techniques of drawing employed in the ancient diagrams, they do testify to the ancient use of rulers. However, since they are not employed systematically in the fragments that we have, it is not certain whether they were used standardly in ancient manuscripts. It is possible either that rulers were used systematically in the ancient manuscripts, or that they were used non-systematically, or that the manuscripts dispensed with using rulers completely. Considering the overall non-technicality and imprecision of the diagrams regardless of whether they are ruler-drawn, it seems necessary to question the importance which has been attached to rulers in determining whether a fragment was part of a manuscript copy of the *Elements*.

---

70 To be sure, there is the ostraca evidence for propositions about three-dimensional objects, but the respective diagram is not extant.

71 For a systematic overview of the types of objects with which the *Elements* deals (based on Netz 1999a) and the constructional requirements which each of them entails see below in the part on the textual construction (Section 4.3.3).

72 *In extenso* on the different possibilities of depicting perspective in relation to the textual information on it, see Section 4.3.3, 256–260, on the textual construction.

The second kind of basic instrument besides rulers which are theoretically necessary for constructing the objects of *Elements* I are compasses. The ancient evidence for the *Elements* does not testify to their use directly, but compasses are generally attested for ancient mathematics, so the use of rulers on papyrus strongly indicates that compasses also would have been used in the same manuscripts in which rulers are applied. They would probably not have been used for book I, because this book does not deal with circles themselves; their construction is implied by the assignment of the construction of objects that are made up solely of straight lines, but, as we have seen, it appears that such auxiliary constructions were not carried out. Thus, while the usability of rulers is proven, that of compasses can only be presumed, and their actual use and its degree of standardization are in either case unsure.[73]

The application of letters

The last aspect by which to compare and, potentially, classify the ancient diagrams, besides the drawing ground and the use of technical instruments, is the application of letters. From the small ancient sample available, the use of letters appears as little standardized as that of the ruler: three fragments feature diagrams *with* letters (the ostraca set [P. 11999 and 12611], P. Herc. 1061, and P. Fay. 9), two *without* letters (P. Oxy. i 29 and P. [inv.] Berol. 17469). In each case, the presence or absence of the letters in or from the diagram corresponds to the presence or absence of the letters in the respective proposition text. That is, the two papyrus fragments which feature unlettered diagrams transmit, at least in the surviving part, only the per se unlettered enunciation of the respective theorem.

The state of the lettered diagrams themselves is rather fragmentary, especially of the diagram of P. Fay. 9, in which only one letter is readable (α). All others had to be reconstructed, as they must have appeared in the parts that have been torn off. As far as one can tell, the assignment of the letters to objects in text and diagram matches that of the modern editions which are based on the Byzantine manuscripts. The use of the letters differs between the two fragments: the one letter which is readable in the diagram of P. Fay. 9 is a lowercase one, whereas most of the letters of the Herculaneum papyrus are capitalized – all but the α, which means that the lettering of this diagram itself is not unequivocal.

Conclusions from the analysis of the ancient diagrammatic evidence

The ancient fragments testify to a highly diverse diagrammatic practice, as they transmit a small number of most idiosyncratic diagrams. To be sure, all of them have two

73 Cf. Netz 1999a, 19: "Rulers and compasses may have been used."

things in common: they are all drawn on an empty background, which has not been previously organized so as to provide orientation, and they do not display any auxiliary lines which would indicate a technical construction. Despite these commonalities, no two diagrams of the sample equal one another in their outward appearance. They differ in various regards such as size, position, drawing techniques, precision, and lettering.

We have taken the material side of the ancient diagrammatic practice into account in order to assess the generalizability of the features of these several diagrams. Some features could be traced back directly to a specific unusual materiality and are not characteristic of the ancient diagrams in general, others seem to be regular features of the ancient diagram. It has turned out that papyrus, the most common material for longer texts, may account for small differences in the look of the diagram rather than determine certain common features or cause major differences, and that ostraca, though putting a number of constraints on the drawing of diagrams, nevertheless do not lead to diagrams which differ essentially from the ones on papyrus.[74] Thus, the above-named major differences of the transmitted diagrams are, by and large, independent of the material aspect.

Even if minor stylistic differences that are due to the respective material are ignored, our synopsis of the ancient diagrams does not show a common diagrammatic phenotype. However, one can draw tentative conclusions on some elements which qualify as possible features of a common form of the diagrams as they could have appeared in ancient manuscripts of the *Elements*: first, the blank background on which they are drawn; second, the non-technicality of the drawing, i.e., the fact that it is not technically constructed with the help of the auxiliary constructions that are suggested by the text; and (maybe) third, the letters that seem to be absent only from diagrams which accompany likewise unlettered texts, although we cannot be sure in the case of P. Oxy. i 29 whether its unlettered diagram indeed accompanied only an unlettered enunciation or rather a full proposition text. By contrast, the use of a ruler is non-standardized in the fragments, and its use there does not overlap with the use of letters. Moreover, whether the diagrams have been drawn by means of a ruler or not, they are not technically constructed and are therefore imprecise throughout. This does not necessarily mean that the use of basic instruments such as a ruler (and compass) is *not* a common and decisive feature of the common diagrammatic form; we simply cannot tell from the ancient evidence whether such a feature would have possibly characterized the diagrams, or not.

---

74 Cf. Netz 1999a, 16 on the general independence of the diagrammatic presentation from the material: "None of these [i.e., "small-scale acts of communication, limited by a small set of media, from the dusted surface, through wax tablets, ostraka and papyri, to the whiteboard"] is essentially different from a diagram as it appears in a book."

4.1.4 Synthesis of the results from the textual and diagrammatic analysis

We have seen, in a first step, that texts of the ancient fragments are not unanimous regarding their form and presumed function. In this way, they do not present a reliable basis for evaluating the diagrams as to their significance regarding the question of the presence and form of diagrams in Euclid. There is, at most, one fragment which can with some certainty be regarded as part of an actual manuscript copy, but even this assumption is far from certain as its text deviates greatly from the Euclidean one. In a second step, we have argued that the diagrams do not display a common form but are highly diverse. In this way, they cannot indicate a standardized presence of diagrams in the ancient manuscripts for the *Elements* which would be imitated in the papyri regardless of the actual purpose of these papyri. The transmitted diagrams do have some features in common, but since we cannot be sure that at least one of the fragments was part of a manuscript copy, the common features could be a consequence of their diverse, but possibly non-professional reception of the *Elements*. At any rate, the letters as they can be found in the diagram P. Fay. 9 (of which the text comes closest to being part of a manuscript copy of the *Elements*) will have been part of the diagrams of the ancient manuscripts, if there were any, although we cannot be sure of this in view of P. Oxy. i 29.[75] However, the use of a ruler in the same diagram does not necessarily indicate the general use of technical instruments in the ancient manuscripts, given the 'non-technicality' of all diagrams of the ancient fragments. The omnipresence of this feature could indicate that the ancient manuscript copies, counterintuitively, do not differ from student notes and other types of secondary reception on account of the technicality of the construction of their diagrams, but it could also indicate that no manuscript copy of the *Elements* is among the evidence.

In the last step, I shall now compare the combinations of text and diagram that the papyrus fragments display in order to see whether it is at all possible to distinguish certain types and draw conclusions on the functions of the respective papyri. I focus on the three pieces that seem to represent direct evidence. P. Fay. 9 is the only fragment which clearly indicates that it contains a full proposition text with an adjunct lettered ruler-drawn diagram. As it combines a text which bears resemblance to that of *Elements* I 39 and the lettered diagram of this proposition, it has been considered part of an ancient manuscript. However, because of its divergence from the text which the Byzantine manuscripts transmit, other functions have been considered.[76] P. Oxy. 29 is the only other fragment which might be qualified for originating from a manuscript copy of the *Elements*, but it is not clear whether it indeed provides the text of Euclidean propositions as a whole or just their enunciations. If it did contain whole propositions and if it were considered a manuscript copy of the *Elements*,

75 See above, p. 212.
76 See above, 198 f.

P. Oxy i. 29 and P. Fay. 9 would present highly divergent diagrammatic evidence for elements texts: on the one hand, an unlettered free-hand drawn diagram on poor-quality papyrus and, on the other hand, a lettered diagram drawn with a fine pen on well-prepared material. The diverse diagrammatic form alone should not prevent one from considering the fragments from parts of (formally diverse) manuscript copies, but the unlettered form of the diagram of P. Oxy i. 29 hardly fits a lettered proposition text. If, on these grounds, one considered P. Oxy. i 29 part of a collection of (per se unlettered) theorems and unlettered diagrams in the manner of P. Berol. [inv.] 17469, then these two fragments would again exhibit highly different diagrammatic forms: both diagrams are unlettered, but the one of P. Oxy. i 29 is roughly drawn, whereas the diagrams of P. Berol. [inv.] 17469 are drawn with a ruler and look precise (despite their mathematical imprecision). The different looks of the diagrams may (but do not have to) be explained by the differing functions of the respective papyri. The diagram of P. Oxy. i 29, which seems to have been redrawn several times, might be the result of practicing the theorems, and P. Berol. [inv.] 17469 might be part of a "memorandum" of Euclidean theorems.[77] In any case, the absence of letters indicates a non-Euclidean text form: in the *Elements*, as in the whole genre, diagrams can only be drawn with regard to a concrete case of an abstract theorem or problem, which is why their textual construction is always lettered, whereas theorems and problems themselves cannot be presented directly by means of a diagram.[78] This means that the non-Euclidean text form of P. Oxy. i 29 and P. Berol. [inv.] 17469 cannot be expected to display the diagrammatic conventions of the ancient manuscripts of elements texts. Neither do the fragments exhibit a uniform way of drawing a diagram from which the form of manuscript diagrams could be distinguished.

As was expected in view of the diversity of both the texts of the ancient fragments and of their diagrams, none of the combinations of textual and diagrammatic form can be found twice among the ancient evidence. Thus, the material basis is next to zero and does not even allow for a comparison on the smallest possible basis of two manuscript fragments that can safely be considered as such.[79] Neither is it possible to differentiate

---

77 Netz 1999a, 19, n. 31.

78 This is the mathematical-philosophical explanation of the strict twofold division of the proposition into a concrete core part and abstract frame (for which see Section 1.2, 23). The practical explanation for introducing letters in the first place is that it is otherwise impossible to perform the proof both economically and understandably, but the standardized adherence to a 'logically' unnecessary (Section 5.1, 418 f.) abstract dimension can be explained only by the philosophical background, and, as I shall suggest below (Section 5.3, 432 f.), an aesthetic motivation. As far as I can see, the paradox of the unlettered diagrams has as yet not been discussed.

79 Fowler's above-quoted verdict about P. Fay. 9 "that the surviving papyrus fragment that most closely resembles a professional scribe's copy differs so much from the received text" (see above, p. 198), seems to suggest that we already have a specimen of a 'scribe's copy' with that our fragments could be compared. We do have ancient papyri from other authors so that we could judge the general appearance and text form regarding its resemblance to a 'scribe's copy.' However, we do not have

text types which could be associated each with different functions. Rather, some of the fragments resemble one another in some ways and differ from one another in other ways. It is rather unlikely that the diversity of the diagrams in the ancient fragments mirrors the Euclidean practice, since the strong textual standardization of the *Elements* seems to imply that the use of diagrams was standardized as well. Likewise, the form of the diagrams which might have been diverse prior to Euclid, and which might have differed according to the respective source text, would have been unified in Euclid and during his transmission.[80] The non-standardization of the diagrams in the small ancient evidence seems to be due to the differing purposes of the copyists or writers and to the scope and look of their source texts. For instance, they could have aimed at producing a precise copy of a manuscript copy or simply made notes including a sketched copy of the diagram, or they could have attempted to do the construction on account of the instructions of the text alone, either as part of a reproduction of the proof for the sake of learning the proof method, or because of the cognitive challenge, or because the original did not contain any diagram. To conclude, even if one assumes that the papyri are based directly on an earlier copy of the Euclidean text, the diagrams included in them are not in any case indicative of the diagrams' regular inclusion in the original, nor of what the original diagrams, if there were any, would have looked like, nor of whether their appearance would have been standardized.

4.2 Diagrams II: the Byzantine diagrams for Euclid's *Elements* transmitted in the manuscripts

The earliest testimonies to transmit diagrams extensively for Euclid's *Elements* (and for all other elements works, indeed, for all ancient scientific works that use diagrams) are six Byzantine manuscripts. These are far-off in time and liable to reflect Byzantine or, if at all, late ancient drawing conventions. But since diagrams are a standard part of all of the Euclidean manuscripts, they will be considered as potential evidence of the use and form of diagrams in the Euclidean text. This section is devoted to reviewing two of the manuscripts closely. In so doing, my purpose is to look for indications as to whether the manuscripts shed any light on the existence of dia-

contemporary papyri of mathematical texts which contain diagrams or texts which refer to diagrams, so our only (and anachronistic) object of comparison would be the Byzantine manuscripts for Euclid and other mathematical authors.

80 For a different view as to the stylistic repercussions of the different sources of parts of the *Elements* cf. Fowler's characterization of the *Elements* and its "compiler" Euclid: "It deals with a variety of topics in a recognizably varied range of mathematical and literary styles. Euclid is the compiler, not the author, of the work: he is believed to have taken source works by other mathematicians and edited them, adapting and rearranging the material, perhaps even inserting new material of his own, to make the complete treatises" (1999, 205). On the debate of Euclid's being the 'author' or 'compiler' of the *Elements* see above, Section 1.1, 8, n. 45.

grams and, possibly, their characteristics in Euclid and in the ancient manuscripts of the *Elements*. To this end, I examine all characteristics of the Byzantine diagrams in view of the question of how and why they originated: whether they seem to build on an ancient tradition or whether they are due to certain Byzantine scribal conventions or the individual style of the copyist. As in the preceding part on the ancient evidence of Euclidean diagrams, my starting point for choosing textual features will always be the Euclidean text rather than modern standards of mathematical beauty and precision. Where a comparison proves productive, I compare the features of the diagrams of the Byzantine manuscripts to those of the ancient diagrams.

The Byzantine manuscripts that transmit the *Elements*, including those of the Arabic tradition, all show similar tendencies regarding the appearance of diagrams. Therefore, it suffices to compare two of the most important Greek manuscripts: the Vatican (hereafter: P) of the ninth or tenth century, which is the only one that does not stand in the Theonine tradition, and the Bodleian (hereafter: B),[81] which is dated to AD 888 and, thus, the oldest of the manuscripts that depend on the Theonine recension. Of this manuscript, only folios 15–118 and 123–387 which contain I 14 "from about the middle of the proposition" until the end of book VI and books VII–XV have been written "in one and the same elegant hand" from the ninth century, while leaves 6–14 (which contain the beginning of the work until the middle of the I 14) have been "carelessly written by a later hand on thick and common parchment," presumably in the thirteenth century.[82] Indeed, the style of presentation differs quite visibly between the later added part and the original one. This concerns, besides the writing, also the position and drawing style of the diagram. I focus primarily on the large original part, as it is much earlier and written by one and the same hand, whereas the part until the middle of I 14 will have depended on other Byzantine manuscripts. However, I compare the practices of the later hand to highlight the specifics of the diagrams as they are presented by the original hand of the manuscript.

The focus of my review of the diagrammatic practices of the two manuscripts will be my sample book I of the *Elements*, the propositions that are also transmitted on the

---

81 For an overview of the main Greek manuscripts used by the standard edition by Heiberg & Stamatis to which the sigla refer, and of the Arabic tradition of transmission, see above, Section 1.2. The Vatican manuscript P (Vat. gr. 190.1, Vat. gr. 190.2) is available in the Digital Vatican Library (http://digi.vatlib.it/view/MSS_Vat.gr.190.pt.1, https://digi.vatlib.it/view/MSS_Vat.gr.190.pt.2), the Bodleian manuscript B (MS. D'Orville 301) in the Digital Bodleian (https://digital.bodleian.ox.ac.uk/objects/d4a23501-0b98-4aff-acd6-fe06fe9b62e3/surfaces/06cfa3b7-2aad-465e-88ac-0ebe3f2b5d13/).

82 Heath 1956, 49. The "Description" of the papyrus in the Digital Bodleian mentions the valuable note by Arethas, the Archbishop of Caesarea, who bought the manuscript, according to which Stephanus clericus is named as the copyist of the manuscript; it mentions that later hands of the tenth to the fourteenth centuries have added notes and diagrams in the margins, as well as corrections to the text, but does not mention that a whole part, which has been lost and supplemented later, is not transmitted in the version of Stephanus Clericus (http://digital.bodleian.ox.ac.uk/inquire/p/d4a23501-0b98-4aff-acd6-fe06fe9b62e3 and Heath 1956, 48).

ancient papyri (*Elements* I 9, I 10, I 39, and II 5), and propositions that include the construction of three-dimensional objects and which in this way pose a challenge for drawing the diagram. First, I consider the position of the diagrams in relation to the text; second, the techniques of drawing the diagram, primarily regarding the use of ruler and compass; third, the lettering of the diagram; and fourth, what I call the 'geometrical configuration' of the diagram, that is, how the objects are related in the diagram. Unlike in the analysis of the ancient diagrammatic evidence, I do not explore the ramifications of the materials that are used. This is because the two manuscripts are written on parchment that allows for a regular fine character of both the writing and drawing.[83] In each of the following sections, I look at the two manuscripts separately, but wherever I do not specify a manuscript, I refer to a practice which is common to either manuscript.

4.2.1 The position of the diagram in relation to the text

As far as the position is concerned, the diagrams of the manuscripts standardly do not appear *in margine*, but within the text itself. Their integral position is especially notable as the margins of the manuscripts contain many further diagrams that are often part of the scholia. Apart from these diagrams in the margins that supplement and explain the text, there is always *one* diagram which appears inside the column and is fully or partly embedded within the actual text.

In P, which features two columns on each page, the diagram invariantly appears below the respective proposition text, with the text of the following proposition beginning only below the diagram of the preceding proposition text. This means that the diagram is conceded the whole width of the respective column, but mostly appears in the center of the column without taking up the whole width. In any case, there is not written any text on either side of it.

Differently, in B, which has only one large column on each page, the diagram never appears below the proposition text, but is always partly embedded in it. This is part of an overall tendency to save space which we can observe in B in comparison with the more generously laid out P. The end of the text of a proposition is always followed rather immediately by the beginning of the subsequent one, with no, or at most one or two lines left blank in between. In the original part of the manuscript from the middle of I 14 onward until the end of the book, the diagram appears invariantly at the right-hand side of the column, reaching slightly into the respective margin, with its bulk, however, being framed on two or three sides by the text (but it

83 However, I have mentioned the poor quality of the parchment that is used by the later hand of B and results in a different outcome of the drawing and writing (see above, p. 217).

never appears in the middle of the text, fully surrounded by it).[84] In relation to the course of the proposition, the diagram appears at varying places. Six times, it is drawn at the very end of the proposition,[85] i.e., in the lower right-hand corner of the proposition text. In these cases, the diagram either lines up precisely with the end of the right-hand side of the column and with the last line of the conclusion, or it extends somewhat beyond both ends or even into the space of the following enunciation. In the vast majority of cases (29 times), however, the diagram appears on the right-hand side somewhere along the proposition, often next to the proof, but also next to other proposition parts, but, in any case, with at least one line of text encasing it from above toward the beginning of the proposition or from below toward the end of the proposition.[86] Interestingly, while the diagrams which appear in the bottom right corner of the proposition text may or may not happen to appear, at the same time, at the bottom right corner of the *page*,[87] most diagrams (25 out of 29) from I 14 onward which do *not* appear in the bottom right corner of the proposition text do appear at the bottom right corner of the *page*.[88] This indicates that, by and large, not the contents of the proposition, but rather the design of the page seems to have been the decisive criterion for placing the diagram. An aesthetic motivation for the flexibility of the diagram's position regarding the course of the proposition could be to have each

84 It is only in the first part which has been supplemented by a later hand that the diagram is sometimes drawn at the left-hand side of the column (in the diagrams for I 7, 9, and 12), though it appears predominantly at the right-hand side here, too.

85 I 17–18, 20, 30, 40, 48.

86 I 14–16, 19, 21–29, 31–39, 41–47 (in I 14, the diagram is not visible anywhere in the second part of the proposition, i.e., the part which is transmitted by the original ninth-century hand, so one may conclude that it must have appeared in the lost first part somewhere on the right-hand side along the proposition, since it does not appear at the end of the proposition). By contrast, in the first part of the text of *Elements* I which has been added to the manuscript by a later hand, the dominant position of the diagram is at the end of the text, either in the lower left-hand (I 7, 9) or lower right-hand corner of the proposition text (I 1–4, 10), whereas the position somewhere on the side (but not at the end) of the text bloc (again, either the right-hand: I 5–6, 8, 11, 13, or the left-hand side: I 12) is less frequent; once, even, the diagram occurs at the beginning of the text (I 6). This is however a peculiar case, because there is another free space for the diagram at the more common place (at least according to the presentation of the first 14 proposition of book I) in the lower right-hand corner of the proposition text. To this, the case of I 11 can be added, where the full diagram appears on the right-hand side somewhere toward the middle of the proposition text, while a half-finished one, not yet containing the line whose construction constitutes the goal of the proposition, appears, as in I 6, besides the enunciation). Thus, in the part that has been added to the manuscript later, although it itself stems from one hand, the position of the diagram is much less standardized than in the subsequent original section.

87 Two of them do (I 17 and 48), four do not (I 18, 20, 30, 40).

88 Only the diagrams of I 27–28, 39, and 41 do not (I 14 has been included in the count, as it is most probable in view of these results that the diagram would have appeared at the bottom of the lost original page, since it does not appear at the originally transmitted end of the proposition).

page, ideally, contain at least one diagram at a standardized place. That is, the layout of the pages seems to follow a 'one diagram per page-principle.'[89] Maybe it was, at first, a convenient way to place the diagram at the bottom of the page to avoid the problems which might arise if the drawing of the diagram was left until the end of the proposition, and from this, an aesthetic preference for the place of the diagram was developed.[90] The standardization of the diagram's position, for the most part, in the lower right-hand corner of the page, or alternatively, in the lower right-hand corner of the proposition is complemented by the way in which the space of the diagram at these standardized positions is standardly demarcated from the text which encases it: an invisible rectangle seems to define the diagrammatic space that is surrounded by the text.[91]

To conclude, our two manuscripts standardly show (at least) one diagram per proposition within the space of the textual column. They are each individually quite standardized as to the position of their diagrams, but they deviate from one other in *how* they standardize: in P, the diagram is drawn standardly below the respective proposition; in B, it tends to appear at the end of the page rather than at the end of the proposition, and its place is in either case within the column at its right-hand side. This means that the position of the diagram within the column is not standardized beyond the individual manuscript. It is overall standardized only insofar as it appears *at the end* (of something).[92] The divergence of our two manuscripts regarding the

---

89 Indeed, the manuscript's presentation of book I from I 14 onward features only four pages which do not contain a diagram (fol. 20v–fol. 21r, fol. 29v, fol. 31v). In all cases, this is the case due to the length of the respective propositions.

90 In some cases, the remaining part of a proposition text that follows the page which has the diagram of this proposition at its end is so small that, if the diagram was not drawn on the preceding page, the text would fit almost or entirely onto that preceding page. In this way, the text would either fill up that page entirely, or it would leave some space but not enough to accommodate the diagram, or it would extend onto the next page with only a few lines. In either case, the diagram would have to be drawn onto the next page, where it would be almost or entirely encased by the enunciation of the subsequent proposition rather than by the proposition text to which it belongs. However, this can only be the reason for the way of placing the diagram in some of the cases where it appears at the end of the page. In many others, the proposition text goes on for many lines on the following page and would be able to accommodate the diagram there.

91 By contrast, at the beginning of the manuscript, which has been ascribed to a later hand, the space left free for the diagram is not as clear-cut as in the ninth-century presentation of the later propositions, though it seems to imitate somewhat the way of standardization which we find later in the manuscript.

92 It is not quite clear what Fowler (1999, 215) means by "at the end" when he says this is where the diagrams are placed "universally in our minuscule Byzantine manuscripts, at the end," unlike in P. [inv.] Berol. 17469 (cf. Section 4.1.2, 197), which "is the only Greek text I know of where the figures are placed before the enunciation," and unlike in the other Greek papyri, where it is placed "alongside the text" (P. Herc. 1061, P. Oxy. i 29) or "in the middle" (P. Fay. 9). The reference to the Berlin papyrus seems to suggest that Fowler refers to the position of the diagram in relation to the course of the respective proposition when he says "at the end" with regard to the diagrams in the Greek manuscripts.

place of the diagram could have to do with the different layout of the pages in each case: P features two narrow columns on each page, whereas B only has one broad column. Placing the diagram below the proposition text in the center of the column of the Bodleian papyrus without it being embedded in the text would mean a great waste of space. If, then, the idea was to have the diagram surrounded by text, the diagram either would have to be drawn before the text was written, or some space on the page would have to be allocated to the diagram so that it could be drawn afterward. In either case, the necessity of drawing the diagram or defining its space in advance could explain its standardized place at the bottom right corner of the page.

What does this tell us about the presence and position of diagrams in Euclid? Since our two sample manuscripts show different standardized positions of the diagram, they do not indicate that the Byzantine mathematical manuscripts relied on a previous tradition in which the position of the diagram was standardized. However, the overall standardized place of the diagram within the space of the textual column suggests that its inclusion into the text was planned from the start – by contrast with the diagrams *in margine*. This could indicate that the diagram was already present in the underlying manuscript copies and that its presence may go back to a longer tradition of including it within the text, even if we do not know at which standardized position, if any, it was included. Of course, this does not necessarily mean that the practice goes back to Euclid. The practice of transmitting the diagrams as part of the text could also be the result of Byzantine conventions or of the development of the

However, his other references to the ancient papyri point in another direction: "alongside the text" seems to refer to the position of diagram within the column but on one side of the text rather than to its position in relation to the course of the proposition, considering that Fowler tends to believe that P. Oxy. i 29 only contains enunciations and diagrams, and considering that the diagram for II 5 appears, to be sure, on the side, but rather toward the end of the enunciation of II 5, which here equals 'at the end of the proposition text.' Fowler's description "in the middle" with regard to the diagram of P. Fay. 9 clearly refers to where the diagram is placed on the page (whereas its position in relation to the course of the proposition is 'at the end'); "in the middle" could also aptly describe its horizontal orientation within the column, but if we understand it in this way, one could say the same about the diagrams of the Berlin papyrus, from which Fowler's discussion started, but Fowler considers them as being placed, uniquely, "before the enunciation." Lastly, in order to illustrate his claim that the "minuscule Byzantine manuscripts" always have the diagram "at the end," Fowler seems to want to refer to the position of the diagrams for II 5 and 6 on folios 35v and 36r of B, both of which are placed at the end of the *page*, not at the end of their respective propositions (he actually, by mistake, refers to his Plate 6, which is folio 14v of the Vatican manuscript that only contains definitions and axioms of book I). By contrast with this implication, Fowler, for further support of his statement that Greek manuscript diagrams are always placed "at the end," adduces his quotation of Heiberg's statement, "dans nos manuscrits mathématiques, la figure se trouve régulièrement à la fin du théorème" (Fowler 1999, 122 with reference to Heiberg 1900, 148). Heiberg, in turn, seems to use "théorème" in the sense of "proposition," but he is wrong in any case, as B mostly places the diagram at the end of the page rather than at the end of the proposition.

text in the long period of time between Euclid and the Byzantine tradition, especially in the context of the *Elements*' late antique didactic reception.

4.2.2 Drawing techniques

I shall now explore the use of technical instruments for drawing diagrams in the Byzantine manuscripts. By contrast with the ancient evidence, the manuscripts unanimously testify to the use of ruler and compass.

In P, ruler and compass are standardly employed, although the results of this use vary in terms of precision. The lines that are drawn by means of a compass are precisely circular, whereas the supposedly straight lines that are drawn with the help of a ruler show some irregularities: for the most part, the ruler-drawn lines are indeed straight, but some lines look a bit crooked, as if they were drawn free hand (cf., e.g., the diagrams of I 47 and II 5). This could be due to several reasons: the scribe's writing surface might have been uneven and able to be felt through the thin parchment, or the edge of the ruler along which the pen was drawn might not have lain precisely on the parchment, or the ruler might have slipped, or the pen slid.

In B, just like in P, the use of ruler and compass is standardized, at least in the original part of the manuscript from I 14 onward.[93] By comparison with the presentation in P, the appearance of the circular and straight lines that are produced by means of ruler and compass is even more standardized. It is for the most part not hampered by any drawing mistakes and therefore conveys a precise and orderly impression. There are only a few thick or crooked lines visible in the photographs, but the thickness is due, e.g. in I 33, to a correction by a later hand (the line that has been deemed wrong is crossed out), and the crookedness results from the warping of the paper due to moisture or folding (I 36, 39, 40), but this probably happened only after the manuscript had been written.[94]

The two manuscripts thus demonstrate in an exemplary manner that the diagrams are standardly drawn with the help of ruler and compass, but that the use of these instruments, as we have seen in the ancient fragments, leads to different results as far as the precision of the drawings is concerned. The diagrams of B look more precise than those of P, in which many supposedly straight lines are not exactly straight, but variously crooked, extremely thick, or having thick terminal points. According to the comparison

93 By contrast, in the preceding non-original part of the manuscript, only three diagrams unambiguously display the use of a ruler (I 1, 3, 12), whereas seven are clearly drawn free-hand (I 5–8, 10–11, 13), and in the remaining ones (I 2, 4, 9), some lines look as if they were drawn free-hand, others as if drawn with a ruler. Analogously, the writing from the rest of I 14 onward is much more refined and regular, than before I 14, both in its style and because it seems to follow imaginary straight horizontal lines.

94 However, whether the parchment has been damaged before or after the event cannot be told for sure from the photographs.

of the two sample manuscripts, the realization of the use of technical instruments differs from manuscript to manuscript. In this way, the diagrams of the manuscripts testify to a range of different looks of ruler-drawn diagrams.

As in the ancient papyrus fragments which contain ruler-drawn diagrams, the use of the basic technical instruments effectuates only a seeming technicality. It does not serve the technical construction of the objects in the way which is suggested by the deductive system. The diagrams do not display any additional lines that reveal the auxiliary constructions which are necessary for constructing them according to the constructional requirements of the proposition. In this way, they only produce an image that imitates the objects and object relations described, as it were, as the diagrams are not strictly proportional, let alone metrical. As a result, they fail to depict the actual geometrical configurations. For example, the diagram for I 9 on the Vatican manuscript does not represent the geometrical situation correctly, although its copyist seems much more anxious to do so than those who accomplished the drawing of the same diagram on either of the two extant papyrus fragments:[95] the line ΔE seems askew, which means that AΔ and AE are not, as requested, of the same length; the lines ΔZ and EZ do not actually meet in Z; and the supposedly equilateral triangle is not actually equilateral. This is the general impression which arises from the diagrams of the manuscripts that we have reviewed: although they overall appear zealous to represent the geometrical situation precisely, they do not represent the geometrical situation correctly because the depiction is non-constructional and non-technical. This means that the diagrams imply a sharp contrast between their seeming technicality and actual non-technicality and accordingly, between their seeming correctness and actual incorrectness. Unlike the ruler-drawn Byzantine diagrams (and the ruler-drawn diagrams of the ancient papyri), the free-hand drawn diagrams of the ancient papyri do not raise any expectations of precision, technicality, or correctness in the first place. What, then, is the function of using technical instruments, if not to construct in a geometrically correct way? The application of ruler and compass seems to serve the purpose of an emblematic fiction, what Netz refers to as 'make-believe:'[96] The fiction the instruments 'make believe' is precisely the technicality which they lack.

A different aspect of the seeming technicality and actual non-technicality of the ruler-drawn diagrams becomes apparent in the depiction of three-dimensional objects for which we have to turn to other books of the *Elements* beyond our sample book I. Drawing three-dimensional objects requires representing depth in order for the objects to be shown correctly and recognizably at all. In the subsequent section on the textual construction (Section 4.3), I discuss different possibilities of depicting depth and the complete silence of the text on how to construct three-dimensionality. In this section, I only consider the use of technical instruments for representing three-

95 The diagram on the Bodleian cannot be compared as it is from the spurious part.
96 Cf. Netz 1999a, 54.

dimensional objects. Both P and B display the most basic of three-dimensional objects, i.e., cubic and rectangular solids, by means of parallel projection, that is, by means of foreshortening the parallel projection lines and presenting the angle of those lines to the projection plane at a skew angle. The diagrams do not, however, indicate any standardized foreshortening factor or any standardized angle (e.g., in the diagrams for XI 34). The parallel projection is neither metrical nor technically constructed, but rather drawn by the eye, albeit quite carefully so. Here, the non-technicality does not really cause geometrical incorrectness as the proposition texts do not talk about 'depth.' A straight line seems to be drawn incorrectly if it turns out to be crooked rather than straight, although the 'correctness' also in this case depends on the respective cultural perception of what is a correctly drawn straight line. But what is a correctly drawn cube or parallelepiped? The proposition texts do not tell us how to realize three-dimensionality in a diagram. In fact, they do not even tell us to draw three-dimensionally at all: 'depth' is only implicitly required by the name of the three-dimensional object. It is the Byzantine diagrams themselves which show that cubes and parallelepipeds were expected to be shown three-dimensionally by means of parallel projection (however, as we will see shortly, not all three-dimensional objects are represented three-dimensionally in the Byzantine manuscripts). Regarding this expectation, the non-technical depiction of perspective by means of parallel projection evokes a seeming technicality, and the use of the ruler aids in the fiction of the seemingly technical representation.

While the Byzantine manuscripts thus testify to an awareness of the necessity of showing depth in the depiction of three-dimensional objects as well as of a valid method of so doing, they seem to avoid the method of parallel projection in the case of other rectilinear solids.

They find alternative, i.e., non-standardized ways around it, either by making no attempt at depicting three-dimensionality at all, or by mixing different kinds of perspective. For example, pyramids are sometimes depicted by means of central projection, but only if prisms are to be depicted as being contained by them. However, the prisms are drawn as if viewed from a different perspective than the pyramid, that is, by some kind of parallel projection, as for instance in the extant diagrams for XII 3, XII 4, and XII 5 (cf. Fig. 11, which reproduces the drawing of P). The reason for this could be either the non-technicality of the drawing, that is, it is difficult to simultaneously foreshorten a pyramid as well as two prisms that are contained by it by means of a free-hand, non-metrical, and non-technical drawing. Or the prisms were drawn by means of parallel projection on purpose, because they would not be recognizable as prisms if drawn by means of central projection.[97] In the latter case, non-technicality is what enables the

97 The diagrams for XII 3, 4, and 5 of P, the mix of perspectives seems to be clearly visible, the corresponding diagrams of B rather seem display the (ultimately unsuccessful) effort to find a common perspective.

kind of depiction in the first place. In any event, these diagrams are not only non-technical, but even geometrically wrong because of the deliberate mix of different perspectives or the inability to foreshorten within a free-hand drawing.

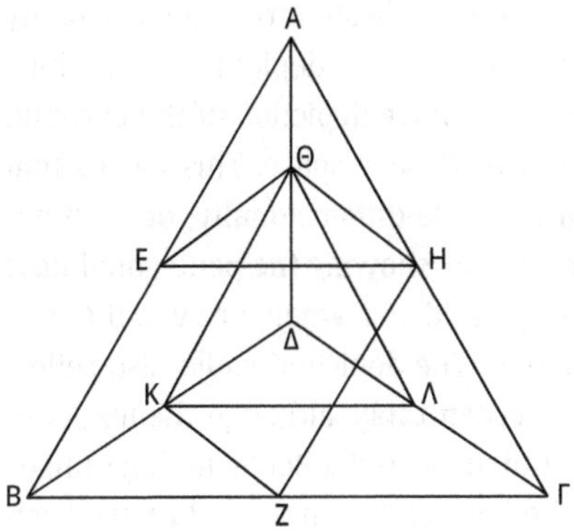

Fig. 11: XII 3 according to Vat. gr. 190.2, fol. 217r.

The mix of perspectives and the resulting lack of (correct) perspectivity seem to be purposefully accomplished in those pyramids which do not contain an internal figure. The reason why we do not find it in the drawings of pyramids with internal prisms seems to be that these prisms could not be represented by the following kind of depiction: the preferred method of drawing pyramids is to draw the polygonal base in plan view, i.e., as a two-dimensional figure without any foreshortening. The rest of the figure, that is, the apex of the pyramid and the triangular lateral faces that are formed by the apex and each of the base edges, is drawn from a different perspective, as one can see, e.g., in the diagrams of both manuscripts for proposition XII 6 (cf. Fig. 12).[98] The apices of the two pyramids each appear on the right-hand side of the respective pentagonal base so that the figure, apart from its pentagonal base, leans to the right. Although it is possible for the apex of a pyramid not to be centered over the base according to the broad definition of the figure in the *Elements* (XI, Def. 12) and although the proposition text of XII 6 does not preclude this possibility, the diagram does not seem to display such oblique pyramids. Rather, this kind of depiction is the conventionalized one for pyramids (and for cones) in the manuscripts.[99] The oblique position could indicate perspective and might imitate the projection of cubes and parallelepipeds. Alternatively, it could be a non-perspectival way of showing the part of the figure which rests on the base so that one can recognize that the figure as a whole is a

98 The respective diagram of P is similar, only here the pentagonal base is non-regular – because the proposition allows for it to be non-regular as well as regular, it seems, not because it has been projected.

99 The diagrams for XII 10 in both manuscripts only show the respective bases of a pyramid, cone, and prism, because the rest of the figures does not play a role in the propositions in terms of letters (the proposition does refer in general terms to the same height of the three figures).

pyramid. For if the axis of the pyramid was drawn from the same perspective as the base, it would as a mere point in the diagram (provided that it is a right pyramid) and would no longer be recognizable as its axis. The same applies to the depiction of the lateral faces: the lines which connect the apex to the base edges would divide the pentagonal base into five triangles so that the drawing would not indicate three-dimensionality anymore. In the latter case, it might have been more intuitive to depict the part which rests on the base in an upright position, considering the above depiction of the pyramid of XII 3. Maybe the reason for not doing so simply was to save space. This means that the diagram either shows a general refusal to draw three-dimensionally, or it shows only part of the figure three-dimensionally. The reason for showing the pentagonal base in plan view could be the difficulty of projecting a pyramid as a whole or even the pentagonal base without applying a technical construction. The depiction could also reflect the attempt to show the defining part of the figure recognizably, although the text does not prescribe whether a regular pentagon is to be drawn, as in the Bodleian diagram, or a non-regular, as in the Vatican diagram. Finally, the kind of depiction might have been chosen by analogy with the standardized depiction of the circular base of cones in plan view which I explain in the following paragraph. While the depiction is obviously incorrect,[100] it has a seeming technicality about it because of its close imitation of the base figure and its use of a ruler. The depiction appears to aim at seeming technicality quite independently of the eventual aim of geometrical validity. In fact, the technicality and correctness that are suggested by the use of the ruler even seem to be at odds with the actual geometrical incorrectness of the figure. For a similar kind of depiction, cf. the octahedron of XIII 14, in which the base of both the pyramids which form the solid is drawn in plan view, whereas the rest is drawn in central projection (Fig. 13).

The same tendencies of representation can be observed with regard to the representation of cones, which, together with cylinders, include an additional problem with regard to the depiction of perspective, that is, the perspectival depiction of circles. Cones are, again, like pyramids, represented by a mix of two different perspectives: as one can see in the diagrams of the manuscripts for XII 11, the base is shown in plan view as a circle, and the rest of the figure is only indicated by an oblique line that represents the axis of the cone, while the slant height (which is the distance from a point on the circle of the base to the apex) is not depicted so that the lateral surface is not indicated (Fig. 14). In the diagrams for the subsequent proposition XII 12, the depiction is similar, but additionally, two straight lines represent the slant height in either of the two cones, because the proof relies on the triangle that is formed by one

100 The falsity is particularly compelling in one of the diagrams for XIII 13. In this proposition, a pyramid is to be enclosed by a sphere, which is represented in both manuscripts by a circle that, however, encircles only the base of the pyramid, while the tip of the pyramid protrudes above the edge of the circle.

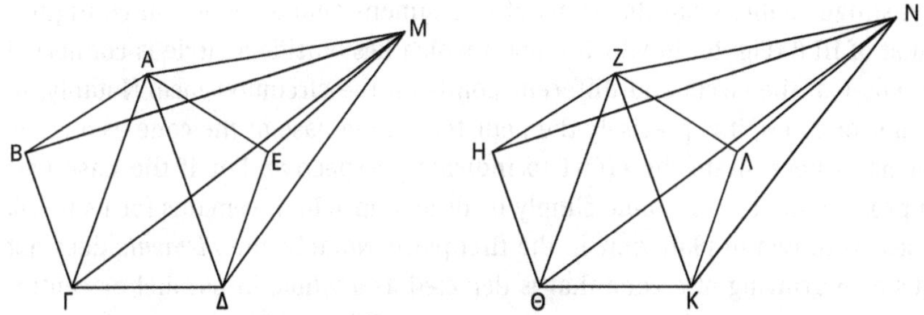

Fig. 12: XII 6 according to MS. D'Orville 301, fol. 315r.

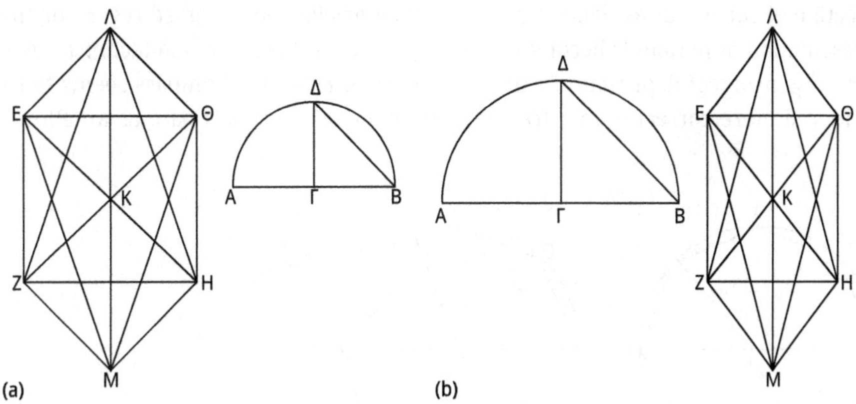

Fig. 13: (a) XIII 14 according to Vat. gr. 190.2, fol. 241r; (b) XIII 14 according to MS. D'Orville 301, fol. 350r.

of these straight lines, the radius of the circle, and the axis of the cone (Fig. 15).[101] Unlike pyramids, which can be oblique according to the definition of the *Elements* (XI, Def. 18 f.), cones are always thought to be right circular, not oblique circular, that is, the axis connects the apex with the center of the circular base at right angles to the circular base. This means that the oblique depiction of the axis of the cones of our diagrams does not imitate the geometrical obliqueness of the axis. As in the case of pyramids, the drawing of the cones would not indicate any three-dimensionality if the axis was drawn from the same perspective as the base (that is, if it was drawn as a mere point). Thus, the purpose of the oblique representation of the axis (and of the slant heights in the diagrams for XII 12) could be to indicate depth so that the figure's three-dimensionality can be recognized. Again, the perspective of the axis seems to imitate the projection of cubes and parallelepipeds. However, in this way, the depic-

101 It also relies on the triangle formed by the two straight lines which lead from the apex to different points on the circle of the base and the straight line which connects them as part of polygonal base of the pyramid that is inscribed into the circular base of the cone.

tion is indistinguishable from diagrams of two-dimensional geometrical configurations as that of III 8 (Fig. 16) in which a point which lies outside a circle is connected with the center of the circle and different points on the circumference. Notably, in our diagrams of cones, it is precisely the depiction of the base of the cone as a circle which seems to necessitate the effort to indicate perspective. For if the base was drawn in projection, the axis would simply be drawn upright. It remains for us to ask why the base is drawn in plan view in the first place. Notably, the *Elements* does not contain a single drawing of a cone that is depicted as a whole in parallel or central projection, such as we find in the case of pyramids. This could be because if a cone was projected, its base would have to be presented by means of an ellipse which cannot really be drawn by means of ruler or compass. It might be for this reason that the projection of cones was avoided as a whole. What has just been argued regarding the representation of pyramids becomes even more obvious here: the non-technical, geometrically incorrect depiction of cones by means of ruler and compass seems to be considered more correct than a free-hand drawing of the cone's base as an ellipse.

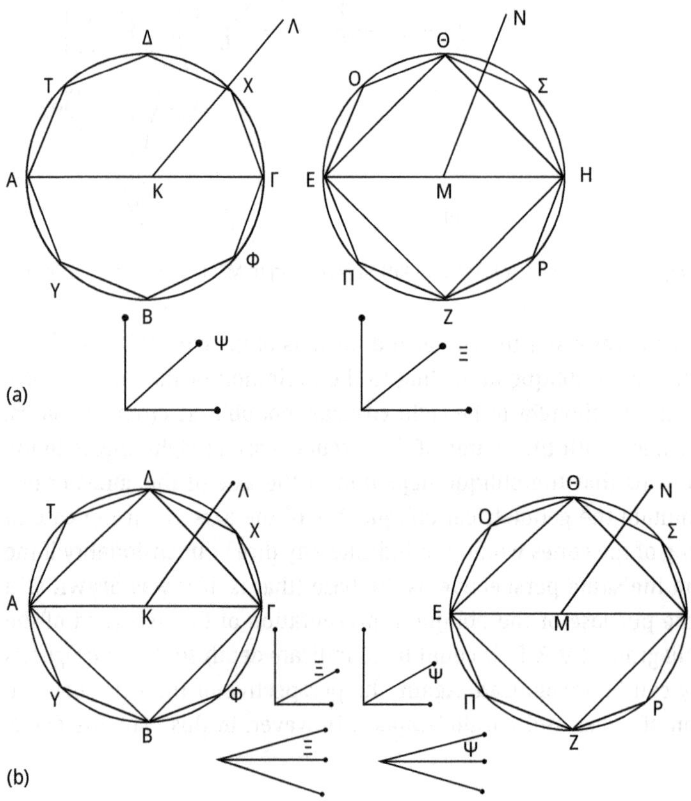

Fig. 14: (a) XII 11 according to Vat. gr. 190.2, fol. 223v; (b) XII 11 according to MS. D'Orville 301, fol. 322v.

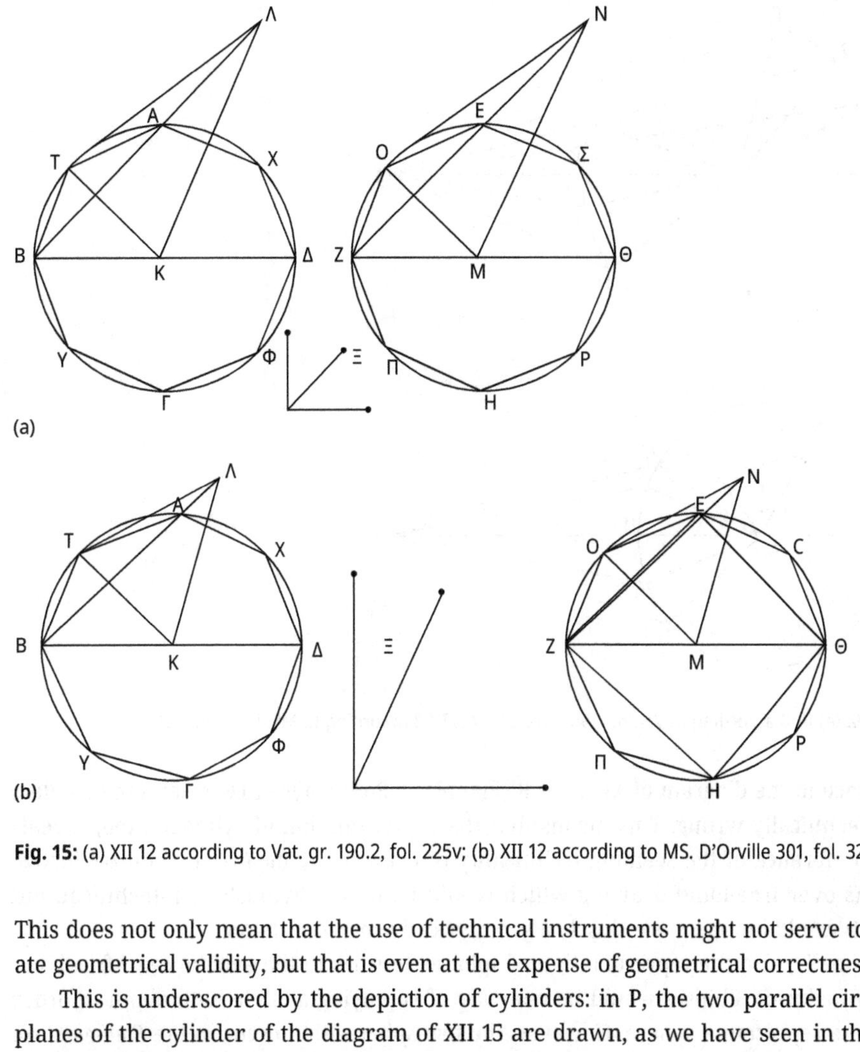

Fig. 15: (a) XII 12 according to Vat. gr. 190.2, fol. 225v; (b) XII 12 according to MS. D'Orville 301, fol. 325v.

This does not only mean that the use of technical instruments might not serve to create geometrical validity, but that is even at the expense of geometrical correctness.

This is underscored by the depiction of cylinders: in P, the two parallel circular planes of the cylinder of the diagram of XII 15 are drawn, as we have seen in the depiction of cones, as circles, whereas the cylindrical surface is indicated by two parallel straight lines which connect the circles (cf. Fig. 17). Since the two straight lines are drawn vertically, the diagram displays a mix of perspectives that is similar to the one which we have seen in the depiction of cylinders and cones. However, in the present case, the drawing does not even attempt to imitate depth by showing parts of the figure in oblique perspective. The aim of the depiction seems to be to make visible all relevant parts of the figure by means of ruler and compass. The standard way of drawing cylinders of *Elements* XII, however, is to draw them three-dimensionally. For this purpose, the circular planes are drawn as what are supposed to be ellipses. Ironically, it is the use of a compass which prevents these drawings from looking like proper ellipses, because they each consist of two joined compass-drawn arcs, as for

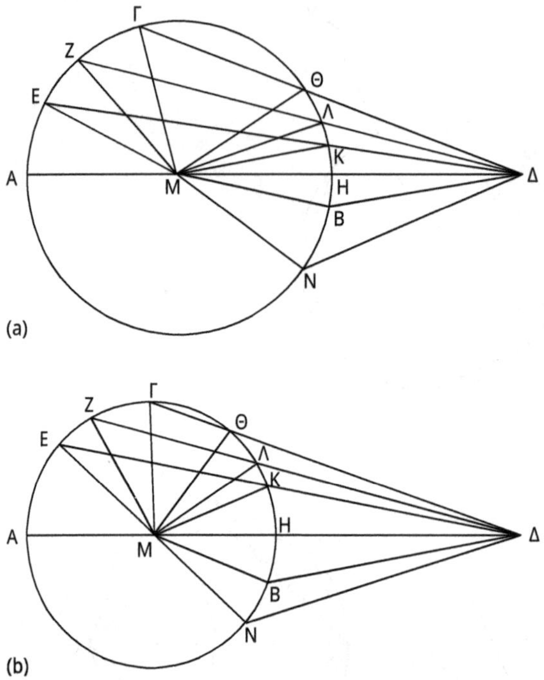

Fig. 16: (a) III 8 according to Vat. gr. 190.1, fol. 53r; (b) III 8 according to MS. D'Orville 301, fol. 48r.

instance in the diagram of XII 15 of B (Fig. 18).[102] Both ways of representing cylinders are technically wrong. This means that the representation of cylinders, too, reveals the preference of the seeming technicality of drawing by means of technical instruments over free-hand drawing which would be more obviously non-technical, but which would imitate more closely a geometrically correct version.

Finally, also the representation of spheres testifies to the avoidance of free-hand drawn ellipses. Ellipses would be one way of supplying a circle with depth in order for it to represent a sphere. However, the few spheres that occur in the *Elements* are shown differently. Sometimes, the sphere is not drawn at all: e.g., in the diagram of XIII 14, the sphere which is to comprehend the octahedron that we have considered

102 For the same kind of depiction of cylinders, cf. the diagrams of XII 13 and 14 in both manuscripts. The representation of ellipses by means two joined arcs also occurs in propositions about two-dimensional configurations: in III 10, which is meant to show by means of a proof by contradiction that one circle cannot meet another at more than two points, the second circle that seems to be represented as an ellipse is drawn in the described way by means of a compass. In this example, the schematically drawn ellipse is not drawn for the sake of representing three-dimensionality, but 'impossibility,' that is, the impossibility of there being a second circle.

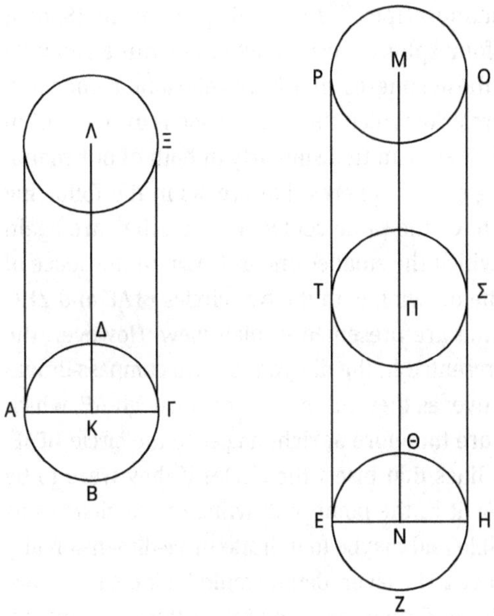

Fig. 17: XII 15 according to Vat. gr. 190.2, fol. 227r.

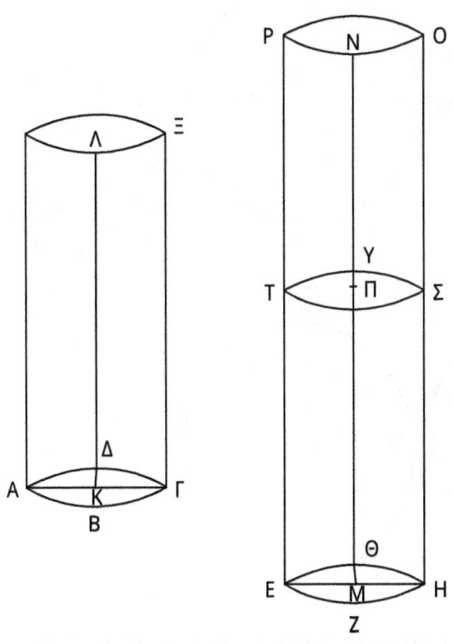

Fig. 18: XII 15 according to MS. D'Orville 301, fol. 328r.

above is simply left out in both of our manuscripts.[103] In the diagram of XII 18, as it can be found in both manuscripts, the four spheres are depicted two-dimensionally by means of circles (cf. Fig. 19). Two of these spheres that have the same center and are of different sizes are shown as two circles of which the smaller one is drawn within the larger one. The diagram of XII 17, as it is transmitted similarly in both of our manuscripts, shows a non-perspectival drawing of two spheres that are, as in the following proposition XII 18, of different sizes and have the same center (Fig. 20): they are again depicted two-dimensionally as circles of which the smaller one is drawn in the space of the larger one. These circles coincide in the drawing with the two circles ΒΓΔΕ and ΖΗΘ that cut the spheres through the center and are presented in plan view. However, the two semicircles ΒΞΔ and ΚΞΝ that are represented in the diagram by two compass-drawn arcs are depicted from a different perspective: as they cut the spheres through ΑΞ, which is the perpendicular on the center Α, and are therefore at right angles to the circle ΒΓΔΕ, they would have to be drawn as straight lines that bisect the circles if they were to be drawn in plan view just like the sphere. As it is, the point of drawing the semicircles as arcs seems merely to be to make them visible and maybe to indicate three-dimensionality regardless of the resulting mix of perspectives. However, depth would be indicated more correctly and recognizably by means of ellipses drawn free-hand rather than by compass-

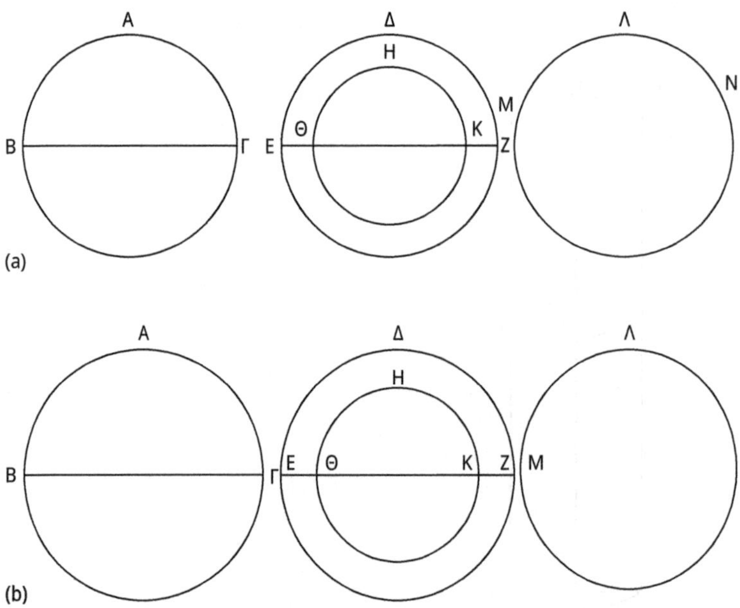

Fig. 19: (a) XII 18 according to Vat. gr. 190.2, fol. 230v; (b) XII 18 according to MS. D'Orville 301, fol. 333v.

103 See above, 226 f. The same thing happens in the diagrams of XIII 15 in which a cube is to be comprehended in a sphere, of XIII 16 (icosahedron), and of XIII 17 (dodecahedron).

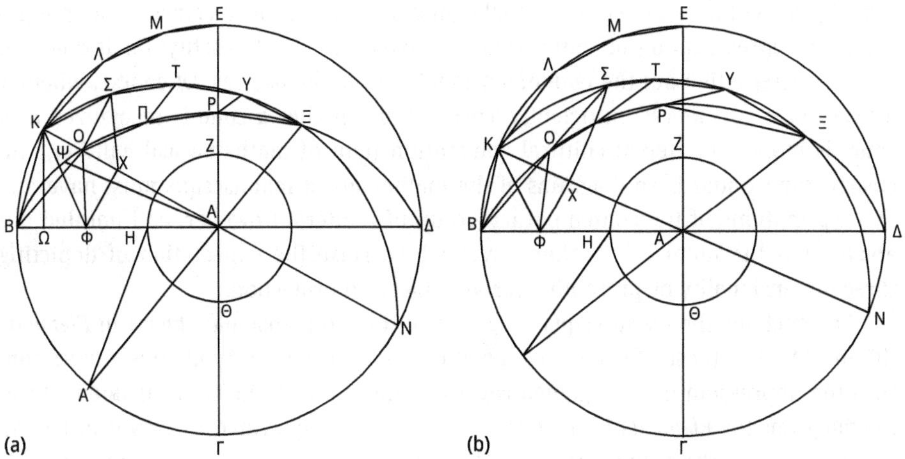

Fig. 20: (a) XII 17 according to Vat. gr. 190.2, fol. 230r; (b) XII 17 according to MS. D'Orville 301, fol. 332r.

drawn (parts of) circular arcs.[104] Thus, also the representations of spheres show that free-hand drawing (of ellipses) is avoided by all means for the sake of the seeming technicality that is evoked by the use of compass. As a result, the depictions lack three-dimensionality and in some cases even geometrical validity.

As I have said before, the necessity of showing depth is only implied by the construction of objects that are by definition three-dimensional. This means that the *Elements* does not explicitly deal with ellipses or other non-circular and non-rectilinear figures as objects of construction. A four-book work on conic sections by Euclid is lost. However, the manuscripts of Apollonius' *Conics* provide diagrammatic evidence for the depiction of ellipses, parabolae, and hyperbolae, i.e., figures which themselves cannot be drawn by means of ruler and compass (and not only if they occur in three-dimensional space). However, the matter is complicated even further by the fact that conic sections originate from a three-dimensional object. Therefore, if they are – as is not always the case – treated as part of cones, the problem arises of whether, and if so, how to draw them three-dimensionally. Just like ellipses in the manuscripts of the *Elements*, the non-circular forms that result from conic sections are always drawn by means of a compass in the manuscripts of the *Conics*, rather than free hand or on the basis of measurements. This practice yields diagrams which do not even approximate geometrical correctness (if that was at all possible), but are mathematically simply invalid. This is in accordance with what we observed regarding the diagrammatic representation of three-dimensional objects in the manuscripts of the *Elements*. But since the falsity in those cases always concerns perspective and either results in the incorrect representation of a certain view on an object, or view on the relation of sev-

104 For a realization of the diagram that is consistently three-dimensional, cf. Thaer 1970, 380.

eral objects within a three-dimensional configuration, the incorrectness that emerges from the representation of conic sections is even more noteworthy: for it does not concern perspective but the two-dimensional objects themselves. Of course, whether a diagram is considered sufficiently 'correct,' or whether a cone is recognized as a cone, is more a matter of cultural convention than of mathematical validity, and clearly, the standardized diagrams of the mathematical manuscripts must have met the expectations of the contemporary group of readers. However, as I pointed out above, it is the manuscripts themselves which raise the expectation of depicting three-dimensionality by presenting certain objects in projection.

To conclude, the manuscripts' diagrams of three-dimensional objects in *Elements* XII and XIII and those of conic sections in the manuscripts of Apollonius' *Conics* confirm the impression that we gained regarding the use of technical instruments from the diagrams for *Elements* I: the diagrams are standardly drawn by means of technical instruments. The only instruments that have been applied are ruler and compass. As has been noted, the copyists may have restricted themselves to using these two means for aesthetic or philosophical reasons (e.g., minimalism).[105] However, while they used only two instruments, they always used one of these or both and avoid free-hand drawing by all means. The use of ruler and compass is not connected to any method of technical construction, nor is it metrical. Therefore, it is only seemingly technical, but actually non-technical. It does not add to geometrical correctness or precision, nor does it seem to be applied for this purpose to begin with. Especially in the depiction of three-dimensional objects, it is used even at the expense of geometrical correctness: free-hand drawing is avoided even where it would be more flexible and lead to a better depiction. Ruler and compass seem to be used precisely for the purpose of creating an impression of technicality. If measured against the mathematical necessity of using technical instruments, the use of ruler and compass is not only standardized but even overstandardized, as it is applied more consistently than mathematically necessary – even at the expense of the correct representation of content.[106]

The two Byzantine manuscripts are much more standardized regarding the use of technical instruments than the non-representative, tiny sample of the few ancient papyrus fragments. The diagram of the one fragment which most likely originates from an ancient book manuscript, as far as the style and structure of its text are concerned, shows a similar style in terms of its use of technical instruments as do the Byzantine diagrams. The similarity, on the one hand, could suggest a continuity between the ancient and Byzantine diagrams on the thinnest possible material basis. The Byzantine manuscripts would, then, put a different complexion on the impression that we received from the diverse ancient fragments which display no fundamental difference between ruler-drawn and free-hand drawn diagrams regarding mathematical correct-

---

105 On the possibility and use of νεῦσις cf. Asper 2007, 112, n. 136.
106 The subsequent Chapter 5 will deal with the *Elements*' overstandardization *in extenso*.

ness or precision: the diagrams of the manuscripts show that the copyists found ruler and compass indispensable despite their geometrical ineffectiveness and virtual non-technicality. It is precisely this paradoxical importance of the instruments which could indicate that their use was already standardized in an earlier source – maybe even in the ancient manuscripts.

The obligation to use ruler and compass is felt especially in the Byzantine diagrams of three-dimensional objects. However, since there are no ancient diagrams of solid geometry extant, the practice of the manuscripts cannot be verified and must be treated with caution as regards any conclusions about the standardization of ruler and compass use on them. That being said, the depiction of three-dimensional objects in the manuscripts could indicate a previous standardization of ruler- and compass-drawn diagrams in view of the different ways of presenting each kind of three-dimensional objects. We have seen that the text underspecifies the construction of three-dimensional objects to the extent that it does not say how to present depth.[107] With regard to this, it is first of all notable that the manuscripts nevertheless display one preferred way of drawing a certain kind of three-dimensional object, e.g., cubes by means of parallel projection or spheres two-dimensionally by means of a circle. Moreover, while this standardization is not 100% and while the manuscripts mostly testify to several ways of drawing a solid object, it is remarkable that the realization of specific diagrams is often (though not always) similar in both of our manuscripts. For instance, the manuscripts from the two traditions would unanimously depict a pyramid by means of a plan view of its base in the case of one proposition and by a mix of perspectives in the case of another. Thus, it is not unreasonable to entertain the possibility that these diagrams may have been standardized in this way in at least certain earlier sources.

The standardization of the use of rulers could, of course, be a convention of the Byzantine manuscripts that had no precedent in the ancient predecessors, or was only developed in the didactic contexts of late ancient times. Thus, the question of whether rulers and compasses were used standardly or not cannot be finally decided on the basis of the evidence. But the unanimous standardization allows for at least the possibility that the use of rulers for diagrams in book manuscripts was adopted from Euclidean time.

4.2.3 The use of letters in the diagrams of the manuscripts

Turning from the pictorial to the literal side of the diagrams, I now review briefly the use of letters in the manuscript diagrams. I consider *if* letters are standardly employed, and if so, what letters, and whether the lettering is standardized as to the

107 On this, see Section 4.3.3 on the practicality of the textual construction, 256–260.

kinds of letters. First, the diagrams of the manuscripts are lettered without exception. Second, the lettering is non-standardized regarding the use of capital and lowercase letters. In P, the text is written entirely in minuscule except for the fully capitalized terms for the structural parts of the proposition, which were inserted later.[108] This includes the letters in the text that refer to the diagram. By contrast, the diagrams that accompany the propositions are invariantly lettered with capital letters, while the use of capital and lowercase letters varies in the marginal diagrams, even within one and the same diagram, e.g., in the marginal next to I 6. There is only one lowercase β to be found in the inside-text diagrams, which however appears in a darker handwriting, presumably stemming from a later hand. B is, like P, written in minuscule, however, differently than in P, excluding the letters that refer to the diagram which are capitalized throughout. The letters in the diagrams are capitalized as well, both in the main (that is, inside-text) diagrams and in the marginal ones. This means that the letters of the main diagrams are standardized in either manuscript. The standardization is somewhat more rigorous in B because of the corresponding standardization of capitalized letters in the text and in the diagram. However, the kinds of letters are not standardized *among* the manuscripts of different manuscript traditions.

Compared with diagrams of the papyrus fragments, the lettering of the Byzantine manuscripts is much more standardized. Of the four fragments, only two contain lettered diagrams, whereas the diagrams of the other two are unlettered. However, as I have said above, the absence of diagrammatic letters from the two fragments appears to have to do with the respective abridgment of the proposition text and the corresponding absence of letters from this text, because the other fragments unanimously include diagrammatic letters in both the text and the diagram. P. Fay. 9 (Fig. 1[a]) is one of them (though only one letter is recognizable because of the fragmentary state); this is the fragment that most closely resembles the respective proposition text of the *Elements*. The standardized appearance of diagrammatic lettering in the manuscripts for Euclid's *Elements* indicates strongly that also previous manuscript copies of the *Elements* contained diagrams that were lettered, but in any case, that the diagrams (if there were any at all) were lettered. While the diagrams of the Byzantine manuscripts are lettered throughout in capital letters, it is notable that the two ancient lettered diagrams each contain one lowercase letter (in diagram: α). P. Fay. 9, in which only this α is transmitted, might have been lettered in lowercase letters throughout, while the diagram of the Herculaneum papyrus is lettered in capital letters apart from the α. This non-standardization of the ancient fragments regarding the kind of diagram-

108 These are definitely not genuine, as they are of the explicit, non-Euclidean type of standardization avoided throughout in the *Elements*. See Section 5.2, 424–426 for what this practice, on the one hand, reveals about the implicitness of the work's standardization, which provokes such explicit standardization (cf. Saito 2009, 804), and, on the other hand, about the extremeness of standardization which allows for inserting such terms easily in the first place (provided that the text did not have to be rewritten for fitting the Proclean scheme).

matic letters can hardly be explained based on the scarcity of the evidence. However, it can remind us not to generalize from the Byzantine evidence and assume that the diagrams of the ancient manuscripts standardly contained capital letters if lettered diagrams were standardly included at all. The lowercase α of the Fayum papyrus is especially remarkable as the text of the fragment is written in majuscule, by contrast with the minuscule of our Byzantine manuscripts.

4.2.4 A brief outlook on the standardization of the geometrical configuration of the manuscripts' diagrams

A question that has not been considered in the case of the papyrus diagrams because of the scarce evidence is how the construction assignments are realized in the diagram. This question is not about smaller differences which are due to the drawing style which we have discussed before, but about what form the geometrical objects take and how they are arranged in the diagram.

In general, the diagrams for one and the same proposition resemble one another quite closely in terms of the geometrical figuration which they depict. This can be seen from the extant comparable diagrams: those which are transmitted for the same propositions on both papyrus and manuscripts or between manuscripts of different traditions (that is, between the only non-Theonine testimony P and one of the manuscripts of the Theonine transmission). This is the case wherever the construction assignments leave only one possible way for arranging the objects. Indeed, to a large extent, the object forms and relations are predefined by the construction assignments as well as the definitions and conventional understanding of geometrical terms. Here it only happens that the same diagram appears mirror-invertedly in different manuscripts.

Despite the overall unanimity of the diagrams, there are cases in which different manuscripts contain different diagrams for the same propositions. The diversity of the diagrams seems to be due to the text's lack of specificity regarding visual representation. For example, the text does not specify how to depict three-dimensionality. We have dealt with this issue in the section of drawing techniques, and we have seen that the different diagrammatic ways of dealing with three-dimensionality have to do with the obligatory use of ruler and compass. We have seen that the different ways of depicting three-dimensional objects have an influence on what form the objects take in the diagram and how they are related, that is, on the geometrical configuration. Although I pointed out that the diagrams of particular three-dimensional configurations nevertheless often coincide in different manuscripts,[109] they sometimes do not. For example, it happens that the same kinds of three-dimensional objects are depicted by means of different methods of showing depths or, if one diagram shows the object

109 See above, p. 235.

three-dimensionally, the other one shows it with a mix of different perspectives, for example in the above-mentioned case of the diagrams for proposition XII 15: in B, the cylinders are drawn three-dimensionally (Fig. 18), whereas in P, the circular planes are drawn from a different perspective than the rest of the figure (Fig. 17). Apart from the way of dealing with three-dimensionality, there are more ways in which the diagrammatic configurations in the manuscripts differ. The reason seems to be again that the text leaves different ways for realizing a proposition diagrammatically. Consider proposition I 27 which, in the two Greek manuscript traditions, contains diagrams which differ regarding how they deal with its proof by contradiction. While B does not depict the contradictory part of the proof at all (except in marginal diagrams), P presents a mix-up of the conditions and the proof construction which turns out to be impossible in view of these conditions.[110] This proposition seems to be a case in point since it is mostly not clear from the impossible proof construction how to draw the diagram(s).[111]

In view of these differences that seem to be due to lack of specification of how to translate the text into a diagram, it is all the more surprising that the two manuscripts often display the same diagrams for the same proposition: for example, the diagrams for XII 13 and XII 14 show the cylinders three-dimensionally in the same way as the cylinders of XII 15 look in B (Fig. 18).[112] This is conspicuous since the manuscripts testify to at least one other way of depicting cylinders.[113] Apart from propositions about solid figures and such that contain proofs by contradiction, some propositions do not make unambiguously clear where exactly to draw a particular point in the diagram. Again, it is noticeable if the transmitted diagrams nevertheless turn out to be the same, as in the case of proposition I 9. Both in P and on the Herculaneum papyrus (Fig. 2) as well as on the Berlin papyrus (Fig. 3[a]), the diagram of I 9 has its point Z lying below ΔE, although the text does not specify whether it is to lie above or below it: the proof would work either way, as the diagram both times would yield the equal angles ΔAZ and EAZ that are necessary for proving that the angle BAΓ has been bisected.[114]

Such consistencies in the transmission of the diagrams are rather conspicuous and could be indicative of underlying standardized diagrams, although this standardization might have come about much later than Euclid. However, if we proceed on the assumption that the diagrams are standardly contained by the manuscripts and simply redrawn

110 See below, Section 4.3.3, 266–269 for a detailed discussion of the diagrams for I 27 in our two manuscripts (Fig. 29–32).

111 On the special relation of proof and construction in propositions with proofs by contradiction cf. below, Section 4.3.3, 261–263.

112 See above, p. 230, n. 102.

113 See p. 229 on XII 15 in P (Fig. 17).

114 The diagram transmitted for I 9 by the later hand of the Bodleian papyrus which restores the lost first part of *Elements* I depicts the equilateral triangle in the same way but betrays its unoriginal provenance by containing an extra line (BΓ) not specified by the text.

in the process of transmission, we should be surprised by those above-mentioned cases in which the diagrams deviate from one another. This non-standardization of the depiction of the geometrical configuration could indicate, in turn, that the diagrams were not standardly included, although it could also be explained by deliberate changes that were made to the diagrams in the course of their transmission and enabled by the ample scope which the text leaves for depicting the geometrical configuration. As it is, the comparison of the configurations of the diagrams in the manuscripts only provides tentative indications regarding the original presence of the diagrams in the text.

4.2.5 Conclusion

While the tendencies of the standardization of the manuscripts' diagrams could point to an ancient regular inclusion of them, especially as some of them can be observed both in the ancient fragments and in the Byzantine manuscripts, they could also be the result of mainstreaming the diagrams, whether they have been transmitted from ancient into Byzantine times or not, into a contemporary representational tradition, with the remaining non-standardization of the depiction of particular objects being more decisively indicative of the lack of diagrams in the ancient 'originals.' To conclude, owing to the great temporal, and, by consequence, cultural distance of the manuscripts from the ancient production and distribution of the work, and because the direct ancient evidence is too thin to allow the construction of a development of the diagrams in terms of continuities and discontinuities, we cannot infer any valid information from the Byzantine diagrams about the regularity of their original inclusion and appearances. We will have to turn to the proposition text to determine the extent to which it enabled and predefined drawing the diagram, and to which the text is predetermined by an underlying geometrical configuration (Sections 4.3 and 4.4). In so doing, we will reconsider the diagrams from the Byzantine manuscripts.

4.3 Diagrammatization I: the 'textual construction'

The analysis of the few ancient diagrams in Section 4.1 and of the diagrams of the Byzantine manuscripts in Section 4.2 has shown that the textual practice alone, if anything, is able to give implicit but potentially conclusive information about how often diagrams were included and whether their form was standardized. Therefore, we need to consider the *textual* form of the proposition as to whether it was shaped by the diagram and contains traces of it. I refer to these traces as 'diagrammatization.'

4.3.1 Forms of diagrammatization

Diagrammatization takes manifold forms in the *Elements*. That is, the text is shaped and organized by the diagram in many ways, both obvious and subtle. Not all of them, however, indicate conclusively that the text has been written on the basis of a diagram. I now briefly review the different ways in which the propositions seem to implicate a diagram and their ramifications as to the standardization of the diagram. Broadly speaking, the textual indications of a diagram can be summarized by two main features: the textual construction of the diagram, and the letters which characterize the whole concrete part of the proposition. These are the two most obvious manifestations of the diagram on the linguistic and structural-semantic level of the text. I first point out the palpable connection of those features to the diagram. Then, I am going to discuss the deeper repercussions of each of them with respect to whether the text offers unanimous evidence of its relying on a diagram. As is my approach throughout, I start from the features of the text rather than from its mathematical logic.

The first textual feature which unmistakably evokes a diagram is the construction part, the section of the concrete (i.e., lettered) part of the proposition which is regularly devoted to assigning the construction of geometrical objects. The construction part occurs, though not in all, in the great majority of the *Elements*' propositions.[115] One of the reasons why the occurrence of the construction is not 100% standardized is that the main objects have already been introduced before: they have been mentioned first by the foregoing exposition (and sometimes by the definition of the goal). Those propositions which do not deal with more objects than have already been set out can do without a construction part.

Linguistically, the construction part is marked by third-person perfect passive imperatives of 'construction verbs,' that is, verbs denoting some kind of diagrammatic action that is to be accomplished by the person who draws the diagram.[116] These verbs thus semantically suggest diagrams and the drawing thereof. What is more, in certain proposition types, this language of diagrammatic construction permeates even larger parts of the proposition, since in them, the construction verbs occur also outside what is according to Proclus' scheme the actual construction part.

For one thing, construction verbs dominate large parts of those propositions which employ the method of 'proof by contradiction.' There are nine of these propositions in *Elements* I, of which all but one are theorem propositions. Propositions of this kind may or may not contain a regular construction part. Additionally, in a frequent subtype of this kind of proposition, the proof is accomplished mainly by means of a (second) construction, so the proof part contains construction verbs in the form in which they occur

115 It may be missing from problem propositions (Section 3.4.1, 178–180).

116 I have distinguished above this class of verbs from verbs in the form of *present* active imperative, which denote the diagrammatic action which geometrical objects do to each other, and pointed out the minimal overlap between the two (Section 3.3.2, 161–165).

mostly in the construction part.[117] To be sure, along the lines of my above argument that 'parts' of the proposition are defined stylistically rather than by their place in the proposition, the propositions with proofs by contradiction which contain a (second) construction merely witness a coincidence of construction and proof part and, thus, strictly speaking, not an occurrence of construction verbs outside the construction part. But in the context of highlighting the importance of diagrammatic construction, it seems worth noting the use of construction verbs in propositions with proofs by contradiction because it is especially prominent and extensive.

Additionally, the language of the 'construction verbs' occurs in problem propositions which account for a considerable part of *Elements* I.[118] By contrast with theorem propositions which prove a hypothesis on the basis of the construction of a diagram, the very goal of problem propositions is to solve a problem of (diagrammatic) *construction*. Accordingly, in such propositions, construction verbs do not only occur in the construction part, but even in the non-concrete passages of the proposition: on the one hand in the enunciation in the form of an aorist infinitive which is used to assign the problem to be solved, and on the other hand in the conclusion, in the form of perfect passive indicative which is used to acknowledge that the problem of construction has been solved.[119]

The second feature which is indicative of a diagram are the letters: as I have pointed out above, the 'letter words' – lexical units of one to up to four letters – quantitatively account for a large portion of the text. They occur exclusively within what I have called 'letter phrases,' that is, composite terms referring to particular geometrical objects. What is more, they occur regularly in almost all parts of the proposition (namely the 'concrete' ones which deal with particular objects). Indeed, they give the large core part of its characteristic look, as they standardly can be found in the exposition, the definition of the goal, the construction, the proof, and, in the case of problem propositions, also in the conclusion.

It is by their mere occurrence that the letters point to a diagram, as they implicate a pictorial plane of reference in which they recur.[120] If the text did not presuppose or envisage a diagram as an external referential plane, the inclusion of letters would not make any sense. To be sure, it has been rightly argued that from a certain point of complexity onward, a proof cannot be accomplished without a diagram, but in the case of many of the easier proofs, especially of our first book, it might as well have been possible to do so. In any case, if the idea was to prove without the help of the diagram, the text would dispense with letters altogether (except if it used them as names that standardly signify certain objects).[121] As it is, the letters provide the largest

117 See Section 3.3.1, 145–147 for this type of proposition and its subtypes.
118 See above, Section 1.2, 16, n. 89.
119 See above, Section 3.3.2, 166.
120 Cf. Netz 1999a, 31: "the diagram sets up a world of reference, which delimits the text."
121 On the latter possibility, see Section 4.4.1, 283.

part of the proposition with a concrete dimension: except for the beginning and (in theorem propositions) the ending, the propositions revolve around a diagram. This is true for all propositions of the *Elements* no matter their contents or difficulty, because we find letters in every single one of them. This means that if we assume on the one hand that the more complex propositions have to be based on diagrams, and if on the other hand all propositions are invariantly lettered, then all propositions of the *Elements* should invariantly refer to diagrams.

From what I have recalled about the most visible and tangible aspects of both the 'construction' as well as the letter words from the foregoing analysis of textual standardization, the diagram appears to be omnipresent in the propositions of the *Elements*: the standardized occurrence of the two prominent features seems to be directly due to the standardization of a diagram. It therefore seems out of the question that the author of the propositions has a diagram in mind which sooner or later must come into being. The crucial question, however, is when and by whom it is created; whether beforehand by Euclid *as part of the proposition*, so that the diagram, in a material sense, would have been part of the text, or only subsequently by the reader according to the instructions provided by the text.

In approaching this question, I am going to sound out the two overtly diagrammatic features of the text – the construction and the letters – for deeper resonances of the preexistence of the diagram. I have just recalled from Chapter 3 some aspects of their standardization: the structural part of the proposition that assigns the drawing of the diagram, the construction imperatives, the letter phrases into which the letters are embedded, and the division of the proposition into an abstract and a concrete part depending on the use of the letters. I have analyzed these aspects in the previous chapter with respect to the standardization of the *text*. In the following, I focus on the various characteristics of the textual construction and the use of the letters regarding their repercussions for the standardization of the *diagram*.

In this section, I explore the textual construction and examine its structural, linguistic, practical, and functional side. The underlying assumption of investigating the textual construction with respect to the standardization of the diagram is the following: if the diagram did not materially underlie the proposition text, the text would have to describe the construction in such a way that the diagram could be constructed on the basis of it. If, however, a diagram is already there and does not need to be constructed, the text would not actually have to enable the construction of the diagram. Therefore, a 'non-constructional' construction would imply that a diagram must have already been there. This assumption, however, presupposes that the construction of the diagram is at all a necessary element of the proposition. Therefore, I will consider whether the diagrams of the *Elements* are relevant for the proof at all, and if so, in what form.

In a shorter first part of the section, I consider the structural implications of the section of the proposition that is concerned with the textual construction of the diagram and explore the implications of the dominant linguistic feature of the textual construction, that is, the third-person perfect imperatives. In a second part, I ask how practical the construction assignments are regarding the actual construction of a diagram: whether they enable the construction of a diagram at all, what kind of diagram one can construct on the basis of them, and how such a diagram relates to the implications of construction assignments in terms of unambiguousness, technicality, and precision. With regard to the drawability of the diagram, I reconsider briefly the previously surveyed diagrammatic evidence. In a third part, I consider how functional the implied diagram is with regard to the proof. That is, I ask whether the text sufficiently enables the construction of the diagram in order for the proof to be accomplishable on the basis of it. This entails the question of whether the diagram forms a basis of the proof and whether the diagrammatic form is proof-relevant.

The letters that refer to the diagram are a constitutive feature of the construction part and play a crucial role in assigning the construction of the diagram. However, since the subject of the letters is a topic in itself, and one that is not limited to the textual construction, I exclude them from my analysis of the textual construction and consider them extensively in Section 4.4 as the second main subject of diagrammatization. At various points, however, I anticipate the significance of the letter use for shedding light on a problem arising from the analysis of the construction assignments regarding the practice of construction. But I save precisely the combined analysis of letters and construction for Section 4.4.

4.3.2 The 'constructing section'

As I have pointed out above, not only the construction part proper, but indeed the whole first section of the proposition introduces and constructs the geometrical objects. This is why I have called this first part 'constructing section' as it is – beyond the boundaries of the Proclean proposition parts – concerned with the coming into being of a diagram. Whereas the following proof part draws freely on the objects that have been introduced before, jumping back and forth between them as the deduction requires, the first part systematically 'builds' the diagram. It begins with those objects that are the most basic and central ones – not only in a local but also in an argumentative sense – and moves on to the ones that are attached to them.

The structural implications of the section

The description follows roughly chronologically the order in which an actual construction of the objects would make sense. That is, it presents the larger objects which consist of several points in the order of their construction, whereas it does not necessarily refer to the points themselves according to the expected order. For example, I 47 first sets out a right triangle, then it assigns the construction of the square on its hypotenuse, then of the squares on both of its legs, and finally of the lines that connect the points of the triangle with points of the respective opposite squares (Fig. 21).[122]

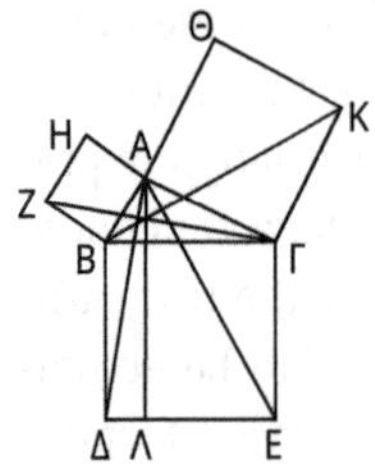

Fig. 21: I 47 according to Heiberg & Stamatis.

In the case of problem propositions, the evolutionary presentation of the diagram in terms of its construction emerges from the goal of this kind of propositions which is to solve a problem of construction. Here, the construction itself fulfills the goal. The subsequent proof only proves that the construction fulfills the goal. By contrast, in theorem propositions such as I 47, the proof shows that a statement about a certain geometrical situation is true, and the foregoing construction merely brings the geometrical situation into being. With regard to these propositions which represent the majority of the propositions of the *Elements*, one may ask whether such an evolutionary 'constructional' presentation would be at all necessary if the diagram preexisted, or whether it would suffice to mention the central objects and their relations toward one another (maybe by means of the ἔστω of the exposition). For example, in I 47 it might suffice to mention the right triangle and the three squares on the sides of the triangle, and where the triangle is meant to have a right angle and where the hypotenuse. The 'oblique' objects (as in I 47 the lines that connect the points of the triangle with the squares) would not even have to be mentioned as they could easily be identified in the adjunct diagram. To push this thought even further, it might suffice to provide a key to the diagram which featured the technical terms for the objects. The point is not that the objects and relations around which the proof revolves need not be specified *prior* the proof in some way or another, but that this need not necessarily happen by means of a *construction*.[123] This notwithstanding, even theorem propositions take the trouble to introduce (almost) all objects by constructing them (that is,

122 It will turn out later that the actual order of construction is indeed only sketched out 'roughly' by the text, and that certain changes to the order would be more practical or even necessary for accomplishing it (Section 4.3.3, 270 f.).

123 Of course, the inclusion of a construction part is a standardized convention in elements texts, and other forms of specifying the objects such as a 'key' provided next to the diagram are inconceivable within this genre. The purpose of the above mind game is simply to think about the origin of this standardized feature and what it reveals about the question of the existence of an adjunct diagram.

by using a language of construction).[124] From these considerations, the chronological and evolutionary presentation seems to suggest that Euclid did not perhaps intend for a diagram to be supplied with the text, but rather that the readers should construct it on their own.

The implications of the construction imperatives

The impression conveyed by the structure of the first section of the proposition is reinforced by the lexical feature of the so-called construction verbs, whose presence in different parts of the proposition I have recalled at the beginning of this section. These verbs account for the strong emphasis which is put on geometrical practice in the axiomatic-deductive propositions of elements texts.

The question now becomes: to whom is the act of construction assigned? The construction verbs make the most prominent appearance in the construction part in the form of perfect imperatives. These perfect imperatives of construction verbs are the constitutive feature of the construction part. By their standardized appearance in the third-person passive, they are decidedly underdetermined as to their addressee and even seem to negate agency because their grammatical subject is always the geometrical object, never a personal agent. However, the imperative mode suggests that they are directed to *someone*, precisely in order for this person to perform the construction. Naturally, this must be someone but the author himself. This only leaves the recipient of the text because the author would hardly assign something to themselves instead of just doing it, least of all retrospectively if they had already done it.

It is only seemingly that these verb forms display a discrepancy between the imperative mode which "always implies future time"[125] and their standard perfect tense.[126] For the future sense of the imperative applies regardless of the tense which "do[es] not refer to differences of time, and denote[s] only the stage of the action."[127] Thus, the perfect imperatives of construction verbs simply denote that the (future) construction is to be "complet[ed] with permanent result."[128] This means that the construction imperatives do not imply counterintuitively (considering that they are *imperatives*) merely by their tense that the diagrams are standardly presented to have

124 The section on letter specification below will show that indeed many but by no means all of the letters are introduced by the 'constructing section' (Section 4.3.2).

125 Smyth §1864.

126 These perfect imperatives recur standardly in other mathematical texts and within geometrical contexts in other scientific texts, but the use has been adopted most probably from the early elements tradition in which Euclid himself stands. Third-person perfect passive imperatives in Attic prose are limited to a small number of non-constructional verbs, for which see Smyth §713. In Aristotle, they appear of verbs of saying and definition (Asper 2007, 125, n. 223 with reference to Lengen 2002, 99).

127 Smyth §1864.

128 Smyth §1864.

already been accomplished by the author prior to the text, that is, as if they were perfect passive *indicatives*.[129]

The other main line of interpretation of the perfect tense connects it with the impersonality of the third-person perfect passive imperatives which, in the case of this mode, is somewhat paradoxical, since one expects an instructional order to be directed to some specific agent-to-be.[130] Some understand the third-person perfect passive form of the imperatives as a repercussion of the 'contemplative' and 'anti-operational' attitude of Greek mathematics.[131] This, above all, Platonist view[132] conceives of the geometrical objects and their relations, and, by consequence, also of proofs about these objects as eternal and already existent, and therefore non-constructible but only discoverable by the mathematician. According to this, the impersonal perfect imperatives have been interpreted as depicting the construction of the geometrical objects as essentially preexisting, and their presence, thus, as uncaused.[133] This means that unlike Netz's interpretation, which takes the perfect tense to refer to the preceding actually drawn construction of the diagram, i.e., within the text, the 'Platonist' interpretation of the perfect tense has no bearing on the question of when the diagram is drawn, as it understands the perfect metaphysically in terms of the ontological priority of the geometrical figure. Just like the first interpretation, however, it understands the perfect falsely as having a temporal meaning, as expressing anteriority, whereas the perfect imperatives regardless of their resultative aspect still point to future time. Moreover, the second interpretation produces an irresolvable paradox: on the one hand, the aim of the perfect tense and of the objectifying third person seems to be to conceal that the geometrical situation comes into being

---

129 *Pace* Netz 1999a, 25: "The text takes the diagram for granted. [. . .] This, in fact, is the simple explanation for the use of *perfect* imperatives in the references to the setting-out – 'let the point A have been taken.' It reflects nothing more than the fact that, by the time one comes to discuss the diagram, it has already been drawn."

130 Asper 2007, 125.

131 Asper 2007, 126 with reference to Piaget 1972, 260 f.

132 For Plato's critique of the language of mathematical procedures or their ontological preconditions cf. Asper 2007, 126 f. He takes the critique to refer to the recipes of practical mathematics rather than to the Euclidean tradition of theoretical mathematics, the language of which, according to him, is a reaction precisely to an objection like the one raised by Plato. While Euclid's own Platonist dependence is uncertain, the transmission and reception of his work have taken place largely in the Platonist tradition, which is why the original Platonist legacy and the later Platonist shaping are hard to distinguish from one another. In Section 5.3, 432, I shall return to the question of the role of a Platonist philosophy of mathematics in forming the textual features of elements texts in general and the extreme standardization of the Euclidean *Elements* in particular.

133 Asper 2007, 125 f.: "Bedeutsam an diesen unpersönlichen Imperativen einer resultativen Aktionsart ist vor allem, daß sie die Konstruktion der jeweiligen Betrachtungsgegenstände als bereits vorfindlich präsentieren, also beanspruchen, eine unverursachte Gegenwart zu beschreiben, einen unpersönlichen und damit objektiv gegebenen Zustand." Cf. also Lachtermann 1989, 65–67.

and is produced by an individual mathematician; on the other hand, the language of construction and the imperative form seem to suggest precisely that.[134]

From this, the question arises why the construction, if it was to be concealed by the objectifying language, would be made a subject of (theorem) propositions at all and, at that, would end up taking up such a large part of the proposition, and why, if it had to be described at all, it would be presented throughout as to-be-done by means of imperatives. This paradox would be especially compelling if the diagrams were included in the text from the beginning. Therefore, it seems that from a 'Platonist' point of view, the paradoxical construction imperatives can only be explained by the absence of diagrams from the text of the original *Elements*: for then there would be no way around including a construction even in theorem propositions (by which the reader is enabled to produce the diagram), and the impersonal style of presenting it would reduce the impression that the construction is idiosyncratic and therefore to some extent random. Nevertheless, the paradox between the assignment of a construction which must be actively accomplished by someone and the third-person passive remains with this interpretation which regards the mathematician's construction as something which the text aims to play down.[135] For the *Elements*' propositions, including the large number of theorem propositions, center around the diagrammatic *construction* rather than a readymade diagram.

If one, however, accepts the construction imperatives as what they are (i.e., imperatives), the third-person passive seems to deliberately leave open our initial question of the addressee of the imperatives. In this way, they could refer to any recipient in a decontextualized context, or even more generally to anyone who would tackle the problem or theorem. Rather than addressing back personally the one who pronounces it, the third-person passive imperative would be similar to an impersonal phrase such as δεῖ with infinitive, and express the necessity of performing the respective construction for anyone trying to accomplish the proof. In this way, the imperatives are characterized by a certain timelessness, but relatively refer to future time from the point of view of the one doing the proof.

The perfect, again relatively, prescribes (in the above-mentioned standard sense of the resultative aspect) that the respective construction is to be accomplished completely, definitively, and permanently. While the requirement of completion is directed to the performing mathematician, the implied reference to the permanence of the construction could, in the sense of enallage, be more of a statement about the status of the diagrammatic construction: the permanence envisaged would be due not so

134 Cf. Asper 2007, 126: "Die von Proklos referierten Ausführungen des Speusipp und des Amphinomos zum Begriff des θεώρημα zeigen, wie wichtig es ihnen war, jeden Anschein von γένεσις zu vermeiden, also trotz mathematischer Konstruktionsbeschreibung den ewigen Status der mathematischen Objekte zu wahren. Genau diese Intention drückt, scheint mir, auch der unpersönliche Imperativ aus."
135 Asper 2007, 126, n. 224 with reference to Netz's interpretation which does not address the peculiarity of the third person.

much to the quality of the construction than either to the supposed nature of mathematics (in the Platonic sense described above) or the quality of the proof (in the sense of an authorial strategy of persuasion). Just like the proof, the construction, once it has been accomplished, is ironclad and holds forever. On the same note, the standardized use of the third person which is unusual with imperatives could evoke objectivity not only on philosophical grounds as described above. Rather, the implication of objectivity by the third person could also be socially motivated, as a means of enhancing the persuasiveness of the mathematical account in a decontextualized situation of communication.

While the tense and personal number of the construction imperatives thus seem to point beyond the literal sense of the assignment of a geometrical construction pure and simple, they do not prevent the imperatives from accomplishing precisely this act of assigning the construction. However, the construction imperatives do not unequivocally delegate the construction to the recipient in such a way as to imply that the proposition texts did not come with an adjunct diagram. Rather, by their impersonality and timelessness – they always imply future time, but their point of reference is not fixed in time – they leave open by whom the construction is to be accomplished. In so doing, the imperatives of construction are ambivalent as to the question of the standardization of the diagram: on the one hand, the imperatives *as* imperatives, if taken at face value, suggest that a construction is yet to be done and that, consequently, a diagram is not necessarily underlying. Moreover, if one understands the impersonal resultative form of the imperatives as indicating an endeavor to avoid the impression that an actual construction is being assigned to the recipient, the point of assigning the construction if not for the very sake of construction becomes all the more questionable. On the other hand, the impersonality itself opens up the possibility that the diagram has been constructed prior to the reception of the text. In the event that the diagram underlies the text, the point of a textual construction could be the reenactment of the actual practical construction; or the point could be to provide for the possibility that the construction can be reenacted later in unforeseeable, because decontextualized, circumstances of reception.[136] This means that the textual indications of a diagrammatic construction do not necessarily exclude a standardized inclusion of diagram, but also do not conclusively indicate it.

To conclude, while the constructional perspective of the first section of the proposition indicates a construction yet to be accomplished, the defining stylistic feature of the construction part, i.e., the most cumbersome and peculiar imperative forms,[137] is ambivalent as regards the question of the standardization of the diagram.

---

136 Cf. Section 4.5, on the explicit reflection by the geographer Ptolemy about the necessity of preparing for such circumstances and his own methods of map description in view of this.

137 I shall return to the remarkable form of the imperatives below: in Section 5.1, 418 I shall think about their excessive standardization and their resulting repetition and dominance in the *Elements* in terms of 'unnecessary' standardization from a cognitive point of view. Finally, I shall attempt a syn-

4.3.3 The practicality of the textual construction

I now consider the drawability of the diagram on the basis of the construction assignments. Specifically, I ask how useful they actually are for someone who tries to draw the diagram according to them, and what this tells us about the standardization of the diagram.

I start from the extreme limitation and standardization of the textual means that the text employs for describing the construction. It only uses the aforementioned imperative constructions that consist of three main elements. First, these are the standard terms for geometrical objects that each entail a definitive set of constitutive properties, which is partly explicitly specified in the definitions, partly deducible from the standardized usage. Sometimes the noun terms are complemented by adjectives that add specific standardized properties. The second element is the standard terms for geometrical operations which are verbs or phrasal verbs that are standardly connected to certain geometrical objects. In many cases, they denote no more than the 'drawing' or 'constructing' pure and simple of the respective object. In some cases, they mention very basic and synoptic geometrical actions which affect one or several objects but reveal nothing about the constructional act itself. Third, there are the letters which denote the constitutive points of the objects and, in so doing, enable referring to these individual points (and not just to larger objects as a whole) and which enable objects sharing one or more points to be textually related to one another. I defer the subject of the letters for Section 4.4 and focus on the standardized terms of the construction assignments.

As they only consist of standardized terms for geometrical objects as well as basic and synoptic geometrical operations, the construction assignments do not specify information which might be necessary for accomplishing an unambiguous and precise construction. Of course, 'precision' with regard to (geometrical) drawing is a somewhat subjective notion which is liable to anachronism. I continue to connect it to the notion of unambiguousness. The necessity of the diagrammatic construction to be unambiguous is suggested precisely by the extreme unambiguousness of the language of the construction assignments. The properties of the geometrical objects and activities are (almost) completely standardized, so that the diagram can be expected to match this description and depict the objects exactly, not just vaguely. This means that the text itself, not modern expectations, suggests a non-sketchy, precise, and, by implication, technical drawing. Ironically, while these qualities are suggested by the standardized language, it is precisely this standardized language which seems to prevent a standardized, unambiguous construction. Let us consider what the construction as-

thesis, including both the peculiarity of the verb forms and the observation that the text indulges in them, and offer another, that is, an aesthetic explanation of their use apart from objectification for philosophical or social reasons.

signments do *not* specify and how this missing information could be necessary for the construction of the diagram in accordance with the textual construction.

Underspecification regarding materials and instruments

The text does not make any specification as to what kind of writing medium to use in order to ensure that the respective geometrical configuration assigned may be drawn so as to achieve an unambiguous result. That is, it does not specify what quality of papyrus should be used so that any writing materials that are necessary for accomplishing the result may be applicated. Further, the text does not specify the expected dimensions of the figure in terms of the number of drawings constituting it, or in terms of width and height or area, so that the figure could be accommodated but still display its decisive properties. Moreover, the text does not specify whether and how the diagrammatic space is to be prepared: in terms of smoothening, or in terms of demarcating the drawing space, or organizing it by means of an orientating pattern. For the latter two points, contrastive examples from Ptolemy may be adduced: he, on the one hand, explicitly specifies squares into which diagrams of a greater width than height are to be drawn,[138] and, on the other hand, advises preparing the space demarcated for his maps with orientating lines representing the meridians and lines of latitude. What is more, the text does not specify any of the materials with the help of which to draw the figures onto the writing medium. That is, it does not recommend a kind of pen or ink, nor does it mention the technical means necessary for accomplishing the construction, such as ruler and compass, more sophisticated instruments like goniometer, triangle ruler, and means for measurement.

The lack of such information is in line with the overall approach of the *Elements* which dispenses with any kind of metadiscourse about its mathematics or how to do it, but which, instead, simply does it – by means of a standardized language which does not leave any room for methodological information. However, in the case of the construction part, which purports to result in a diagrammatic practice, the lack of information is particularly conspicuous. This central generic characteristic which entails a general silence on practical issues could have different implications for the time and kind of construction of the diagram. Either (1) the text does not talk about material questions because it already includes the diagrams and does not need to enable their reproduction – this would conspicuously contradict the imperative nature of constructions, however impersonal it is; or (2) mathematical experts are the only intended addressees who would know what materials were suitable or conventionally used; or (3) the material questions are not of special importance for the construction

138 This does not prevent the Byzantine copyists from drawing the diagram into the narrow space between two columns, but this only shows that one should question the reliability of the Byzantine manuscripts regarding the question of the standardization of the diagram in ancient times.

of the diagram. The question of which of these implications holds true shall come up time and again in the analysis of the different aspects of the practicality of the textual construction; at this point, at least, it cannot be decided.

Most of the information on materials might indeed be dispensable, as the result of the drawing would only differ by degree. In the case of the technical instruments, however, the third possibility – that it actually does not matter whether any instruments were used, which instruments were used, and how they were used – would be quite a significant one. For it would mean that the text accepts the possibility that the diagram is merely drawn free hand rather than technically constructed. As we have seen in the analysis of the Byzantine diagrams in Section 4.2, the question of the (theoretically) necessary technical instruments is closely tied up with the envisaged method of construction. Therefore, the possibility that the construction allows for dispensing with technical instruments may only be discussed in view of the methods which were available. I shall consider this possibility more systematically under my next point which deals with the lack of information on the methodology. At present, I only anticipate the possible reasons for and implications of both the non-mention and the eventual non-use of certain technical means.

The two basic technical means which are in any case essential for technically constructing and unambiguously representing a large portion of the figures of the *Elements* are ruler and compass. The fact that these instruments are not mentioned is non-conclusive regarding their use or standardization: possible reasons for not mentioning them may include that the text does not talk about technical issues in general or that the use of these instruments was so obvious and prevalent, because they would have definitely been known and in use by the time the *Elements* was written.[139] Other instruments that are known from modern geometry classes such as goniometer and triangle ruler may not be mentioned either on account of their being unknown or unavailable to ancient Greek geometers, or for their being undesirable in their eyes. The use of anything other than ruler and compass may have been precluded for aesthetic reasons, even though the elegant confinement to these two entails in many cases a great number of auxiliary constructions.[140] Goniometer and triangle ruler or something of the kind may not have been employed because they imply a radically different concept of 'geometrical construction' than one which gets by with ruler and compass only. They involve measurements which may not have been considered part of a 'proper' construction. Crucially, measurements and geometrical calculations do not form part of the text at any point. As a rule, the text does not make any reference to specific dimensions. Instead, it only mentions simple relations between objects regarding size ('is equal to,' 'is larger than') and position ('is parallel to,' 'is orthogonal

139 Rulers are used in some of the ancient diagrammatic evidence for the *Elements* itself, the use of compasses is uncertain (cf. Section 4.1.3, 212 with n. 73).

140 I shall give an example in the subsequent section on the methods of construction.

to') in the geometrical books, and proportions in the arithmetical books. This indicates strongly that instruments that presuppose measurements were not put to use, even though measurements also might have been made up in accordance with the text's merely proportional specifications.

The objects of *Elements* I can be constructed in a technical way by means of ruler and compass with or without the help of such imaginary measurements. By contrast, some of the three-dimensional objects that occur in later books of the *Elements* involve circular lines. They seem to require measurements for the depiction of three-dimensionality and perspective unless they are to be drawn free hand. Making up those measurements would presuppose calculations on the basis of rather complex mathematical formulae. Although experts would have been familiar with such calculations, it would seem unlikely that the text presupposes them. The text's silence on both measurements and the necessary means for using them such as the preparation of the drawing area with a coordinate system seems to indicate that such objects would not have been technically constructed but rather drawn free hand if they were to be drawn three-dimensionally.

Let us now briefly recall the results of our analysis of the diagrammatic evidence in Sections 4.1 and 4.2 regarding drawing materials and instruments. The evidence is in line with the textual construction insofar as it does not display any kind of preparation of the drawing area. Also as regards technical instruments, the material evidence largely corresponds to the expectations raised by the textual construction. Indeed, in none of the transmitted diagrams has any instrument but ruler or compass left any traces. The diagrams of the papyri have been drawn either by compass and ruler or free hand, whereas the manuscript diagrams have been drawn invariantly by means of ruler and compass. The diversity of the papyrus diagrams may be due to their different (partly non-manuscript) functions, as the only fragment which we suspected to have been part of a manuscript copy features a ruler-drawn diagram. For this reason, it does not necessarily indicate that the original diagrams were sometimes drawn by means of ruler and compass, sometimes not. Conversely, the standardized use of ruler and compass in the Byzantine manuscripts could be the result of formal smoothening in the aftermath, so we cannot decide on this basis how standardized the use of ruler and compass had been in Euclid's own time.

However, the manuscripts have brought up one important new aspect regarding the use of technical instruments. They indicate that the standardization of the use of ruler and compass might have been even stronger than could have been assumed on the basis of the textual construction only. Compasses are applied standardly even in those cases in which free-hand drawing had seemed to be the only possibility in the supposed absence of other technical means. Circular planes that are part of three-dimensional objects are always drawn by means of a compass: either as ellipse-like figures that consist of two circular arcs,[141] or two-dimensionally as circles. That is, the manuscripts do not only display a diagrammatic practice which makes use only of ruler and compass, but they even do not feature diagrams which do *not* testify to either or both of these instruments. Whether this practice has ancient predecessors is not sure, but the standardization of this oddity in the manuscripts could point out a (maybe late) ancient tradition.

Underspecification regarding the method of construction

The most important way in which the textual construction is underspecified is regarding the respective method that is needed for accomplishing the construction of a certain object. The method is not mentioned, let alone explained. This kind of underspecification is closely bound up with the silence on technical questions. I start with some general considerations concerning the constructions of book I. Then I deal specifically with the challenges posed by the different kinds of objects, also beyond *Elements* I.

As I pointed out, it is the standardized language of construction itself which suggests that the construction be unambiguous and precise. However, it does not divulge how to erect a square on the hypotenuse of a right triangle, or how to draw a straight line through a point parallel to two other ones, to cite two examples from the construction of I 47.[142] That is, the construction assignments are extremely precise regarding the result to be achieved by the construction – objects that are defined unambiguously by certain properties and related unambiguously to other objects – but they do not inform about how to achieve this result. In order to be able to construct these objects in a geometrically precise way, one would have to know which intermediate constructional steps are necessary for doing so. As it is, the construction assignments of each proposition taken by itself enable drawing a sketch which imitates the central properties of a figure by approximation rather than properly constructing the figure.

In the absence of explicit information in the respective proposition itself, the axiomatic-deductive nature of the text suggests that it implicitly does contain the information which is necessary for the construction, and it also indicates where to look for it. For according to the deductive system of the *Elements*, the constructions of the more complex propositions presuppose construction problems that have been previously solved. In this way, the technical constructability of each geometrical configuration is theoretically guaranteed. However, the deductive principle also crucially entails that the presupposition of earlier results is always implicit. The text does not repeatedly explain previous proofs or constructions, but only mentions the respective

141 Netz 1999a, 17 notes that the "manuscripts do not show any technique for drawing non-circular curved lines, which are drawn as if they consist of circular arcs," without, however explaining this fact by the avoidance of drawing free-hand and the according obligation to use ruler and compass.

142 *Elem.* I 47 (vol. 1, p. 63.11–13 Heiberg & Stamatis).

result or construction assignment, which, by its standardized language, implicitly points back to the respective foregoing proposition. In the modern editions, the numbers of the underlying propositions are specified in brackets within the proof itself or in synoptic tables to compensate for the implicitness. Not so in our ancient manuscript evidence.

To technically construct a figure from further on in the book such as that of I 47 solely based on the implications of previously solved construction problems will be very difficult and tiresome, not only because of the implicitness of the constructional information, but also because of the highly implicatory character of almost all propositions except for the very first ones of the work. For each proposition presupposes a whole 'deductive tree' which entails a number of propositions which themselves, in turn, implicate further underlying constructions, which, again, presuppose more basic constructional knowledge. This means that this whole underlying 'tree' would have to be reconstructed from memory or from going through the previous propositions again in order to do the construction properly, that is, in the way that is implicated and envisaged by the deductive system.

What is more, the nature of the deductive system is such that the construction problems of *Elements* I are not about the systematic construction of the range of 'rectilinear figures' ('those which are contained by straight lines'), as defined from definition I, Def. 19 onward,[143] even though these constitute the majority of the objects which appear in the diagrams of *Elements* I. Rather, they are about constructions of relations of straight lines which both entail and enable the construction of larger rectilinear figures. That is, the construction of a square on a straight line that is necessary for the figure of I 47 does not constitute explicitly a foregoing construction problem, but one has to revert to those propositions which solve the construction problems of letting fall a perpendicular and drawing straight lines of the same length. Constructing a square on a straight line in this way is theoretically possible, but certainly not the easiest way to do it. It is, however, the only way suggested by the system of the *Elements*, and, therefore, the one which must be assumed to be underlying, but it is debatable whether the text would have suggested a more economical solution if it specifically assigned the construction of a *square*.

Considering the wide range of possible recipients of the impersonal construction assignments, lay readers, on the one hand, theoretically would be able to accomplish the construction on the basis of the indications of the text. However, it would be very difficult for them because first they would have to remember the implicitly underlying assumptions and previous constructional results and then perform a very complicated construction. On the other hand, expert recipients would know the constructions that are suggested by the *Elements* and might even be able to think of easier ways of con-

143 *Elem.* I, Def. 19 (vol. 1, p. 3.9–12 Heiberg & Stamatis).

struction according to the principles of the *Elements*.[144] However, if the construction was properly done in the way which is implied by the text, it would be quite cumbersome even for experts in many, even seemingly simple cases would and seem disproportionately complex in the case of auxiliary objects or even the main objects on which the proof is based. Thus, the question is whether the author or recipient really was supposed to go through it, and if not, why the text does not suggest otherwise, but rather explicitly assigns the *construction* of an object with clearly defined properties whose construction is clearly solved elsewhere.

If a proper construction in the way which is implied by the structure of the *Elements* was indeed envisaged, the auxiliary constructions that are necessary for constructing most of even the simplest geometrical objects would find standardized expression in auxiliary lines. Such lines would be proof of the technicality of the diagrams. We have seen, however, that neither the ancient nor the Byzantine diagrams display such underlying constructions, but rather feature non-technical diagrams, which is just what the construction assignments, taken by themselves, allow for. Thus, the non-constructional nature of the construction assignments seems to be mirrored and confirmed by the whole material evidence, both ancient and Byzantine. For we have seen that it is strongly indicative of the non-technicality of the diagrams (as not a single transmitted diagram is technically constructed). However, since there may be only one single ancient diagram that originates from a real manuscript, the general practice could still differ and be technical, which the deductive system of the *Elements* does not only allow for, but even implies.

It remains a remarkable fact to be kept in mind, however, that the assignments of *construction*, i.e., of performing a geometrical action which in almost all cases entails a whole set of actions, only mention synoptically the main action to be accomplished, which denotes the result to be achieved by the construction rather than how to get there. In so doing, they are stereotyped and non-informative as regards the construction. They posit objects rather than facilitating their being brought into being.

As a result, the construction assignments turn out not to differ essentially from the present imperatives which merely 'set out' some of the objects beforehand as if they were to appear out of thin air. This is conspicuous as both types of imperatives and resulting 'stylistic parts' of the proposition are formally clearly demarcated.[145] This raises the question of how the objects of the exposition are meant to come into being. While they are all constructible in the same way as are the objects that are introduced in the construction, the modern editions, notably, never specify the numbers of the underlying propositions, apparently on the assumption that the objects of the exposition are merely posited. This would, of course, have consequences for the

144 Ancient and modern commentators have discussed alternative ways of proof for individual propositions, which would result also in different diagrams, but they have not discussed alternative ways of technical construction.

145 See above pp. 161 and 179 f.

nature of the whole construction: why would the 'construction' of certain objects of the proposition be technical, if the objects that are first set out were merely 'drawn' or even sketched out?

The absence of more detailed explications of construction and of technical information may not seem all that significant with regard only to *Elements* I. For this book exclusively revolves around two-dimensional objects which are made up only of straight lines and circles and theoretically can be constructed, because the constructions previously solved presuppose only ruler and compass on the material-technical side. As has been pointed out in the foregoing section on the silence of the text on technical means, the question of the method of construction is closely connected to that of the instruments which it requires. In the case of *Elements* I, the constructability of the objects and relations is not only guaranteed methodologically, but also practically, as the method entails the use of instruments which are available at the time and enable a technical construction. Surprisingly, the review of the material evidence has revealed that, in spite of the possibility of a technical construction, ruler and compass – in the evidence we have – are never even used for the technical kind of construction which they generally enable.[146] Instead, they have been shown to purport a seeming technicality. In view of this use of ruler and compass which is non-technical, but seems to serve precisely the diagram's technical *appearance*, the two instruments appear to be significant, even though their application has turned out, counterintuitively, not to be indispensable from the point of view of the accomplishment of a technical construction.

While *Elements* I only deals with two-dimensional rectilinear and circular objects, some of the subsequent geometrical books deal with three-dimensional objects, both non-spheric and spheric ones. Their description implies difficulties regarding the depiction of perspective which they require on account of their three-dimensionality. Notably, the construction assignments never specify how foreshortening is accomplished. The representation of three-dimensionality is nowhere assigned as a construction problem and so cannot be presupposed. Thus, the reader either would have to know how to do it or would not be able to do it.

In modern geometry, there are numerous recognized techniques of foreshortening that are recognized as the most unambiguous, useful, or correct ones depending on context, audience, and purpose. If an ancient reader knew of the existence of different techniques, they would not know which one to choose on the grounds of the constructional description only. The most common techniques today which we have also seen in our two Byzantine manuscripts are central projection and parallel projection. In the first of these, which most closely imitates the vision of the eye or a camera

146 To be sure, this is only true for the construction of (straight or circular) lines in relation to other (straight or circular) either for its own sake or for the sake of the construction of a larger object, while straight or circular lines simply drawn by themselves by means of ruler and compass, are, of course, indeed 'technical.'

lens and therefore makes a natural and realistic impression, the figure is oriented toward one or more vanishing points on the horizon line using vanishing lines which connect all points of the object to the vanishing point(s). That is, all parallel lines in space eventually run together in this point. This can be realized as one-point perspective, that is, with one vanishing point, as two-point perspective, or as three-point perspective, either as worm's eye or as bird's eye view.[147] By contrast, in the simplest type of parallel projection, that is, oblique pictorial projection, two-dimensional objects are projected onto a projection plane by means of oblique parallel lines which correspond to real parallel lines in space of the three-dimensional object depicted in this way. While the front side and the projection plane keep their real dimensions, the plane of the parallel lines in space is usually foreshortened (cabinet perspective).[148] Here, both the factor by which they are foreshortened as well as the angle made by the oblique and the horizontal line may be set individually. Besides, there are a great number of other ways of producing the illusion of three-dimensionality on a two-dimensional surface. It is important to note that neither type is more 'perspectival' than the other, but that each of them proves useful depending on the purpose of depiction.[149] Since the text does not mention perspective at all, let alone specify the most suitable method, it proves highly impractical especially as regards the depiction of three-dimensional objects.

Apart from the question of the general method by which to present the three-dimensionality of an object, the question arises how to put the chosen method into practice. In the case of the most basic of rectilinear solids, i.e., parallelepipeds, this includes the construction of the projection lines either as parallels which make a certain angle with the horizontal line or by orientation to a specific point on the projection plane. Moreover, one has to determine the length of object lines with the help of the foreshortening factor in relation to their real length in order to map them onto

147 Additionally, there are even more sophisticated varieties of central projection using more than three vanishing points, but these are usually not relevant for the purposes of common geometry.

148 Other types of oblique pictorial projection include the cavalier perspective, which is produced like the cabinet perspective except for the oblique lines' not being foreshortened, and the military projection, in resembling the bird's eye perspective, in which the ground plane (not the front side) is the plane projected by means of oblique lines.

149 Such a classification seems to be implied by Netz 1999a, 17 in claiming that "three-dimensional objects do not require perspective in the strict sense, but rather the practice of foreshortening individual objects." It is not quite clear from this what exactly "perspective in the strict sense" refers to, but I take it that it means central projection as explained above by contrast with oblique pictorial projection, that is, "the practice of foreshortening individual objects" in Netz's distinction. This can also be a problem of English vs. German terminology: whereas the latter regards both types in terms of 'perspective,' the former, along the lines of Netz, seems to reserve the term for types of central projection. Interestingly, Netz does not only claim that foreshortening suffices, but also, with reference to the opinion of M. Burnyeat (1999a, 17, n. 24) that it is even more effective in elements texts than perspective proper, without, however, saying why. Indeed, foreshortening, or oblique pictorial projection, is the prevailing method in the Byzantine diagrams, on which see further below.

the projection lines. In the case of more complex rectilinear solids such as pyramids or prisms, the same actions apply, but the construction includes the additional difficulty of drawing the base in parallel projection, which may be more complex than that of a rectangular solid or cube. This is the case especially if the base is a regular polygon which turns into an irregular one in projection. Theoretically, all of this could be accomplished by means of the construction of angles of a certain size in relation to other angles, and of straight lines of a certain length in relation to other straight lines. However, the more complicated the base of the object which is to be projected, the more difficult it would be to do the projection on the basis of proportions only and do without calculations of sizes of angles and lengths of lines. Also, the technical construction proper of a polyangular prism or pyramid will be quite complicated to begin with, and the question arises whether the text would indeed intend it.

A more serious problem of construction, and one that is unsolvable by the means provided by the *Elements*, arises with the technical representation of such three-dimensional objects which include circular lines, i.e., cones, cylinders, and spheres. No matter the exact method by which to show depth in their representations, a technical construction of them from a perspective which displays these objects unambiguously would always involve ellipses. In the three-dimensional depiction of cones and cylinders, the circular bases would have to be presented as ellipses. If spheres are to be drawn three-dimensionally, one or several circular planes which are the result of cutting through the sphere would have to be indicated by means of ellipses in order to evoke an impression of depth.[150] Thus, ellipses present the central construction issue with such objects. It is first of all conspicuous that the text does not even mention ellipses as an object that is so central even though it is only virtually part of the solid. It is, however, impossible to draw ellipses in a technical way unless one undertakes calculations on the basis of specific figures and maps them onto a coordinate system, which in turn requires measuring.[151] However, neither calculations nor measurements form part of the propositions in the geometrical books of the *Elements*, nor are coordinate systems nor points mapped onto them attested in any of the ancient diagrams, which suggests that such a practice was not common even if experts knew it. Moreover, again, such a construction would be extremely complicated and would not seem worth the trouble starting from the actual aim of the proposition which is to prove on the basis of deduction.[152] However, without applying such methods and means, it is impossible to make

150 The general possibility of depicting three-dimensionality in the case of spheres in this 'geometrical' way is not mentioned by Netz 1999a, 18, n. 28, who – besides the practice visible in the transmitted diagrams which I shall discuss below when comparing the constructional descriptions of the *Elements* with the ancient and Byzantine diagrammatic evidence – only mentions "shading," a more graphic and 'non-geometrical' way of depicting depth, which I do not consider here.

151 See above p. 252.

152 On the relation of the diagrammatic construction to the purpose of the proposition and to deduction, see Section 4.3.4 on the 'functionality' of the textual construction.

anything but sketchy drawings according to the construction assignments of such objects. Thus, the textual description of three-dimensional objects that involve circular lines seems to accept or even presuppose that ellipses are not drawn technically but in an improvised manner.

Thus, three-dimensionality does not constitute a problem of construction in the *Elements*, neither in the case of spheres nor in that of non-spherical objects. This seems to be for the simple reason that the depiction of it is about the subjective act of viewing and not about the essential properties of an object. Such a view, however, raises questions about the meaning and relevance of diagrammatic representation. For the depiction of depth serves precisely the preservation of the properties of the parts of an object if they appear in space. Apparently, or so it seems from the silence on three-dimensionality, the correct depiction of the properties of objects in a diagram does not play such a vital role for Euclid and other elements' authors. This, then, has consequences also for the depiction of two-dimensional objects, as it more generally calls the relevance of the technically correct construction of the diagram into question.

Our review of the diagrams of the manuscripts in Section 4.2 seems to confirm the assumptions made about the diagrammatic practice on the basis of the textual construction, and it adds another aspect to the picture. In general, we have seen that the depiction of the individual kinds of geometrical objects is in some cases non-standardized. This non-standardization is possible because of the textual underspecification of the methods of construction, and is possibly also *due* to this underspecification. As a rule, the variation regarding the depiction of certain kinds of objects seems to be the stronger the more complex an object and the more varied therefore the possibilities of its representation. The diagrams of the most basic objects, i.e., those that are easily drawn in an unambiguous way, cubes and rectangular solids, only differ with respect to the foreshortening factor, the direction of the projection lines, and the side from which they are viewed. By contrast, more complex objects testify to different kinds of representation that deal differently with the problem of representing depth. Some even avoid depicting it altogether.

The non-standardization of the depiction of three-dimensionality might not only have to do with the underspecification of the text, but also with the standardized use of technical instruments that we have observed in the manuscripts. The application of ruler and compass has turned out to be a constant feature of all methods of depicting three-dimensional objects, whether three-dimensionally or not. The standardized application of those instruments underscores the simultaneous non-standardization of whatever is to be done with ruler and compass regarding the challenge of representing three-dimensional objects. No means other than ruler and compass are employed, and, conversely, ruler and/or compass are employed in each and every diagram. The use of compass, which is more conclusive than that of ruler in this respect, leads to essentially different results regarding three-dimensionality than free-hand drawing would do, and therefore seems to have been employed meaningfully. These results are not superior in terms of the representation of depth but rather the opposite. No free-hand example is extant for comparison, but the ruler- and compass-drawn diagrams of three-dimensional objects often do not even show the attempt of depicting three-dimensionality but present these objects two-dimensionally or by an irreal mix of perspectives. I have argued that the purpose is to produce a seeming technicality even at the expense of geometrical correctness. In this way, the applicability of ruler and compass determines the respective method rather than the other way around. As we have seen, the evidence of the *Elements* for this is supplemented by the diagrams that have been transmitted for conic sections which show the same practice – drawing non-circular lines by means of compass[153] – even more clearly. To be sure, we do not know whether this relation of the use of ruler and compass and the method of construction in the depiction of three-dimensionality in the Byzantine manuscripts goes back to an ancient practice. However, its standardization in different manuscript traditions despite its pecularity could be indicative not only of such a (standardized) ancient practice, but even of the standardized application of the diagrams themselves.

Apart from the methodological challenges of depiction which have to do with properties of the kind of geometrical objects to be depicted, there are other issues of presentability which do not have to do with the presentability of the objects themselves. The following two issues, the representation of objects which need to be moved as part of the proof and the representation of the construction proofs by contradiction, already come up in *Elements* I and continue to be relevant throughout the work. To be sure, the first one always occurs as part of the second one, but as 'movement' is a particular feature which is non-representative of some proofs by contradiction, I review it separately first.

Several proofs in the *Elements* involve some kind of movement of their objects. I now look briefly at the two propositions from book I which present instructive examples for the diagrammatic representation of such a situation. Both I 4 and I 8 require that two objects be laid on top of each other (ἐφαρμόζεσθαι ἐπί τι, 'to be applied to something'), but they do not specify whether and how this movement is to be represented in a static two-dimensional drawing. Strictly, movement is not representable at all by means of a drawing; moreover, it is not formulated as part of a construction assignment, but in the language of the proof. Even in proofs by contradiction which standardly require some additional constructional action, this action is usually assigned by the standardized imperatives.

Two improvised ways of representation are conceivable: (1) the text could either be taken literally, with the movement not being an actual part of the drawing but being indeed performed. This would require that the drawings of the respective objects are cut out of the drawing material and then put on top of one another, which is hardly imaginable because of the cost of papyrus and the impossibility of a precise

153 Netz 1999a, 17.

cutting of alternative materials such as clay tablets. (2) Or ἐφαρμόζεσθαι could be understood metaphorically: either in the sense that one of the objects, after having already been constructed, is to be constructed again on top of the other one; or, since the action of laying the objects on top of each other is not even part of the construction part, but of the beginning of the proof proper, in the sense that it must be merely imagined.

However, the diagram for I 4 in B shows a different, non-intuitive solution which recurs several times in the work: the second one of two triangles does in fact carry some sign of the movement of the object: an arc below one side of that triangle which is to be moved onto the other (Fig. 22).[154] The arc even seems to be drawn by means of a compass, even though the rest of the diagram does not seem to be drawn with the help of the ruler as it stems from the unoriginal part of the manuscript whose diagrams are mostly drawn free hand. The way of representing movement in this diagram, however, does not seem to be the standardized way to do it, for the analogous diagram of P does not carry any traces of it. Neither do the diagrams for I 8 in P and B which each contain two and three figures, respectively: in P, a triangle and another two triangles that are drawn on top of another and overlap only by one of their three lines and part of the plane that is comprised by either of them (Fig. 23), and in B two separate triangles and the overlapping triangles. The overall non-standardization seems to be due to the fact that the text is not unambiguous about what to do, and, thus, highly impractical as regards the construction. It may also indicate the originality of the diagrams: for a depiction as in the diagram of I 4 would be expected to be fully standardized if it did not go back to the respective source.

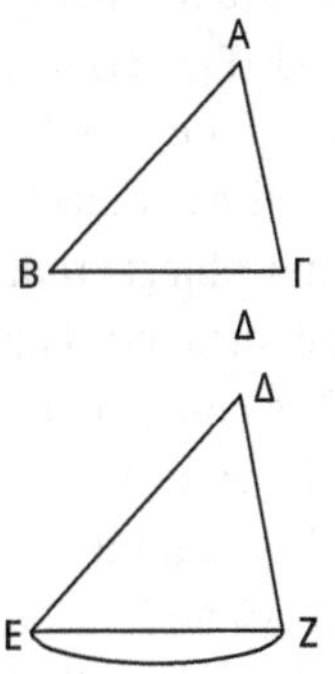

Fig. 22: I 4 according to Heiberg & Stamatis.

The second issue of presentability concerns a specificity of a subtype of propositions which contain a 'proof by contradiction,' that is, those in which the proof involves not just a hypothetical assumption of the opposite theorem which it proves to be contradictory, but an impossible construction which shows that the opposite of the proposed

154 The editions of Heiberg & Stamatis and of Acerbi 2007 both print the diagram of I 4 as in B.

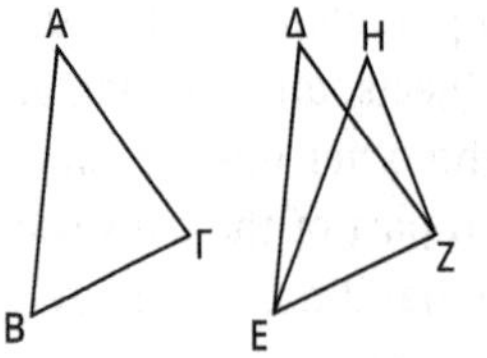

Fig. 23: I 8 according to Heiberg & Stamatis.

theorem is contradictory (cf. I 6, 14, 26, 27, 29, and 39). For such proofs would theoretically necessitate that two versions of the diagram be drawn for the same proposition – one which is possible and a second one which looks similar but turns out impossible to construct at all or unable to contain all of the characteristics of the assigned figure.[155] However, in all of these cases, the text does not specify how many diagrammatic figures to construct in order to present the textual construction unambiguously by means of a diagram, or, alternatively, how to combine the two constructions so as to produce a single unambiguous and meaningful figure. Following the textual construction, one will have to assume that a single figure indeed is to suffice, and that a second one is not intended.

This is because, first, the text does not explicitly assign the repeated construction of *all* objects so far introduced, that is, of the whole diagram. Instead, the proof part which in this kind of propositions consists mainly of a construction proceeds in one of two ways: either it only assumes the opposite of or a particular deviation from the characteristics which the definition of the goal ascribes to some of the objects of the original setting out and then goes on to assign only the construction of additional objects. For example, in I 39, the non-parallelism of AΔ and BΓ is assumed without the implied re-setting out or re-construction of the original configuration actually being mentioned before the additional object (the straight line AE) is assigned (Fig. 24); or, after assuming the opposite of the characteristics that are assigned to the objects by the definition of the goal, the text explicitly re-sets out or re-constructs those objects by attributing to them different properties or a different relation to other objects that have already been constructed before additional objects are being constructed. For example, in I 6, AB is newly set out as being longer than AΓ before ΔB is constructed on it and ΔΓ is drawn (Fig. 25); and in I 27, the straight lines AB and ΓΔ shall be extended so as to meet in point H, which is constructed additionally to the original setting out (Fig. 26). It is noteworthy that the proof does not demand that AB and ΓΔ be re-drawn altogether despite their original parallelism, but that it merely assigns that they be prolonged (ἐκβαλλόμοναι). Notably, in all of these examples, the proof ends

155 One could, additionally, consider here I 4 and I 8, which deal with two diagrams each from the start, but eventually require a third diagram each to depict the movement of one onto the other. However, since the description of movement is a special problem whose intended realization is uncertain, as we have seen in the preceding section, I shall not deal with them again with regard to the present problem.

up showing that the originally set-out objects indeed possess the qualities that have been ascribed to them in the first place and that they appear in this way in the original diagrammatic configuration. Thus, the three examples show that the propositions of this sort all differ slightly as to the way and explicitness with which part of the old configuration is changed in the course of the proof. But they all have in common that the respective remaining part of the configuration is not mentioned again regarding a re-setting out. This suggests that the figure as a whole is not to be redone as a redrawing of the same figure, or, more conveniently, as a second figure. This is, on the one hand, logical, as the proof still works with the original setting out and proves it correct in the end, whereas the reconfiguration will be proved wrong; on the other hand, if one follows the procedure of the proposition, it must strike one as odd because the changes that are presupposed for individual objects of the original configuration in fact imply changes to the whole of it. For example, if in I 6 AB is prolonged, then the angles of both AB and AΓ with BΓ have to change in order for the three lines to form, again, a triangle.

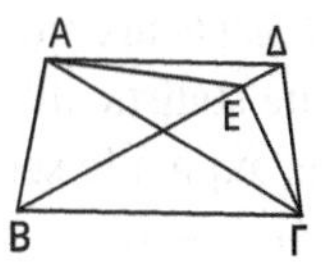

Fig. 24: I 39 according to Heiberg & Stamatis.

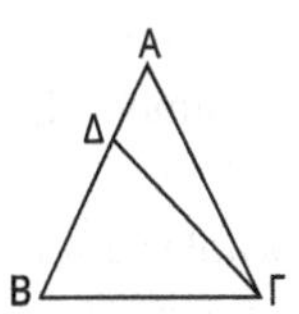

Fig. 25: I 6 according to Heiberg & Stamatis.

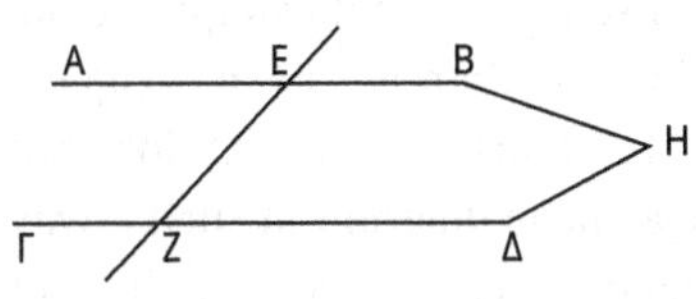

Fig. 26: I 27 according to Heiberg & Stamatis.

The second point which indicates the construction of only one figure instead of two apart from the incomplete re-construction of the original diagram is that, in any case, the proof part does not attribute new letters to the objects when changing their properties according to the opposite of the proposed hypothesis. For example, AB remains to be called AB in I 6 even in the second part of the construction; the same is true for AB and ΓΔ in I 27, and for AΔ in I 39.

Third, it is important to note that the respective assumption of the opposite of the claim defined as the goal of the proof is standardly done by an indefinite conditional clause (εἰ γὰρ μή in the present cases, or εἰ γὰρ δυνατόν in those cases in which the possibility of an impossible claim is considered).[156] While the subordinate εἰ clause is elliptic, the indefiniteness of the εἰ clause is clear from the main clause, which is in the future tense or in the imperative which is equivalent to the future tense in the temporal respect, so that ἐστί must be supplied in the εἰ clause. This means that the construction is not in the irrealis mode, and does not, therefore, betray from the start that it will turn out wrong and therefore is not meant to be done in the first place while the claim of the definition of the goal will be proven correct. This shows that although the proof by contradiction is marked by the standard introduction by εἰ as a hypothetical scenario, its validity is truly considered, and it should therefore be expected to find resonance in the diagram. To be sure, within the part of the proof by contradiction, the construction assignments which are reserved for additional objects are linguistically differentiated from the supposition of a different case by means of εἰ; that is, they appear in the usual form of the construction imperatives. This difference could indicate that the case which is envisaged by the εἰ clause is not reflected in the diagrammatic construction. It might explain why the original configuration is not being redone and why the original letters of those objects of which two contradictory claims are made are not changed in the proof by contradiction: these objects are not supposed to be constructed, only the additional objects are. However, the stylistic differentiation is not always as clear-cut. In our I 6, as we have seen, AB is explicitly set out in the proof part as being longer than ΑΓ, and the construction of ΔΒ is explicitly assigned to take place on top of this "longer" AB. According to our previous argument, this construction would have to be reflected in the diagram, along with the corresponding changes to the whole configuration. This again indicates that a single diagram emerges from the textual construction.

In any case, such a diagram would be impossible with regard to the whole sum of assignments made, because it would either combine two contradictory constructions and therefore would be contradictory and impossible itself, or it would simply be obscure if the combination ended up depicting something else than has been assigned. Even if it was presupposed that only one figure was to be drawn, the text does not specify how to depict the constructions together in one single diagram in such a way that the aim of the proof would be at all recognizable, i.e., the impossibility of the second construction and the resulting validity of the first. Both cases would have to be represented correctly according to the respective constructional assignments. The reason why the text does not indicate how to do it is probably that it is simply not possible to do it without annotating the diagram with clarifying explanations. Methodologically, however, this is not an option that agrees with the *Elements*' principle of being self-explanatory. The main problem besides the necessity of accommodating two identically lettered versions of the same object(s) would be that, in many cases, the object to be additionally constructed would have to take the very same place of

156 See above, Section 3.3.1, 145–147, on the standardization of the syntax.

(what is assumed to be, by the definition of the goal, and finally proven to be) the original position of the now re-positioned object. For example, in I 39, AE is to be made parallel to ΒΓ and so takes the place of ΑΔ which was originally set out as being parallel to ΒΓ but now is assumed to be *not* parallel to it. This means that the same straight line would have to be lettered twice.

In our three examples, the underdetermination of the textual construction has the following consequences for the construction of the diagram: In I 6, which proves that one side AB of a triangle ABΓ equals another side AΓ, the construction of the proof by contradiction assigns the construction of AB as longer than AΓ if it is assumed that AB is unequal (ἄνισος) to AΓ, and of an additional line ΔΓ which equals AΓ, so as to form ΔB on AB. This would require drawing AB twice (for in the original configuration, AB is not longer than AΓ but equals it, as will be subsequently proven), but this would mean that one would have to draw two different triangles. Otherwise, i.e., in one single drawing, AB could only either equal AΓ (and ΔB) or be longer than AΓ (and ΔB). So if AB was drawn only once as having the length of AΓ, this would represent the conclusion, but not the two-time construction of AB.

In I 39, which proves that the straight line ΑΔ is parallel to the straight line ΒΓ which forms the base line of the two triangles ABΓ and ABΓ, the proof assigns the construction of an additional line AE parallel to ΒΓ, in the case that ΑΔ is not (εἰ γὰρ μή), as has been set out, parallel to ΒΓ. If, however, AE was drawn indeed parallel to ΒΓ in the very same diagram, it would have to overlap partly with ΑΔ, as it was first assigned, i.e., parallel to ΒΓ. So there would have to be either two diagrams: one showing ΑΔ parallel to ΒΓ and one not, or one and the same diagram would have to feature ΑΔ twice, which would make it ambiguous. Otherwise, if ΑΔ was drawn only once in accordance with the setting out, there would be no contradiction ensuing, as both AE and ΑΔ would be parallel to ΒΓ, and their respective parallelism would not be mutually exclusive, by contrast with what the proof purports to show.

In I 27, which proves that the two lines AB and ΓΔ, on which the line EZ makes the alternate angles AEZ and EZΔ equal to one another, are parallel to one another, the proof assigns, in case that the two supposedly parallel lines are not parallel (εἰ γὰρ μή), that they meet in the direction of B and Δ in H. Again, two figures would be necessary, since, as it turns out, AB and ΓΔ can only be *either* parallel (as is claimed and finally proven with regard to the original configuration) *or* meet in another point. Otherwise, if depicted in one single figure, two versions of the two lines AB and ΓΔ would have to be drawn and lettered, or the figure would only be able to show either the version proven correct or the one proven false, or an impossible mixture of both.

I have used these examples to show that the textual construction of propositions with proofs by contradiction is highly underdetermined, partly because of the ambiguity of the textual construction – not in logical terms, but as regards the realization of the diagram as described – partly because of the ambiguity of the letters. Many of them are effectively used twice, with one and the same letter referring to two different points. As a result, the diagrammatic figure to emerge from the description of construction is unpredictable on the basis of the textual construction alone and can hardly turn out meaningful so as to represent the text unambiguously. In Section 4.4, I will elaborate on the ambiguity of the letters on account of which the construction of the diagrams of the present type of propositions is textually underdetermined.

At present, I shall return to the material evidence for our three sample propositions, which precisely because of the unpredictability of the diagrammatic construction should be promising regarding the issue of the standardized inclusion of the diagram. I shall review the diagrams for the three propositions in the context of the manuscript evidence for the propositions that contain proofs by contradiction. The question is first of all whether the evidence displays any patterns in spite of the ambiguity of the textual construction. Afterward, the origin of the identified patterns must be assessed.

Indeed, the representation that is invariantly opted for in the diagrams of the manuscripts is that of one single diagrammatic figure. In the overwhelming majority of cases, it consists of only one figure which combines the possible and the impossible part of construction. This combination forms a hybrid of both which neither depicts the proven hypothesis adequately nor the opposite of the proven one. There is only some variation among propositions as to how, and in what ratio to the first part, the second, impossible part is accommodated. This variation depends, on the one hand, on the specificity of the construction problem; on the other hand, it simply has to do with the scope of variation enabled by the underdetermination of the textual construction.

The evolving standardized tendencies of hybridization are that, first, the objects which play a double role appear only in the form in which they have been originally set out and in which they are in the end proven to be in line with the original claim. By contrast, their new position does not get depicted, although it is in many cases explicitly set out or constructed, but in any event implied by the formulaic assumption of the falsity of the hypothesis. That is, not only the objects which play a double role but even the whole original configuration remains the same. Second, in many cases, the diagrammatic construction of the additional objects introduced in the proof part paradoxically does not match their textual construction, which would have them, impossibly, coincide with the textually but not diagrammatically transposed object of the original setting out: e.g., in the diagram for I 39 in our two manuscripts, AE is to be drawn parallel to BΓ, but in this place, there is still AΔ, so AE is drawn falsely as being non-parallel with BΓ and so as to fall on the diagonal BΔ below AΔ. Thus, the diagram depicts the claim made in the definition of the goal instead of the construction that is proved wrong: AΔ rather than AE is parallel to BΓ and takes the respective spot (Fig. 27). In a comparable manner, the manuscripts' diagrams for I 6 present AB as being equal to AΓ, i.e., in accordance with the proof result, while ΔB, which is textually constructed as having the same length as AΓ on the longer version of AB, in the diagram appears as being shorter than AB and AΓ (Fig. 28).

While the diagrammatic transmission for I 6 and I 39 is roughly unanimous, the diagrammatic representation of I 27 in the manuscripts is more variant but, by and large, displays the same principles as the diagrams of the type we have just examined

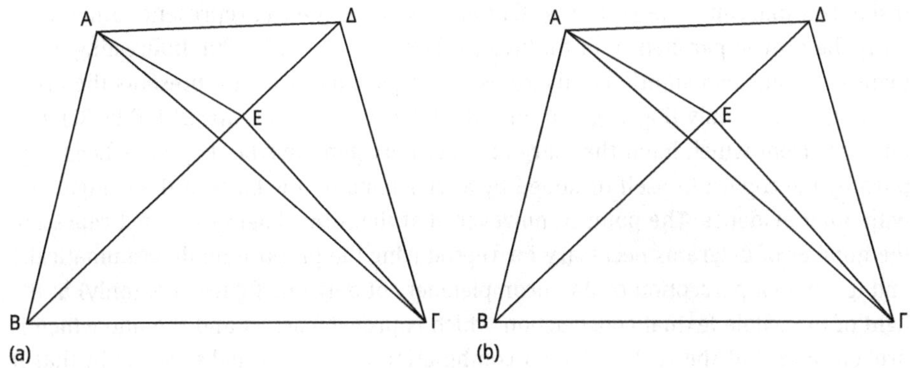

Fig. 27: (a) I 39 according to Vat. gr. 190.1, fol. 33v; (b) I 39 according to MS. D'Orville 301, fol. 27r.

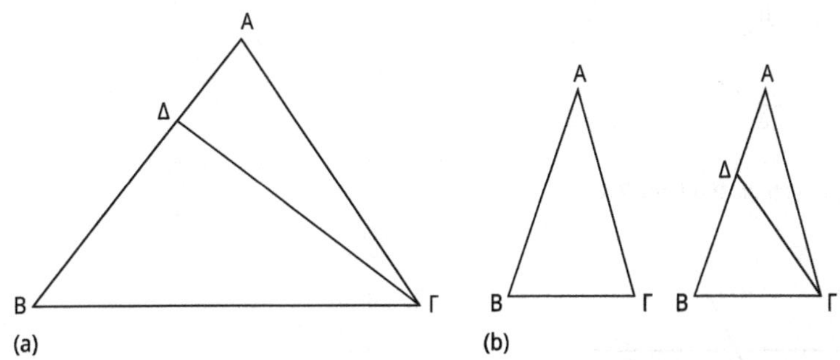

Fig. 28: (a) I 6 according to Vat. gr. 190.1, fol. 19v; (b) I 6 according to MS. D'Orville 301, fol. 11r.

more closely. The version which appears in P and is printed in our editions is one which, in the manner of the diagrams of I 6 and I 39, contains something of both contradictory parts of the construction. In so doing, it only depicts the proven version fully, while the proof by contradiction is not represented completely, so that it does not become clear from the diagram why it is impossible (Fig. 29): AB and ΓΔ appear as the parallels which they will be proven to be, and from B and Δ two straight lines are drawn so as to meet in H, with the whole resulting in a kind of rocket shape. What the proof by contradiction actually does, however, is to assume the non-parallelism of AB and ΓΔ and *prolong* them in direction of H. It is only in this way that they can be shown not to make equal alternate angles with EZ as required by the setting out. However, in the diagram, the two originally parallel lines AB and ΓΔ have not been prolonged (because as parallels they would never meet), but instead, two new straight lines, BH and ΔH, have been drawn (not so, of course, named by the text) which connect the parallels as they are to point H. Again, thus, the diagram only hints at the actual textual construction which cannot be assessed on this basis. The Bodleian manuscript goes even further as it does not present the proof by contradiction at all in the inside-text diagram. The resulting diagram, by consequence, represents unambiguously the proven parallelism of the two straight lines only (Fig. 30). Interestingly, the manuscript has two additional diagrams *in margine*: one which represents the proof by contradiction only (Fig. 31), and one which has the rocket shape of P (Fig. 32). It is not clear from which hand the marginal diagrams stem; maybe they have been supplied by the copyist himself or added by a later hand on the basis of the comparison with other evidence. The point is, however, that the three diagrams in fact represent the number of diagrams necessary for reproducing the proposition diagrammatically and testify to a perception of the incompleteness of only one figure. It is (only) in the light of the whole textual construction which is proven correct and the one which is proven false that the rocket-shaped combination of the two makes sense in that it shows the incompatibility of the two.

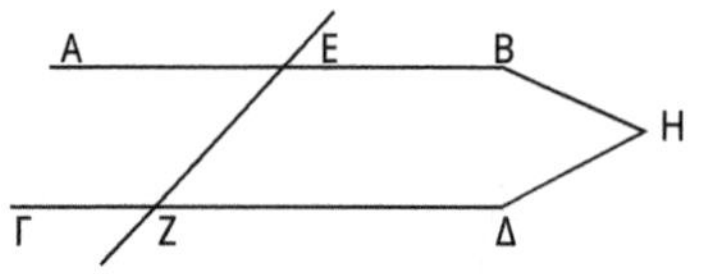

Fig. 29: I 27 according to Vat. gr. 190.1, fol. 27v.

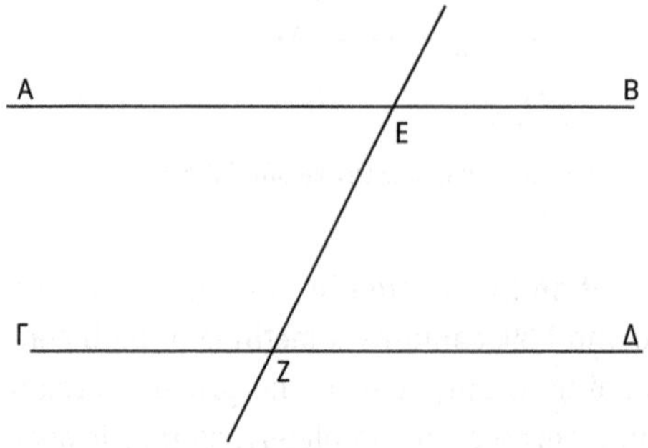

Fig. 30: I 27 according to MS. D'Orville 301, fol. 21v.

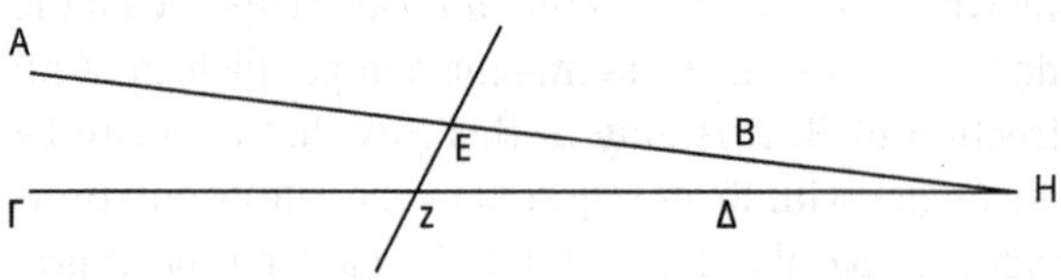

Fig. 31: I 27 according to MS. D'Orville 301, fol. 21v *in margine* (1).

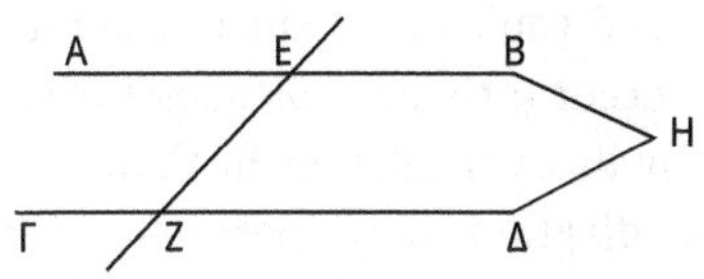

Fig. 32: I 27 according to MS. D'Orville 301, fol. 21v *in margine* (2).

Thus, in all three cases, leaving the marginal diagrams of B of I 27 aside, the actual construction assigned by the text is ironically not accomplished in the manuscripts, and the diagrammatic construction does not correspond to the text of the proposition. This is because the (textual) construction in propositions with proofs by contradiction takes place partly in the proof, which, however, ends up proving the falsity of this construction. The formal peculiarity of such propositions lies, on the one hand, in the divergence of the resultative diagram and the processual proof text, whereas in 'normal' propositions, proof and diagram are congruently resultative as regards the construction; on the other hand, it lies in the non-linearity of the proposition: first, something is set out and defined as the goal, then the text goes back toward the beginning and starts (almost) anew, which is then proven to be the false way, so the text goes (again) back, this time to the original beginning. This is different from the strictly linear and forward way in which the normal type of proposition proceeds.

Considering that the diagram standardly deviates from the text in propositions with proof by contradiction, the question is what the point of such a diagram would have been. Rather than helping to follow the text at all or more easily, the cryptic diagram has to be illuminated by the text and must be disentangled first with regard to the two versions somehow contained by it before becoming a useful referential basis for following the text – but only once the text is self-sufficient anyway. That is, not only is the text here underdetermined regarding the realization of the diagram, but also the diagram is underdetermined regarding what it is to show, or even meaningless. Crucially, it is the textual underdetermination regarding the diagram which provokes the underdetermined diagrammatic representation in the first place.

However, while the underdetermination would have enabled all kinds of representation, there is a certain kind of diagram which is used in a rather standardized way. As has been shown, the diagrams for this type of proof by contradiction all vary somewhat from proposition to proposition but generally display the same principles and are standardized per proposition in the different manuscripts (except in the case of I 27). This standardization is quite conspicuous. For although experts no doubt would have been able to interpret this type of diagram, it does not seem to be the most natural way of representation: it is cognitively not helpful and would not have naturally suggested itself to whoever supplied the diagram-less texts with figures and, thereafter, have become standardized. Rather, it seems to be indicated by the unexpected standardization in view of the textual ambiguity regarding the diagrammatic construction that the diagrams of this form, and, by consequence, also all other diagrams, might already have been included in the text and copied ever since. The diagrammatic standardization could imply that it did not occur (primarily or necessarily) for cognitive reasons but may have been pursued for its own sake: even though it would have been impossible to produce unambiguous diagrams in propositions with proofs by contradiction, they nevertheless seem to have been standardly supplied, maybe simply because diagrams were standardly included in elements texts. Moreover, it seems that these diagrams took the form of the other diagrams rather than deviating from the formal requirements of the genre for producing a result which would be cognitively more useful, e.g., by containing more figures or additional lettering. The lack of practicality which characterizes the textual construction in these propositions, in combination with the nevertheless standardized material evidence, may indicate a standardized form and inclusion of the diagrams from the beginning. However, as in other cases of the diagrammatic standardization of the manuscripts, this one, too, does not necessarily have Euclidean predecessors but may go back to a late ancient or Byzantine tradition.

Our review of different kinds of geometrical objects and of special types of textual construction in *Elements* I and beyond the first book, and of possible ways of constructing them in a geometrically precise and informative way has shown that the text confines itself to assigning the construction in a standardized way, rather than providing any information about the *how* to do the construction, that is, about the methods of construction. In this way, it proves to be highly impractical. We have seen at some points how the impracticality of the textual construction raises the question of its functionality as it seems to provoke diagrams that do not really represent the proof or help understanding it.

Underspecification regarding the act of drawing

Finally, as a last point to be made on the question of the practicality of the construction, the text does not make any reference to the act of drawing itself. It has been shown that the construction proceeds cursorily: the assignments only summarize the actual construction which would actually comprise several acts. In so doing, they name resultatively the appearance of the steps of construction in the diagrams rather than explaining the way they are produced. It has further been pointed out that the order of naming the objects to be constructed is roughly chronological so as to purport the impression of the diagram's gradually coming into being, but that the textual construction does not quite follow this order consistently.

This situation seems to implicate that it is not possible for the construction assignments to elaborate on how the act of drawing itself is to be accomplished with regard to the smallest object unit, i.e., the (straight, curved, or circular) line. For, on the one hand, even if an assignment is made of only one single straight line, its construction comprises a set of constructional actions which lead toward the construction of the line in question. However, the text does not specify the construction in such a detailed manner. On the other hand, if one takes the text seriously in not assigning 'full,' i.e., technical constructions and simply attempts to sketch or draw the objects in the order of their being mentioned, one will be not able to do so in many cases because of the impracticality of this order: it is roughly chronological from the point of view of description, but turns out to be not strictly chronological regarding the practical drawing.[157] In this way, the construction assignments can hardly inform about the manner of drawing every single line.

The most significant lack of information regarding the act of drawing is that the assignments do not specify the starting and end points of the drawing of straight lines or circles. I 47 (Fig. 21), e.g., assigns to draw through A AΛ, instead of saying that a straight line should be drawn 'from A to Λ,' which would explicitly indicate the direction of construction. The example is especially significant in this respect as the text mentions A first anyway, so the assignment could have been more conveniently rephrased for the purpose of accomplishing the construction. This information would be expectable if the objects were simply to be drawn rather than technically constructed according to the text. It would greatly facilitate the non-constructional drawing of more complex figures as the directions of drawing the lines would indicate where to start drawing the respective next line so as to enable the drawing of larger objects at one go and ensure the quality of the figure by describing the most convenient way of drawing it. To be sure, while this lack of specification shall be taken as another indication of the impracticality of the construction, the resultative rather than processual way of mentioning the lines and the more complex objects could also be regarded as a sign that the objects are not to be simply drawn one by one. Contrary to what has been concluded from the analysis of the textual construction so far, it could indicate that the objects have to be constructed in a technical way according to the implied constructions. However, in the face of the various implications of non-technicality, it seems more probable that the language of the textual construction is utterly impractical as it is not orientated toward the accomplishment of an actual drawing.

Conclusions on the (im)practicality of the textual construction

To conclude, the construction assignments prove highly impractical regarding precisely the construction which they assign, since, owing to their lack of methodological and technical information, they do not enable a precise, unambiguous, and 'technical'

157 The order of construction is recognizable from the order in which the letters are mentioned in multi-lettered letter words (see Section 4.4.4). In this way, the letter order provides an unambiguous basis for determining the order of the construction and its implications for the question of whether the text is based on a diagram or not. Therefore, I shall consider the (im)practicality of the order of construction only in Section 4.4.

construction which they themselves suggest by their own precision and unambiguousness. In fact, they have proved to be no more effective regarding accomplishing the diagrammatic construction than the imperatives that posit geometrical objects as part of the exposition. The general non-practicality which characterizes the construction assignments regarding all kinds of objects that are treated in the *Elements* could be interpreted as an indication that the diagram was physically included in the text. The text assigns a diagrammatic construction and presents it as a precondition of the proof, since the whole part which precedes the proof (which we have called 'constructing section') serves the construction of the diagram. However, the text does not really enable what it assigns. From this, one might assume that the actual correct construction is already available, while the construction imperatives merely sum up the main steps of the construction by mentioning formulaically the geometrical objects and their relations roughly in the order of their appearance. In this case, the imperative mode would point, as presumed above, to the necessity of the respective constructional step. Even so, it would be remarkable that the text did not ensure the unambiguous construction of the diagram in order to provide against the danger of the diagram's becoming separated from the text and getting lost in the course of its decontextualized situation of transmission.[158]

To this, one might object that mathematical expert readers, just like the author, would have known from the standard names for objects as well as operations how to do the constructions: which methods and means to employ, and in which cases to make up exemplary dimensions, e.g., for accomplishing basic relations like sameness or for drawing ellipses. On the assumption that the primary addresses were expert readers, the impracticality would be no longer suggestive of a preexistence of the diagram; for the lack of constructional detail would be due to the expectable knowledge of the recipients and therefore would not contradict the apparent reliance on a construction. Any expert reader would be able to construct the diagram or to reconstruct it in case of its loss. However, if experts are indeed the intended readers, this does prevent the text from providing a lot of specification and disambiguation in many

158 Such is the motive which Ptolemy explicitly ascribes to his own detailed description of the construction of the maps in his *Geography*. He explicitly mentions the danger of the corruption of the maps and on these grounds wants to make sure that the reader may reproduce the maps on the basis of text: the construction is therefore much more detailed and contains a number of specifications that do not appear in our elements texts. To be sure, the maps are less self-contained than mathematical diagrams to begin with, with the general appearances and exact properties of the geometrical objects being clear from the standardized use of the terms, and the description of their construction therefore is in need of greater precision; nevertheless, one notes a remarkably greater emphasis on and examination of the construction in Ptolemy. While I argue above that the less detailed account of the construction in the *Elements* indicates the authorial production of the diagrams, the detailed account by Ptolemy, in turn, does not implicate that he did not himself provide diagrams nor that the recipients necessarily would have had to reenact the construction as a vital part of receiving the work. It only makes sure that they would have been enabled to do so in certain circumstances of transmission.

other ways which would not be necessary for such a readership.[159] In view of this, it is rather conspicuous that the text is not quite so unambiguous as regards the practical construction of the diagram. So even if expert readers are presupposed, it remains remarkable that the textual construction of all things – as opposed to the proof proper – leaves open so much. The only explanation for the lack of information about the practical dimension of drawing the diagram and the resulting poor constructional quality enabled by the text at this point seems indeed that the diagram was physically part of the text, and that the possibility of its loss was not considered nor provided against by a more specific and unambiguous account.

The underspecification and ambiguity of the text regarding the diagrammatic construction should result in quite diverse diagrammatic outcomes, had the diagrams not been standardly included from the beginning. As we have recalled from our review of the manuscript evidence, the Byzantine diagrams do show differences which seem to be due to the underspecification of the text, but they also testify to rather striking similarities between diagrams that are unexpected in the face of the ambiguity of the textual construction. The similarity of the depiction of specific figures in different manuscripts could be interpreted as a sign for the standardized application of the diagram, and, conversely, as an indication for the manuscripts’ relative reliability regarding the diagrammatic depiction. However, we do not know whether the relative unanimity of the diagrams really reaches back to Euclid’s time.

In any case, before being able to conclude from the impracticality of the construction assignments directly about the standardization of and reliance on a diagram, we have to review the kind of diagrammatic ‘constructions’ which the text enables in terms of their functionality: that is, how far they are able to contribute to fulfilling the aim of the propositions, whether and how far they are deficient in this regard, or whether they actually turn out to be sufficient. This brings up the question what the function of the diagrammatic construction in the *Elements* is anyway, that is, what they effectively contribute to fulfilling the aim of the propositions.

4.3.4 The functionality of the diagrammatic construction

The relevance and function of the diagram for the mathematical proof, both ancient and modern, have been the subject of a long debate. Euclid has been famously criticized by Hilbert for the ‘implicit assumptions’ in his proofs which had to be understood from the diagram,[160] who thereby suggested that in Euclid, the proof had some undesirable dependence on the diagram, which made the *Elements* deficient in the

159 That a lot of the unambiguousness caused by standardization is unnecessary both from a cognitive and from a communicative stance if an expert reader is assumed as addressee will be my central argument of Chapter 5. There, I shall think about the reasons for overambiguousness.

160 Netz 1999a, 26–29, especially 27, n. 39.

measure that its proofs, owing to this precarious dependence, actually lacked self-sufficiency. Modern research on Euclid has ignored both the diagrammatic evidence as well as the diagrams' relation to the text for a long time, but has developed some considerable interest in (not only Euclidean) diagrams in recent years. Those who actually have taken the diagrams into account have, among other things, inquired into the cognitive functions of the diagrams.[161] For the most part, scholars have not questioned that the propositions of the *Elements* are to a certain extent based on the diagram, but have put forward different conceptions of coming to terms with this dependence.[162]

The question is how far the diagram is relevant to the proofs. This concerns both the conclusiveness of the diagram, that is, the question whether it can be conclusive at all from the point of view of elements' mathematicians, and, on the other hand, the necessity of the diagram for the proof: the question whether it *has to* be taken into account as a conclusive part of the proof, because the text presupposes it. Further, the question is to what extent the diagram is proof-relevant, that is, in virtue of what (part of it) the diagram is proof-relevant, and what is the relation of the diagram to the text regarding proof-relevance.

Netz has argued that the proof, rather than always depending only on the text, in some ways depends on *both* text and diagram, and sometimes even *only* on the diagram.[163] With regard to the first kind of dependence, he mentions the "synopticity" enabled by the diagram.[164] I do not consider this visualizing function of the diagram for the proof, because I am concerned with the question of the standardization of the diagram, and the existence or non-existence of the diagram cannot be proved with regard to the function of visualization.[165] As regards the second kind of dependence of the proof on the diagram (i.e., on the diagram only), Netz has shown that the text depends on the diagram as a referential basis for the letters employed in the text, as the letters

161 E.g., Manders 2008; Mumma & Panza 2012; Panza 2012; Krämer 2016. The works of Saito and Sidoli focus, differently, on the form of the transmitted diagrams, mostly in terms of their alleged "overspecification" and "visual inaccuracy" (e.g., 2012, 157).

162 Of course, the "interdependence" is "mutual" (Netz 1999a, 19 and 33), but while the diagrams have long been regarded as an insignificant addition to the proposition texts, with their dependence on the text not being questioned, it is the converse dependence which is at stake.

163 Netz 1999a, 26–29.

164 Cf. also Krämer 2016, 92: "Als textbegleitende Diagramme haben sie die 'panoptische Funktion,' geometrische Relationen – in der Vogelflugperspektive – anschaulich zu vergegenwärtigen – das ist ihre Visualisierungsaufgabe."

165 As Netz himself admits, "[. . .] such cases [sc. in which "assertions derive from the combination of text and diagram"], while ubiquitous, are difficult to pin down precisely" (1999a, 28). He goes on to explain that in his example from Apollonius, the argumentative value of the diagram compared with the text lies in its synopticity, i.e., its ability to present all detailed information contained by an assertion all at once (29), which is indeed a somewhat vague argument for the diagram's being the source of proof-derivation.

are not consistently specified in text. He mentions that the ways in which both media interdepend in the Euclidean proposition are manifold and differentiates between diagram-based, text-based, and text-and-diagram-based assertions. However, he acknowledges that most of the types of interdependence are ramifications of the textual dependence on the diagram in terms of letters (for which cf. Section 4.4). He does not systematically consider the conclusiveness of the diagrams as drawings but holds that because of the diagram's non-metricality and resulting schematic nature, one cannot draw conclusions about properties such as parallelism from the diagram.[166]

By contrast, Kenneth Manders has argued, specifically with regard to Euclid, that the diagrammatic representation of the objects itself is in large part conclusive, that is, regarding the "co-exact attributes" of the diagram ("triangles," "*alternate* angles") as opposed to the "exact attributes" ("that an angle is *right*," that lines are *straight*").[167] By this distinction, he offers an instrument for quantifying the conclusiveness of the diagrammatic construction. However, the use of this distinction presupposes dichotomizing the terms of the *Elements* in a way which is not suggested by the work itself. Moreover, the resulting dichotomy of features regarding which the diagram can be conclusive and features regarding which it cannot implies conversely that the text needed to be conclusive with regard to the "exact" features, whereas it could have allowed itself to be inconclusive regarding the "co-exact" features, as the diagram would have been able to represent them conclusively. This does not seem to be a practical basis for composing propositions: for either the proof would have to be wholly self-sufficient, or it would be able to base itself on the diagrammatic figure generally.

I shall regard the conclusiveness and proof-relevance of the diagram from the question of the diagrammatic standardization. For it is only relative to the kind and degree of diagrammatic conclusiveness and proof-relevance that the impracticality of the textual construction may serve to judge the standardization of the diagram. If, on the one hand, the diagram was wholly irrelevant for the proof since the text functioned completely independent of it, the impracticality of the textual construction would be strange in view of the prominent role of the construction in the text. However, it would not be able to reveal anything about the original inclusion of the diagram, for it would not matter whether the diagram was present in the original text or not. If, on the other hand, the proof was based to whatever degree on the diagram, either the text would have to provide everything that is necessary in order for the diagram to be constructed in such a way that it would be able to be conclusive regarding proof, or the diagram would have to preexist in such a way that it contributes its share in conclusiveness.

While previous studies have considered the relevance of the work's diagrams mostly in absolute terms regarding whether the propositions are based on the dia-

166 Netz 1999a, 18.
167 Manders 2008, 91–94.

gram or not, I shall, more narrowly, as the term 'proof-relevance' indicates, focus on the diagram's relevance for the purpose of the proof, that is, on whether the *proof* is based on the diagram or not. In so doing, I follow the text's own suggestion that the diagram is only just to be constructed prior to the proof part, and, thus, must not underlie from the beginning. Indeed, except in some cases is the diagram's construction fully completed before the proof begins.[168] Accordingly, in order to evaluate the standardization of the diagram, I shall regard only the proof proper in terms of whether the diagram (and what part of it), the text, or both are the basis of its conclusiveness, while the foregoing part shall be taken into account only as to whether it enables the kind of conclusiveness which the subsequent proof suggests. Consequently, I take the standardization of the diagram to be its readiness at a certain point of the proposition, namely, by the time the proof begins. If the diagram turns out to be presupposed by the proof but is not completed in time, this will imply its physically coming with the text.

Apart from the chronological dimension of the question of the diagram's relevance, we must consider which aspects of the diagram, if any, are conclusive and proof-relevant. Previous scholars have ascribed conclusiveness to the Euclidean diagram either based on its consistent provision of a reference system of letters, or in terms of the depiction of the geometrical objects and object relations. While I shall consider the ramifications of the letter use regarding the conclusiveness and standardization of the diagram in Section 4.4, I shall now focus on the proof-relevance of the depiction of the properties and relations of the geometrical objects and, in light of this, reconsider the impracticality of the textual construction regarding the diagram's standardization.

The conclusiveness and proof-relevance of the diagrammatic construction

The relevance of the construction for the proof is suggested structurally by its forming one of the two major parts of the proposition and by its immediately preceding the proof. The logical significance of the construction for the following proof is also suggested by the fact that the proof relies linguistically exactly on the foregoing construction. It is, therefore, conversely, from the proof that we must judge whether the construction is indeed relevant for it, and whether the proof is based on it.

The question that arises from the structural and linguistic implications is: are deductive assertions made partly or fully on the basis of the diagram as it is enabled by the textual construction? For this to be the case, the assertions about object properties

168 These are the propositions with 'proofs by contradiction,' which standardly include the (impossible) construction, and more complex propositions, in which further objects are to be constructed after a certain part of the proof; here the evaluation of the conclusiveness of text and diagram regarding the proof pertains only to the proof accomplished up to a certain point, that is, until another construction part is inserted.

and relations would have to be not (fully) explicable by referring to the textual construction alone, and they would have to be verifiable directly in the diagram. This means, at least according to what the precision of the language which refers to the diagram suggests, that they would have to be drawn in such a way that they do not merely give an idea of the general shape of the figure or relation mentioned but picture it in an exact way so as to match the exactness of the language of the construction assignments. That is to say, they would have to be properly constructed instead of drawn by the eye. The precision and unambiguousness of the drawing could only be verified either by the existence of certain auxiliary constructions or by doubling the construction, i.e., by applying a second version on top of the first in the way of control experiments in the modern sciences, or by measuring the result. However, a drawing on the basis of the *Elements*' construction assignments would most probably not be technical in the sense that it would not be the result of a construction that is accomplished with the help of auxiliary construction as suggested by the text. Therefore, any subsequent re-construction or measurement would bring forth the result that the drawing does not, or does not unambiguously, depict the object properties' relations that are so unambiguously pronounced by the text, with the degree of deviation from what the text assigns differing between object kinds and their respective complexity.

As we have seen, the constructability of the objects of the *Elements* on account of the means which the work itself provides differs. While the rectilinear and circular objects of book I may be constructed with the help of previously solved construction problems, the depiction of three-dimensional objects is nowhere specified in the *Elements*. Whereas it is possible to depict rectilinear three-dimensional objects with the help of the underlying, i.e., previously solved construction problems, the construction of spherical objects or such that involve circular lines would require the construction of conic sections with which the *Elements* does not deal. However, we have seen that the three-dimensional depiction of three-dimensional objects might not even be considered as an objective of the construction in the first place, as it seems to be regarded as a matter of subjective viewing and not proof-relevant, for it is nowhere itself dealt with as a problem of construction. Moreover, it has been shown that it is in general highly questionable whether a technical construction is actually indicated, for such a construction would be extremely complex and time-consuming and therefore unlikely. The manuscript evidence does not show traces of technical constructions either.

The supposed non-technicality of the depiction entails a deviation from the geometrically correct version, i.e., the one which would match exactly the implications of the geometrical terminology. This deviation differs according to the objects to be represented and according to the properties and relations ascribed to these objects. In the case of two-dimensional objects consisting of circles and straight lines only, the deviation from the textual specification could be, in very few cases, none at all, e.g., if a simple triangle without further specification is to be drawn. However, in most cases, the deviation is a matter of precision. The properties and relations to be drawn are recognizable from their mimetic depiction, but not exactly depicted. In most of these cases, to be sure, some of the features of an object, as Manders has it, will be indeed represented precisely, whereas others will not. However, the Euclidean conception of definitions and the standardized use of the geometrical terms do not seem to allow for making such a division regarding the unambiguous representation of a single object: either a square is depicted, including all the characteristics which make it a square, or not. We have seen in the ancient fragments and Byzantine manuscripts that the use of ruler and compass does not change the non-technicality of the drawings; nor does it change, as a consequence, the overall lack of geometrical unambiguousness of the diagram. In some basic cases, e.g., of the representation of a straight line pure and simple, or a simple triangle, the use of a ruler indeed induces a depiction which equals the definition of the object. In the case of objects, by contrast, which require the drawing of conic sections, the application of a compass leads to a kind of depiction which would not only have been imprecise, but which would not even imitate the definitive properties and prescribed shape of the objects.

Thus, a drawing which is accomplished strictly on the basis of the proposition text in a non-technical way, for which there are several strong indicators, would deviate by different degrees, but in any case qualitatively, from the unambiguousness that is predetermined by the text. Therefore, it does not seem able to serve as a valid basis on which the text could make deductive assertions (which presuppose the unambiguousness of the language). It is in this sense that the diagrammatic constructions cannot be conclusive and proof-relevant. That is, assertions such as 'lines AB and CD are parallel' or 'triangles ABC and DEF are the same' cannot be proven or unambiguously shown to have been proven by means of the diagram.[169] What the diagrams might be able to do is to produce evidence of the assertions in combination with the assertions themselves. That is, one would recognize what they depict because the text itself disambiguates it, and the depiction, in turn, would then have the effect of confirming the text.[170] Moreover, the diagram is "synoptic" in that it makes all "facts [. . .] simultaneously available."[171] In this way, the diagrams would have certain cognitive, proof-*related* functions which help accomplishing and understanding the proof. However, at least in certain cases in which the depiction on the grounds of the transmitted diagrams may be expected to deviate essentially from the text, the diagrams seem to obscure the evidence of the text rather than to reinforce it, e.g., in the representation of three-dimensional objects by means of a mix of perspectives.

To be sure, the question is whether diagrammatic drawings in general, even the most technical and unambiguous ones, could be at all 'proof-relevant,' or whether they can only ever represent clearly defined objects and object relations by approximation. Regardless of whether modern mathematics actually considers diagrams as conclusive,

---

169 Netz 1999a, 29.
170 Similarly, ibid.
171 Ibid.

which is still considered an essential problem of the philosophy of mathematics, diagrams generated with the help of computer programs certainly could realize all of the constructions of the *Elements* perfectly unambiguously, as opposed to all drawing by hand. However, they would do so on the grounds of calculations performed by the software and, thus, not by means of what both according to the *Elements* and even today would be considered a geometrical construction. In any case, it would be anachronistic to suppose that diagrams in general could not have been considered conclusive since only analog drawing was available to Euclid and his fellow mathematicians, for the respective most exact method could still have been considered conclusive.

The deductions appear to be mostly text-based as far as the predicates of the assertions are concerned. They appear to depend, in the main, on the implied knowledge about the constitutive properties of the geometrical objects and object relations. This is, essentially, the nature of the deduction: the basis of an assertion is another previously proved assertion or an unproved principle which is verbalized as part of the ἐπεί clause of the deductive clause. To be sure, the text does not mention whether this underlying conditional assertion is an axiom or where it has been formerly proven. While the latter can be determined with some certainty by looking at the preceding proofs in the *Elements*, the former can only be found in the list of axioms at the beginning of the work which is part of the introductory material. Axioms appear as basic claims which are mentioned within deductions but nowhere proven in the text. Here, the question is whether those predicates which contain such basic claims indeed refer to unproved principles or whether they are rather based on the diagram.

While the Euclidean proof, thus, appears to be largely text-based, the question, from the point of view of our interest in diagrammatic standardization, is whether it is exclusively text-based. The *communis opinio* is that it is not, and various reasons have been brought forward to support this claim. The alleged diagrammatic share in the proof has been viewed either as a deficiency or a simple fact, and has been, again differently, ascribed primarily to the letter specification accomplished by the diagram, or to the figurative specification, which is my current interest. Whereas previous studies have focused on the logical consistency of the proof part in dealing with the question whether the proof is only text- or both text- and diagram-based, the present focus shall be extended according to a broader definition of 'text-based' which includes the foregoing construction, not only the proof proper. That is, if an assertion is based on an assumption which is not explicitly deduced in the proof part but relies on an explicit specification of the underlying construction, I count it as a text-based assertion rather than a diagram-based one. Similarly, if an assertion is based on an *implicit* consequence of the textual construction which does not depend on the exactness of the presentation, it will be counted as text-based. Even these cases of the proof-relevance of the *textual* construction are, however, hard to distinguish from the proof-relevant letter specification, which shall be dealt with in the next part. Most of the logical gaps of the proof text will be filled by taking the textual construction in terms of letter specification into account (for which cf. Section 4.4.3), whereas the textual construction in terms of figuration and quality of the drawing turns out to be both incapable of possessing proof-conclusiveness and, at the same time, unnecessary for proof.

The moments identified by Netz in which the proof seems to depend on both text and diagram can almost all be classified either as incidents in which the diagram contributes a synoptic impression to an argument which cannot be produced by a textual assertion, but which relies completely on the letter specification of the construction, and thus on the text, or in which the diagram simply provides evidence for what the text asserts. In view of the present question of the standardization of the diagrammatic construction, they are irrelevant.

The implication of the impracticality of the textual construction reconsidered

As it is, it seems that the diagrammatic drawings indeed do not have any proof-relevant function. Therefore, the impracticality of the construction assignments is no longer conclusive regarding the standardization of diagram, for there is no need for a more practical textual construction that contains all information necessary for an unambiguous diagrammatic construction if such an unambiguous construction is not necessary for the proof. This means that we cannot presuppose a technical and unambiguous construction merely based on the impracticality of the textual construction. The underdetermination as far as the drawing of the diagram is concerned strikes one as odd in the face of the otherwise extreme unambiguousness of the work in general and especially of the linguistic unambiguousness of the very construction assignments which appear to assign an unambiguous construction: the text is completely unambiguous as regards the properties and relations of the geometrical objects, but it leaves open how they must be drawn and especially, whether they are meant to be drawn unambiguously or not.

The discrepancy between the standardization of the language of construction and the diagram which can hardly match the textual construction in terms of unambiguousness becomes apparent in the proof part, when the diagram, being sketchy in the way that we must assume it to be, simply does not show what the proof asserts, e.g., the sameness of two triangles, or a right angle.[172] The most unsettling manifestation

172 Krämer 2016, 92 differently, emphasizes the cognitive value of the sketchy diagrams which the reader draws on the basis of the text: "Als Eigenkonstruktionen der Leser haben sie – in der Teilnehmerperspektive – die Aufeinanderfolge der notwendigen Konstruktionsschritte [. . .] deutlich zu machen. Erst in dem *Zusammenwirken* beider diagrammatischer Funktionen ereignet sich dann [. . .] der Erwerb, das 'Aufleuchten' einer neuen geometrischen Erkenntnis. Es ist gerade dieses vom offiziellen, textbegleitenden Diagramm unterschiedene Diagramm 'Marke Eigenbau,' das einen wichtigen Schritt hin zur Einsicht in die *Allgemeingültigkeit* des gewonnenen geometrischen Wissens verkörpert. Denn so kann eingesehen werden, dass die im Text gegebene Konstruktionsanweisung in ganz unterschiedlich gezeichneten Figuren realisierbar ist." However, the supposedly sketchy diagram which might be included in the text is precisely *not* "Marke Eigenbau," that is, it is not drawn by the reader, so it is not clear how the reader is supposed to recognize the diagram. Saito & Sidoli 2012, 156 assume on the

of this discrepancy is the conclusion of problem propositions: They state that 'x has been constructed' – however, it has *not* been constructed, at least not on the basis of the foregoing textual construction. Even though the conclusion of problem propositions (just like in theorem propositions) seems to be reached by means of a deduction that is based on axioms and previously proved hypotheses, i.e., on the text rather than on the diagram, it raises the question of proof-relevance of the drawing anew. This is because the very goal of problem propositions is a successful construction, whereas the goal of theorem propositions is to prove a statement on the basis of a construction. The proof of problem propositions is just as theoretical and deductive as that of theorem propositions. However, unlike in the latter, the conclusion is still related to the concrete case of the constructed diagram instead of abstracting from it. Moreover, the conclusion consists in the result of a diagrammatic construction: the imperative infinitive of the construction verb from the enunciation is turned into a perfect indicative so as to indicate the task's completion. If, however, the diagrammatic construction, at least in problem proposition, turns out to have some relevance for the proof, this will change the evaluation of the impracticality of the textual construction completely.

Thus, the question of the consequences of the impracticality of the textual construction for diagrammatic standardization cannot be settled on the basis of the analysis of the practical implications of the language of construction, for the function of the drawing remains unclear. While the text, on the one hand, does not focus on the drawability of the objects and object relations and thereby seems to imply a foregoing construction of the diagram, it does not, on the other hand, seem to employ the drawing as such for accomplishing the proof, although this is not entirely certain.

4.3.5 Conclusion

I have discussed the diverse constructional possibilities, technical and mathematical preconditions, and the difficulties of geometrically depicting the objects mentioned in view of the almost complete silence of the text on the method(s) and technical issues of construction. The point of this has not been to criticize the non-technicality and impracticality of the *Elements*' construction descriptions anachronistically from a modern point of view, by applying culturally different standards of technical precision to the ancient mathematical texts, as far as practices of description and construction are concerned. Rather, the point has been to show a striking discrepancy: on the one hand, the construction of a concrete diagrammatic case is emphasized as if it was the main basis of the proof (this is also implied by the contemporary metonymic use

basis of later ancient mathematical authors that the ancient diagrams were more correct than the medieval ones.

of the term διάγραμμα for the whole proof). The structural part of the 'construction' takes up a large portion of the proposition and is, on the face of it, full of seemingly practical assignments and implications. The construction assignments linguistically prefigure the assertions of the proof. On the other hand, the small effort which is put into describing the construction and the effective impracticality of the 'textual construction' appear to correlate with the fact that the propositions are utterly theoretical and proof-, not construction-centered. The diagrammatic construction appears to be irrelevant for the proof, as far as the quality of the drawing is concerned. This discrepancy prompts the question of what the point of assigning and describing the construction of a diagram is in the first place; but the only answer to this which may be safely provided on the basis of the foregoing analysis of the 'practicality' of the textual construction is that the point of the 'textual construction' cannot have been to enable and guarantee drawing the geometrical situation in a precise, proof-matching way. The exclusion of this possibility implies that the dimension of lettering, rather than the dimension of drawing, determines the form and function of the textual construction. As regards our question of the standardized inclusion of the diagram, the result of this section is that the impracticality of the textual construction does not necessarily suggest that the diagrams are already there, for it would only be indicative of the regular inclusion of the diagram if the proof presupposed a certain quality of the drawing. However, as we have seen that this is not the case, the diagrams (even those that might have been originally included in the Euclidean text) may well be imprecise, non-technical, and even non-geometrical, as is confirmed by the Byzantine diagrams.

4.4 Diagrammatization II: the implications of the use of letters

I have pointed out that the constant presence of the letters as such suggests the presence of a diagram. Now, I shall investigate the use of letters with regard to whether it conclusively indicates the foregoing construction of the diagram, after we have seen that the indications of the textual construction are ambivalent regarding this question. The use of the letters, as of all other textual features, is highly standardized. I have not dealt with the standardization of this textual feature in Chapter 3 but saved its analysis for this chapter, for it is only with regard to the standardization of the diagram that its analysis is at all possible. I shall analyze it according to four of its main characteristics which have emerged from Netz's insightful study: (1) the syntactical embeddedness of the letters, (2) the alphabetical order of the introduction of the letters, (3) the 'specification' of the letters in terms of the fixation of their reference to points in a diagram, and (4) the standardization of the letter words as regards the order of the letters. As I shall make use of the terminology which Netz has coined with regard to each of these aspects, and as I draw fundamentally on his results, I shall start from his conceptualization each time. Then I shall adjust his terms to the aim of my own study and point out the ways in which the aspects of letter use will have to be analyzed with a focus on diagrammatic standardization (which Netz touches upon, but which is not at the center of his interest). It is, above all, the subjects of letter specification and of letter ordering which seem to be decisive regarding whether diagrams have been included in the text and accordingly on which my study focuses.

4.4.1 The syntactical embeddedness of the letters

I have already alluded to the letters' being indicative of diagrams because of their indexical nature. At present, I shall specify the ontological ramifications of the syntax of the phrases in which letters are embedded with regard to the question of the actual presence of the diagram. Netz rightly claims that the function of the letters is that of "indices,"[173] that is, of ad hoc designated pointers to points in a diagram rather than of "symbols"[174] conventionally naming the points of certain geometrical objects.[175]

The indexical nature cannot be read into the syntax of every single letter phrase. Rather, Netz concludes it from the use of letters within expressions that involve spatial prepositions in the earliest surviving mathematical text by Hippocrates and in Aristotle; evidence of these expressions can still be found in Archimedes. They establish a spatial connection between objects and the letters which, e.g., stand "on" these objects.[176] With regard to the historical roots of the letter phrases which play such an important part in the *Elements* and its generic tradition, we must assume that the Euclidean text at least envisages a diagram. In fact, the letter phrases are even more suggestive as regards the relation of text and diagram: As pointers, the letters would be more likely to point to objects of an adjunct diagram rather than to objects yet to be constructed.

To be sure, it might be that the letter phrases preserve a 'dead' terminological tradition which no longer matches the mathematical practice of the time. However, it more likely seems to be the other way around: the terminology indicates that, from the beginning, elements texts have been written on the basis of a diagram, and nothing in the text suggests that this had changed by the time of Euclid. Still, the semantic implications of the syntax of the letter phrases are suggestive rather than conclusive as regards the question of the material inclusion of the diagram in the Euclidean text.

---

173 Netz 1999a, 47, 49, 50, 56, n. 118, 80.

174 Netz 1999a, 80 and *passim*.

175 Cf. Netz 1999a, 69 on what letters stand for in arithmetical contexts: "What do letters stand for? The baptised object in Greek mathematics is most often a point. In non-geometrical contexts, the baptised object is often the term directly under discussion, e.g. a number. However, through the assimilation of arithmetic to geometry, numbers, which are represented by lines, may sometimes be baptised indirectly. Their 'extreme' points will be baptised instead. Thus, in Euclid's Elements IX.20, the letters Α, Β, Γ, Η represent numbers, while the letters Δ, Ε, Ζ represent extremes of line-segments (in turn representing numbers) [. . .]. The decision on which approach to take is not arbitrary. In Euclid, letters represent 'extremes' if and only if the represented objects are meant to combine."

176 Netz 1999a, 43–51.

4.4.2 The alphabetical principle in the introduction and reuse of the letters

The relation between letters and objects

With the letters being indices rather than symbols, the relation between the specific letters and the specific kinds of geometrical objects is generally non-fixed[177] and potentially differs from proposition to proposition.[178] An exception is presented by K, which in some elements texts designates the center of the circle, though not in the *Elements*, where it is used in the same way as any other letter:[179] its denotation must be established anew in each proposition.

The alphabetical attribution of the letters to objects

The "attach[ment]" of the letters to the objects around which a particular proposition revolves does not happen for its own sake, as Netz's term "baptism" may, though is not intended to, suggest,[180] but takes place mainly alongside the setting out, definition

---

177 Netz 1999a, 79 is right in claiming "there is no regular relation between signifier and signified in Greek mathematics: specific letters do not represent stereotypical objects [. . .]." From the example he provides ("no 'r for radius'"), however, it appears that he is mixing up 'non-regular' (in the sense 'non-fixed' or 'non-standardized') with "arbitrary" (80). Similarly, he adduces the example of the use of the letter Ψ for both circles and cones in Archimedes' *On Conoids and Spheroids*, even though there is "nothing Ψ-ish about circles or cones (which are rather K-ish, if anything)" (80). However, the point about the letters is that their application to the objects which they happen to signify is in any case 'arbitrary,' but also 'non-regular;' for it could be 'arbitrary,' but still regular – as is the case with the rest of the *Elements*' lexicon (for which see Section 3.2) and indeed with the use of almost all words of the Indo-European languages (although the relation to the signified, e.g., in Greek, is almost never as strictly regularized as in the case of elements texts). Netz's mix-up of arbitrariness and non-regularity seems to be caused by his opposing "indices" with "symbols" (which, at least according to some theorists, are not purely conventional signs), rather than with "names" or simply "denotations." In any case, "car plates" are cited by Netz somewhat infelicitously as a parallel for the arbitrariness/non-regularity of the letters; for both in Germany and in the US, the letters and numbers may be and are chosen by the car owners more often than not as a 'symbol' for their own initials, dates of birth etc. The fact that this is so even though this costs an extra fee shows how people are thus opting against the arbitrariness of their car plates. As we shall see in this section, the application of the letters does follow certain standardized principles even though the relation between specific letters and specific kinds of objects is not regularized in the sense that it is not fixed.

178 Netz 1999a, 79 about the situation that letters may sometimes stick with the same kind of objects over the course of a few propositions, which, however does not change the fact that the relation between letters and objects is generally non-fixed: "It may happen that, for a stretch of a few propositions, a local regular relation between signifier and signified is established. But this is strictly local."

179 Netz 1999a, 80.

180 Netz 1999a, 68.

of the goal and construction, thus being confined mostly to the first part of the concrete section of the proof, though new letters can come up in the proof part and in later, additional construction assignments. The letters are assigned to the objects rather mechanically in roughly standardized alphabetical order: the point which is mentioned first is usually called 'A,' and if it is mentioned as part of a line, this one would be called 'AB,' etc.[181] Every letter is only assigned once in each proposition,[182] so if an aforementioned point is mentioned again as part of another object, it keeps its aforesaid letter, with the respective due letter of the alphabet being used only for the next to be newly introduced point in line.

The mere existence of the principle of the alphabetical attachment of the letters to the objects and the resulting textual introduction of the objects in alphabetical order does not yet reveal anything about whether a diagram has been underlying the text or not. The letters either could have been attached ad hoc alphabetically to the objects as they were introduced according to "the sequence of the text," with no diagram underlying at all, or with the letters being attached simultaneously to an underlying unlettered *drawing*; or the "lettering may be structured following the sequence of geometrical actions (instead of the sequence of the text),"[183] with the construction and the corresponding letter sequence only being envisaged or with the letters being quoted, as it were, from an underlying diagram in which the objects have been lettered in the order in which they have actually been drawn. The question whether diagrams were coming with the text or not, and if so, at what point of the production of the propositions they were drawn, and at what point they were lettered, remains to be answered precisely on the basis of the analysis of the letter use, as it has not been possible to answer it as yet on the basis of the material evidence and analysis of the textual construction.[184]

We cannot conclude on the origin of the alphabetical order from the mere existence of such an order per se. However, the recognition of a principle that underlies the textual attachment of the letters to objects provides a starting point for determining when the diagram would have (been thought to) come into being. It is the way in which the principle is applied, whether it is adhered to or varied, which could provide a clue as to the presence of a diagram. A perfect abidance by the alphabetical order would be inconclusive as to its orientation toward the text or the diagram, since it

181 Netz 1999a, 71. 'I' is usually not used (ibid., n. 7).

182 An exception to this rule is presented by propositions with proofs by contradiction, which sometimes attach the same letter twice or redefine the object to which it refers, respectively.

183 Netz 1999a, 85.

184 Netz 1999a, 85, on the basis of his letter analysis, provides a theory of the order of the steps of drawing the diagram, lettering, and writing the preliminary and final version of the text. My own interest, by contrast, is not in how the proof was discovered and how its formulation came into being, but rather, from the point of view of formal standardization, in the form of the ready product of the *Elements*. Regarding diagrams, this means that my interest is whether they standardly came with the text or not and whether they came standardized in form or not.

could either indicate that the text is wholly self-sustained in describing a diagram or suggest, with regard to an underlying diagram, a complete congruence between it and the text in terms of letter distribution. Rather, it is any deviations from the order of the alphabet in terms of incompleteness or inconsistency which would suggest that the text relies on a lettered diagram. It could be that the description of the construction of the diagram closely imitates the order of the construction though it departs from it sometimes where the logic of the text demands it or where the textual standardization is stronger than the diagrammatic one. Indeed, although the alphabetical order is the main principle to organize the introduction of the geometrical objects in the majority of cases, the text deviates quite often from it, both in the very introduction of the objects and in the later repeated references to them. I shall now briefly explain the practices of deviation from the alphabetical order according to Netz and their ramifications regarding the existence of a diagram, and I shall further unfurl them when dealing with the specification of the diagrammatic letters and the order of the letters in multi-lettered letter words.

Deviations from the alphabetical order of letter introduction

For the most part, the text of the *Elements* roughly adheres to the alphabetical order of letter introduction, even where it does not follow it consistently. That is, it does not, as a rule, deviate from it completely so as to display a 'non-alphabetical' order of letter introduction. Such practices are regularly visible in Archimedes and Apollonius, even though there, too, they present exceptions from the overall rule of alphabetical letter introduction. Wherever they occur, they do not seem to do so by accident, but they usually convey some meaning by deviating from the apparent rule of alphabetical letter introduction.[185]

Since such cases are not prominent in the *Elements*,[186] I shall concentrate on explaining the much more frequent category of deviation from the alphabetical order which Netz calls "permutation."[187] It covers, with regard to letter introduction, all

185 As examples for "non-alphabetical" letter introduction, Netz 1999a, 76 mentions the Archimedean practice of introducing "auxiliary objects which are not part of the main geometrical configuration," or, put more generally, his "mark[ing]" "discontinuities in the same proposition," or, conversely, the practice of Apollonius of "mark[ing]" "continuities between different propositions, e.g., in a case where a certain object which keeps recurring in successive propositions is always referred to by the same letter sequence in accordance with alphabetical lettering, but in one proposition, "where the sequence of construction is different," turns out to be lettered non-alphabetically by the very same letter sequence.

186 Netz does not mention non-alphabetical letter introduction for Euclid, and it does not occur in our sample from the *Elements*.

187 Netz 1999a, 71 and 83. Netz applies the term both to according practices in the introduction of letters, which he calls "baptism," and in the reuse of multi-lettered letter words, on which cf. pp. 290–303.

cases in which the "string" which one can produce from the letters in the order in which they are introduced is alphabetical in the sense that it consists of letters which, in any case, in the alphabet come before all letters that are *not* part of this string,[188] but in which this set of letters is not present in strictly alphabetical order, although the principle of the latter always remains recognizable. I have reviewed *Elements* I by Netz's string method with the following result: of 48 propositions, only 6, that is, only 12.5%, lead to a string which features one or more permutations of the alphabetical order,[189] while all others are strictly alphabetical.

These 'permutations' can either take the form of certain letters or letter sequences referring to points or larger objects that are mentioned later than they would be due alphabetically, or they can occur in the form of inversions of the letters in multi-lettered letter words. In the following short review of the practices of deviation from alphabetical lettering, I shall use the term 'permutation' only for the latter, while the former shall be called 'omission.' This is because I shall, on the one hand, regard these two forms of 'permutation' as distinct from one another. On the other hand, I shall connect either of them with specific practices of letter use which can be perceived only when letters begin to be reused after the textual introduction of most of them has been accomplished. In a first, shorter part I shall explain the phenomenon of 'omission,' in a second, larger part I shall discuss the repercussions of 'permutation.'

Omission

'Omission' is the term that I shall use for those 'permutations' of the alphabetical order in the sequences of introduced letters which are not due to the permutations of the alphabetical order in multi-lettered letter words, but which are caused by the situation that certain letters or letter sequences referring to points or larger objects are mentioned later than they would actually be due alphabetically. Netz does not conceptualize this as a type of its own (in fact, he does not discuss reasons for 'permutation' outside multi-lettered words at all), neither does he use the term 'omission,' even though his method of extracting strings of introduced letters could be useful as a heuristic tool for identifying such deviations from the alphabetical order. His 'permutations,' in any case, would only cover those omitted letters which will be introduced belatedly at some point in the proposition, but not those which are utterly missing from it. To be sure, Netz mentions the case that letters are "missing" ("usually no

188 Netz 1999a, 71: "they employ *n* letters, which are the first *n* letters of the alphabet."

189 These propositions are I 9, 23, 30, 43, 45, 47. Netz (1999a, 71 f.) has reviewed 100 sequences from Apollonius' *Conics* and Archimedes' *On the Sphere and the Cylinder* I and arrived at the conclusion that they contain 15% deviation from the alphabetical order, which, however, refers to "the average percentage" of instances of permutation; he does not give the number of propositions of the 100 reviewed which are actually affected by permutation, and to what extent these are, on average, affected by permutation, only mentioning the proposition (*Sph. cyl.* I 1) with the highest number of permutation ("above 50%").

more than a few") as a possibility besides 'alphabetical letter introduction' and 'permutation of the alphabetical,' but treats this as an exception which does not deserve any conceptualization of its own.[190] By contrast, in my view, letters omitted from the alphabetical order of letter introduction, whether they occur in permuted position or whether they are not explicitly mentioned in the proposition at all, are part of the same phenomenon.

This is because the reasons for omitting letters from the alphabetical order seem to overlap in either case. Whether the omitted letters remain unmentioned until the end of the proposition or whether they are introduced outside of their expected alphabetical order at a later point, they seem to be omitted in most cases for one of two reasons. Either they are left out if it suffices to mention only some of the letters which constitute the full reference to an object in order to refer to it: for example, a square may be referred to by two or three of its four letters only. Or, alternatively, letters may be omitted if the order of the textual construction appears to be different from the order of the diagrammatic construction and lettering. Since letters thus are omitted for similar reasons whether they appear belatedly in permuted alphabetical order or whether they are not mentioned at all, the analysis in terms of strings is only useful for detecting omitted letters if it serves to register all letters which have been omitted from the alphabetical sequence.

The interesting point regarding the practice of 'omission' is that the text, far from ignoring the alphabetical order of letter introduction, to the contrary, standardly pays meticulous attention to it, as it bypasses letters which have already been silently 'spent' instead of automatically applying alphabetically due letters. This practice is very suggestive regarding our question of diagrammatic standardization. It strongly indicates that the text from the start refers to an underlying diagram which has been constructed and lettered beforehand. For the text appears to follow roughly the order of the diagrammatic construction as it emerges from the likewise alphabetical distribution of letters to the diagram. However, since the lettered diagram is already present, the text is free not to mention every occurring letter in order of appearance, either because it is not necessary to mention all letters that constitute an object for referring unambiguously to it, or because the text deliberately chooses to deviate slightly from the diagram's construction sequence.

However, the omission of letters according to the alphabetical order is not in any case a conclusive argument for the preexistence of the diagram. In the case of letters omitted from object terms, the versed reader would know the practice of implicitly 'using up' letters. Even to the lay reader, the strong (though not complete) standardization both of the terminology of the 'letter phrases' and of the principle of alphabetical letter introduction provides a guideline for supplementing the missing letters in the diagram if readers were to draw and letter the diagram themselves. In fact, it

190 Netz 1999a, 71.

would be this very standardization which allows for the practice of omission.[191] Of course, this can be different from case to case. Often, it is unequivocally clear where to supplement a letter in the diagram when it is missing from a multi-lettered name; sometimes, however, e.g., if two letters of an object's name have been omitted, the question arises not only where to supplement which of the omitted ones, but also which points are to be lettered by the letters that are mentioned in the text. This would be clear mostly from the context and from the principles according to which letters in multi-lettered words are ordered. Hence, it would again depend on the recipients' knowledge of these principles. However, the question eventually would have to be left open until the letters explicitly occur in the text – if they do so at all. This means that the 'omission' of letters due to abridged multi-lettered references is in most cases strongly indicative of an underlying diagram, but not a secure sign of its preexistence, as the letters are sufficiently specified, that is, localizable in the diagram in spite of their omission. However, in some instances in which the omitted letters cannot be supplemented for a very long time or not at all, and in which their omission temporarily or permanently obstructs the further process of construction, letter omission becomes a very strong sign of an underlying diagram.

In the case of letters which are omitted because the textual and diagrammatic constructions deviate from one another, the question is whether they can be supplemented from the context or remain in need of explicit specification; that is, whether the letter is identifiable at all, and whether the one drawing the diagram from the text is able to identify it. In any case, an expert first of all would know about the fact that a letter has been implied by omission, and second, might be qualified to draw conclusions on where it belongs, whereas a learner or lay reader would have more difficulty in recognizing the need of supplementing the omitted letter and in accomplishing the supplementation. Thus, in cases in which the omission of a letter from the alphabetical order seems to be only explicable by a deviation of the textual and diagrammatic constructions from one another, the question of the conclusiveness of the 'omission' of a letter comes down to the question of the specification of the letter.

'Specification,' regarding which I shall review the use of letters in Section 4.4.3, is the unambiguous identification of a letter, not merely with an object, but with an object *in the diagram*. For an omitted letter to become specified, it will have to be supplemented explicitly at some point in the proposition; for it to be supplemented so as to enable the construction of the diagram, it will, as shall be explained in more detail in the subsequent section, have to be specified *in time*, that is, until the end of the construction part. Conversely, this means that an omitted letter only indicates conclusively the presence of a diagram, if it is *not* specified in time (i.e., either only in the proof part, or not at all). Strictly speaking, this is the rule applying to all omitted let-

191 See Section 5.1 for more examples of less-than-perfect standardization enabled by and even provoked by complementary (over)standardization.

ters. However, one has to qualify it insofar as specification which is accomplished too late or not at all is only conclusive regarding the text's reliance on a diagram under two conditions: first, if the specification of the omitted letter in question is at all necessary for the proof,[192] and second, if it is not self-explanatory, as it is mostly the case with letters that are omitted from the references to multi-lettered objects, where the omitted letters are implicitly specified by the underlying rules of letter distribution (which the recipient must know). The analysis of letter specification in Section 4.4.3 will include many examples of letter omission and give an idea of the extent of the phenomenon in the *Elements* and its suggestiveness of the presence of diagrams.

Permutation

I shall use the term 'permutation' proper only with respect to the order of letters *within* multi-lettered letter words. In the context of letter introduction, I take it to denote the deviation from the expected alphabetical order within a multi-lettered letter word, while the letters which form the letter word are the alphabetically due ones; e.g., a line may be called ΔΓ instead of ΓΔ, after the letters A and B have been introduced.

This practice, however, is not confined to the process of letter introduction. Besides, also the order of the letters in letter words which have already been introduced may be permuted in comparison with the original order when these words recur in the same proposition. The order may be alphabetical or inversely alphabetical at first, then it may get permuted, and again permuted at any further recurrence of the letter word. Netz has coined the terms "repetition" for the reappearance of the introduced letter order, "switch" for its inversion, and "reswitch" for the reversion of the switched order to the original one.[193] These practices are enabled by the fact that the letters within the same letter phrases are apparently understood as referring to the same objects whether they are being ordered alphabetically or inversely alphabetically.[194]

The two practices of multi-lettered letter words, i.e., permutation in the introduction and in the reappearance of such words, are, in fact, part of one and the same practice of 'permutation.' For permutations within letter introduction could be regarded, just like permutations in the recurrence of multi-lettered letter words, as 'switches,' namely of the underlying, originally expected alphabetical order of lettering. For both of these practices rely, on the one hand, on the principle of alphabetical ordering, and, on the other hand, on the awareness that the alphabeticality of the

192 See below p. 336–338 on the proof-relevance of the diagram as a whole and on proof-relevance as a criterion for judging the necessity of textual and diagrammatic standardization and unambiguousness.

193 Netz 1999a, 71.

194 See Section 4.4.3 on letter specification. In the few cases of letter phrases which are potentially ambiguous, it is precisely the order of the letters which may be used to indicate the kind of geometrical object; for this, see Section 4.4.4 on the order of the letters in multi-lettered letter words.

order does not matter geometrically. Netz, by contrast, though he uses the term "permutation" widely to cover both deviations from the alphabetical in the introduction of the letters and reappearances of multi-lettered words, conceptualizes "baptism," i.e., the introduction of letters, including those which designate (part of) many-lettered words, and the change of "many-lettered names" as two distinct practices.[195] However, the two practices overlap in the case of the 'permutation' of multi-lettered letter words: namely, with regard to the form in which these words are introduced.

In the twofold conceptualization of 'baptism' and 'change of many-lettered names,' however, the introduction of many-lettered words does not play a role as such regarding the adherence of these words to or their deviation from the expected alphabetical order. Rather, the letter sequences are only considered as part of the 'strings' of introduced letters – that is, if the letter words form part of the sequences at all. For if the letters of which the multi-lettered words consist have already been introduced as referring to single points or as part of other objects, they would not make it into the string of newly introduced letters, and, therefore, would not be accounted for *as newly introduced letter words*.

For example, the first 20 propositions of the *Elements*, the sample on which I have based my own analysis of the statistical behavior of multi-lettered letter words,[196] in-

195 In the introduction to his chapter, 1999a, 69, it seems that the distinction is between the practice concerning individual letters and that concerning many-lettered words (which is Netz's term for what I call 'multi-lettered letter words'): "The practices with regard to letters fall into two kinds. The first is what I call 'baptism' – the process of attaching individual letters to individual objects. Second, given the distribution of individual letters to individual objects, many-lettered names can be made [. . .]." By contrast, his subsequent specifying distinction of practices (71) indicates the distinction outlined before: "I follow two practices: (a) The sequence of baptisms in a given proposition can be plotted. It turns out to be often, though not always, alphabetic [. . .]. (b) Objects may change their many-lettered names [. . .]." According to this second distinction, the analysis of "baptism" in terms of the sequencing of introduced letters covers many-lettered names as well, as they are made of individual letters, while the second kind of analysis deals with the *changes* which such many-lettered names undergo. It is for such changes that Netz reserves the term "switches," while he does not extend it in the above-explained sense of 'permutation' onto multi-lettered letter words which have not occurred in the same proposition before but are 'switched,' as it were, in relation to the expected alphabetical sequence.

196 It is for two reasons that I have limited the sample for my own preliminary assessment of the deviation from the alphabetical order in multi-lettered letter words to the 20 propositions of book I: first, this sample, by containing 585 multi-lettered words in total (that is, including all reappearances), of which 223 are first appearances in the respective propositions, appears sufficiently large for providing a representative picture of how multi-lettered letter words behave throughout the proposition. Second, the sample contains not only the most basic form of multi-lettered letter words, namely two-letter words referring to straight lines, which can only be presented in two possible ways, either alphabetically or inversely alphabetically. It also features three-letter words which allow for a considerable permutability, since they can be used for referring to different kinds of objects, in relation to which, in turn, they are varyingly permutable: they either allow for two different forms (in the case of angles which are, in our sample, standardly referred to by the three letters for the three points constituting the two straight lines which form the angle, of which the letter referring to the point in which the two lines run into each other is non-permutable and takes the middle position of the three-letter

troduce, in total, 223 such words. Of these, 178 (80%) are introduced alphabetically, and 45 (20%), conversely, non-alphabetically. However, if the introduction of such words is reviewed by proposition, it turns out that only 6 out of 20 propositions, that is, 30% only, introduce all of their multi-lettered letter words alphabetically. That is to say, though the majority of such words is introduced alphabetically, their non-alphabetical introduction is a regular feature, occurring in the majority of propositions. This situation is not mirrored by the review of strings produced from newly introduced letters for I 1–20 according to Netz's method. Only one of the strings (for I 9) contains a permutation which is due to the allegedly non-alphabetical introduction of a multi-lettered letter word; for I count the letter word ΒΑΓ, which is responsible for the permuted introduction of the first two letters of the alphabet, as a multi-lettered letter word which is alphabetically introduced, as it is part of the designation for an angle which could only have been referred to by either of the two combinations ΒΑΓ or ΓΑΒ, of which I regard the first as alphabetical, the second as non-alphabetical, that is, on the grounds of the order of the first and third letter, which are the only permutable ones. A similar result emerges from taking the whole book into account: of 6 propositions (of 48 in total) which yield strings that contain permutations of the alphabetical order, only 2 besides that of I 9 feature such permutation that is due to the way a multi-lettered word is introduced. Of those two, I 23 is a case like I 9, with the permutation within the string ΑΒΔΓΕΖΗ resulting from the introduction of the designation for an angle by the letter sequence ΔΓΕ which is, in fact, as alphabetical as it can be. This leaves only I 34 which indeed returns a permuted sequence (ΑΓΔΒ) due to the permutation of the letter order within the multi-lettered word ΑΓΔΒ. Thus, the strings are non-indicative regarding the degree of alphabeticality that characterizes the introduction of multi-lettered letter words. This is because the strings contain only such newly introduced multi-lettered letter words that are responsible for the introduction of letters into text. However, many of the letter

word, while the other two technically are exchangeable); or they allow for six different forms (in the case of triangles and circles). Thus, the sample, by being restricted to two- and three-letter words only, contains sufficient variety and changeability of the order of the letters, without involving more complex cases as are presented by the four-letter words which occur in the second part of *Elements* I. These complicate the picture since they do not only quantitatively allow for even more variation in permutation, but also include a different kind of variability (and resulting permutability), that is, by the possibility of the respective objects to be represented by less than the four letters constituting them. Here, the practice of 'omission,' besides 'permutation,' affects the deviation of the form of multi-lettered letter words from the alphabetical. While four-letter words shall be included in the subsequent study of the principles of letter ordering according to which it deviates for what reasons from the order of the alphabet (Section 4.4.4), a clearer picture of the extent and kinds of permutation of the alphabetical order which I analyze at present is gained by limitation to the letter words of I 1–20. However, in the case of the strings of introduced letters produced according to Netz's method for the sake of methodological comparison, a larger sample of 48 strings covering the whole first book shall be provided besides the smaller one for the first 20 propositions.

words are only introduced when the letters of which they consist have already been introduced and, thus, have already been covered in the letter 'string.' Thus, the introductory form of multi-lettered letter words does not play a role in the analysis of permutations in the process of 'baptism' as reviewed by Netz; consequently, the degree of alphabeticality of the introduced letter words and the forms and meaning of the deviation from it do not emerge from this method.

Neither, however, is the introductory form of multi-lettered letter words considered in Netz's assessment of permutations as 'switches' of the originally introduced form of multi-lettered letter words. In his survey, the introduced form, whatever it itself is, is only taken into account negatively as the foil against which the respective first switch of a letter word may be detected. According to this method, the first 20 propositions of *Elements* I contain 362 'reappearances' of multi-lettered letter words, of which 63, i.e., 17%, are 'switches' compared with the previously occurring form.[197] The number is 108, i.e., ca. 30%, if 'switch' is understood in the sense in which I have redefined it, that is, as including changes from the supposedly *expected* letter order that already occur in the introduction of multi-lettered letter words.

In fact, the method of counting switches, on the basis of which Netz discusses the reasons for the changes of the order of letters within multi-lettered letter words, does not say anything about the dominant forms of such words in terms of letter order; that is, it does not only not betray the dominant form of the letter words as they are being introduced, but neither, as they are being reused, whether they are predominantly used in the alphabetical form or in permuted order, and in the case of more than two-letter words, in what kind of permuted form. Further, we do not get to know how often a word is changed from the one to the other form, i.e., how the number of switches in a proposition is distributed onto the individual multi-lettered 'names.'[198] The reason why all of this is not clear from the count of switches is, first, that the basis of the switches of each multi-lettered 'name,' i.e., the form in which it first occurs in the proposition and the resulting directions of the switches, are not taken into account, and second, that the distribution of switches is not considered with regard to each letter 'name.'[199] That is to say, it does not become clear how many switches

197 This result confirms for the *Elements* the results which Netz's assessment of "switches" in works by Archimedes, Apollonius, and Autolycus has yielded: "20% or less" (1999a, 72).

198 By letter 'name,' I mean the particular combination of letters which occurs in a particular letter phrase throughout a particular proposition in reference to a particular object and takes various forms regarding the order of the letters, as opposed to the individual instance of this letter 'name;' of course, as has been pointed out above, the letter words are essentially not 'names,' but indices, and thus not tied to an object beyond the course of the proposition, but always newly attributed to it at the moment of reference, but by their keeping the same letter combination it is nevertheless possible to track their behavior during the proposition.

199 Netz indicates that besides having counted the overall number of switches in different texts in relation to the theoretically possible number of switches, he has also considered the number of switches per proposition ("I have not come across any proposition where the percentage is above

occur per 'name' in relation to the number of reappearances, and at which one of the occurrences the switch(es) take(s) place with regard to the number of reappearances.

For example, it makes a difference regarding the switches per 'name' and the resulting kind(s) of reuses whether the same number of switches occurring within a proposition (e.g., four) affects four-letter words (as in I 11, with one switch occurring per word)[200] or only two-letter words (as ΒΓ and ΑΒΓ in I 14, where one of the two words, ΒΓ, is switched three times). Further, it makes a difference regarding the dominant overall letter order in which a 'name' is used, whether the switch – if, e.g., a word is switched only once within the proposition – occurs after the first of two appearances, as is the case with the references to the triangle ΑΒΓ/ΑΓΒ in I 6, so as to produce one alphabetical and one permuted form, or rather at the last instance of a comparably large number of reappearances; and here, it is crucial which letter order is the one used from the start until before the very last occurrence: whether the alphabetical one, as with the letter word ΑΕΔ in the designation of an angle in I 15, which occurs nine times before appearing once in permuted form, or rather the permuted one, as with the word ΔΕΒ that refers to another angle in I 15 and occurs four times in this same form before being switched to the alphabetical form for its very last occurrence.

Rather than considering, in this manner, *what* letter order is switched, and *when*, Netz's approach toward the behavior of multi-lettered letter words is confined to counting the overall number of switches within a work and within propositions of a work. This is because his focus is on the cognitive processes involved in writing up the proof, whereas our own interest is, differently, in the standardization of the text with regard to the diagram. From this latter perspective, the average number of switches has no significance except for indicating that the order of the letters was variable without changing the validity of the reference. The number of switches emerging from our own count of switches in propositions of the *Elements* according to Netz's method simply shows that changes of the alphabetical order of letter introduction are regularly to be expected, and that the standardization of the letter words extends to the number of letters and to certain rules of their order depending on the object denoted by the respective letter phrase, but generally allows for variation as to which letter comes when.

However, from the point of view of the standardization of the diagram, the alphabetical order and the deviation from it within letter words are not merely of interest in suggesting in general the fact of the changeability of the order of the letters. Rather, it is the extent, the principles and functions of the deviation from the alphabeticality of the letter order within multi-lettered letter words which are significant with respect to the question of the presence of a diagram.[201] The question is how decisive the

50%. Above 30% is the highest level still common," 1999a, 73). But he gives no indication of having assessed the number of switches per word.

200 The affected letter words are, in the form in which they have been introduced: ΓΔ, ΓΕ, ΓΖ, ΕΓΖ.

201 See above, Section 4.4.4.

alphabetical order is in determining the order of the letters in letter words: whether it is just the underlying principle of letter introduction which happens to characterize the introduction of multi-lettered letter words insofar as they introduce new letters and letter combinations into the text, with the subsequent use, however, deviating from it either irregularly or according to certain principles; or whether the alphabetical order is actively pursued as the prime principle of letter organization throughout the proposition, from which the text deviates only inadvertently or if it for whatever reasons has to. The standardization of the alphabetical order of letter words can only be analyzed on the basis of assessing both: the order of the letters in the first mentions of multi-lettered letter words in a proposition, and the subsequent behavior in terms of 'switches' in the reappearing words. Both practices seem to be intricately bound up with one another regarding the analysis: first, by similarity, because they both show variation between the same types of behavior, namely, adherence to and deviation from the alphabetical, and they both potentially rely on the same principles; second, conceptually, because the permutation of a letter word *as permutation* depends on the expected or underlying form of the letter word.

I shall now briefly present my results of such an analysis of the order of the letters of letter words in the first 20 propositions of *Elements* I. This analysis has been undertaken, just like the previous analyses of the text of the *Elements*, solely on the basis of the text and in the way of statistical evaluations. It is only in Section 4.4.4 on 'the order of the letters,' when dealing with the reasons for and functions of the practices of letter ordering in the letter words of the *Elements*, that the order of the letters shall be discussed on the basis of the diagram.

Of all appearances of multi-lettered words in the sample (585), 76% are ordered alphabetically. As has been already pointed out above, of 223 of those 585 instances which are first appearances of the respective letter word, 178, that is, 80%, are alphabetically ordered. As for the reappearances of the introductory forms of multi-lettered words (362), 266 (74%) display an alphabetical order. This means that the degree of alphabeticality among introduced letter words is only somewhat higher than among reused words; it is, in fact, quite stable. While the overall number of alphabetically lettered words points to the general predominance of the alphabetical principle of letter ordering, the comparison of the quantities for first and re-appearances suggests that this principle is equally pursued as the dominant one throughout the proposition, with the slightly smaller number among reappearances being tentatively explicable by a decrease in discipline or as slips of memory on the part of the author or writer.[202] Besides indicating that the order within letter words is standardized to a considerable degree, however, the above quantities also suggest a rather conspicuous and stable amount of deviation from the alphabetical order both among introduced

202 For a deeper discussion of such factors see Netz 1999a, 74 and Section 4.4.4, 344–360, on the functions of letter ordering.

and reappearing words. This calls for a closer investigation of those deviations, since, as has been pointed out, what could be, if at all, conclusive is not the adherence to alphabeticality, but precisely the deviation from it.

On the basis of the text only, the distribution of the deviations from the alphabetical principle, both in the introductory forms and in the reappearances, could be observed according to parts of the proposition with regard to certain standardized tendencies, which, in turn, could be indicative of diagrammatic standardization. For new letter combinations get introduced into the text not only in the exposition and construction part where most of the individual letters are first mentioned either individually or as part of multi-lettered letter words, but throughout the proposition, and letter combinations get repeated from the exposition part onward.

Indeed, the analysis of the deviation from the alphabetical order in the introduced forms of multi-lettered words yields that the number of multi-lettered letter words which are introduced in permuted letter order differs according to the different parts of the proposition. Notably, the exposition part is almost completely alphabetical regarding the introduction of multi-lettered letter words; while 17 of 19 expositions do not feature any permutation in introduced letter words, the order in the reference to one angle, respectively, is permuted in both I 13 and I 19.[203] By comparison with this, the definition of the goal is less alphabetical regarding the introduction of multi-lettered words, with 5 of 19 propositions[204] containing altogether 5 multi-lettered words introduced in permutation in this part of the proposition. In 6 of 11 constructions which appear as a distinct part before the beginning of the proof, 10 letter words introduced in permutation occur,[205] and in 10, that is, half of all proof parts, 26 letter words with permuted letter order occur. Thus, the number of multi-lettered words, the letters of which are not alphabetically ordered, increases in each proposition part in absolute terms and regarding the propositions affected, the relative occurrence of permuted letter order in newly introduced multi-lettered words (i.e., relative to the overall number of introduced letter words per proposition part) is extremely low (4%) in the exposition part and thus significantly lower than in all subsequent parts, where it is between 21% in the definition of the goal and in the construction, respectively, and 27% in the proof. This means that words begin to be introduced by not adhering to the alphabetical principle regarding their own letter order in a significant way only from the definition of the goal onward, and that their number from there even tends to increase in the course of the proposition. Conversely, there is a strong adherence to alphabetical ordering in the exposition part.

The alphabeticality of the exposition and the steady increase of the permutation of the letter order in the first appearances of multi-lettered words could suggest that the

203 I 7 does not have a proper exposition (nor a definition of the goal).

204 I.e., I 15, 16, 18 and 20.

205 I.e., in I 1, 2, 5, 11, 16 and 20.

alphabeticality of the order of the letter words is tied to the introduction of the letters themselves into the text – that the introduced letter words are themselves only secondarily ordered alphabetically, depending on whether they contain newly introduced letters. However, there does not seem to be any such overlap: of all words introduced with alphabetical letter order (178), only 63 (35%) actually contain newly introduced letters; of those 50 introduced alphabetically in the exposition, only 27 do, of 19 in the definition of the goal, only 2, of 38 in the construction, 25, and of 69 in the proof, 9. Of all introduced words whose letters appear in permuted letter order (45), most do not contain new letters, but 3 actually do: 2 of 10 in the construction part, and 1 of 26 in the proof part. Although the numbers regarding words introduced with permutation are not very significant since the number of permuted multi-lettered words itself is quite low, this suggests that increase of permutation in the introduction of letter words over the course of the proposition does not depend on the newness of the letters.

I shall now focus on the permutation of the alphabetical principle in the reappearances of these words in the context of the relation of the reappearances of the letter words to the forms in which they have been introduced. Four such relations are possible: either the introductory letter order – which is alphabetical or permutation of the alphabetical – may be permuted so that the reappearance turns out to be alphabetical or permuted, respectively, by contrast with itself; or the introductory letter order may be retained. I shall concentrate on those reappearances which constitute changes from the alphabetical order of the letters in the words when they first appeared, as my interest, at present, is in how the text deviates from the dominant alphabetical principle. As far as the other types of reappearances are concerned, I take it that neither the alphabetical reappearances of alphabetically ordered introduced forms, nor the permutations of the alphabetical that are reappearances of likewise permuted introductory forms are suggestive in any way from the point of view of formal standardization. By contrast, the alphabetical order in the reappearances of originally non-alphabetically ordered multi-lettered words could be usefully compared to the reverse practice of letter order change which is my main focus.

I shall start from the quantitative relations of all types of reappearance before going into the permutations of the introduced letter orders, and especially the permutation of the alphabetical one. At this point, I shall not regard the permutation in the reappearances in terms of 'switches' from the immediately preceding form of the word but take into account all reappearing deviations from the (non-)alphabetically ordered introduced letter combination (and likewise, all reappearing instances in which the respective introductory forms have been preserved). In the first 20 propositions, I have counted 51 such permutations in reappearances where the first instance of the word is alphabetically ordered, compared with 241 alphabetical reappearances of alphabetically introduced words. That is, ca. 17% of all reappearances of alphabetically introduced multi-lettered words are permutations of the alphabetical. As far as the permutation of introduced words which present a permuted version of the alphabetical is concerned, that is, the alphabetical reoccurrence of non-alphabetically ordered words, 25 are alphabetical reappearances, compared with 39 reappearances which stick to the permuted version. Thus ca. 39% of all reappearances of non-alphabetically introduced letter words are permutations of the non-alphabetical, that is, become alphabetical, whereas 61% remain permutations of the alphabetical. This means that, while alphabetically introduced words strongly tend to be reused alphabetically, with the number of deviations nevertheless being considerable, non-alphabetically introduced words are more likely to appear in the permuted form than alphabetically introduced words are. This is true even though the number of non-alphabetical reuses still prevails in the case of non-alphabetically introduced words. This underscores the significance of the alphabetical ordering principle and the predomination of alphabetical reappearances in absolute terms: it is not only because the reappearances stick to the order of the dominant alphabetically introduced forms that the alphabetical order numerically dominates also the reappearances of multi-lettered words despite the general possibility of the change of letter order; for this could be interpreted in terms of mechanical, and therefore insignificant repetition. Rather, also the reappearances of non-alphabetically introduced forms show some considerable inclination toward alphabeticality.

At the same time, it is notable that almost two thirds of all reappearances retain the permutations of the alphabetical order of the introductory forms, while roughly one fifth of all reappearances of alphabetically introduced forms present changes from the dominant form which could as well have been retained. With regard to the latter situation, it is notable that the relative number of the changes *from* the alphabetical is already smaller than the above-mentioned number of the changes *to* the alphabetical. The absolute number of permutations of the original forms to non-alphabeticality is more than twice as high (51) as that of permutations to alphabeticality because of the absolutely much higher number of reappearances of alphabetically introduced forms. This adds a considerable number to 'non-alphabetical reappearances.'

I shall now consider the number of changes from the originally introduced letter order in the reappearances per part of the proposition in order to see where deviations are most likely to occur. Almost the same picture emerges for both the reappearances which deviate by permutation from the alphabetically ordered introductory forms, and for those which deviate alphabetically from the non-alphabetically ordered introductory forms. Of the 51 changes from the original alphabetical order, 47 occur in the proof part, while only 2 can be found already in the definition and the construction, respectively. Similarly, of 25 alphabetical reappearances deviating from the originally permuted (i.e., non-alphabetical) form, 22 occur in the proof part, 1 in the conclusion, and only 2 already in the definition of the goal. On the face of it, these results could indicate that the order of the letters in the reappearances changes significantly only in the proof part, and that the change of letter order, consequently, has to do with the goal of this part of the proposition. However, the more likely explanation of them is that the majority of letter words, whether they change regarding the original letter order or stick to it, tend to reappear only in the proof part to begin with – which is, then, also the natural place to change for those which do. Of all reappearances (362), 47, that is, only 13%, occur before the proof part, while the rest of them occur, for the most part, in the proof (304) or in the conclusion (11). Thus, no general explanation of the change of the letter order regarding the alphabetical principle in the reappearances seems to be possible. In the part devoted to the functions of the order of the letters (Section 4.4.4), I shall nevertheless attempt to explain the letter order comparatively with respect to the supposed aim of the construction and proof part respectively, in order to show the specificity of the letter ordering in the *Elements* and the role of the alphabetical principle therein.

I shall now take into view the switches among the reappearances *as switches.*[206] That is, I shall consider the changeability and frequency of change of the letter order in the reappearances. This has been, as I pointed out above, the main agenda of Netz's approach to the order of letters in multi-lettered letter words. He has argued on account of the number of 'switches' and 'reswitches' in the reuse of letter words which he has counted per author/work and individual proposition that the kind of letters of a multi-lettered word are the ones which are due in the introduction of the letter words according to the alphabetical principle of letter introduction, but that the order of the letters is, wherever they occur, non-fixed.[207] He has pointed out that the number of switches is limited per work/sample and proposition, but since he has not taken into account the number of switches and reswitches per letter word but quantified the switchability of the letter order only more generally, he has insinuated that the order of the letters may change back and forth in the reappearances of a letter word over the course of a proposition.[208] An analysis of *Elements* I 1–20 yields that, by featuring 64 switches compared with 362 reappearances, that is, possible switches, the sample has 18% switches, and, conversely, a switch every 5.66 reappearances.

In accordance with this method, I shall now quantify switches independently of the relation of the switched and non-switched words to alphabeticality.[209] However, my own focus in analyzing the behavior of switches is the individual letter word and the standardized patterns which eventually emerge from a statistical evaluation of the individual cases. This is because it does not become clear from the abovementioned results how switches are related to reappearances in every single case, and in general per reappearing word.

206 Here, I shall regard the notion of 'switches' in Netz's original sense, that is, as referring only to the change of letter order in the reappearances, rather than extend the term in order for it to refer also to the introductory forms which have a permuted letter order, as I have suggested earlier with the purpose of showing the interdependence of the change of letter order in the introductory forms and in the reappearances of letter words.

207 Netz 1999a, 81.

208 Netz 1999a, 71–73.

209 It is only the conclusion from the reconsideration of the nature of change in the order of the letter words (p. 304) which shall shed some light again on the significance of the ordering principles, alphabetical or permuted, in multi-lettered letter words and form the basis of the investigation of the functions of these principles under point four of the present study of letters (Section 4.4.4).

First of all, of all multi-lettered letter words that are introduced in I 1–20 (223), a good amount does not reappear at all in the course of the respective proposition (80). This leaves 143 multi-lettered letter words which do reappear. Of these, more than two thirds (101, i.e., 71%) are not switched at all, with only 42 switched words remaining (29% of reappearing multi-lettered words, and 19% of all 223 multi-lettered words), on which the above number of 64 switches is concentrated. These numbers are only partly explicable by the general shortness of the propositions which does not allow for an endless repetition of letter words nor for an accordingly frequent repetition of switches. For there is a total of 362 reappearances, and, thus, 2.53 reappearances per reappearing word on average. Depending on their actual distribution, these reappearances do not only allow for more regular but also for more frequent switches of reappearing words. It seems, in spite of the overall stable number of switches in our Euclidean sample as in the samples from other contemporary authors analyzed by Netz, that the phenomenon of switches is rather confined in two respects: both to certain reappearing words and within switched words. This shall be confirmed with the help of some more detailed analysis.

We have seen that 101 words of all reappearing words are not switched at all. Of those 42 letter words which *are* switched, 28 words are switched once, 8 are switched twice, 4 are switched three times, and 2 are switched four times. This means that, while it is by far the most common behavior of a recurring word not be switched at all, the most common behavior of switched words is to be switched only once, as it happens in the majority (two thirds) of cases. Having more than two switches is extremely rare, with the maximum number of switches, namely four, occurring only in I 1 with AB and in I 17 with AΓB. This is the case even though words with more than two reoccurrences, which thus provide the possibility of more than two switches, are actually quite frequent, though not in the majority. The low number of switches per reappearing letter word is, in the face of the overall not inconsiderable number of switches, thus compensated for by the high number of words which are switched only once among those which are switched at all.

It is conspicuous that the number of switches per reappearing word, as a rule, tends to be smaller than the number of reappearances of the same word, and even tends to become a lot smaller the more reappearances a word shows. For example, while multi-lettered words with 1 reappearance are switched once 11 times compared with 51 cases of zero switches, and words with 2 reappearances are switched twice only 3 times, but more frequently once (7 times) or not at all (14 times), words with 3 reappearances and more never reach the maximum number of switches. The proportion of zero switches becomes gradually smaller the more reappearances there are, which means that it becomes more likely for a word to be switched at all the more frequently it recurs. However, the fact that many letter words with such frequent recurrences considerably score under the possible number of switches illustrates that it is especially in the case of words with many reappearances that many reappearances remain unswitched. For example, of words with six reappearances, three are not switched at all, while the only one remaining is switched three times; of words with seven reoccurrences, one is not switched, and another one is switched twice; the two words with eight reappearances are switched once and three times, respectively, those with nine, once and four times. This becomes even clearer from the reverse statistic which looks at the number of reappearances per switches. Here, the number of words with the minimum number of reappearances required for a letter word to be switched is in all cases smaller than the number of words which exceed the minimum number by its reappearances and in this way enable more switches than are actually affecting them. For instance, of all those words which are switched only once, only 10 reappear only once, while 18 reappear at least twice, and more than half of these 18 more than twice, that is, up to nine times;[210] similarly, for words with 2 switches, the ratio is three to five, for words with 3 switches, zero to four, and for those with 4 switches, zero to two.

From these considerations of the relation of switches and reappearances with regard to individual letter words in I 1–20, it tentatively emerges that the appearance of individual letter words regarding their letter order is quite homogeneous over the course of a proposition rather than switching back and forth between the introduced and the permuted version. In this regard, the letter words with one reappearance (which are in the majority), in allowing only for one switch or none, are less conclusive than words with many reappearances: here, one can see that the case of the ΑΓΒ which is used in I 17 in reference to an angle is absolutely exceptional in being repeated once in this same form, before being switched to ΒΓΑ, then switched back to ΑΓΒ, then again to the permuted ΒΓΑ, and finally once more to the original alphabetical version ΑΓΒ. As has been pointed out, there is only one more word with four switches (ΑΒ in I 1), whose use, because of the higher number of reappearances, is more straightforward, with two permuted forms occurring toward the middle of the string of reoccurrences, and a number of repetitions of the alphabetical form occurring at the beginning and end (ΑΒ, ΑΒ, ΑΒ, ΒΑ, ΑΒ, ΒΑ, ΑΒ, ΑΒ, ΑΒ, ΑΒ). With the majority of switched words being switched only once (28) but reappearing more than once (31), it is common for reappearing switched words to display block-like numbers of repetitions of one or two forms or of both before and/or after the switch.[211] This suggests that the letter words tend to be both used with a particular letter order and

210 More precisely, of those 18, eight words have eight reappearances, four words three, two words four, another two words five, and one each have eight and nine, respectively.

211 Cf. the following strings of introductions and subsequent reappearances for switched multilettered letter words that reappear more than once and display block-like repetitions: ΑΒ, ΑΒ, ΒΑ, ΒΑ (I 8), ΔΕ, ΔΕ, ΕΔ, ΕΔ (I 8), ΑΔ, ΑΔ, ΔΑ (I 8) ΑΕ, ΑΕ, ΕΑ (I 9), ΓΔ, ΔΓ, ΔΓ (I 11), ΓΕ, ΓΕ, ΕΓ (I 11), ΖΓ, ΖΓ, ΓΖ, ΓΖ, ΓΖ (I 11), ΓΘ, ΓΘ, ΘΓ, ΘΓ, ΘΓ, ΓΘ (I 12), ΓΒΑ, ΓΒΑ, ΓΒΑ, ΓΒΑ, ΓΒΑ, ΑΒΓ, ΑΒΓ, ΑΒΓ, ΑΒΓ (I 13), ΑΒΔ, ΑΒΔ, ΑΒΔ, ΔΒΑ, ΔΒΑ, ΔΒΑ (I 13), ΕΒΔ, ΕΒΔ, ΕΒΔ, ΕΒΔ, ΔΒΕ, ΔΒΕ, ΕΒΔ, ΕΒΔ (I 13), ΑΒΕ, ΑΒΕ, ΕΒΑ, ΕΒΑ (I 13), ΑΒΓ, ΑΒΓ, ΑΒΓ, ΓΒΑ, ΓΒΑ, ΓΒΑ (I 14), ΔΕΒ, ΔΕΒ, ΔΕΒ, ΔΕΒ, ΒΕΔ (I 15), ΑΕΔ, ΑΕΔ, ΑΕΔ, ΑΕΔ, ΑΕΔ, ΑΕΔ, ΑΕΔ, ΑΕΔ, ΑΕΔ (I 15), compare these with the only other two instances of back-and-forth switching besides the ones mentioned from I 1 and I 17: ΒΓ, ΓΒ, ΒΓ, ΓΒ, ΓΒ, ΓΒ (I 14) and ΒΓΔ, ΔΓΒ, ΒΓΔ (I 20).

switched for a reason, and to be used in the switched form until another reason for switching the order arises, rather than to be switched back and forth randomly. Thus, there seems to be, despite the possibility of switching the letter order all the time, some standardization in the application of the letter order. This conclusion, which is possibly significant regarding the question of the standardization of the diagram, is based on the individual consideration of the number of switches and reappearances in switched reappearing words, whereas the average numbers of switches, e.g., per reappearing word (0.45) or per switched reappearing word (1.52), would not be very informative.

The above-raised question of when the switch(es) occur(s) in relation to the number of reappearances, i.e., at what reappearance(s) the switch(es) occur(s) in relation to the number of instances of the word preceding in pre-switch form and following in switched form, has become obsolete in view of the above consideration of the time and place at which switches occur in the proposition: they have turned out to occur almost invariantly in the proof part. In any case, as one can see from the above examples of strings of reappearances, the switches occur at various times relative to the presence of a letter word in the proposition, either right after the introduction of the word, at the last instance of a number of reappearances, or after the word has occurred a couple of times and before recurring again once or more times.

To conclude from the analysis of the 'permutation' of the letter order in the introduction and reappearance of multi-lettered letter words: first, we have noted a strong adherence to the alphabetical order both in the introduction and in the reappearances of the multi-lettered words, with the proportion of words alphabetically ordered being similar in either case. While the alphabetical ordering principle is the predominant one, both introductions and reappearances of multi-lettered letter words display a considerable, stable, and similar number of words deviating from the alphabetical principle regarding their letter order. From this rough congruence of the numbers of alphabetical and permuted letter order in the introduction and in the reappearances of letter words, however, one cannot conclude that the introduced letter words keep their original form when reappearing. Rather, alphabetically introduced words for the most part retain their alphabetical order although they get switched in nearly one fifth of all cases, while non-alphabetically introduced words are switched so as to recur alphabetically in their reappearances in a much larger fraction; in the process, the evolving similar ratio of alphabetical and permuted forms in the reappearances of multi-lettered letter words is due to the fact that the overall number of alphabetically introduced words and reappearances of those words is larger than that of non-alphabetically introduced words and their reappearances.

Second, regarding the distribution of alphabetical and permuted multi-lettered words along the proposition, it has turned out, as far as the introductory forms are concerned, that their deviation from the alphabetical order in the introductions is seen to take place only after the exposition, where the introduction of letter words is mostly alphabetical. Further, it has appeared that there is no overlap of multi-lettered words which newly introduce letters and such that are ordered alphabetically. So the difference between the exposition part which for the most part contains alphabetically ordered letter words, and the subsequent parts, which contain both alphabetically and non-alphabetically ordered letter words, cannot be traced back to the presence or absence of letters to be introduced. As far as the reappearances of multi-lettered words are concerned, deviation from the alphabetical order is mostly confined to the proof part, which, however, is also the part where multi-lettered letter words tend to recur at all, so the mere occurrence of most deviations in the proof part is in itself not significant.

In any case, one cannot align the comparison of the introduction of letter words and their reappearances regarding their deviation from the alphabetical principle with a comparison of the different parts of the proposition (e.g., exposition vs. the following parts or the constructing section vs. the proof part) regarding the deviation of letter words from the alphabetical principle, as the occurrence of introductory and reappearing forms overlaps in the different parts of the proposition: letter words continue to get introduced beyond the exposition and even beyond the construction part, and begin to reappear as soon as they have been introduced – although, as has just been emphasized, by far the largest amount of reappearances is in the proof part.

Third and finally, the frequency and manner of change in the reappearances of introduced letter words, as evolving from an analysis of the number of switches and reappearances per word, have been considered: while the minority of words which reappear gets switched at all, many of those that are get switched only once, or else only few times (up to four), while the number of reappearances tends to exceed the number of switches. The result of this is that the reappearances, if followed chronologically through the proposition, are often repeated in the same forms before and/or after being switched, but are not switched back and forth all the time. This seems to suggest a special importance of the change of letter order, and of the application of the particular types of letter order.

Having thus outlined under the present point how far the order of the letters in the multi-lettered letter words observes the alphabetical principle and where, when, and to what extent permutation takes place, both when the letter words are first introduced into the proposition text and when they are repeatedly referred to, it is under point 4 of my study of the use of letters in *Elements* I (Section 4.4.4) that I shall return to the subject. There, I shall comprehensively consider possible reasons for and functions of the quantitative standardization of the alphabetical order and deviation from it, and of the limited number of changes to the letter order overall as well as regarding the individual word. The focus of examining the reasons and motivations for applying a certain letter order will be whether the application and change of the letters betrays an underlying diagram. The question will be whether the diagram is described on the basis of an underlying diagram, or whether the text is supposed to enable the construction of a diagram, and whether it takes the perspective of construction and description throughout, or whether it changes once the construction has been accomplished.

Conclusion

To conclude, 'omission' (of letters from the alphabetical order of letter introduction) and 'permutation' (of letters in the introduction of multi-lettered letter words) appear to be distinct practices of deviating from the alphabetical attribution of letters to objects in the *Elements*. While the first, omission, describes deviation from the alphabetical only as it seems to ensue on the explicit level of text, the second, permutation, refers to the explicit change of the alphabetical order within the confined limits of multi-lettered letter words; implicitly, however, the text always pays attention to the alphabetical order. Both practices of deviation from the alphabetical letter attribution, omission and permutation, are part of larger phenomena which I need to consider within a larger conceptual framework in view of diagrammatic standardization.

The 'omission' of letters from the sequence of alphabetically introduced letters has turned out to be part of the overall problem of letter specification. In this context, the omission of alphabetically due letters is one possible indicator for the underspecification of letters, although, as has been pointed out, initially omitted letters may be specified in time or may not be in need of specification. In any case, omission provides a soft argument for the reliance of the text on a diagram; besides, also letters which have not been omitted from their alphabetical position may turn out to be underspecified. Thus, while the omission of letters from the string of alphabetically introduced letters provides a means for retrieving one kind of potentially underspecified letters, the analysis of the sequences of introduced letters regarding alphabeticality provides only one starting point for the investigation of letter specifications regarding diagrammatic standardization in the following section. The 'permutation' of letters in multi-lettered letter words has been shown to be not restricted to such words as they are being introduced into the proposition, but to affect their application throughout the proposition. The relation of the form in which they are introduced and the form(s) in which they are subsequently employed is yet to be determined. It is in the section on the order of the letters that I shall take possible motivations of letter ordering in multi-lettered letter words into view and analyze the principles of letter ordering in introduced and reused letter words regarding the question of diagrammatic standardization.

The alphabetical principle of letter introduction crucially underlies both subjects that we will explore in the following parts of the chapter: the specification of the letters and the order of the letters within letter words. This is because the deviation from the alphabetical principle could be apt to generate meaning in either context and in this way might be inadvertently indicative regarding the question of the standardization of the diagram.

4.4.3 The specification of the letters

Netz's concept of letter specification

I shall now take into view what Netz calls "fixation of reference" or "specification" of the letters,[212] that is, the establishment of the relation between the *letters in the text* and the *objects in the diagram*.[213] As we have seen, the diagrammatic letters in the text evoke a diagram. Since their introduction into the text adheres to an alphabetical principle, we were able to see that not all letters of the envisaged or physically included diagram are mentioned in the text, but that all of them are alphabetically implied. Those letters that are mentioned in the text refer to a particular object in the diagram. The question now is whether the text unanimously locates these objects and letters as part of this diagram.

In order to answer the question of the origin of the diagrams it is crucial to determine whether the text alone standardly undertakes the specification of all letters which are to occur in the diagram that it describes. For if the diagrams were not standardly included the *Elements*' propositions, it would be necessary for the proposition *texts* to do all relevant specifications so that the diagram can be drawn and lettered on account of it. Netz has shown that, by and large, this is not the case.[214] Rather, he claims that a great number of letters do not become fully specified. However, he does not say what he understands by "specification" or how he thinks it occurs. In any case, it does not happen "per se,"[215] but through the introduction of the letters.[216]

Netz mentions three methodological prerequisites for what, in his view, "counts as a possible moment of specification."[217] First, he claims that derivative assertions cannot, at the same time (i.e., simultaneously with deriving), specify, as specifying assertions would present definitions instead of derivations. He concludes that, "[r]oughly speaking, specifications occur in the imperative, not in the indicative."[218] This sounds rather vague: on the one hand, it does not seem to preclude that specifications ever occur in another mode but in the imperative, e.g., in the indicative or by means of non-finite verb forms; on the other hand, it does not say anything about the possibility of specification in the definition of the goal, which fulfills the criterion of being non-derivative, but which usually happens in the indicative.

212 Netz 1999a, 19 f.

213 Netz himself does not explicitly define "specification," but his invariant use of the terms "specification" and "fixation of reference" (*passim*) suggests this latter, general meaning.

214 His study of the letters, just like all his statistical evaluations in his 1999a book, does not comprise the whole of the *Elements*, nor is it confined to this work, but takes into account a greater sample of parts of elements texts by different authors (1999a, 68–88).

215 Netz 1999a, 25.

216 For the necessity of connecting conceptually the two processes of alphabetical introduction and specification of letters, which Netz does not undertake, see below pp. 311 f.

217 Netz 1999a, 21.

218 Netz 1999a, 21.

The second claim regarding specification is that "letters are specified by other letters."[219] Again, it is not clear whether this apodictic statement claims exclusive validity or whether it is the focus of Netz's approach.[220] According to it, to specify a letter would be to describe unanimously how it (and the object it refers to) is related to other letters (and their corresponding objects). However, this is not always the case; especially in the simplest diagrammatic configurations letters are usually specified with regard to other objects, which do not necessarily have to be lettered themselves.[221] More problematically, Netz bases his method of judging the specification of a letter relative to its preceding letters on a false assumption: that all preceding letters are fully specified.[222]

The third methodological claim that Netz makes about the analysis of letter specification is that the benchmark for determining the degree of specification is, in any case, the "purpose of the proposition," not absolute unambiguousness.[223] The idea of '(un)necessary unambiguousness' in relation to the proof's functioning successfully will play an important role in Chapter 5, but I shall already in the subsequent part of the present chapter consider whether the letters are specified beyond the purpose of the proof with regard to the question of the standardized reliance on diagram.

Netz distinguishes four categories of letters with regard to their degree of specification, which he, again, refrains from defining. I shall repeat his simple made-up examples, which he provides in the place of a definition,[224] in order to present his conception of letter specification, although it should be noted that with regard to actual examples from the *Elements*, the categories do not appear as clear-cut as they do from the following out-of-context letter uses. First, "complete specification" appears to be the case if a letter unambiguously points to a specific point in the diagram, e.g., if 'A' is given as the center of a circle. Second, "underspecification" can be observed if a letter is partly determined as to its position in the diagram in relation to other letters,

219 Netz 1999a, 22.

220 Cf. ibid.: "I have concentrated on *relative* specification, specification of a letter relative to the preceding letters."

221 Cf. Netz's own fictitious example of a fully specified letter: "Let there be drawn a circle, whose center is A" (1999a, 20). In this case, A is fully specified only with regard to the unlettered circle.

222 His justification of the method is not persuasive: on the one hand, he seems to suggest that, since a letter can always gain more specification in the course of a proposition, it can, though it is not fully specified the moment it is relevant for a subsequent letter's specification, retrospectively specify it: "I have been like a very lenient teacher, who always gives his pupils a chance to reform" (1999a, 22), in which light the whole point of determining the specification of a letter at a particular point – its introduction – is no longer clear. On the other hand, the assumption that, in this way, "the statistical results are more interesting," because "[o]therwise, practically all letters would turn out to be underspecified in some way" (1999a, 22), cannot be the reason for choosing an approach that has the letters appear more specified than they really are. Besides, I do not believe that one ends up with a lot of unspecified letters; for my own approach toward letter specification, see below, pp. 307–314.

223 Netz 1999a, 20 and 22 f.

224 Netz 1999a, 20 f.

but cannot be located exactly, with a limited number of possibilities remaining, e.g., if the line 'BΓ' is specified as being the radius of a circle, but if, by this specification, it is not clear which of the two points referenced by B and Γ, respectively, is the center of the circle, and which is on the circumference, and where. Third, "complete unspecification" seems to mean that a letter is mentioned without any information as to its whereabouts, e.g., if, in addition to the situation just mentioned of BΓ being the radius of a circle, ΔB is given as a line double the size of BΓ, in which case it is completely unclear where Δ is supposed to lie in relation to the circle and in relation to Γ.[225] The fourth category applies if letters "change their nature through the proposition" in terms of increasing specification, with their status regarding specification changing from unspecification to underspecification, or from underspecification to complete specification, or from unspecification to complete specification.[226]

In Netz's sample from *Elements* XIII, "about 47% of the letters are completely specified, about 8% are underspecified, about 19% are completely unspecified, and about 25% begin as completely unspecified or underspecified, and get increased specification later."[227] According to Netz, the results "have little quantitative significance," as they vary greatly even within parts of the same text.[228] However, the number of fully specified letters is quite stable, amounting to less than half of all cases.[229] This, in turn, means that elements propositions regularly seem to contain a considerable number of letters which are not fully specified. From this, Netz concludes that in many propositions, "the diagram is not directly [in fact, not at all] recoverable from the text,"[230] for it contains a considerable amount of letter information not specified by the text. Therefore, according to Netz's classification, unspecified or underspecified letters provide a strong argument for an underlying diagram. As Netz concludes, "while the text is being worked through, the diagram is assumed to exist. The text takes the diagram for granted."[231] It would be then precisely because of the standardized underlying diagram that the text would be able to dispense with complete specification and would be, for once, non-standardized and underdetermined.

Letter specification and the standardization of the diagram

As to the specification of the letters, Netz's systematization of degrees of letter specification is directly relevant to the present question of what the use of letters reveals about the standardization of the diagram in the *Elements*. The gist of his analysis is

225 Netz 1999a, 21.
226 Netz 1999a, 21.
227 Netz 1999a, 23.
228 Netz 1999a, 23.
229 Netz 1999a, 24.
230 Netz 1999a, 19.
231 Netz 1999a, 25.

clear: the fact that the *Elements* contains a large portion of un(der)determined letters points to the preexistence of a diagram. However, at a closer look, the analysis is not conclusive in this regard. For Netz directly concludes, from the average number of not-completely-specified letters evolving from a statistical evaluation across propositions, the standardized existence of diagram. This conclusion, however, is far from evident. This is because it is not clear how the numbers for the different classes of letters which deviate from text to text should indicate the standardization of diagram. Even though the number of unspecified or underspecified letters has been said to be quite stable, it is not clear why this number of all things (slightly more than 50%) indicates a regular use of the diagram, and, conversely, how much higher the number of completely specified letters would have to be for it to imply the opposite, that is, the non-standardization of the diagram. On the same note, it is not clear of what use the distinction between unspecified and underspecified letters is regarding the standardized preexistence of diagram, and what the letters which change as to their degree of specification imply about the standardization of diagram. For Netz has pointed out that the numbers for these subtypes vary quite a lot not only between authors, but also between works or books of the same author.[232] In view of this subclassification of underspecified letters, the question of what the greater or lesser standardization of highly differentiated letter types means for the standardization of the diagram is even more urgent.

To be sure, the present study of standardization in the *Elements*, too, uses a loose definition of 'standardization,' which does not tie the existence of 'standardization' necessarily to a definite regularity of whatever percentage, but instead observes tendencies of unification and detailedness.[233] However, this applies above all to the analysis of the standardization of linguistic, stylistic, and structural, i.e., in a narrow sense of the word, textual phenomena: accordingly, I have argued in Chapter 3 that statements about standardization by degree on the basis of overall quantifications will be meaningful. In the case of diagrams, the situation is different, as we cannot assume that the diagrams materially exist to begin with and scrutinize them directly regarding the standardization of their application and form. Rather, we have to analyze them through the text, which, ironically, we have to expect to be conclusive regarding the standardization of the diagram not only on account of its own standardization, but also in some respects on account of its own non-standardization.

However, such an analysis, if pursued for the sake of determining diagrammatic standardization, is not conclusive if done purely on the basis of overall quantification by degree. This is because my primary question of the *existence* of the diagram, which is the one which could be answered with the help of letter analysis,[234] cannot be an-

232 Netz 1999a, 23 f.

233 For my conception of the term, see pp. 2, 30, 64 f., 108–110.

234 If at all, the analysis of the use of the letters can divulge something about the *existence* of diagrams; it cannot betray anything about the standardization of the *form* that the diagrams had if they existed. I had hoped to answer this question by the analysis of the textual construction, which how-

swered in general 'by degree,' but only with respect to a particular proposition in terms of 'either or' in view of a certain proposition: i.e., either the letters of a particular proposition are sufficiently specified so as to enable the diagram to be drawn, or the letters are not sufficiently specified for drawing the diagram and indicate that the diagram is not meant to drawn but is already included in the text. It is only on the basis of determining the letter specification of every single proposition in this way that it makes sense to quantify the specification of the letters by degree, that is, by determining the number of propositions whose letters are (not) sufficiently specified rather than the overall number of specified and un(der)specified letters. Even then, the result in terms of percentage would have to be judged differently than in the case of textual standardization. In order for the letter specification to be indicative of underlying diagrams, it is not even necessary that the letters of the majority of propositions (however that is defined) are not sufficiently specified. This is because the inclusion of the diagram is a crucial decision which I expect to be made once and for all so that a lack of specification should indicate that diagrams were generally included, not only in the respective propositions. If the application of the diagram should vary, then supposedly not randomly, but only with regard to certain types of propositions: e.g., the diagram might have been lacking in the first, rather simple propositions of the work but supplied from a certain degree of complexity onward. It is hardly possible to define a number of propositions with insufficiently specified letters which would unequivocally indicate that diagrams were present. However, we may recall that the standardized application of letters to the text indicates a standardized practice, either in terms of a standardized practice of material inclusion or construction on the basis of the textual letters. Thus, it would be a somewhat significant number of proposition cases with underspecified letters which are not limited to a certain type of proposition, which would be indicative regarding the standardization of diagrams. The number would be significant if it excluded the possibility of randomness. The necessity of analyzing the specification of the letters per proposition is also clear from the fact the letters themselves are not used with a standard meaning across propositions but have to be specified anew in each proposition. The specification of the letters is therefore only conclusive if viewed individually per proposition, and if, subsequently, the overall evaluation is made regarding the number of propositions which are unambiguous regarding the letters in them.

The other main problem with drawing conclusions about the standardization of the diagram from the statistical evaluation by Netz besides its overall, cross-propositional approach is the insufficient focus on the question of the possibility of constructing the diagram on the basis of letter specification, which necessitates a restriction of the analysis to parts of the proposition. The focus within the individual proposition must be what

ever, as we have seen, turned out to be inconclusive both regarding the standardization of the existence and of the forms of diagrams.

we have called the 'constructing section.' For our central question is whether the text alone unambiguously locates the objects referred to by the letters it mentions, in order for them to fully determine the diagram, so the construction of the diagram and, *on account of this,* the proof can be accomplished unambiguously based on the diagram, or whether the letter specification does not allow for a timely diagrammatic construction. Whereas we have debated the question of whether the diagrammatic *construction* (i.e., in terms of the kind and quality of the drawing) is relevant the proof, I am currently presupposing the proof-relevance of the diagram *as a letter configuration.* This seems to be evident from how the proof is accomplished: the deductive assertions are made with regard to a particular diagrammatic letter configuration, and they can only be made and understood with the help of a diagram that is physically present. Some of the simplest propositions might be accomplished just by imagining the letter configuration, but even they presuppose that one is able to imagine the place of each letter and point based on its specification.

It follows that from our perspective of the standardized inclusion of the diagram, Netz's notion of letter specification must be readjusted regarding its definition, its necessary scope and degree, its linguistic means, its time and place within the proposition, its object, and the method of testing it:

(1) Regarding its definition: a letter is (fully) specified if the point it refers to can be drawn or located unambiguously and if thus the letter can be attached unambiguously to the respective point in the diagram. That is, 'specified' means nothing more than 'localizable.' The diagram as a whole, in turn, can be drawn unambiguously, if all of its letters are 'localizable.' The ultimate proof of their specification or 'localizability' is that the subsequent proof functions on the basis of that letter's specification.

(2) Regarding the necessary scope of specification, i.e., the admitted percentage of not (fully) specified letters per proposition: one single unspecified letter suffices for the diagram to be impossible to be drawn and for the proof to fall apart.

(3) Regarding the degree of the specification of each letter: the focus on the feasibility of the proof entails that it does not matter whether a letter is a little or completely underspecified, but only whether it is fully specified so that the diagram can be drawn, or whether it is not so the diagram cannot be drawn. Unspecification and underspecification cannot be distinguished from one another in this respect. For this reason, I am going to apply the two terms synonymously. Netz's fourth category of letters (those which become gradually more specified) is relevant for the question of diagrammatic standardization insofar as it depends on whether and if so when the letter becomes fully specified, whether the letter is to be regarded as fully specified or not. For regarding the place and time of specification, it is vital that all letters be (fully) specified before the proof, insofar as the proof part presupposes them. That is, specification by means of later additional construction assignments within the proof section is possible from this point of view, as long as its information is not needed in

the foregoing part of the proof. With regard to gradually specified letters this means that if a letter that is introduced as unspecified or underspecified according to Netz's classification becomes fully specified only as late as the proof part, although its specification is needed by the beginning of the proof, it should be regarded simply as unspecified and, consequently, as strong evidence for the preexistence of diagram. But if it becomes fully specified in time, that is, by the end of the first, 'constructing' part of the proposition succeeded by the proof proper, it is, effectively, (fully) specified. Its former underspecification does not conclusively indicate diagram reliance, though it could provide some soft evidence for it. Letter specification which happens only gradually impedes the step-by-step construction of the diagram; thus, in such cases, the diagram cannot be constructed according to the individual construction assignments, but only after the whole construction has been perused once completely. However, in order to regard specification which happens fully, but only gradually before the proof part, as conclusive evidence, one would have to know precisely when the diagram would be expected to be constructed, that is, as a whole by the end of the first part or rather simultaneously step by step. Though it will not be possible to answer this question definitively, a close reading of the respective propositions will offer some useful hints. At present, it would seem that our method allows for gradual specification *until a certain point*; that is, a letter which is not (fully) specified right away does not necessarily count as unspecified, and the category of letters that are gradually specified is only relevant temporarily.

(4) Regarding the verbal mode of specification: from the considerations about the time and place of the kind of specification which would enable the actual performance of the diagrammatic construction assigned, that is, relative to the proof, the assertion that specifications take place in the imperative turns out inadequate in two ways: first, imperatives do not necessarily contribute to (relevant) specification which *must ensue* before the proof builds on it; second, as specification *can happen* anywhere before the proof builds on it, it can also happen in the definition of the goal, which usually does not take the imperative.

(5) Regarding the object of specification: the method of analyzing specification must take all letters into account, both those that are explicitly mentioned as well as those that are omitted but alphabetically implied. Of the letters which have been implicitly used, those that will not reappear later will be irrelevant as regards specification; but those which do reappear must count in any case as unspecified from a perspective which only takes into account the letters one after the other as they materialize in the text. This is true at least as regards the uninformed perspective of a lay reader. However, by the versed reader who is familiar with the standardized principle of alphabetical letter introduction, the omitted letter will be understood as (fully) specified in most cases, even if they occur as late as the proof part. To be sure, as far as those of the omitted letters are concerned which are not unambiguously specified by their 'implicit introduction,' the above-mentioned rules apply: they only count as fully specified if they explicitly and localizably (re)appear before the beginning of the proof part. Thus,

an analysis of the specification of the letters with regard to the 'drawability' of the diagram as a letter configuration would benefit from regarding the introduction of the letters as a part of their specification and from taking as a starting point the letters in the order in which they effectively occur before the background of the alphabetical principle of letter introduction. The majority of the omitted letters, that is, those which are omitted in a standardized way and thus can be supplied, do not present hard evidence for the standardization of diagram, but they nevertheless point in the direction of the preexistence of diagram, as has been explained in the previous section.

(6) Regarding the means of specification: it has already been pointed out above that the concept of 'letters that are specified by other letters' on the premise that all previous letters be regarded as fully specified is problematic. It becomes clear now why: the only criterion for determining whether a letter is specified, that is, 'localizable,' is whether it can be attached unambiguously to a point in the diagram before the proof begins, and within this time frame, preferably at the very moment that they are introduced. For this, it is crucial that the underspecification of foregoing letters is not ignored; rather, the underspecification of a letter may lead to the underspecification of all or some of the subsequent letters, and in many cases, it is only just the lack of localizability of one letter which indicates the unspecification of a previous letter as it is precisely due to this unspecification. To be sure, this subsequent letter may also, if it is not itself affected by previous underspecification, 'specify back' previous underspecified letters. But as has been said, the development of a letter's specification is only relevant for the central question of this section until the proof refers to that letter, so if a letter's localizability depends on that of the previous letters, the localizability of these letters within this time frame has to be ascertained. Further, omitted but alphabetically implied letters contribute to a subsequent letter's specification, which means that letters are specified on the one hand by previously specified or implicitly specified letters and unlettered objects, and on the other hand by letters or unlettered objects that are mentioned later.

(7) Regarding the method of testing: from the previous points it follows that instead of considering abstractly the specification of letters as if in a vacuum, as it were, the whole concrete situation of specification which surrounds a letter has to be taken into account for determining its localizability. Ultimately, this can only be done by trying to locate the letter and draw its corresponding point in a diagram, a process which has to be repeated for every single letter that (implicitly) appears in a proposition.

To sum up: in order to investigate conclusively what Netz calls the 'specification of letters' with regard to the question of whether the *Elements*' propositions standardly came with diagrams or not, the propositions have to be analyzed individually with regard to the localizability of all of their letters and the drawability of the diagram (i.e., as a configuration of letters) on the basis of which the proof can be performed. This means that the focus of the analysis has to be whether the text enables the construction of the diagram on account of the specification of the letters, and the focus as regards the text would have to be the relation of letters and objects in the first, 'constructing' part of the proposition. With regard to this first part, again (and only with regard to it), one may attempt to distinguish between a 'summary constructability' and a 'step-by-step constructability,' with the lack of the former conclusively pointing to the standardization of diagram, and the lack of the latter providing at least soft evidence for it.

The determination of the localizability of the letters would have to be orientated toward the demands of mathematical necessity which Netz has usefully set as the benchmark for letter specification. However, while only the lack of localizability regarding this benchmark is conclusive, the deviation from absolute, non-necessary localizability constitutes soft evidence for an underlying diagram and should therefore be problematized: for if the readers are meant to construct the diagram themselves, they must locate the letter *somewhere* and might have to wonder whether the exact location matters or not. If, however, the diagram is already there, it makes perfect sense for the text not to localize exactly all of the letters if their exact location is not relevant at that moment or in general.

It is only with the focus on individual constructability that the analysis of letter specification will be conclusive regarding the question of the standardization of diagram. That is, it will be conclusive only with regard to those propositions which feature at least one unspecified letter which prevents the drawing of the diagram: those definitely must have come with a diagram. Another group of propositions will be indicative but not conclusive regarding the presence of a diagram: i.e., those which supply all information that enables the construction of the diagram somewhere in the first, 'constructing' part but which require rereading this part as the specification does not happen in the expected order. Finally, propositions which are fully specified as regards their letters do not provide any evidence as to the existence of a diagram: they do not necessitate a diagram, but nevertheless might have come with one.

Once we have classified the *Elements*' propositions in this way, how should interpret our results? First, if the number of propositions that belong to the first two groups is considerable compared with that which contain only fully specified letters (though not necessarily higher), this might suggest that an underlying diagram was common practice rather than utter exception. Moreover, a considerable number of propositions in which not all letters are fully specified could even indicate that the practice of supplying diagrams was standardized: since all propositions work by the same method, that is, by employing letters which refer to a diagram on which the proof-by-deduction is based, it seems unlikely that Euclid and his fellow mathematicians decide from proof to proof whether to materially supply a diagram or not.

An analysis of the 'constructing section' regarding letter specification

I shall analyze the specification of the letters in *Elements* I according to the above criteria and test each proposition as to whether a diagram (as letter configuration) is drawable on the basis of the localization of the points and corresponding letters before the proof part or before the letters become relevant for the proof. In doing so, I shall leave out all issues of technical construction which concern the exactness of the depiction of the properties and relations of the objects and focus exclusively on the exactness of the localizability of the points and letters by which the diagrammatic configuration may be set up. I will assess the specification of the letters with respect to the degree of specification which is necessary for the proof; besides, I will note even such underspecification which does not prevent the construction of the diagram that is relevant to the proof, but which nevertheless seems to point to the underlying presence of the diagram. The question of 'localizability' implicates the role of the recipient. I shall conduct the general evaluation without presupposing knowledge about drawing conventions on the part of the reader; subsequently, however, I will consider to what extent the presupposition of expert knowledge changes the results. I shall classify the results into three groups: propositions with (1) diagrams which can be drawn before the proof begins on account of the specification of the letters, (2) diagrams which can be drawn before the proof begins but which contain letters that are specified only gradually or not precisely according to the order of the construction, and (3) diagrams which cannot be drawn before the proof part because of the underspecification of one or more letters. Propositions that contain proofs by contradiction will not be included in the statistics but will be classified separately into propositions with (a) diagrams which can be drawn on account of the letter specification and (b) those which cannot because of the underspecification of the letters. Considering these propositions separately is necessary because they cannot be assessed by the same criteria: in propositions that contain proofs by contradiction, the letters do not need to be specified before the proof part because the construction mostly takes place only within the proof which is accomplished essentially by the construction; moreover, letters are underspecified for different reasons in proofs by contradiction.[235]

Let us now turn to the results of our survey. First, I shall focus on the majority of the propositions (38 of 47)[236] which do not prove by contradiction before turning briefly to those 9 which do. Of the 38 propositions without proofs by contradiction, more than half (21) contain diagrams which can be drawn before the proof part, as all of their letters are sufficiently specified in time.[237] In most of these, in turn, the specification of the letters leaves little room for variation as far as the attachment to a specific point is concerned. (It is because of the nature of the textual construction rather than of the insufficient letter specification that the general orientation of the diagram – on which side of the diagram all the points of a certain side of it are supposed to lie – and the proportion of the objects

235 Here, too, however, it is vital that the letters are specified within the proof part before any non-constructive deductive step builds on them.

236 That is, excluding the presumably spurious I 40.

237 I 4, 5, 9, 10, 13, 15, 17, 18, 19, 23, 24, 31, 32, 33, 34, 36, 41, 42, 44, 45, 48.

are mostly non-fixed.[238]) In 3 of the 21 propositions, there is more than one possibility of locating the letter and the respective point; the different possibilities result in a different look of the diagram but do not have consequences that are relevant to the proof: in I 9, Z could also lie above ΔE, i.e., between A and ΔE rather than below it (Fig. 33). In I 15, the orientation of the two lines AB and ΓΔ in relation to one another is not specified; the letters of one of them could be switched in the diagram without changing the validity of the proof (Fig. 34). In I 24, ΔH could lie on the side of ΔE rather than, as in our evidence, on the side of ΔZ (Fig. 35). The underspecification of the respective letters in these three propositions itself is not indicative of the underlying presence of the diagram, since all of them are sufficiently specified from the point of view of the proof. However, the uniform transmission of these diagrams is indicative in view of the constructional variation allowed by the letter specification. This is especially notable in the case of I 9, for which we do not only have the diagrams of the manuscripts but also, uniquely, two ancient ones on papyrus, all of which show the diagram in the same way as regards the position of Z.[239]

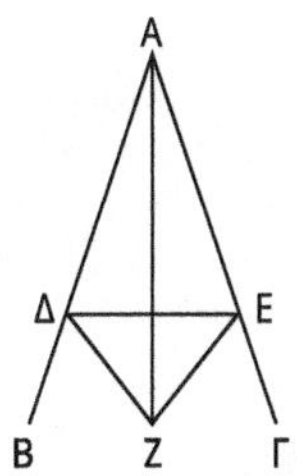

Fig. 33: I 9 according to Heiberg & Stamatis.

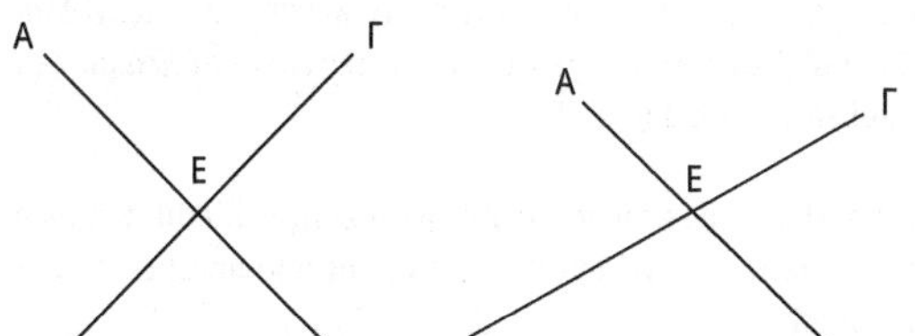

(a) (b)

Fig. 34: (a) I 15 according to Vat. gr. 190.1, fol. 16r; (b) I 15 according to MS. D'Orville 301, fol. 15r.

238 On the indeterminacy regarding the general orientation of the diagram cf. the part on the order of the letters below (Section 4.4.4, 411), on the silence of the textual construction about matters of size and proportion cf. above Section 4.3.3, 251 f.

239 See pp. 193 f. with Fig. 2 and 3 (a) for the depiction of I 9 on P. Herc. 1061 and P. Berol. [inv.] 17469 and Section 4.2.2, 223 for my description of the diagram in P.

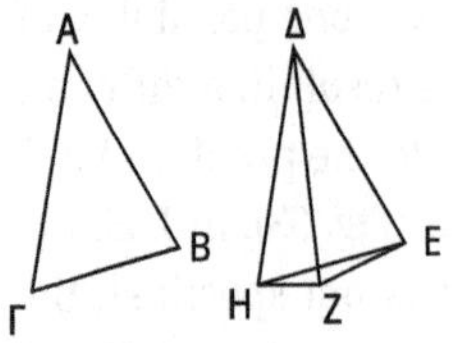

Fig. 35: I 24 according to Heiberg & Stamatis.

Another 7 of the 38 diagrams of our first group can be drawn and lettered unambiguously before the proof, but the localization of one or more of their letters is made difficult by their belated specification.[240] The reason for the belated specification of the letters is in most cases that the letters are not fully specified at the moment they are introduced, but only gradually in the further course of the construction. I am going to discuss the following example from I 1 in some detail, before referring to the remaining instances of gradual specification more briefly (Fig. 36):[241]

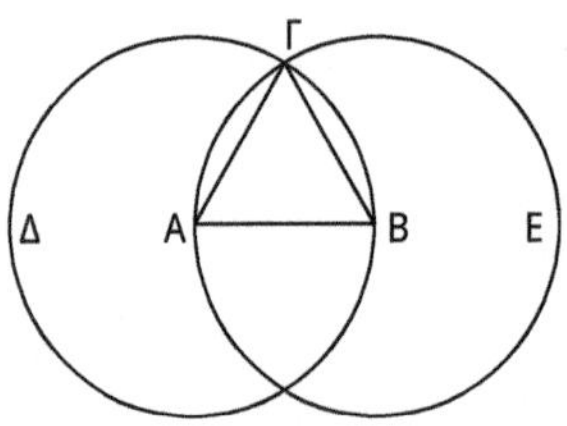

Fig. 36: I 1 according to Heiberg & Stamatis.

> Κέντρῳ μὲν τῷ Α διαστήματι δὲ τῷ ΑΒ κύκλος γεγράφθω ὁ ΒΓΔ, καὶ πάλιν κέντρῳ μὲν τῷ Β διαστήματι δὲ τῷ ΒΑ κύκλος γεγράφθω ὁ ΑΓΕ, καὶ ἀπὸ τοῦ Γ σημείου, καθ' ὃ τέμνουσιν ἀλλήλους οἱ κύκλοι, ἐπὶ τὰ Α, Β σημεῖα ἐπεζεύχθωσαν εὐθεῖαι αἱ ΓΑ, ΓΒ.
>
> With center A and distance AB let the circle BΓΔ be described; again, with center B and distance BA let the circle AΓE be described; and from point Γ, in which the circles cut one another, to the points A, B let the straight lines ΓA, ΓB be joined.

Point Γ is first introduced as part of the first circle BΓΔ, without further specification as to where on the circle it is supposed to lie, then mentioned again as part of the second circle AΓE, which implies that it must be one of the two intersection points, and only then explicitly referred to as a point of intersection (it is still not said which of the two it is, but this is the kind of non-proof-relevant variation allowed, as explained in the previous paragraph). The underspecification of Γ at its first mention is indicative of an underlying diagram, because the circle BΓΔ cannot be lettered simul-

240 I 1, 16, 20, 22, 37, 38, 43.
241 *Elem.* I 1 (vol. 1, p. 7.7–13 Heiberg & Stamatis).

taneously with the perusal of the construction – nor, from the point of view of the emergence of the text, could it have been lettered simultaneously with writing the text. Point Γ would have to be put *somewhere*, but only by the following assignment of the construction of the second circle ΑΓΕ it becomes clear that it could not have been put just anywhere but that it can only be one of two specific points. At the moment of its introduction, Γ is as underspecified as Δ of the same letter word (and as E of the following ΑΓΕ). But whereas the position of Δ and E on the respective circles turns out to be indeed free, the one of Γ is not. The case of Δ and E is also indicative although their specification is not needed, because at the moment of their introduction it is not yet clear that they really can be put anywhere. However, they would have to be put somewhere if the lettering of the diagram were to follow the order of the text. (If, in turn, it was assumed that all letters of a circle which are not further specified at the moment of their introduction can be applied anywhere on the circle, the immediate specification of Γ appears to be all the more urgent.) Thus, from the point of view of both the production and the reception of the text, the likeliest explanation for the belated specification of Γ and the belated 'specification' of Δ and E (i.e., in the sense that it becomes clear that they do not require further specification) is that the description was made on the basis of an already lettered diagram which was included in the text and was at the disposal of its first readers. In this way, it seems that while the text was being formulated, the letter 'names' for the circles were simply read from the diagram, and the need for specifying Γ and the other two letters was not felt. Nevertheless, the case of I 1 is inconclusive as we cannot presuppose that the reader was expected to accomplish the construction of the diagram indeed step-by-step simultaneously with the 'constructing section.' It could also have been done only after the perusal of the whole construction in terms of what we have called 'summary construction;' and as a whole, the construction in any case provides all the necessary information for constructing the diagram before the proof which presupposes the exact location of Γ – indeed, the letter already becomes specified later in same sentence in which it is introduced.[242] From the point of view of the production of the text, too, the case is inconclusive: the writing of the text could have been accomplished without the lettered diagram's underlying by envisaging the geometrical situation, which is possi-

242 The proposition has been mentioned as the notorious example for the hypothesis that the Euclidean proofs rely on the visualization by the diagram because of their logical incompleteness of their deduction (cf. Netz 1999a, 26 f.: "As is well known, the very first proposition of Euclid's Elements contains an implicit assumption based on the diagram – that the circles drawn in the proposition meet [. . .]. There is a whole set of assumptions of this kind, sometimes called 'Pasch axioms'"). My own interest, by contrast, is not in the independence of the proof from the diagram in logical terms, but in the self-sufficiency of proof in terms of letter specification. This means that I regard the proof as independent of the diagram if it works on the basis of the diagram constructed on account of the sufficiently specified textual construction. Accordingly, in terms of specification, the proof of the present I 1 does not rely directly on a diagram, but on the textual construction which states explicitly that the circles cut one another and enables the diagram which shows their intersecting one another.

ble in the case of I 1 because of the simplicity of the figure. However, this is a very theoretical scenario, and unlikely to have been the case.[243] The example, though inconclusive, supports our previous hypothesis in view of the impracticality of the textual construction that the textual construction introduces the relevant 'names' and objects and the object properties and relations on the basis of an already constructed and lettered diagram, rather than explaining how the diagram is to be accomplished. A similar example of gradual but timely letter specification, again concerning the intersection points of two circles, occurs in I 22 – with the same implications for the presence of a diagram (Fig. 37): the points K and Λ are first introduced as part of the circle ΔΚΛ but not specified as to where they lie on the first circle; it is only the construction of the second circle ΚΛΘ which makes it clear that K and Λ are the points of intersection of the two circles.[244]

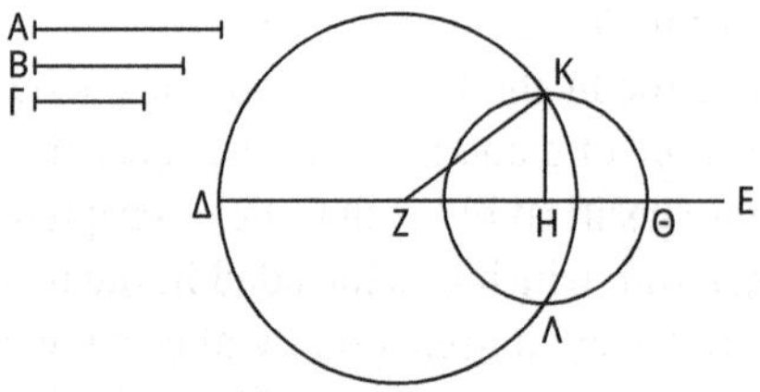

Fig. 37: I 22 according to Heiberg & Stamatis.

Two other examples of gradual specification concern the extension of a straight line toward another point. This point is only specified by a subsequent construction assignment which specifies the length of the new segment of the straight line: compare from the construction of I 16 (Fig. 38): καὶ ἐπιζευχθεῖσα ἡ ΒΕ ἐκβεβλήσθω ἐπ' εὐθείας ἐπὶ τὸ Ζ, καὶ κείσθω τῇ ΒΕ ἴση ἡ ΕΖ ("and let BE be joined and produced in a straight line to Z; let BE be made equal to EZ").[245] Similarly, the construction of I 20 (Fig. 39) demands: Διήχθω γὰρ ἡ ΒΑ ἐπὶ τὸ Δ σημεῖον, καὶ κείσθω τῇ ΓΑ ἴση ἡ ΑΔ ("For let BA be drawn through the point Δ, let AΔ be made equal to ΓA").[246] Points Z (in I 16) and Δ (in I 20) cannot be applied to a specific point of the diagram until after the subsequent construction assignment. These examples show that the boundaries between immediate and gradual specification are fluid, and therefore underline that a gradual specification within the construction part does not present conclusive evidence for the presence of a

243 Another possibility is even less likely: that the text was written on the basis of a sketch-like lettered diagram which, however, was not included into the text, for why would it have been written on a different material than the text or left out by a copyist?

244 καὶ κέντρῳ μὲν τῷ Ζ, διαστήματι δὲ τῷ ΖΔ κύκλος γεγράφθω ὁ ΔΚΛ· πάλιν κέντρῳ μὲν τῷ Η, διαστήματι δὲ τῷ ΗΘ κύκλος γεγράφθω ὁ ΚΛΘ ("With center Z and distance ZΔ let the circle ΔΚΛ be described; again, with center H and distance HΘ let the circle ΚΛΘ be described," *Elem.* I 22 [vol. 1, p. 31.11–14 Heiberg & Stamatis]).

245 *Elem.* I 16 (vol. 1, p. 25.6–9 Heiberg & Stamatis).

246 *Elem.* I 20 (vol. 1, p. 28.15 f. Heiberg & Stamatis).

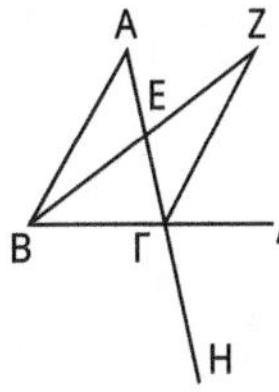

Fig. 38: I 16 according to Heiberg & Stamatis.

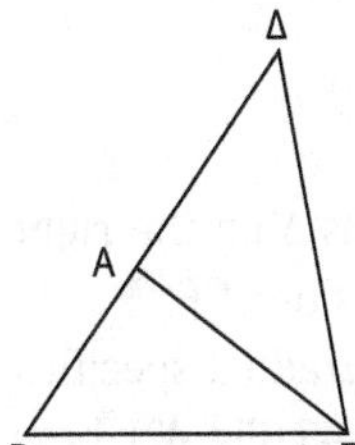

Fig. 39: I 20 according to Heiberg & Stamatis.

diagram: in both I 16 and I 20, the specification of the underspecified letters happens, as in the example from I 1, in the very next main clause. But whereas in I 1 the specification by means of the construction of the second circle is only implicit and requires recognizing that Γ is the intersection point of both circles, the lengths of the straight lines EZ and AΔ in the present cases are specified precisely for the purpose of specifying the exact location of Z and Δ, respectively. Therefore, these two points are specified, though gradually, in fact almost immediately.

Our fifth example (from I 37, Fig. 40) presents a case of gradual specification within the construction which must be assessed differently depending on the assumed readership. The objects introduced in the exposition are two triangles ABΓ and ΔBΓ which lie between the same parallels AΔ and BΓ. Subsequently, the following construction is assigned: Ἐκβεβλήσθω ἡ AΔ ἐφ᾽ ἑκάτερα τὰ μέρη ἐπὶ τὰ E, Z ("Let AΔ be produced in both directions to E, Z").[247] From this, it is not clear on which sides of A and Δ the new points E and Z are to lie, respectively. It is only indicated by the next double assignment of drawing a straight line BE through B parallel to ΓA, and a straight line ΓZ parallel to BΔ.[248] This shows that E must lie on the side of A, and Z on the side of Δ because there is no way how BE could be parallel to ΓA if E was situated on the side of Δ, nor could ΓZ be parallel to BΔ if Z was situated on the side of A. In this case, one may well assume that a reader familiar with the conventions of lettering the diagram would know that the objects were lettered preferably alphabetically

---

247 *Elem.* I 37 (vol. 1, p. 51.1 Heiberg & Stamatis).

248 καὶ διὰ μὲν τοῦ B τῇ ΓA παράλληλος ἤχθω ἡ BE, διὰ δὲ τοῦ Γ τῇ BΔ παράλληλος ἤχθω ἡ ΓZ ("through B let BE be drawn parallel to ΓA, and through Γ let ΓZ be drawn parallel to BΔ."), *Elem.* I 37 (vol. 1, p. 51.2 f. Heiberg & Stamatis).

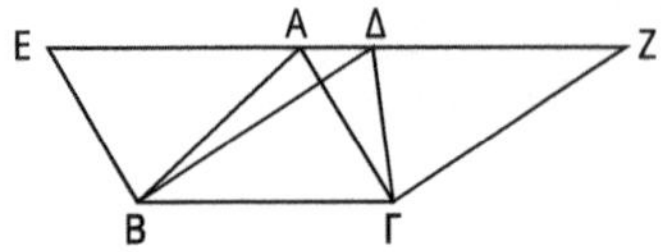

Fig. 40: I 37 according to Heiberg & Stamatis.

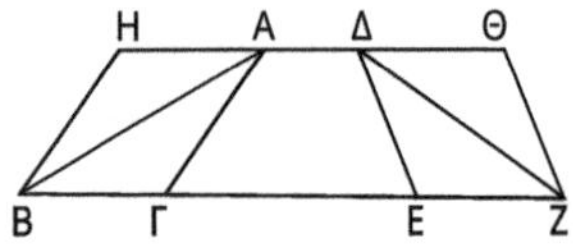

Fig. 41: I 38 according to Heiberg & Stamatis.

from left to right.[249] Accordingly, the diagram transmitted for I 37 has Δ on the right-hand side of A, Γ on the right-hand side of B, and Z on the right-hand side of E.[250]

The preceding five cases of belated specification all concern the gradual specification of letters. A different case of belated specification is presented by I 43 (Fig. 42): here, the respective letters are immediately rather than gradually specified, but they indicate a foregoing step of the construction in the course of which they have already been applied to the diagram. As this step is not mentioned, the letters are introduced only subsequently as part of the parallelograms. It is in this sense that their introduction can be called 'belated.' The letter words in question are EΘ, ZH which refer to two parallelograms. They are introduced in the exposition as follows: Ἔστω παραλληλόγραμμον τὸ ΑΒΓΔ, διάμετρος δὲ αὐτοῦ ἡ ΑΓ, περὶ δὲ τὴν ΑΓ παραλληλόγραμμα μὲν ἔστω τὰ ΕΘ, ΖΗ, τὰ δὲ λεγόμενα παραπληρώματα τὰ ΒΚ, ΚΔ· ("Let ΑΒΓΔ be a parallelogram, and ΑΓ its diameter; and about ΑΓ let EΘ, ZH be parallelograms, and BK, KΔ the so-called complements").[251] The letters of the letter words EΘ, ZH are immediately specified when they are introduced: they presuppose the knowledge that the two points in each case name the diagonal of the parallelogram; from this, it is clear that neither of them can be ap-

249 Here, I am anticipating one of the main results of Section 4.4.4 on the directions of the mutually related practices of lettering the diagram and of ordering the letters in multi-lettered letter words (Section 4.4.4, 405–411). On the reliability of the general orientation of the transmitted diagrams see Section 4.4.4, 411.

250 I 38 is an analogous case (Fig. 41). The diagram is the same as in I 37 except that the two triangles ΑΒΓ and ΔΕΖ which lie between two parallel straight lines BZ and AΔ do not share the same base. Here, the letters H and Θ are at first underspecified as to which one lies on which side of AΔ, and again, it is the construction of two straight lines BH and ZΘ parallel to ΓA and ΔE, respectively, which clarifies the position of H and Θ (*Elem.* I 38 [vol. 1, p. 52.3–5 Heiberg & Stamatis]).

251 *Elem.* I 43 (vol. 1, p. 57.7–10 Heiberg & Stamatis). Acerbi's edition follows the wrong text of Heiberg & Stamatis, which has ZΘ, ZH as references to the two parallelograms instead of EΘ, ZH, while referring to the two "complements" in the above way as BK, KΔ (i.e., the parallelograms that emerge as part of the area of the parallelogram ΑΒΓΔ besides the parallelograms which are "about" the diameter; on the term "complements" [παραπληρώματα] cf. Heath 1926, 341). The reference ZΘ cannot be correct as it refers to the same parallelogram called KΔ, but the parallelograms which are "about" the diameter and the "complements" must be different objects.

plied to the point which is subsequently named K. EΘ, ZH are only underspecified as regards the side of AΓ on which each of the two letters of the parallelograms are intended to lie, but either position does not impair the validity of the proof. Thus, the two references EΘ, ZH are unambiguous; however, they become comprehensible only from assuming that the diagram has been constructed and lettered beforehand. This is because in the diagram, other objects appear to have been lettered by the letters E, Z, H, Θ before than are introduced into the text by the same letters in the present passage. This is clear from the principle of the alphabetical lettering: the objects which apparently have been lettered in the diagram are the two straight lines EZ and HΘ from the intersection of which the two parallelograms EΘ, ZH emerge. The four letters are specified belatedly with respect to their apparent application to the diagram; the two references EΘ, ZH indicate the belated specification of their letters by ordering the four letters non-alphabetically (instead of EZ, HΘ), even though they newly introduce all four letters into the text. The implication of a constructional step by the exposition of the two parallelograms is notable in view of the overall non-constructional character of the 'constructing section' which almost never indicates how the objects introduced by the exposition and construction are supposed to come into being. Even here, the construction of the two parallelograms is not explained, but from the comparison with the diagrammatic lettering we get an idea of how it was accomplished[252] – or how it was meant to be accomplished: the case of EΘ, ZH strongly indicates an underlying diagram, but not conclusively, because, as in the previous examples, the lettering situation is manageable enough to have been envisaged by a knowledgeable reader.

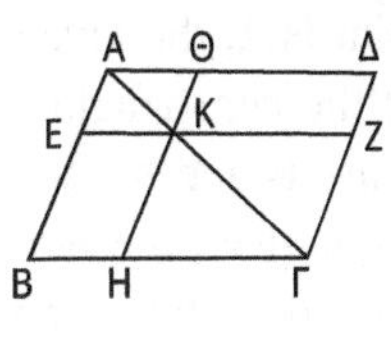

Fig. 42: I 43 according to Heiberg & Stamatis.

252 The present rare implication of a concrete constructional step, however, does not include any practical information about how the construction is to be accomplished, that is, whether the two straight lines EZ and HΘ are to be constructed technically or simply drawn parallel to AΔ/BΓ or AB/ΔΓ, respectively (see Section 4.3.3 on the impracticality of the construction). The identification of a constructional step implied by the exposition is noteworthy also with respect to the relation of exposition and the construction part: while the language of the exposition is non-constructional and that of the construction implies some diagrammatic action, the construction had turned out to be equally non-constructional as the exposition (see above, Section 4.3.3). Here, conversely, the exposition implicates a diagrammatic construction. This confirms our earlier hypothesis on the relation of the exposition and the construction part: while their language indicates a different function for each proposition part, the function of both is the mere introduction of the objects and object relations; constructional information is required in neither of the two parts, as the construction has already been done previously to the writing of the text, which is reflected by the occasional glimpses on the diagram's coming into being (such as the present one).

Let us now turn to the conclusive evidence: the diagrams which cannot be drawn before the proof (or before the proof builds on their respective parts) because one or more of their letters are not specified in time. They account for a considerable portion of the diagrams of those propositions which do not contain proofs by contradiction: 10 of 37 propositions contain at least one unspecified letter.[253] They are conclusive indeed because except in one case (I 28), the respective letters are unspecified in the eyes of both an expert and a lay reader; neither would be able to draw the complete diagram on the basis of the 'constructing section.' The unspecified letters only become specified in the proof part or even only in the conclusion,[254] or not at all. In the latter case their exact location could be concluded upon only by a retrospective process of elimination on the basis of the information of the proof. Since the problem of letter specification in all cases becomes immediately solved once the diagrams are taken into account and considered as underlying the text, the propositions containing underspecified letters present decisive evidence that the diagrams were present in the text from the beginning. I now consider four of the propositions which contain unspecified letters in great detail (I 2, 3, 35, and 47). In I 2 and 3, the respective unspecified letters appear within the construction in due alphabetical order; in I 35, the respective letter does not appear at all until the end of the proof, and is not implied by the order of letter introduction in the constructing section, because it would be due only after all other letters that are introduced; in I 47, two of the respective letters are left out from the alphabetical order (but implied are by it), one of which subsequently appears in permuted order within the construction, and one more letter appears unspecified in due alphabetical order.

I 2 is concerned with the problem "[t]o place at a given point (as an extremity) a straight line equal to a given straight line" (Fig. 43).[255] The given point is A, the given straight line is ΒΓ. The letters which remain unspecified throughout the construction, even though they turn out to be central for accomplishing the proof, are H and Λ; they are the letters which are each introduced as part of a circle: καὶ κέντρῳ μὲν τῷ Β διαστήματι δὲ τῷ ΒΓ κύκλος γεγράφθω ὁ ΓΗΘ, καὶ πάλιν κέντρῳ τῷ Δ καὶ διαστήματι τῷ ΔΗ κύκλος γεγράφθω ὁ ΗΚΛ ("with center B and distance ΒΓ let the circle ΓΗΘ be described; and again, with center Δ and distance ΔΗ let the circle ΗΚΛ be described").[256] Of the reference to the first circle, Γ has been specified before, while H and Θ are introduced unspecified. Θ does not play any further role; that its exact position does not matter, however, is only clear by the end of the proof (which turns out not to refer to it unless as part of the circle), and since not all the letters which need to be specified are indeed specified, one cannot be sure at the moment of the introduction of Θ that its non-specification is deliberate. Therefore, this letter gives some indication of the pres-

253 I 2, 3, 11, 12, 21, 28, 30, 35, 46, 47.

254 Cf. Netz 1999a, 158 on the impossibility of specification by means of deductive assertions.

255 Πρὸς τῷ δοθέντι σημείῳ τῇ δοθείσῃ εὐθείᾳ ἴσην εὐθεῖαν θέσθαι (*Elem.* I 2 [vol. 1, p. 8.11 f. Heiberg & Stamatis]).

256 *Elem.* I 2 (vol. 1, p. 8.19–21 Heiberg & Stamatis).

ence of the diagram, even though it is not conclusive. The letter H, by contrast, must lie on the straight line BZ which has been introduced before. However, this does not become clear from the rest of the construction.[257] Of the reference to the second circle HKΛ, all three letters are underspecified: H is the point which the circle shares with the first one but the position of which is otherwise unclear; K (just like Θ in the reference to the first circle) remains underspecified with no further specification needed; and Λ, analogously with H, must lie on the straight line AE, which has been introduced before in conjunction with BZ on which H is supposed to lie. As with H and BZ, however, the relation of Λ and AE remains unspecified. It is only in the proof part that it becomes clear that H must lie on a straight line with ΔB, and Λ on a straight line with ΔA:[258]

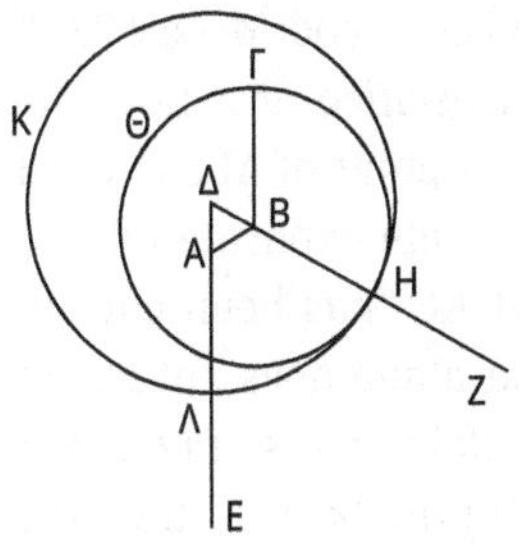

Fig. 43: I 2 according to Heiberg & Stamatis.

Ἐπεὶ οὖν τὸ B σημεῖον κέντρον ἐστὶ τοῦ ΓΗΘ κύκλου, ἴση ἐστὶν ἡ BΓ τῇ BH. πάλιν, ἐπεὶ τὸ Δ σημεῖον κέντρον ἐστὶ τοῦ ΚΛΗ κύκλου, ἴση ἐστὶν ἡ ΔΛ τῇ ΔΗ, ὧν ἡ ΔΑ τῇ ΔΒ ἴση ἐστίν. λοιπὴ ἄρα ἡ ΑΛ λοιπῇ τῇ BH ἐστὶν ἴση.

Then, since the point B is the center of the circle ΓΗΘ, BΓ is equal to BH. Again, since the point Δ is the center of the circle ΚΛΗ, ΔΛ is equal to ΔΗ. And in these ΔΑ is equal to ΔΒ; therefore, the remainder ΑΛ is equal to the remainder BH.

The reference to BH could indicate that this straight line must be part of the given straight line BZ, as it (BH) has not been constructed before. In this case, BZ would have to be long enough so as to cut the two circles (which, as with AE and the circle ΚΛΗ, was not specified by the construction). It is only by the next sentence that the overlap of BH with BZ becomes more cogent: it implies that ΔΑ and ΔΒ are part of ΔΛ and ΔΗ, respectively, and in doing so, that ΑΛ and BH are also part of ΔΛ and ΔΗ, respectively, which is explicitly stated by the next sentence that calls each of ΑΛ and BH λοιπή ("the remainder,"[259] i.e., of ΔΛ and ΔΗ). From this, it is clear that ΑΛ lies on AE, and BH on BZ, provided that AE and BZ extend beyond the circle ΚΛΗ which is still not explicitly

257 Cf. Thaer 1970, 421: "Daß G [i.e., Γ] ein ganz bestimmter Punkt des ersten Kreises, nämlich sein Schnittpunkt mit BF [i.e., BZ] sein soll, muß der Figur entnommen werden" (my own addition in brackets).

258 *Elem.* I 2 (vol. 1, p. 9.1–8 Heiberg & Stamatis).

259 On this translation cf. Heath 1926, 245.

said; but it is again implied by the mention of ΔΗ, ΔΛ and ΑΛ, neither of which was explicitly constructed, but which must be assumed to have emerged, like ΒΗ, from the circle's ΚΛΗ cutting ΑΕ and ΒΖ. The underspecification of Η and Λ is particularly conspicuous considering that the two points are central for solving the problem of the proposition: ΑΛ is in fact the very straight line which had to be constructed as equal to ΒΓ at point Α, and the equality of ΒΗ with the given straight line ΒΓ is decisive for proving the equality of ΑΛ with ΒΓ.[260]

In I 3, "it is thus required to cut off from ΑΒ the greater a straight line equal to Γ the less" (δεῖ δὴ ἀπὸ τῆς μείζονος τῆς ΑΒ τῇ ἐλάσσονι τῇ Γ ἴσην εὐθεῖαν ἀφελεῖν, Fig. 44).[261] For this purpose, a straight line ΑΔ equal to Γ is to be drawn at Α, and a circle ΔΕΖ with the center Α and the distance ΑΔ: Κείσθω πρὸς τῷ Α σημείῳ τῇ Γ εὐθείᾳ ἴση ἡ ΑΔ· καὶ κέντρῳ μὲν τῷ Α διαστήματι δὲ τῷ ΑΔ κύκλος γεγράφθω ὁ ΔΕΖ.[262] This is the end of the constructing section which leaves Ε unspecified: it is not clear that it is supposed to lie on ΑΒ and to be, thus, the intersection point of ΔΕΖ and ΑΒ. Then the proof shows that ΑΕ is equal to Γ, but still does not state explicate that ΑΕ lies on ΑΒ. It is only the conclusion which makes clear that ΑΕ "has been cut off" (ἀφῄρηται) from ΑΒ. However, the preceding proposition text alone does not in any case show this, because the diagram which is necessary for doing so is only one of several possible diagrams to emerge from the construction. On the basis of the present construction, which does not sufficiently specify Ε, it is only possible to prove the equality of ΑΕ and Γ. The initial construction problem, by contrast, can only be solved on the basis of the combination of the text and a previously constructed diagram by which the relation of ΑΕ and ΑΒ is specified. Therefore, this proposition presents conclusive evidence that it has been written on the basis of a diagram.

In I 35, several letters are unspecified or underspecified, and the construction does not necessarily produce the diagram on which the proof is based (Fig. 45). It may

260 Heath 1926, 245 notes that "[t]his proposition gives Proclus an opportunity, such as the Greek commentators revelled in, of distinguishing a multitude of *cases*." These "cases" of theorems and problems correspond to different figures which correspond to different constructions, but to the same method of proof. It is true that many of the Euclidean theorems and problems can be proved and solved by means of other constructions and proofs than the one provided by Euclid; however, it is important to note that the text of the Euclidean propositions generally is meant to allow for only one single construction and proof, even though it, because of the underspecification of its letters, might suggest that there are different possibilities of drawing the diagram and proving on the basis of it (as in I 35 which is among my subsequent examples, cf. pp. 324–327). The text only allows for such variability which has been pointed out above in the discussion of non-conclusive letter underspecification, i.e., slight differences in the look of the diagram which do not have any influence on the text of either the construction or the proof. By contrast, all eight cases distinguished by Proclus for the present I 2 "would require a slight modification in the wording (1) of the construction [. . .], (2) or the proof" (Heath, ibid.).

261 *Elem.* I 3 (vol. 1, p. 9.16 f. Heiberg & Stamatis).

262 *Elem.* I 3 (vol. 1, pp. 9.22–10.2 Heiberg & Stamatis).

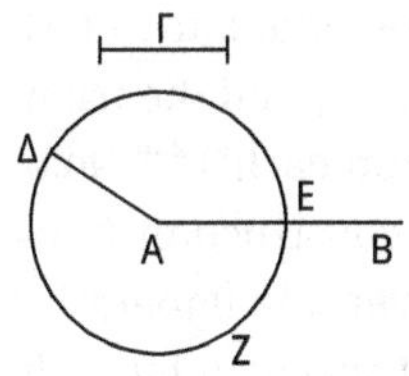

Fig. 44: I 3 according to Heiberg & Stamatis.

instead lead to a different diagram on whose basis the theorem of the proposition could also be proved, however, not by means of the present proof. The proposition claims that "[p]arallelograms which are on the same base and in the same parallels are equal to one another."[263] All letter specification is done by the exposition, as the proposition lacks a construction: Ἔστω παραλληλόγραμμα τὰ ΑΒΓΔ, ΕΒΓΖ ἐπὶ τῆς αὐτῆς βάσεως τῆς ΒΓ καὶ ἐν ταῖς αὐταῖς παραλλήλοις ταῖς ΑΖ, ΒΓ· ("Let ΑΒΓΔ, ΕΒΓΖ be parallelograms on the same base ΒΓ and in the same parallels ΑΖ, ΒΓ").[264] This means that Β and Γ lie on one of the parallels, and all remaining letters (Α, Δ, Ε, Ζ) on the other parallel, and that Δ and Ε must lie between Α and Ζ. However, the exposition does not specify the order of Δ and Ε on the straight line ΑΖ. The order of the letters on ΑΖ could be either Α, Δ, Ε, Ζ or Α, Ε, Δ, Ζ; accordingly, the parallelograms would overlap in different ways. Either, according to the first possibility and the transmitted version of the diagram, their overlap would produce another point from the intersection of the sides ΔΓ and ΕΒ, that is, point Η in the present diagram. This would result in the triangles ΔΗΕ and ΗΒΓ, and in the trapezoids ΑΒΗΔ and ΕΗΓΖ. Or, according to the second possibility (the letter order Α, Ε, Δ, Ζ), the overlap of the two parallelograms would not produce any point in addition to the ones mentioned by the exposition, but create the triangles ΑΒΕ, ΔΓΖ, and the trapezoid ΕΒΓΔ. Thus, two versions of the diagram are possible, but the proof will turn out to match only the former. Therefore, the letters Δ and Ε are underspecified by the exposition (in the absence of a construction).[265] Practically, Δ and Ε as part of their respective letter words could not be specified any more themselves. Rather, the underspecification of the two letters is a consequence of the non-specification of point Η: for by specifying Η, the exposition would have explicated the relation of the two parallelograms, and of points Δ and Ε. However, the point Η first occurs late in the proof as part of the reference to the

263 Τὰ παραλληλόγραμμα τὰ ἐπὶ τῆς αὐτῆς βάσεως ὄντα καὶ ἐν ταῖς αὐταῖς παραλλήλοις ἴσα ἀλλήλοις ἐστίν (*Elem.* I 35 [vol. 1, p. 48.10 f. Heiberg & Stamatis]).
264 *Elem.* I 35 (vol. 1, p. 48.12–15 Heiberg & Stamatis).
265 The order of the letters in the letter words ΑΒΓΔ and ΕΒΓΖ does not help to explain the order of Δ and Ε on ΑΖ. Apparently, the order of the letters in the reference to ΑΒΓΔ was determined first, that of ΕΒΓΖ second and in relation to ΑΒΓΔ. This seems to be the case regardless of the question of the exact relation of the position of the two parallelograms. On the implications of the letter order of ΑΒΓΔ and ΕΒΓΖ for the principles of ordering of the letters in the multi-lettered letter words, for the lettering of the diagrams, and for the presence of the diagram in Section 4.4.4, 362–367, where I discuss this proposition again with respect to the letter ordering in the references to quadrangles.

triangle ΔHE. There, it is used without having been specified beforehand: καὶ τὸ EAB τρίγωνον τῷ ΔZΓ τριγώνῳ ἴσον ἔσται· κοινὸν ἀφῃρήσθω τὸ ΔHE· ("[. . .] and the triangle EAB will be equal to the triangle ΔZΓ. Let ΔHE be subtracted from each").[266] Notably, the use of the perfect passive imperative formally indicates a construction,[267] by which H could be specified even as late as the proof part;[268] however, the imperative clause turns out not to contain any specification. The letter H is not specified but only mentioned. This is because the triangle ΔHE is not brought into being by means of this clause but is assumed to be already there and now 'subtracted' from each of the other two triangles; that is, the meaning of the verb indicates the opposite of a 'construction' and implies that the letters of the already present object are not in need of specification. In the exposition part, the non-specification of H is not in any case felt because the letter is simply not mentioned and is not missing from the alphabetical order of letter introduction as it would have been the last letter to be specified in alphabetical order. Its absence would only strike one as odd if the diagram happened to be constructed indeed according to the first of the above-mentioned possibilities, i.e., in the way in which it is presupposed by the proof, because the diagram would feature an unlettered point in this case, which is not common practice in the *Elements*. However, the question is whether someone trying to draw the diagram from the exposition would not have looked for a way of drawing the diagram which does not involve an unlettered point. The first time that the proof could implicitly indicate that it presupposes the present version of the diagram is where it asserts: διὰ τὰ αὐτὰ δὴ καὶ ἡ EZ τῇ BΓ ἐστιν ἴση· ὥστε καὶ ἡ AΔ τῇ EZ ἐστιν ἴση· καὶ κοινὴ ἡ ΔE· ὅλη ἄρα ἡ AE ὅλῃ τῇ ΔZ ἐστιν ἴση ("For the same reason also EZ is equal to BΓ, so that AΔ is equal to EZ; and ΔE is common; therefore the whole AE is equal to the whole ΔZ").[269] Here, the alphabetical letter order of ΔE could suggest that the two letters appear in that same order on AZ which is the case only in the present version of the diagram, because all other letter words in this passage are alphabetically ordered, tracing (like ΔE) the diagram from left to right, as it were. The other version of the diagram, by contrast, would require the non-alphabetical letter order EΔ in keeping

266 *Elem.* I 35 (vol. 1, p. 49.4 f. Heiberg & Stamatis). Heath 1926, 327: "Euclid speaks of the triangle DGE [i.e., ΔHE] without any explanation that, in the case which he takes (where AD, EF [i.e., AΔ, EZ] have no point in common), BE, CD [i.e., BE, ΓΔ] must meet at a point G [i.e., H] between the two parallels. He allows this to appear from the figure simply." The other diagram which is possible on the basis of the construction which does not yet mention H is considered by Proclus as another "case" of the proposition besides the one which turns out to has been given by Euclid (Heath 1926, 328, my additions in brackets).

267 See Section 3.3.2, 167 f., on the meaning of the verb forms with regard to the structure of the proposition.

268 See above, Section 4.4.3, 310 f. on the requirements for letter specification in terms of verbal mode and position within the proposition.

269 *Elem.* I 35 (vol. 1, p. 48.18–20 Heiberg & Stamatis).

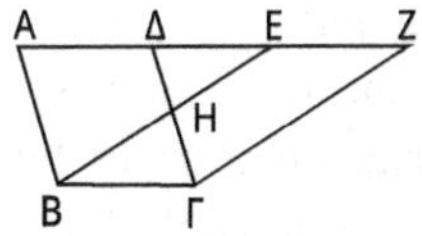

Fig. 45: I 35 according to Heiberg & Stamatis.

with the direction of lettering of the other straight lines.[270] This indication, however, is only recognizable by expert readers who know the practices of letter ordering that I explain in Section 4.4.4.[271]

In I 47 (Fig. 46), too, variably severe and indicative cases of underspecified letters occur in the construction part: Ἀναγεγράφθω γὰρ ἀπὸ μὲν τῆς ΒΓ τετράγωνον τὸ ΒΔΕΓ, ἀπὸ δὲ τῶν ΒΑ, ΑΓ τὰ ΗΒ, ΘΓ, καὶ διὰ τοῦ Α ὁποτέρᾳ τῶν ΒΔ, ΓΕ παράλληλος ἤχθω ἡ ΑΛ· καὶ ἐπεζεύχθωσαν αἱ ΑΔ, ΖΓ ("For let there be described on ΒΓ the square ΒΔΕΓ, and on ΒΑ, ΑΓ the squares ΗΒ, ΘΓ; through Α let ΑΛ be drawn parallel to either ΒΔ or ΓΕ, and let ΑΔ, ΖΓ be joined").[272] The letters which are underspecified are, on the one hand, Ζ and Κ and, on the other hand, Λ. First, Ζ and Κ which are omitted from the alphabetical order of letter introduction: this is recognizable from the reference to the square ΗΒ which introduces Η although Ζ would have been due first, and from the reference to the straight line ΑΛ which makes apparent that Κ has been left out. The omitted letters Ζ and Κ are part of the diagrammatic lettering of the two squares on ΒΑ, ΑΓ, which, however, are referred to not by all of their four letters, but only by two diagonal letters each. This is a common way of referring to squares and parallelograms besides the one which mentions all four of their letters. This means that the letters are not really omitted but actually implied by the references to the squares. However, it depends on the kind of readership and their familiarity with this type of reference whether the seemingly omitted letters would be perceived as unspecified or not: an expert would immediately be able to attach the letters Η and Θ (of the letter 'names' ΗΒ, ΘΓ) to the respective points – these can only be the ones diagonally opposite from Β and Γ – and to supplement the omitted letters Ζ and Κ at the respective remaining points of the squares. By contrast, a reader who is not familiar with the convention would be capable of neither: for them, Η and Θ would remain underspecified because they are not specified any further within the construction, and Ζ and Κ would remain unspecified; although Ζ finally occurs at the end of the construction as part of the reference to the straight line ΖΓ, it does not become specified because its foregoing implicit specification is presupposed by the assignment of the construction of ΖΓ. Κ even occurs only in the proof for the first time explicitly; again, the construction of the two straight lines ΑΕ and ΒΚ there presupposes Κ as

270 The order of the lettering in multi-lettered letter words according to specific directions which I only hint at here shall be discussed systematically in the next part of the chapter (Section 4.4.4).

271 Cf. especially pp. 362–367 for my discussion of the present proposition I 35 there.

272 *Elem.* I 47 (vol. 1, p. 63.11–14 Heiberg & Stamatis).

being already specified (ἐπιζευγνυμένων τῶν ΑΕ, ΒΚ, "if AE, BK be joined").[273] However, even non-expert readers might have been able to localize Z and K on the basis of the construction part because both the convention of referring to squares by means of two letters and the principle of alphabetical letter attribution could have been learned from the preceding propositions of book I. The example of the two letters shows that the perception of letter specification can vary in some cases. Since we cannot preclude that the work was directed to an expert readership,[274] the underspecification of Z and K is in any case not conclusive with respect to the presence of a diagram. As far as the production of the text is concerned, it is not impossible that Euclid would have formed the letter words and taken account of the omitted letters even without a diagram, simply by envisaging the configuration. But it seems to be much more likely that he would 'read' the references HB and ΘΓ from the underlying diagram, because the figure is more complex than many others in book I and cannot be pictured as easily. Therefore, the case of the letters Z and K does give some indication of underlying diagram, even though no conclusive one.[275]

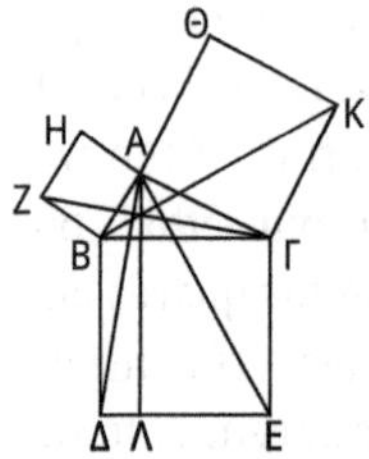

Fig. 46: I 47 according to Heiberg & Stamatis.

The other example from I 47 is indeed conclusive: the letter Λ which is introduced in an underspecified way by the construction of the straight line ΑΛ which has been quoted above.[276] From the construction assignment it only emerges that Λ is supposed to lie on a straight line with A which is parallel to ΒΔ and ΓΕ, but not that Λ should lie on ΔΕ. Strictly speaking, it is not even clear from its construction that ΑΛ should extend downward in the direction of ΔΕ (rather than upward, in which case ΑΛ would still be parallel with the other two lines). This might be guessed from the alphabetical

273 *Elem.* I 47 (vol. 1, p. 64.18 Heiberg & Stamatis). The construction of AE and BK is not assigned in the imperative, which is the usual mode of construction in elements texts, but takes the form of a genitive absolute which substitutes a temporal-hypothetical clause. This is because the construction is done for the sake of a chain of proof steps which is analogous to the foregoing one and which therefore can be related in an abridged manner. The diagrams transmitted for I 47 unanimously testify to the perception that the construction of AE and BK was mandatory.
274 On the context in which the *Elements* has been written and received (that is, as it can be assumed to have been on the basis of our analysis), see below, Section 5.2, 423–426.
275 An additional argument for the presence of a diagram is presented by the order of the letters of the letter words HB and ΘΓ which I am going to discuss.
276 P. 327.

lettering of the straight line analogously with the other two references, which could be interpreted as 'isodirectional' with the letter ordering ΒΔ and ΓΕ – each of the three mentions the letters alphabetically from top to bottom. However, this would only present an additional, rather than a decisive argument in favor of the presence of a diagram, since 'isodirectionality' is often, though not always, based on the viewing of the diagram.[277] In any case, the exact position of Λ is unclear, and the drawing of the diagram cannot be completed on the basis of the construction. The location of Λ remains unclear when the letter recurs in the proof part: καὶ [ἐστὶ] τοῦ μὲν ΑΒΔ τριγώνου διπλάσιον τὸ ΒΛ παραλληλόγραμμον· βάσιν τε γὰρ τὴν αὐτὴν ἔχουσι τὴν ΒΔ καὶ ἐν ταῖς αὐταῖς εἰσι παραλλήλοις ταῖς ΒΔ, ΑΛ· ("Now the parallelogram ΒΛ is double of the triangle ΑΒΔ, for they have the same base ΒΔ and are in the same parallels ΒΔ, ΑΛ").[278] This does still not reveal the position of Λ because the two-letter reference to the parallelogram ΒΛ (which gives us the diagonal letters of the parallelogram)[279] only tells us what we already know: that Λ lies on the line that is parallel to ΒΔ, i.e., on ΑΛ. However, it does not tell us until where ΑΛ is meant to extend and where exactly Λ is intended lie, for while it is clear that Δ is part of the parallelogram (since ΒΛ and ΑΒΔ share the base ΒΔ), the fourth point of the parallelogram (i.e., the one diagonally opposite of Δ) is neither mentioned nor implied. This is because ΒΛ is ambiguously called a 'parallelogram' rather than a 'rectangle' (ὀρθογώνιον),[280] although it is actually supposed to be a rectangle, with Λ lying on ΔΕ and the fourth, unlettered point on ΒΓ. However, the term 'parallelogram' only requires that the respective opposite sides are parallel and of equal length but not necessarily that the four angles are right ones. While the term is not defined in *Elements* I, some of the propositions which precede I 47 even presuppose that it is not understood in the special sense of 'right parallelogram' (i.e., 'rectangle'),[281] e.g., I 35, in which the two parallelograms ΑΒΓΔ and ΕΒΓΖ would be one and the same figure if they were right parallelograms, because they have the same base ΒΓ. Consequently, the proof does not indicate the position of Λ. To be sure, the proof suggests that the two parallelograms ΒΛ and ΓΛ are rectangles which are two parts of the square ΒΔΕΓ, for it states the equality of the parallelograms with the corresponding squares ΗΒ and ΘΓ. This might be obvious to expert readers at least. However, to conclude the position of Λ from this assertion would be circular and invalid: the deductive assertion about Λ cannot specify the position of the letter at the same time because it already presupposes the position of Λ. Therefore, the underspecification of Λ in the construction conclusively indicates that this construc-

277 On the principle of "isodirectionality," originally introduced by Netz, see below Section 4.4.4, 353.
278 *Elem.* I 47 (vol. 1, pp. 64.8–12 Heiberg & Stamatis).
279 As in the above case of ΗΒ and ΘΓ, recognizing the ramifications of the reference "ΒΛ" presupposes knowledge of the conventional use of this type of reference.
280 For the latter term, cf. *Elem.* I, Def. 22.
281 The special case of 'parallelogram' is only defined at the beginning of the second book (*Elem.* II, Def. 1).

tion was written on the basis of a diagram because it would have been needed by experts and lay readers alike.

These four examples should suffice to demonstrate that the underspecification of letters in some of the propositions of the *Elements* conclusively points to the presence of an underlying diagram. In a similar way as these four cases, six other propositions of book I contain at least one underspecified letter which conclusively indicates that the proposition was written on the basis of a diagram.[282] The form of the propositions

282 I give a brief overview of the kind of letter underspecification in each:

In I 11 (Fig. 47), the letter E is underspecified, because it is not clear from the assignment καὶ κείσθω τῇ ΓΔ ἴση ἡ ΓΕ ("let ΓΕ be made equal to ΓΔ," *Elem.* I 11 [vol. 1, p. 19.1 f. Heiberg & Stamatis]) that the point E and the straight line ΓΕ lie on the straight line AB.

In I 12 (Fig. 48), the letters Γ, H and Θ are underspecified: the construction does not betray that all three of them lie on the straight line AB, with E and H being the intersection points of the circle EZH with AB. This is only made clear by the subsequent definition of the goal which locates Θ on AB from which it emerges that also E and H must lie on AB. However, the definition of the goal makes clear that it already presupposes the whole construction because it is in the perfect tense (κάθετος ἦκται ἡ ΓΘ, "ΓΘ has been drawn perpendicular," *Elem.* I 12 [vol. 1, p. 20.15 f. Heiberg & Stamatis]).

In I 21 (Fig. 49), the letter E is underspecified when introduced, and remains so throughout the whole proposition: The construction only says that "the straight line BΔ should be drawn through to E" (Διήχθω γὰρ ἡ BΔ ἐπὶ τὸ E, *Elem.* I 21 [vol. 1, p. 29.19 Heiberg & Stamatis]), but not that E should lie on AΓ. (Cf. Thaer 1970, 423: "Für den Beweis wesentlich ist die der Anschauung entnommene Tatsache, daß E zwischen A und C fällt, wie D zwischen B und E.")

I 28 (Fig. 50) does not contain any conclusively underspecified letters, because the letters in question are only difficult to localize for a non-expert readership. First, the points H and Θ are not introduced as points of intersection of the straight line EZ and the straight lines AB and ΓΔ, but only as part of the references to the angles EHB and HΘΔ. This is especially confusing as the diagram of the preceding proposition 27 also has a straight line EZ cut two straight lines AB and ΓΔ, but there, the points of intersection are E and Z rather than two additional points. But since H and Θ could not refer to any other but the intersection points, they are in fact sufficiently specified, especially in the eyes of expert readers. Nevertheless, the way they are introduced gives some (though not conclusive) indication of the presence of the diagram. More severely, the construction does not specify that A and Γ, and B and Δ must lie on the respective same sides of EZ: only then, the proof works out (unless EZ intersects the other two lines at right angles, in which case the order of A and B, and Γ and Δ does not matter). An expert reader could conclude the necessary order of the lettering from the 'isodirectional' order of the letters in the references to AB and ΓΔ (cf. Section 4.4.4, 353); for this reason, the underspecification of the letters A, B, Γ, Δ is not conclusive with respect to the presence of a diagram, but nevertheless points to it.

I 30 (Fig. 51) presents a similar case, but conclusively indicates an underlying diagram by the underspecification of the letters H, K and Θ. H and K are introduced in the construction as part of the straight line HK which cuts the straight lines AB, ΓΔ and EZ (Ἐμπιπτέτω γὰρ εἰς αὐτὰς εὐθεῖα ἡ HK, "For let the straight line HK fall upon them," *Elem.* I 30 [vol. 1, p. 42.21 Heiberg & Stamatis]). The position of the two letters is not specified; for this reason, one might assume that they do not have a specific position (as in the foregoing propositions 28 and 29, in which the straight line which intersects two other straight lines is named by two letters which are not the intersection points). If one assumes them to be points of intersection as in I 27 (which they turn out to be in the proof), it is not clear which of the three possible ones. The third intersection point Θ is introduced only in the proof.

itself suggests a strict distinction between the construction and the proof that is done on the basis of the construction. According to this distinction, the great majority of letters is indeed specified by the constructing section. It is, thus, the relative standardization of letter specification itself which suggests that the diagram is expected to be drawable on the basis of the specification of all letters. The underspecification of some of the letters can only be explained by the presence of a diagram: if Euclid wrote the text while viewing the ready diagram, the underspecification of some of the letters might simply have escaped his notice, whereas, without having a diagram at hand, he would have specified the letters more meticulously.[283]

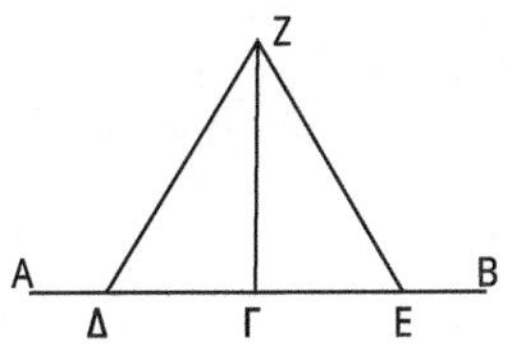

Fig. 47: I 11 according to Heiberg & Stamatis.

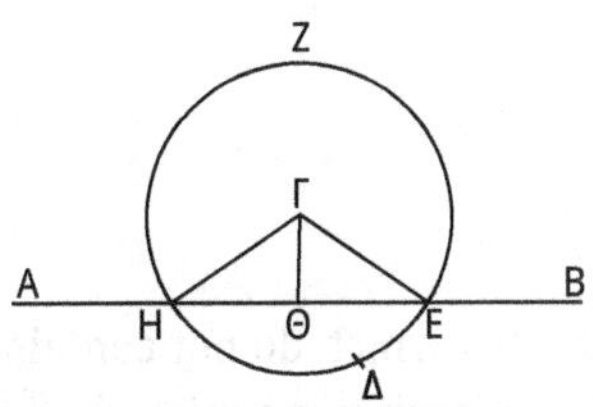

Fig. 48: I 12 according to Heiberg & Stamatis.

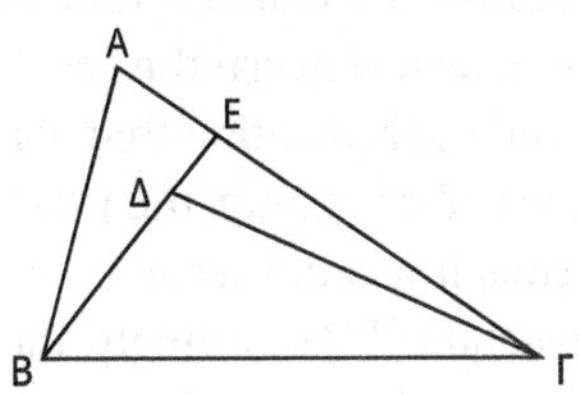

Fig. 49: I 21 according to Heiberg & Stamatis.

In I 46 (Fig. 52), the point Δ is underspecified which indicates conclusively an underlying diagram: the construction assigns that the straight line AΓ is to be drawn at right angles to the straight line AB, and that AΔ should be equal to AB but does not specify that Δ, and therefore AΔ should lie on AΓ. Thus, the construction does not necessarily result in a square (as is the goal of the proposition), but might also lead to a parallelogram.

283 Accordingly, the method by which I have detected underspecified letters has been to try and draw the diagram myself on the basis of the constructing section rather than to analyze the letter specification by means of a comparison of the text and the transmitted diagrams.

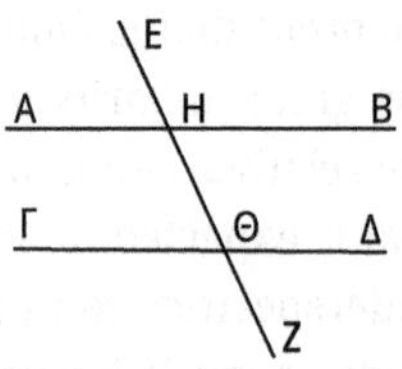

Fig. 50: I 28 according to Heiberg & Stamatis.

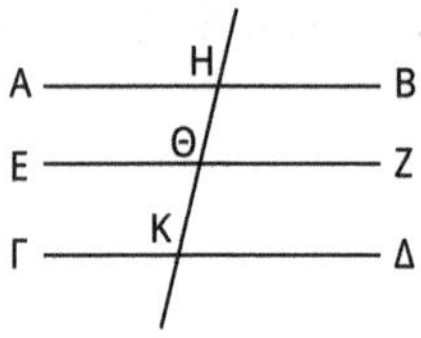

Fig. 51: I 30 according to Heiberg & Stamatis.

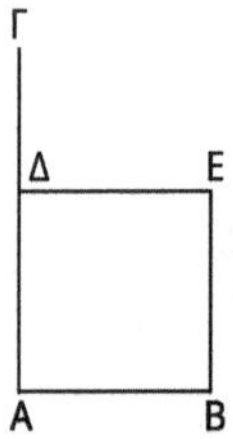

Fig. 52: I 46 according to Heiberg & Stamatis.

Having thus surveyed the letter specification in the propositions which do not contain a proof by contradiction, I shall now turn to those nine propositions which do.[284] These have not been included in the previous survey, because the construction of many of them poses special problems regarding drawability which concern the specification of the letters, but mostly display other reasons for underspecification than we have seen so far. Generally, in the majority of the propositions of this group, the proof is accomplished by means of a construction which either constitutes the second part of the construction of the proposition or the only one construction.[285] Accordingly, the analysis of letter specification in such propositions cannot be restricted to a first 'constructing section' of the proposition, and the proof part cannot necessarily serve as a benchmark for the letter specification of the construction. Of course, as in the case of propositions without proofs by contradiction which involve a construction within the proof, the letters which are introduced by a construction within the proof must be specified before being used subsequently in non-constructional parts of the proof (and can be judged accordingly).

284 I 6, 7, 8, 14, 25, 26, 27, 29, 39.

285 For the types of proofs by contradiction, see above Section 4.3.3 on the 'practicality' of the textual construction, pp. 261–266.

In those propositions which contain (at least) two constructions, the main reason for the underspecification of the letters is that the same letters are applied to two different constructions within the same proposition. In the process, different properties and relations are attributed to objects which are nevertheless referred to by the same letters. These letters are thus doubly specified. I have already discussed this problem and its ramifications in the part on the 'textual construction:' the diagrammatic construction on the basis of such letter use would only be unambiguous if two diagrams instead of the usual one diagram were drawn according to the double specification of the letters.[286] However, the respective propositions do not explicitly mention the necessity of drawing two diagrams; to the contrary, by not repeating the exposition of the configuration which forms the basis of both constructions, they indicate that the diagram was not going to be fully redrawn but that the second construction had to be merged somehow with the first one. The diagrammatic evidence also points into this direction by providing usually only one diagram in such cases.[287] As a result of the limitation to one single figure, the letters which have been specified twice can only be attributed to the diagram in one of the two ways, and letters which are specified only as part of the second, ultimately impossible construction might not be applicable in the specified way.

I 39 (Fig. 53), which we have already adduced in the above analysis of the construction assignments,[288] may again serve as an example: the exposition posits two equal triangles ΑΒΓ and ΔΒΓ which lie on the same base ΒΓ on the same side (Ἔστω ἴσα τρίγωνα τὰ ΑΒΓ, ΔΒΓ ἐπὶ τῆς αὐτῆς βάσεως ὄντα καὶ ἐπὶ τὰ αὐτὰ μέρη τῆς ΒΓ)[289] and assigns the construction of the straight line ΑΔ (Ἐπεζεύχθω γὰρ ἡ ΑΔ).[290] The goal is to prove that ΑΔ and ΒΓ are parallel (λέγω, ὅτι παράλληλός ἐστιν ἡ ΑΔ τῇ ΒΓ).[291] On the basis of the preceding I 37 which has proved that "triangles which are on the same base and in the same parallels are equal to one another,"[292] the two triangles can only be drawn in such a way that ΑΔ must be, when it is drawn, parallel to ΒΓ. Accordingly, this is how the straight line appears in the diagrams transmitted both on papyrus and in the

286 See above, 261–266.

287 See p. 267 f. with Fig. 27 (I 39 in P and B), Fig. 28 (a) (I 6 in P), and Fig. 29 and 30 (I 27 in P and B). The representation of I 6 by means of two diagrams in B (Fig. 28 [b], p. 267) is from a later hand, and the additional diagrams *in margine* in B (Fig. 31 and 32, p. 268 f.) do not necessarily stem from the copyist.

288 See above, Section 4.3.3, 262 and 265.

289 "Let ΑΒΓ, ΔΒΓ be equal triangles which are on the same base ΒΓ and on the same side of it," *Elem.* I 39 (vol. 1, p. 53.1 f. Heiberg & Stamatis).

290 "For let ΑΔ be joined," *Elem.* I 39 (vol. 1, p. 53.4 Heiberg & Stamatis).

291 "I say that ΑΔ is parallel to ΒΓ," *Elem.* I 39 (vol. 1, p. 53.2 f. Heiberg & Stamatis). The transmitted text has another, first definition of the goal already before the construction of ΑΔ; but this one duplicates the second definition of the goal, and so has been deleted in Heath and Thaer.

292 Τὰ τρίγωνα τὰ ἐπὶ τῆς αὐτῆς βάσεως ὄντα καὶ ἐν ταῖς αὐταῖς παραλλήλοις ἴσα ἀλλήλοις ἐστίν, *Elem.* I 37 (vol. 1, p. 50.17 f. Heiberg & Stamatis).

manuscripts.[293] The subsequent proof by contradiction begins with the following construction assignment: Eἰ γὰρ μή, ἤχθω διὰ τοῦ Α σημείου τῇ ΒΓ εὐθείᾳ παράλληλος ἡ ΑΕ, καὶ ἐπεζεύχθω ἡ ΕΓ ("For, if not, let ΑΕ be drawn through the point Α parallel to the straight line ΒΓ, and let ΕΓ be joined").[294] Here, first, the conditional clause assumes that ΑΔ is not parallel to ΒΓ – contrary to how it has been drawn on account of the exposition. This is not an explicit assignment of a different construction of ΑΔ, but the construction which follows conflicts with the diagram as it has been drawn on the basis of the exposition. The construction assignment is unambiguous: ΑΕ should be drawn parallel to ΒΓ; this means that ΑΕ would have to overlap with ΑΔ in the diagram as it has been first constructed, with Ε lying either between Α and Δ or on the other side of Δ (if one extends ΑΔ), or coinciding with Δ. However, in all transmitted diagrams, ΑΕ is not drawn parallel to ΒΓ in accordance with the construction assignment, but as non-parallel with ΒΓ in accordance with the result of the proof by contradiction (οὐκ ἄρα παράλληλος ἡ ΑΕ τῇ ΒΓ).[295] The diagram only shows the result of the proof, but not the second construction in the way in which it has been assigned. This is because one diagram cannot accommodate two contradictory constructions.

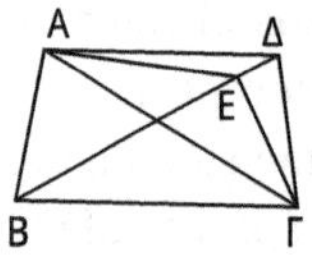

Fig. 53: I 39 according to Heiberg & Stamatis.

The letter Δ is ambiguous: it is first implicitly specified by the exposition, and a second specification, contrary to the first one, is implied by the second construction within the proof. The ambiguous double specification of Δ may be regarded as 'underspecification' in view of the drawing of the diagram: it cannot be constructed because Δ is ambiguously and therefore not sufficiently specified. The letter Ε itself, by contrast, is fully specified: However, it is likewise ambiguous because of the implied same specification of Δ, and it cannot be constructed in the way in which it is specified because the resulting overlap of ΑΕ with ΑΔ would contradict the contrast with ΑΔ as it is intended by the second construction. Accordingly, the transmitted diagram only shows one of the two implied specifications of the letter Δ, and falsely depicts the specification of Ε. This means that the diagram does not, as in the propositions without proofs by contradiction that we previously surveyed, reveal the specification of the letters which have been underspecified by the text; rather, it shows one of several inadequate ways of depicting the ambiguously specified letters in one single diagram (the only adequate way would be by means of two diagrams).[296]

293 See above, Fig. 1(a) (p. 192) and Fig. 27 (p. 267).

294 *Elem.* I 39 (vol. 1, p. 53.6 f. Heiberg & Stamatis).

295 "Therefore ΑΕ is not parallel to ΒΓ," *Elem.* I 39 (vol. 1, p. 53.13 f. Heiberg & Stamatis).

296 See above, Section 4.3.3, 266 f. with Fig. 27.

The previous kind of underspecification of the letters was indicative of an underlying diagram because it prevented the construction of the diagram which otherwise would have been constructible. In the present case, the ambiguity (and therefore underspecification) of the letters, conversely, is itself the result of the impossibility of the construction. Nevertheless, the present underspecification of the letters, too, is indicative of an underlying diagram, though in a different way: the letters which are underspecified because of their ambiguity indicate that the diagram in any case cannot be unambiguously constructed but they do not indicate which kind of depiction to choose. One gets the impression that the construction is not described for the sake of an actual construction but rather because some kind of diagram is underlying and must therefore be referenced by the text. Accordingly, the diagram transmitted for I 39 is not unambiguous nor could it have been constructed unambiguously on the basis of the textual construction and its letter specification: the localization of E in it contradicts the specification of the letter in the text. The diagram therefore strongly suggests that it was included in the text. This is true especially because it is attested in this same manner both on papyrus and in our manuscripts despite the ambiguity of the textual construction.[297] While in the previous cases of letter underspecification, the diagrams helped clarify the position of the underspecified letters, the diagram in the present case does nothing to explain the position of the letters but rather is itself in need of explanation. It even temporarily obscures the text, especially if it was present in the text from the beginning, for it shows the result of the proof as it is based on the exposition and construction, but it contradicts the second part of the construction as it is assigned in the proof part.

The kind of underspecification of the letters described – which results from the ambiguous specification of the letters with regard to two diagrammatic situations – occurs in six of the nine propositions of *Elements* I which contain proofs by contradiction (I 6, 14, 26, 27, 29, 39). The relation of the letter (under)specification and the realization of the two constructions within one single diagram is different in each case. Therefore, one cannot assume that an expert reader would know a standard way of constructing an impossible diagram – only in that case, the ambiguity of the letters would not be indicative of an underlying diagram. Underspecification of letters in the previous sense in which it has been found in ten of the propositions without proof by contradiction, that is, of letters which are presupposed by the proof part, can also be found among the propositions with proof by contradiction, but only twice (in I 8 and in I 26, which shows both this type of letter underspecification and the one which results from the reuse of letters for different constructions as part of the same diagram). Only two propositions of those with proof by contradiction are unambiguously drawable on the basis of the specification of their letters (I 7 and 25). Thus, the ambiguous use of letters for different constructions within one and the same proposition is the main reason for letter underspecification in this group, and it accounts for the comparatively large proportion of

297 See above, Section 4.1, Fig. 1(a) (p. 192), and Section 4.2, Fig. 27 (p. 267).

propositions with unspecified letters among this group (seven propositions with underspecified letters compared to two with only specified letters).

Conclusion

Let us first summarize our results: of the first group of propositions in *Elements* I that we have surveyed (those without proofs by contradiction), twenty-one contain only specified letters, seven at least one gradually or belatedly specified letter, and ten at least one underspecified letter; of the second group (those with proofs by contradiction), two contain only specified letters, six at least one ambiguously specified letter, and two at least one underspecified letter (in the first, proper sense of underspecification that we have defined). This means that of 47 propositions in total, only 23 contain only letters which are unambiguously specified by the textual construction and give no indication of their being based on an underlying diagram; seven give some indication that they have been written on the basis of a diagram, as they contain at least one gradually or belatedly specified letter which prevents the diagram from being constructible step by step, although it is constructible on the basis of the letter specification of the construction as a whole. This leaves altogether 17 propositions – more than one third of the sample – to present highly indicative evidence of an underlying diagram because they contain at least one not sufficiently specified letter and do not allow for drawing the respective diagram on the basis of the construction.

Both types of letter underspecification that we have found – underspecification proper and the ambiguous specification in propositions with proofs by contradiction – are highly indicative of an underlying diagram because drawing the diagram unambiguously is impossible on account of them. Ultimately, however, the conclusiveness of the evidence of letter underspecification depends on the proof-relevance of the specification of the letters. As we argued in Section 4.3, it is the form of the proposition itself which suggests that the second part of the proposition, the proof, depends on the first, the 'constructing section' which therefore must be 'proof-relevant.' In this section, we had considered the proof-relevance of the textual construction with regard to whether it enables a proper or technical construction of the diagram; we have shown that the textual construction prevents such a construction, but that this fact is inconclusive of the presence of the diagram since the technicality of the construction has turned out not to be 'proof-relevant.' That is to say, the proof does not depend on how the objects are drawn. The matter is different in the case of the specification of the letters by the textual construction. Since the proof presupposes the knowledge of the position of the letters, all letters have to be localizable in a diagram in time, i.e., before the proof draws on them. That is, their localizability is, in any case, proof-relevant: this means that the diagram either has to be there from the start or it must be constructible on the basis of the 'constructing section.' The letters must be presumed to be visible in a diagram that is provided by the text or must be specified by the textual construction so that the diagram can be drawn, with the exact degree of necessary specification varying and depending again on the requirements of the proof. This means, conversely, that the textual construction conclusively indicates the presence of the diagram if it does not in a timely manner and adequately specify the letters as they recur in the proof. Therefore, our first, proper type of letter underspecification is proof that the respective proposition text included a diagram, for it prevents the construction of the letter configuration and thereby also prevents accomplishing the proof on the basis of this configuration. In those propositions which contain constructions with underspecified letters of this kind, it is indeed the diagram-as-letter configuration itself which is proof-relevant since the textual construction does not fully accomplish the necessary letter specification (whereas in fully specified propositions, it is the diagram resulting directly from the textual construction which is relevant to the proof in terms of letter specification).[298] We have seen that this conclusive type of letter underspecification occurs both in propositions with proofs by contradiction and in propositions without them. The other type of underspecification is confined to propositions with proofs by contradiction and includes all cases of ambiguous letter specification which do not allow an unambiguous construction of the diagram. Such insufficient letter specification, however, does not prevent the proof from being accomplished nor does it therefore conclusively indicate an underlying diagram. Thus, this type is ultimately not conclusive of the presence of a diagram if we regard it from the point of view of proof-relevance. However, it is indicative of the presence of a diagram because of the contingency of the ensuing diagram. In view of this contingency, the comparative unanimity of the diagrammatic evidence further underlines the indicativeness of this kind of letter underspecification.

Let us consider the overall scope and relevance of letter underspecification. For the most part, letters are specified. In many propositions, they fully enable constructing a diagram-as-reference system and accomplishing the proof on the basis of it. In those propositions which do not enable the construction of the diagram it is only one or few letters that are underspecified. That is, these diagrams are *almost* constructible: they could be constructed if it were not for the underspecification of one letter. Within this setting of extreme specification, the few unspecified letters clearly indicate their own non-specification and imply what they lack in order for the diagram to be constructible and for the proof to be doable on the basis of the diagram. It is possible for us to discover cases of letter underspecification precisely because almost all letters are specified.

298 See above, p. 19 with n. 106 for previous criticism of the dependence of the Euclidean propositions on the diagram (cf. Netz 1999a, 27). This criticism, however, extends to any reliance on the diagram rather than taking as a basis a division of the proposition into a 'constructing section' and a proof part which allows us to consider the reliance of the proof on the assigned diagram rather than on *the* diagram, whatever is meant by that.

While the majority of propositions still are inconclusive regarding the presence of a diagram, the number of those which do present conclusive evidence is considerable and cannot be explained by accidental non-specification. To the conclusive cases, we can add the 'soft evidence,' as it were, that is, the propositions which are suggestive of the presence of a diagram because of their belated and therefore impractical letter specification. It is not only in quantitative terms that the evidence does not seem to be contingent: it includes very diverse cases of propositions and diagrams and is not confined to the most complex propositions which would have necessitated a diagram in any case. The conclusive evidence is conclusive primarily with respect to itself; that is, we can only be certain that a diagram was present in those propositions which feature underspecified letters whose underspecification is relevant to the proof. However, the conclusive evidence also appears to suggest its own generalizability. This is because it is highly improbable that diagrams would have been drawn only for those propositions in which not all letters happened to be specified, for the underspecified letters do not seem to have been left underspecified on purpose. We may safely assume, had Euclid been aware of the underspecification of a letter, that he would have specified this letter right away, rather than having deliberately left it underspecified. This seems to be the case regardless of the question of what motivated formal standardization (which I shall try to answer in Chapter 5), because whatever motivated it, it has been practiced in an extreme way. Even the specification of the letters has proved extreme. The reason why some letters are not sufficiently specified appears to be precisely the standardized presence of the diagram which we concluded from the occasional underspecification of the letters. It is the double standardization of letter specification and diagrams which allowed for the underspecification of some of the letters. If indeed all propositions contained diagrams, those propositions in which letters are fully specified, as well as the remaining propositions simultaneously having underspecified letters alongside otherwise extreme letter specification, would have to be regarded as displaying (unnecessary) 'overstandardization.' This is by no means an argument against the general presence of diagrams – as we shall see, it is, to the contrary, characteristic of the *Elements* to 'overstandardize.'[299]

The underspecification of the letters has finally provided conclusive evidence for the underlying existence of a diagram in a number of cases in *Elements* I. Letter specification will remain the only conclusive criterion to diagnose the presence of a diagram in a particular case. We have suggested that the underspecification of the letters in certain propositions may indicate the standardized presence of diagrams in the *Elements* even beyond the respective propositions. I shall now consider whether another criterion is able to suggest the standardized use of a diagram more generally for the propositions of the *Elements*: the order of the letters in multi-lettered letter words.

299 In Sections 5.1 and 5.2, I shall distinguish cases of necessary standardization and unnecessary (over)standardization in order to elaborate the purpose of standardization.

4.4.4 The order of the letters in multi-lettered 'letter words'

Aim and plan

The fourth characteristic of the use of letters which I shall discuss regarding its implications for the reliance of the proposition text on a diagram is the order of the letters in multi-lettered letter words. I have already analyzed it regarding the principle of alphabetical ordering and the deviations from it in terms of permutation. I have shown that the order of the letters in multi-lettered letter words is predominantly alphabetical but also regularly deviates from the alphabetical order by permutation, with 'switches' between the alphabetical and the permuted order affecting the same letter words within a proposition. This standardized ambivalence does not appear to be random but to follow certain principles which have been discovered by Netz. In trying to determine how the letter order and its change are motivated, I shall first take these principles into account one by one, before singling out one of them, namely, 'isodirectionality' and considering it as part of a larger phenomenon, that is, 'directionality.' I shall explore the range of its manifestations and its ramifications both for the standardization of the diagram and of textual diagrammatization.

As I have claimed in Section 4.4.2, the basic principle underlying the ordering of the letters in multi-lettered letter words is the one of the alphabetical introduction of the letters into the text. It is through this principle that multi-lettered objects receive the letters which they retain throughout the proposition – though not necessarily in the order in which they appear in the references when they are first introduced. While the alphabetical principle determines in almost all cases the set of letters by which an object is referenced (i.e., the alphabetically due ones), the order of those letters themselves may be permuted regarding the alphabetical order already when they are first introduced. Subsequently to their introduction, they may 'switch' from the alphabetical to a permuted form or vice versa, but usually take either of these two, even where more than two ways of letter ordering are possible. All permutations apparently occur on the shared assumption that a letter word within the same letter phrase within the same proposition refers to the same object, no matter the order in which the letters reappear.[300] To be sure, while there is no kind of geometrical object that is referenced by at least two letters which could not be permuted in some way, there are a few limits to the number and kinds of permutation possible depending on the kind of geometrical object. Apart from these constraints, however, it makes no difference from the mathematical point of view of the unambiguousness of the proof whether the letters in multi-lettered words are ordered in one way or another. In view of the mathematical unambiguousness of the letter words regardless of the letter order, it is all the more notable

300 Netz 1999a, 71.

that the alphabetical order has emerged as the dominant one, both of the introductory and of the reappearing forms of letter words.[301]

The role of the quantitatively predominant principle of alphabetical letter ordering with regard to multi-lettered words is as yet to be determined. It has been shown that its predominance is explained only in part by the standardized practice of assigning letters to points newly introduced into the proposition in alphabetical order, which either follows the construction and lettering of the underlying diagram in alphabetical order or determines the construction and lettering according to this order. For the alphabetical ordering of the letters *within* multi-lettered letter words naturally emerges from the practice of alphabetical letter introduction only in those cases in which these letter words introduce new letters into the text. However, as we have seen, many letter words, when they are being newly introduced themselves, consist of letters which already have been introduced into the text and nevertheless many times still adhere to the alphabetical order. The strong abidance by the alphabetical principle throughout the introduction of letter words and in the reuse of the letter words could be the result of aiming for maximal linguistic standardization, as with all other aspects of language and style in the *Elements*. In view of this, however, the degree of deviation from the alphabetical principle is notable because it would have been easy to adhere to the same letter forms without exception. Instead, the text seems to explicitly allow for the permutation of the order of multi-lettered words.[302] While this practice seems already a strong indication that the letter combinations were 'read' each time from a diagram that was already there rather than imagined or remembered as references to particular objects for the duration of the proposition, the point of the deviations has to be examined more closely before we may conclude from them the presence of a diagram. The question is then: what is gained by deviating from the standardized principle? It is, therefore, the permutations which provide a starting point for considering the meaning and functions of letter ordering (just as the practice of 'omission' provided the starting point for analyzing the specification of the letters).[303] The statistical evaluation of the deviation of the letter order from the alphabetical principle has not shown any patterns according to which the letter order is permuted either in the references to particular objects, or in specific parts of the proposition, or in the introductory or reappearing forms. The deviations from the alphabetical order shall now be regarded in their immediate textual context, that is, in relation to other letter words mentioned alongside (and their letter order), and in relation to the diagrammatic representation of the letter word and textually related letter words. That is to say, I take the lettering of the diagram into account. On the basis of the results of the closer inspection of the deviations, the standardized application

301 Cf. the results of the analysis of Section 4.4.2, 304.

302 Cf. Netz 1999a, 78: "The deviations are allowed by the 'but not always' clause [. . .]," i.e., of the "overall structure of the regularity" which is "'usually – but not always.'"

303 See above p. 304.

of alphabetical ordering itself may be reevaluated regarding its motivations, again with regard to the lettering of the diagram.

In order to determine how exactly the alphabetical letter order and its permutation are motivated, I again begin by reconsidering different options put forward by Netz who, however, does not try to explain the coming about of the characteristics of the letter order as they can be found in the *Elements*, above all the adherence to alphabeticality, but only the origin of the practice of 'permutation.' As with all formal standardization in elements texts, he regards the (imperfect) standardization of the alphabetical letter order as the result of "self-regulating conventionality" rather than of "chance"[304] or of deliberate linguistic standardization.[305] In the context of the present comprehensive study of the *Elements*' standardization, my own interest differently is precisely in how the textual standardization of the letter order is to be explained, that is, the standardization of the letter order both in terms of the overwhelmingly standardized alphabetical principle and of the regular deviation from the standardized form. Unlike other kinds of textual standardization, the textual standardization of the letter order shall be regarded not only for its own sake, but primarily with respect to the standardization of the diagram itself. Here, it is, as has been explained above, especially the standardized tendencies of deviation from the dominant practice which can be expected to be informative.[306]

---

304 From the overall standardization, it can be safely concluded that "the results are not random" (Netz 1999a, 73); for in order for them to be random, the ratio of alphabetical and permuted forms would have to be 50:50%. Of course, the question is whether, in a standardized text like this, the letter order could at all be random, that is, whether it could end up with a 50:50 ratio, or whether the text would not stick to one of the ways of ordering, that is, the alphabetical one, more frequently, especially because this one is to some extent predetermined by the regular choice of a certain number of the first letters of the alphabet for designating the points in the diagram, and by the likewise not random alphabetical order of their introduction into the text.

305 See above p. 59 for other instances of "self-regulating conventionality" according to Netz. The way such conventionality is supposed to emerge is not clear: "The conventionality, therefore, is not the result of any explicit attempt to conventionalize a practice. Further, the conventionality is not just the operation of blind, general cognitive forces. It does have a component of intentional agency. This agency, however, does not shape the overall structure of the regularity. Rather, it takes this overall structure and uses it for its own purposes. The overall structure is 'usually – but not always,' and it is possible to use such a structure to convey information by deviation from the pattern." This, however, means that intentionality only applies to the deviation, while it remains unmentioned what it is that "shape[s] the overall structure of the regularity" itself. Apparently, the standardization of the alphabetical principle, according to this, is thought to have come about by itself, since "explicit," i.e., deliberate conventionalization, in the sense of "explicit codification, something like a style manual" according to which the text was written and proofread, is excluded as the driving force (74).

306 Netz 1999a, 77, emphasizes the potential informativity of deviations from the "recognisable pattern." See also 78 (I repeat in part and continue the quote of the preceding note): "The overall structure is 'usually – but not always,' and it is possible to use such a structure to convey information by deviation from the pattern. The deviations are allowed by the 'but not always' clause, and are informative of the 'usually' clause."

I shall thus start from Netz's suggestions for the origin of permutation before considering how the principles which account for permutation determine the practice of letter ordering as a whole. In the first, shorter part of the section, I shall review how Netz explains the deviations from the alphabetical order. Here I shall begin by considering the possibility that permutation came about by mistake, as it were. In this regard, I shall take into account the role of the textual transmission and debate the role of the term "memory"[307] in view of the possibilities of the presence or non-presence of the diagram. Afterwards, I shall explore what function permutation might have according to Netz. Here, I shall first consider the function of the "disambiguation" of kinds of objects (as "intentional, informative deviation[]"), which Netz has discovered in Archimedes and Apollonius. I shall explain the principles which Netz discusses with regard to what he calls "visualisation," namely "linearity" and "isodirectionality,"[308] and consider their relevance for our question of the standardization of the diagram. In the second, larger part of the section, I shall trace the phenomenon which I call by the broader term 'directionality.' That is, starting from Netz's concept of "isodirectionality," i.e., the description of two parallel lines from the same point of view, I consider whether the use of the letters in multi-lettered letter words even more generally reveals a common direction of description and thus could indicate a common perspective on the diagram. In order to determine the ramifications of these principles for the underlying presence of the diagram, I shall examine selected propositions as to whether their perspective onto the diagrams is the same throughout or differs according to certain parts of the proposition.

Possible reasons for deviation

First, I shall consider how permutation might have come about by mistake. Let us start by looking at the role of the textual transmission: what if the order of the letters in multi-lettered words was merely the result of the text's having been copied so often, and, at that, by scribes who either might not have cared about the order of the letters and switched it inadvertently to a lesser or greater extent, or changed it as they thought it best? Once the proof-irrelevance of the letter order had been recognized, which would have been the case especially if there had indeed been a considerable number of changes of the letter order in the original, the letter order would have been particularly susceptible to later change. The influence of the manuscript transmission, however, seems to be neglectable. This is the result of Netz's review of the evidence for a sample of Euclid's *Data* in this respect, according to which there is only one variation between manuscripts regarding the letter order

307 Netz 1999a, 74.
308 Netz 1999a, 77.

per proposition on average.[309] Netz concludes from this that the copiers "copied, slavishly" without a clue of what the text meant, and that the Byzantine transmission was, thus, unsuspicious of change of the transmitted letter order, while the ancient development of the phenomenon until Euclid's time was uncertain. One may assume that the same attitude toward the textual source also characterized the ancient scribes. However, the only way to show that the deviations and the underlying alphabeticality are not the result of random changes made to the letter order in the course of the ancient transmission would be to prove that alphabeticality and permutation have a meaning and function. In other words, we would have to be able to show that the text loses in meaning if the order of the letters within words were either wholly alphabetical or less alphabetical than in the modern editions. In this case, the way of letter ordering could be ascribed to Euclid himself.

Apart from considering the role of the textual transmission, Netz grants the "short-term memory" of the author some influence, that is, more precisely, the natural limitations of it. He does so on the basis of an evaluation of a small sample of propositions[310] with regard to the distances between reappearances. It shows that the "average length of switches" ("about 6 to 7 [Heiberg] lines") is "larger than that of repetitions" ("about 5 to 6 lines"). "Length of switches/repetitions" in terms of lines seems to refer to the amount of text between the appearance of letter word and its reappearance with a switched or repeated letter order, respectively. Netz concludes from this that the order of the letter words tends to be retained whilst remembered over a comparatively short period of time, and gets changed because it tends to be forgotten after some time. However, the correlation between switch and longer distance on the one hand, and repetition and shorter distance, on the other, is not necessarily due to a corresponding memory-caused connection between the distance of letter words and the relation of their letter order. It could also have other reasons: for example, it could have to do with one of the possible functions of permutation which I shall consider shortly.

As I have shown in the part about the alphabetical letter order,[311] switches actually do not occur that often to a particular letter word; instead, larger stretches of repetition of either the alphabetical or the permuted form are much more common. I have suggested on these grounds that the switches could be meaningful. I have not considered then the spatial-temporal distance of the respective reappearances from one another. It could be well be that there is a greater distance between the repeti-

309 Netz 1999a, 73, n. 13. Besides variation of this "AB/BA type," Netz considers "AB/AC variation[s]," which are more common than the former, but which may be identified unambiguously as corruptions and emendated accordingly. Netz does not discuss the generalizability of this survey of the *Data*, but seems to take its results to also hold for the other elements authors and works which he studies, among them the closely related and simultaneously transmitted *Elements*.

310 Netz 1999a, 74. He does not mention from which work or by whom.

311 See above Section 4.4.2.

tions than between a switch and its foregoing form. This could be the case if the occurrence of a letter word with a particular letter order has a certain function within one part of the proposition, or within a part of a part of the proposition, and assumes a different function in a following different part, with the distance between the switch and the last repetition of the non-switched form being due to the occurrence of both in different proposition parts or in different construction assignments or deductive statements. I shall return to this under the section about isodirectionality.[312]

That memory has an influence of significance at all should imply that the text is written without the diagram at hand.[313] Conversely, the presence of the diagram and the resulting possibility of writing the text by constantly referring to it would indicate that the letter words need not be memorized at all, let alone their order, but that they could be 'read' again and again from the underlying diagram. To be sure, with the number of objects being sometimes quite limited, it happens that names get to be remembered anyway; so one would not have to refer to the diagram each time and, consequently, could make memory-induced mistakes when not relying on the diagram for some time. However, the need for constant reference to the diagram would not only and not even primarily be necessitated by the need to recall the references to the objects, but rather by the need to describe relations between objects and formulate deductive assertions about object relations on the basis of it. I imagine that in the course of referring to the diagram anyway for this purpose, the references would be read from it and formulated anew every time. So permutation due to lapses in recollection does not seem to be likely in such a scenario.

From considering the role of the textual transmission and of the memory of the writer in accounting for the switches in the order of the letters, it emerges that it could well be that the permutation of the letter order, for the most part, will not have come about by mistake. I shall now turn to meaningful uses of permutation. The two functions that Netz considers are "disambiguation" and "visualisation."

I shall begin with disambiguation. In the part about the alphabetical introduction of the letters, I have hinted at the fact that the order in which letters are introduced, also as part of multi-lettered letter words, can be shown to be meaningful sometimes in elements texts – e.g., by distinguishing "auxiliary objects" from objects "of the main geometrical configuration," or by marking "discontinuities" within, and "continuities" between propositions – though no instances of this kind occur in our

312 See below from p. 353. Netz 1999a, 75, while attributing switches after long stretches of text to "memory," goes as far as claiming that "[w]hoever switched a name within the very same line did so on purpose" and as calling the "differentiality" of the behavior of such switches "apparently intentional." I should refrain from using the term "intentional," as I prefer to think, instead, about the functions of permutation. It is, however, not clear why switches after long stretches should not be able to happen on purpose and be ascribed a function.

313 Netz, by contrast, although he ascribes some influence to lapses in recollection, throughout argues for the opposite: that the diagram was present while the text was written.

sample from Euclid.[314] Similarly, Netz shows, using examples from Apollonius and Autolycus, that the order of the letter within letter words can be applied in an informative way: mostly for distinguishing objects which, by definition, are referred to by the same number of letters and happen to "share the same letters" in a particular proposition; in such cases, the permutation of the letters appears to serve the purpose of disambiguating the two objects, so they will not be confused.[315] Netz mentions the Apollonian practice of regularly referring to rectangles by three-lettered letter words (which is otherwise possible also by means of two diagonal or all four letters), thus using the very letter phrase reserved for denoting triangles, and of permuting the letters of triangles "in *exactly* the way impossible for rectangles," where references to triangles and rectangles would be homonymous.[316] To be sure, most 'letter phrases' are, by their unique combination of article, prepositions (where applicable) and number of letters unambiguous anyway, and, therefore, in no need of further disambiguation. But Netz shows, by taking the example of the references to circles and to arcs on circles in Autolycus (which use the same letters, though not the same articles), that the letters get permuted nevertheless, apparently for the sake of even greater unambiguousness.[317]

Regarding the *Elements*, it is noteworthy to begin with the observation that homonymous letter phrases are avoided, in line with the overall extreme unambiguousness of the lexicon;[318] e.g., rectangles are regularly referred to by two or four letters in the first place, precisely to distinguish them from triangles, so further disambiguation by means of letter order is made unnecessary. Moreover, it emerges from our sample I 1–20 that in the event that the same letters happen to appear within the same proposition as part of different kinds of letter phrases that are unambiguous anyway by how they frame the letters, the text does not seem to strive for additional unambiguousness by making sure that the letter order of different 'letter names' holds these 'letter names' apart: in most cases, the text appears to embrace the fact that the letter order, by following the alphabetical principle no matter what, repeats itself exactly within letter phrases referring to different kinds of objects. In a few cases where the order of the letters gets indeed switched so as to differ between references to different objects, the permutation does not seem to appear for the sake of disambiguation.

314 Netz 1999a, 76.
315 Netz 1999a, 77.
316 Netz 1999a, 77.
317 Netz 1999a, 77 with n. 17.
318 Netz 1999a, 134 holds that they are not only "avoided" in Euclid, but "*throughout the literature* [his italics]," but, as we have just seen at his own example from Apollonius, there are indeed letter phrases that are, as abstract terms, homonymous, and are only disambiguated in the concrete case by switching the order of the letters, whereas we did not come across any such cases in our Euclidean sample. See above Section 3.2.1, 116, on the exceptionless avoidance of homonyms and the allowance for only very few synonyms in the lexicon of the *Elements*, including the letter phrases.

The only kinds of references in our sample which are at all able to have the same letter number and order are those to triangles and angles. Regardless of the order of their respective three letters, the references in any case unambiguously denote the object of reference, since the three letters, in the case of triangles, are standardly framed by τό (sc. τρίγωνον), and, in the case of angles, by ἡ ὑπό (sc. γωνία). Thus, they are easily distinguishable. As far as the possibilities of letter order are concerned, triangles allow for six different forms (in the case of the triangle ΑΒΓ, these are, apart from ΑΒΓ: ΒΓΑ, ΓΒΑ, ΑΓΒ, ΓΑΒ, and ΒΑΓ), while angles only allow for two: in the case of an angle made by the straight lines ΑΒ and ΒΓ in Β, these would be ΑΒΓ and ΓΒΑ, with the only permutable letters being the first and the last one. This means that the same letter words are a lot more permutable in references to triangles than in those to angles.

References to triangles and angles are most likely to appear together within the same propositions in the sample I 1–20, since this part of book I focuses primarily on the angles of triangles; triangles and angles, naturally, share the same letters, and overlap of letter order becomes inevitable if all three angles are mentioned by adhering to the alphabetical principle. Thus, because of the limited number of objects and the frequent combined references to them, the relevant references in I 1–20 constitute a pool of cases which may be usefully compared. I have only considered propositions and, within them, cases, in which letter words containing the same three letters occur at least twice in references to both a triangle and an angle, that is, in such a way that a coincidence of the letter order is possible. For, by contrast, the letters in the references to two angles of the same triangle which are referred to by the same letters could not possibly appear in the same order.

A specificity of the combined treatment of triangles and their angles in I 1–20 is that the triangle, being the object of central interest, is always introduced first, and the angles are only subsequently introduced. This means that the letter phrase which is more flexible regarding letter order is introduced first, while the less flexible ones come afterward. Naturally, it is the later references which one would expect to take into account the letter order of the foregoing ones and change regarding their own letter order accordingly. It is only if two references appear in very close conjunction that the first one could already be expected to adjust its letter order to the immediately following one. Thus, the first reference to the triangle, which is mostly the very first one of the proposition, is unlikely to have been formed regarding the order of its letters in view of the later mentions of its angles; conversely, it is the letters in the references to the angles which would, if at all, have been ordered in view of the first reference to the triangle. Since this situation even further delimits the number of likely permutations of letter order in the introductions of the letter words, it becomes possibly easier to determine what motivated the choice of letter order, in view of there being only two options in the case of the later introduced angles (alphabetical or permuted), that is, whether it was disambiguation or something else.

As for the results, I have counted five propositions which each feature a triangle ΑΒΓ (or, analogously, ΔΕΖ) plus at least one of its angles, including the one which is capable of being referred to likewise by ΑΒΓ, and in which both the triangle and this angle are referred to homonymously alphabetically throughout – both in the introductions of the terms and in their reappearances – that is, by the letter word ΑΒΓ (I 5, 17, 18, 19) or ΔΕΖ (I 4), respectively.[319] As has been mentioned, with the triangle being introduced into the proposition first, and with it being the very first object to be introduced at that, the letter order of the reference to the triangle would have been chosen according to the alphabetical principle of letter introduction regardless of the order of the letters in the subsequent references to the angle by the same letters. Rather, one would expect the letter word referring to the angle in B, i.e., the ΑΒΓ/ΓΒΑ-angle, to have its letters ordered in the second, disambiguating way in view of the foregoing reference to the triangle. Notably, however, in none of these cases, the letters of the letter word referring to the angle get permuted regarding the alphabetical order, so as not to form the same sequence as the reference to the triangle; but the reference to this angle, just like to the triangle, is always ΑΒΓ. This means that the text here does not aim for disambiguation of the two letter words, although it would have been easily achieved by the permutation of the letters in the reference to the angle, or else in the reappearances of the letter word referring to the triangle. Instead, the adherence to the alphabetical principle in both letter phrases is preferred over (admittedly, unnecessary) disambiguation, and the recurrence of the same letter order in reference to two different objects is accepted.[320]

This adherence to the alphabetical principle in spite of the resulting ambiguity of the letter words in the references to the triangle and to the angle in B is even more telling considering that in three of the propositions mentioned (I 17–19), one of the other two angles of the triangle, or even both of them, get permuted after having been introduced alphabetically (ΑΓΒ and ΒΑΓ in I 17), or are introduced in a permuted way (ΒΓΑ in I 18 and 19) before reappearing alphabetically. In none of these cases, permutation could have been for the sake of disambiguation, since no ambiguity of letter order would have been looming – for no other angle could have been referred to by ordering the letters in the same way as any of the other. It is, of course, the letter word for the triangle which could have taken the form of any of the letter words referring to the angles; but since the reference to the triangle does not reappear (I 18, 19), or reappears alphabetically (I 17), there would not have been any ambiguity in this regard either. Thus, it seems to be for reasons other than disambiguation that the references to some of the angles in I 17–19 contain permutations of the alphabetical

319 Besides, in I 4, the triangle ΔΕΖ and its respective angles appear analogously with the triangle ΑΒΓ and its angles.

320 Neither is the alphabetical letter order in the letter phrase for the angle chosen for distinguishing the reference from those to the other two angles, since the references to the three angles of the triangle ΑΒΓ would not be able to have the same letter order, but are unambiguous in any case.

order. It is, in any case, remarkable that it is the reference to the ΑΒΓ/ΓΒΑ-angle of all three angles, that is, the only one which is capable of featuring the chosen alphabetical letter order of the letter word for the triangle, which is *not* permuted in these three propositions, so as to feature indeed the same letter order as the letter word for the triangle. Thus, in these cases, deviation from the alphabetical, where it occurs, is not for the sake of disambiguation, but rather for the sake of something else; and where it could have been used to fulfil the function of disambiguating two letter words with the same letter order referring to different kinds of objects, it is avoided.

The propositions just surveyed provide strong evidence for the assumption that disambiguation does not present a decisive motivation for permutation of the letter order in multi-lettered letter words in the *Elements*. However, they are not fully conclusive in this respect. For they contain, on the one hand, cases in which the letter order could have been permuted for the sake of disambiguation, but is not, and, on the other hand, cases in which the letter order is permuted, but cannot have been permuted for the sake of disambiguation as the permutation does not effectuate any disambiguation.

Now, two propositions (I 6, 16) shall be briefly examined which involve permutation which could actually have been accomplished for the sake of disambiguation, since, as a matter of fact, it leads to the disambiguation of letter words. These are the only other two propositions in *Elements* I 1–20 which contain letter words capable of or actually having the same letter order in references to different kinds of objects.

First, I 6 (Fig. 54): here, the triangle ΑΒΓ and its angles ΑΒΓ and ΑΓΒ are all introduced alphabetically, with the homonymy of the former two, again, not being avoided though it would have been possible to do so by the permutation of the reference to the angle. In the proof part, the permutation of the letter word in the reference to the triangle which reappears as ΑΓΒ, cannot have the function of disambiguation, although it indeed retrospectively effectuates the disambiguation of the introductory references to the triangle and the angle ΑΒΓ. For the resulting form ΑΓΒ turns out to be homonymous with the second angle, in the same way as the former had been with the first angle. That is, the same permutation, while leading to disambiguation in one case, results in ambiguity in another. That the function of the switch is not disambiguation is further underlined by the close conjunction in which the respective ambiguous forms, rather than the disambiguated ones, are mentioned: the triangle ΑΒΓ is mentioned in the same sentence of the exposition as the equally lettered angle ΑΒΓ, whereas the triangle in the permuted form ΑΓΒ is mentioned in the proof part shortly *after* the equally lettered angle ΑΓΒ has reappeared. This means that the switch of ΑΒΓ to ΑΓΒ has no disambiguating influence on the perception of the angle ΑΒΓ, because this one is not even mentioned again after the exposition; rather, it potentially causes additional perception of ambiguity in the proof part with respect to the formerly unambiguously introduced angle ΑΓΒ.

Therefore, a different function of permutation shall be considered which suggests itself in view of the analogously equally lettered angle ΔΒΓ and triangle ΔΒΓ, the equality of which with the angle ΑΓΒ and triangle ΑΓΒ, respectively, is asserted in the proof

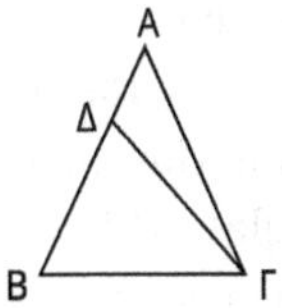

Fig. 54: I 6 according to Heiberg & Stamatis.

part (and proved impossible). First, in stating that the angles ΔΒΓ and ΑΓΒ are equal, the text mentions the two opposite angles, which are made by the straight line ΒΓ shared by the two triangles ΔΒΓ and ΑΓΒ and by opposite sides of the two triangles, ΔΒ and ΑΓ, analogously starting from these opposite sides in the mentions of these angles. (The permuted letter order in the introduction of the angle ΔΒΓ, in turn, seems to be due to the foregoing introduction of the straight line ΔΒ, itself permuted, apparently by analogy with the previously mentioned straight lines ΑΒ and ΑΓ, that is, by starting both times from the line ΑΒ on which both Α and Δ lie, and finishing with the respective point on ΒΓ.) Subsequently, when the supposed equality of the two triangles is deduced from, among other things, the supposed equality of the two aforementioned angles, the respective equal letter order of the references to triangle and angle underscores the analogy of the relation of the triangles with that of the angles. At the same time, it contrasts the only seemingly equal angles and triangles (their actual inequality will be asserted subsequently), because it looks at them from opposite sides of the figure, as it were. I shall elaborate in the course of this section on the idea that the letter order suggests how the diagram is being 'looked' at in the description and on the ramifications of this.

The second proposition to contain a permutation of letter order which is possibly accomplished for the sake of disambiguation is I 16 (Fig. 38): here, the exposition introduces a triangle ΑΒΓ, two of the angles of which are referred to in the definition as ΓΒΑ and ΒΑΓ. That is, the first of the two angles is mentioned by means of permutation of the alphabetical order, so as to have its letters ordered differently than the afore-introduced triangle. Here, in view of the close proximity of the introductions of triangle and angle, it cannot be excluded that the function of the permutation is indeed disambiguation, especially because, at the very end of the proposition, the angle in question reappears alphabetically by displaying the letter order of the triangle ΑΒΓ, with the triangle, this time, not being mentioned anywhere near it, and, thus, no ambiguity looming.

It is, however, again possible that both the non-alphabetical introduction of the letter word and its alphabetical reappearance are motivated by something other than disambiguation, although such a motivation is not as obvious as in I 16. It is notable that the two interior angles ΓΒΑ and the alphabetical one ΒΑΓ are mentioned so as to follow the lines of the triangle consistently in a certain direction, which can be seen from the overlap of the letter order ΒΑ in each of them. If the exterior angle ΑΓΔ (made by ΑΓ and the prolongation of ΒΓ, i.e., ΓΔ) which is claimed to be larger than the interior ones was mentioned after those two, its letter order could be interpreted so as to continue following the lines of the triangle ΑΒΓ in the manner of ΓΒΑ and

ΒΑΓ. However, ΑΓΔ is mentioned first, so the view on the diagram on the basis of its letter order takes a completely different direction than which is suggested by the introduction of the permuted letter order of the subsequently mentioned interior angle ΓΒΑ. In this way, the contrast of the exterior angle with the interior angles, which is the very subject of the definition of the goal, could be emphasized by the contrastive letter order. By contrast, the alphabetical introduction of the first mentioned interior angle in the form ΑΒΓ (overlapping with that of the triangle) would implicate an analogical, non-contrastive view on the two angles which starts, in each case, from point A and leads to line ΒΓ to points B and Γ respectively and from there to Γ and Δ, respectively. This is, in fact, the view taken on the two angles at the end of the proof, where the letter order of ΓΒΑ is switched to ΑΒΓ. The reason for the change of perspective which entails the change of letter order in the case of one of the interior angles could be that the definition of the goal mentions two interior angles each of which is smaller than ΑΓΔ, whereas the proof, having already proved the ΑΓΔ is larger than ΒΑΓ, closes with the assertion that it is (sc. also) larger than ΑΒΓ (= ΓΒΑ): the letter order ΓΒΑ in the definition of the goal would underline the contrast of ΑΓΔ on the one hand, and ΓΒΑ and ΒΑΓ on the other; the order ΑΒΓ in the proof would help visualize the comparative size of ΑΓΔ and ΑΒΓ by presenting them analogically from the same direction.

This review shows that, in our sample I 1–20, there is not a single case in which permutation can only be explained by the text's aiming for the disambiguation of the letter words. What is more, there is only one single case (the angle ΓΒΑ in I 16) in which disambiguation presents an interpretative option besides, however, another, different way to explain it. In another case (the triangle ΑΓΒ in I 6), what could be interpreted as 'disambiguation' in fact leads to ambiguity as much as it leads to disambiguation. In all other propositions which feature letter words capable of or actually showing the same letter order in references to different kinds of geometrical objects, the text dispenses with the mathematically unnecessary disambiguation of letter words which are already unambiguous since they appear as part of unambiguous letter phrases. The reason why it does so seems to be the formal standardization of the alphabetical principle of letter ordering. Even though the disambiguation of the letter order is not necessary from the point of view of mathematical unambiguousness, this is remarkable, because neither is the standardization of the alphabetical principle. The text, which is otherwise unambiguous beyond mathematical necessity precisely through extreme formal standardization,[321] seems to accept the potentially confusing reappearance of the same letters in the same order in different letter phrases which comes about through extremely standardized, mathematically unnecessary alphabeticality. This means that the text seems to prefer formal standardization over clarity of the mathematical content; for even though disambiguation through change of letter order has been pointed out to be unnecessary, it might have added even more clarity to the text. Deviation from the stan-

321 For mathematically unnecessary standardization, see Section 5.1.

dardized alphabetical principle seems to happen, if at all, for reasons which have to do with the 'visualization' or the viewing the objects in the diagram that is expressed by other principles of letter ordering. I shall explore the role of these principles and their relation to the alphabetical one in the subsequent part.

As the second main function of letter ordering identified by Netz besides "disambiguation" I shall now consider what he calls "visualisation." I shall start from two principles identified by Netz according to which the letters in the references to certain objects may be ordered "because of a visualisation:" on the one hand, "linearity," which refers to the linear order in which the letters of multi-lettered letter words are mentioned; on the other hand, "isodirectionality," which is to denote that the letters of the words for two parallel straight lines are mentioned from the same (viewing) direction, as it were.[322] The term "visualisation" for what "motivate[es]" these principles[323] is somewhat misleading because it is not clear from it what is being visualized and why: the term seems to suggest that ordering the letters to these recognized principles serves the illustration or depiction of what can be seen in the diagram or of what the diagram based on the description would show. However, rather than 'visualizing' in this sense, the principles often simply appear to reflect how, that is, from what directions and perspectives, the objects of the underlying diagram are being looked at in their description. This means that the order of the multi-lettered letter words affected by these principles mostly seems to depict the view on the diagram, rather than the diagram itself. In this way, the principles of 'viewing the diagram,' on the one hand, would be able to indicate the very presence of the diagram that is being viewed; on the other hand, they would betray according to what rules the diagram's description is formally standardized, how much it is standardized, and how this standardization relates to the proof content. Netz himself does not consider such ramifications of the two ordering principles for the subject of formal standardization (since, as must be repeated, this is not his subject – he appears to view the evolving standardization again as a "self-regulating conventionality").[324] It is especially the second of his categories, "isodirectionality" which I have found extremely productive for determining the scope of diagrammatic standardization. While he applies the principle only to parallel straight lines (and parallel objects defined by straight lines), I shall analyze how isodirectionality, or, more generally, "directionality,"[325] determines the order of the letters to a much greater extent, both as regards the kinds of objects affected by it, and as regards the number of objects affected at one time.

Let me start from Netz's conceptualization of the two principles, and explain, by using examples from *Elements* I, how they shape the text according to him. I ask what

322 Netz 1999a, 81.

323 Netz 1999a, 82.

324 See above p. 341.

325 Netz also uses the more general term once (1999a, 82). I shall explain below p. 383, why this broader term might be more suitably applied to the phenomena which I have recognized.

can and cannot be concluded from the conceptualization for our own questions, before expanding the concept and adapting it to our own focus and analyzing the ordering of the letters according to it.

The concept of linearity

First, for linearity, which, according to Netz's evaluation, regularly determines the order of letters in multi-lettered letter words. It means that the letters are mentioned in the order in which one encounters the corresponding points in the drawing when tracing the contour of the object.[326] For example, in I 47 (Fig. 46), the letter word used to refer to the square on the hypotenuse ΒΓ, defined by the letters Β, Γ, Δ, Ε, which first introduces the letters Δ and Ε into the text, does not have these letters in alphabetical order at any of its instances; for this to be the case, the letters of the side of the square shared with the triangle would be mentioned before the letters referring to the parallel side ΔΕ. Instead, it permutes the alphabetical order of the letters in order to present them in a linear way (ΒΔΕΓ). The decidedly linear way of presentation is recognizable precisely from the deviation from the otherwise dominant alphabetical principle of letter ordering. In I 47, this letter word is actually the only one which qualifies at all for displaying linearity, and it is of the only kind of letter words (that is, such as refer to quadrangles) which does so in *Elements* I. This is because the letters in all two-lettered 'names' and all three-lettered ones which occur in book I and refer to triangles, circles or angles cannot be but linear.[327] Thus, in order for the letters of a word to be arranged in a linear way specifically, the word has to be able to allow for non-linearity, which, in book I, is the case only with four-letter words referring to quadrangles. A survey of all references to quadrangles in *Elements* I indicates a regular preference of linear over non-linear presentation: of all words referring to quadrangles (53), 52 are described linearly.[328]

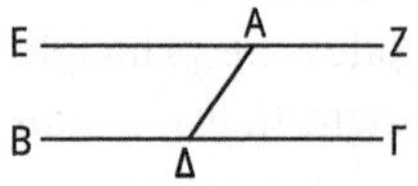

Fig. 55: I 31 according to Heiberg & Stamatis.

326 Netz 1999a, 82: "The sequence of letters in a many-lettered name may be either linear (i.e., corresponding with some continuous survey of the line [. . .]) or non-linear [. . .]."

327 Netz 1999a, 82, n. 24. An exception to this is presented by a straight line which is designated by three instead of the necessary two points/letters. This, in fact, Netz's example for linearity (1999a, 82), but it rarely occurs in the *Elements* (cf., however, I 31 [Fig. 55], which refers twice to the straight line ΕΑΖ). This is because, strictly speaking, three points on a straight line actually constitute two straight lines (or line segments) rather than only one.

328 This confirms the result of Netz's review of all "relevant" "many-lettered names" in Apollonius' *Conics* I and Euclid's *Elements* II (1999a, 82, nn. 24 and 26).

The concept of isodirectionality

The second principle which, according to Netz, shapes the order of the letters of multi-lettered letter words for reasons of visualization, isodirectionality, is supposed to mean that the order of the letters in letter words referring to parallel straight lines is aligned with one another. That is, this principle is about the related presentation of at least two objects by means of their letter order. Two parallel straight lines are "isodirectional," if the "flow" indicated by the letter order "is the same" in the case of both lines, i.e., if it implies that they are being described from the same point of view, whereas they are "counter-isodirectional," if the "flow" is opposite, with the lines being looked at from opposite directions (I shall use the term 'non-isodirectional' instead).[329] All parallel straight lines are, by their letter order, necessarily either isodirectional or counter-isodirectional. Isodirectionality occurs several times in I 47 (Fig. 46), e.g., in the construction part, where it affects three parallel straight lines at a time: καὶ διὰ τοῦ Α ὁποτέρᾳ τῶν ΒΔ, ΓΕ παράλληλος ἤχθω ἡ ΑΛ ("and through Α should be drawn ΑΛ parallel to ΒΔ and ΓΕ").[330] From this example, it is, however, not clear whether the presentation explicitly follows the principle of isodirectionality, or whether it happens to do so merely by following the alphabetical principle. It would require a deviation from alphabetical lettering, as in the above example for linearity, by at least one of the letter words, in order for the pair to present conclusive evidence for the explicit application of the principle of isodirectionality in the description of the objects. In mere quantitative terms, in any case, isodirectionality seems to be almost as much preferred over its alternative, i.e., non-isodirectionality, as linearity has been shown to be preferred over non-linearity: of 57 instances of (mostly two, only occasionally three) parallel straight lines mentioned in conjunction with one another in *Elements* I 27–38 – a string of propositions which centrally revolves around parallel straight lines – 52 are isodirectional.[331]

Linearity, isodirectionality, alphabeticality, and diagrammatic standardization

From our point of view which is focused on the standardization of the diagram and of the text on account of it, the main question arising from Netz's conceptualization is how far and how consistently the principles of what he calls "visualisation" determine the letter order in *Elements* I. A first review of the phenomena of linearity and isodirectionality has returned overwhelming results. However, the exact scope of the application of these principles and their conclusiveness regarding diagrammatic standardization, even where they are regularly applicated, remain to be determined. The two main questions in the face of the subject of formal standardization are: (1) how are each of the two 'prin-

329 Netz 1999a, 82.

330 *Elem.* I 147 (vol. 1, p. 63.12 f. Heiberg & Stamatis).

331 Thus, the results confirm those of Netz's survey of a set of cases from Archimedes and Apollonius.

ciples of viewing,' as I conceive of them, related to the other principle we have seen to shape the letter order regularly, namely alphabeticality, and (2) how relevant are the principles for the description of the diagram in general, as they seem to concern only a small number of objects overall and at a time. I focus on the first one, the relation of linearity and isodirectionality to alphabeticality, now; the second will emerge from my considerations and be pointed out in due course. First, I explain the problem and the methodology required for determining how far the principles of viewing shape the description of the diagram and how far they are conclusive regarding the question of the standardization of the diagram. Second, I focus on 'linearity and alphabeticality,' before turning to 'isodirectionality and alphabeticality.'

Both alphabeticality and the 'principles of viewing' have been shown to characterize the order of the letters regularly in the – albeit, different – samples which have been surveyed. Of all multi-lettered words of *Elements* I 1–20, 76% had turned out to be ordered alphabetically; of all letter words in references to quadrangles in *Elements* I 34–47 (which is to where the occurrence of these objects is confined in book I), 98% have displayed linearity; and in *Elements* I 27–38, around 91% of all pairs (or trios) of parallel straight lines have been shown to be isodirectional. Provided that the numbers gathered for each principle are representative, the strong prevalence of both alphabeticality and the two viewing principles suggests a strong overlap of both alphabetical and isodirectional ordering and of alphabetical and linear ordering in the description of the respective objects. The regular adherence to alphabetical-linear and alphabetical-isodirectional description indicates, in any case, that the letter order of multi-lettered letter words is, as a matter of fact, standardized with regard to both types of letter ordering principles. However, as has just been pointed out regarding the examples for linearity and isodirectionality from I 47, the alphabetical-linear and alphabetical-isodirectional description, respectively, does not betray which principle motivates the letter order in the first place, and is, thus, not conclusive of the explicit application of one of the principles. Rather, it is the deviation from the alphabetical principle resulting in linearity or isodirectionality, and, vice versa, from linearity and isodirectionality resulting in alphabeticality which would seem to indicate that the respective type of letter ordering has been practiced in a particular case, and may have been the one standardly pursued.

Further, in order to conclude from the standardization of principles determining the letter order in multi-lettered words about the standardized presence of the diagram itself, the principles of lettering the diagram itself must be taken into account. For, if one considers the possibility that the text describes an already present diagram, one must be aware that the alphabetical distribution of the letters to the objects in the diagram will already have followed certain ordering principles. As far as multi-lettered objects are concerned, these might be the very principles the influence of which on the order of letters in the multi-lettered *words* is presently under discussion: linearity in the case of certain three- and more-lettered objects such as quadrangles, and isodirectionality in the case of parallel straight lines. The letters in the *letter words* could be ordered either alphabetically or, in view of the (emerging or existent) diagram, in a linear/isodirectional way, or both alphabetically and isodirectionally, with four combinations being possible in the references to the respective objects.[332] Linearity and isodirectionality in the diagram itself would be indicated accordingly by the alphabetical-linear and alphabetical-isodirectional allocation of the letters. Thus, alphabetical-linear or alphabetical-isodirectional description would correspond to (alphabetical-)linear and (alphabetical-)isodirectional lettering of the diagram.

If linearity and isodirectionality should indeed turn out to have determined the lettering of the diagram itself, alphabetical-linear and alphabetical-isodirectional letter ordering must be presumed to be the dominant forms of letter ordering in the letter words. In this case, linearity and isodirectionality would be no longer conclusive of either of the principles' having shaped the letter order of multi-lettered words *as principles of viewing an already existing diagram*. For these kinds of letter order in the letter words could have been formed by imitation of the letter order of the objects in an underlying diagram – in which, if it preexisted, the lettering of objects described in an alphabetical-linear or alphabetical-isodirectional way could not be but alphabetical-linear or alphabetical-isodirectional as well. Alternatively, alphabetical-linear and -isodirectional description could also have been the conventional ways of referring to polygons (and other objects capable of linearity) and parallel straight lines, respectively, in which case the letter order would simply indicate how to draw and letter these objects in a diagram yet to be constructed. The latter is easily conceivable since the linear and isodirectional way of lettering and referencing, e.g., quadrangles and parallel straight lines, respectively, is, though not the only possible and though a geometrically not necessary, yet an obvious way to do it, as it imitates the geometrical situation on a formal level: linear letter ordering traces the contour of a figure, isodirectional referencing of parallel straight lines presents analogously constructed objects in an analogical way.

It is, thus, if at all, the deviation from the in itself inconclusive alphabetical-linear and alphabetical-isodirectional ordering in letter words which could be indicative of an underlying diagram. However, of the three possible kinds of deviation, two in each case are likewise not indicative: on the one hand, non-linear- or non-isodirectional alphabetical description would not be indicative of an underlying diagram; for alphabetical ordering within letter words is a principle which does not depend on viewing a diagram which is already there but which suggests itself even as a purely textual means. On the other hand, a letter order which does not pay attention to any of the ordering principles discussed so far (non-alphabetical *and* non-linear/non-isodirectional) would not be indicative of the presence of a diagram unless, in conjunction with the lettering of the diagram, it displayed some meaningful imitation of or deviation from the latter. This leaves only non-alphabetical linear and non-alphabetical isodirectional letter order to be poten-

332 I.e., alphabetical-linear, non-alphabetical linear, alphabetical-non-linear, and non-alphabetical non-linear, and alphabetical-isodirectional, non-alphabetical isodirectional, alphabetical-non-isodirectional, and non-alphabetical non-isodirectional ordering.

tially indicative, because these combinations might indicate that an underlying diagram is being viewed and described from a perspective which does not coincide with the one from which the diagram most obviously has been or is to be lettered.

However, even in the instance of non-alphabetical linear or non-alphabetical isodirectional letter ordering, which has been, thus, singled out as being in general potentially indicative of an underlying diagram, the deviation from the alphabetical ordering is not necessarily indicative of an underlying diagram. The question, here, is, first of all, why, and on the basis of what corresponding diagrammatic lettering, the letter order in the letter words deviates from the alphabetical order. In the case of non-alphabetical isodirectional description, the corresponding lettering in the diagram can only be non-alphabetical isodirectional as well. Non-alphabetical-linear description of, e.g., quadrangles, by contrast, can correspond to a kind of diagrammatic lettering which is either equally non-alphabetical linear (i.e., non-linear), or which is, divergently, alphabetical-linear. In the latter case, the deviation from the corresponding alphabetical-linear lettering by means of starting the letter word with one of the letters which, of all the letters which constitute the letter word, is not the first in the alphabet, could indicate that the diagram has already been there: for if indeed the alphabetical-linear type of ordering the letters in letter words is (one of) the standardized one(s), deviation from it, even though the standardized type would have been possible on the basis of the diagram's lettering, seems to be due to a particular perspective taken on the diagram. For the adherence to the linear type of ordering suggests that the letter order is not random but follows a certain viewpoint or direction, the starting point of which, however, in such cases, could only be explained by the diagram's underlying.

The other kind of diagrammatic lettering which may correspond to a non-alphabetical linear letter ordering in the letter words, and the only kind which may correspond to non-alphabetical isodirectional letter ordering, is non-alphabetical-linear (= non-linear) or non-alphabetical isodirectional (= non-isodirectional) lettering, respectively. A non-alphabetical linear/non-alphabetical isodirectional letter order in combination with non-linear or non-isodirectional lettering will, however, not in any case present conclusive proof of the presence of a diagram, and it will be in most cases impossible to determine definitively if they do so. To be sure, conditions under which such a letter ordering situation will be conclusive may be defined, but it will be difficult to decide whether they apply in each particular case. Broadly speaking, the non-alphabetical linear letter order will be conclusive if it implies a decision, that is, either against another equally or more plausible letter order on the basis of the corresponding diagrammatic lettering, or against another lettering of the object in the diagram.

In accordance with these preliminary considerations, I will proceed as follows: I shall ask first whether the non-alphabetical linear letter order immediately suggests itself at the point at which the letter word is introduced into the diagram, or whether there would be another way of referring to the transmitted lettering of the diagram which is more evident at this instance, while the present one is elusive. Apart from other ways of ordering the letters non-alphabetically linearly, especially the possibility of alternative alphabetical non-linear ordering is worth considering, which may have been especially plausible if the diagram was not present. Alphabetical-linear ordering, by contrast, will not be a possible alternative for referring to an object which can also be referred to by means of non-alphabetical linear ordering. An explicit, non-intuitive decision for the non-alphabetical-linear variant appearing in the text as opposed to other variants which would be more obvious at first sight would seem to imply that the linear description is due to a certain perspective upon the already present diagram. In this way, it is possible to compare the possible letter order in the case of linearity and test the non-alphabetical linear letter order as to its conclusiveness in a way which is not possible in the case of isodirectionality.

Since, however, I shall not presuppose that the text and its letter ordering have been shaped on the basis of the diagram and its lettering but rather aim to prove that has been so, I shall, when analyzing linearity and isodirectionality, take into account not only the kinds of letter order which are possible on the basis of the transmitted lettering, but indeed all variants of letter order which are possible on the basis of all kinds of diagrammatic lettering which are possible. In so doing, I shall take a detour via the lettering of the diagram itself and how it came about, in order to retrieve other forms of textual letter order which are conceivable and compare them to the transmitted one.

Thus, I shall ask, second, why the diagrammatic *lettering*, which is transmitted as the one which corresponds to the non-alphabetical linear letter order in the letter word, deviates from the alphabetical-isodirectional/-linear principle to begin with. I shall consider precisely how the object(s) in the diagram would have been lettered if the diagram was lettered before the text had been written, and how, consequently, the letters would have to be ordered in the corresponding letter word according to this. The aim of this is to determine how flexible the lettering and letter ordering had actually been in each case, and what the lettering possibilities imply about the origin of the transmitted non-alphabetical linear order: whether it can be concluded to have been formed on the basis of the diagrammatic lettering or not.

There are two possibilities: on the one hand, it could be that the lettering in any case has to be non-linear/non-isodirectional because of when it is due in the order of construction, which might entail that some points of the object have already been named by previously attributed letters, leaving fewer choices or none of applying the rest of the letters and eventually precluding the possibility of linear/isodirectional lettering altogether (as, e.g., in the case of a square which is drawn onto a straight line that has already been lettered, with the 'new' points of the square being lettered separately from the ones of the straight line). On the other hand, it might turn out that the attribution of the letters is to a considerable extent or even completely free and that a linear lettering, resulting in alphabetical-linear letter ordering in the corresponding letter word would have been possible.

If the diagrammatic lettering only deviates, according to the first possibility, from the linear or isodirectional principle because it has to, with the principle being adhered to wherever possible, and, thus, being recognizable as the preferred one, the deviation from the alphabetical ordering in the text – in the form of non-alphabetical linear or isodirectional lettering – would not be indicative of the diagram's being viewed and, thus, underlying. For then, the resulting letter order would be either the only possible one or at least the most natural one conceivable which would suggest itself to anyone drawing and lettering the diagram. To be sure, it depends, here, on the complexity of the diagrammatic figure, whether this only one possible letter order could be extemporized without looking at the diagram, or whether its underlying lettered drawing would have to be presupposed in any case.

If, however, according to the second possibility, the non-linear/non-isodirectional lettering transmitted in the diagram according to the letter order of the text is not the only possible type of lettering conceivable, but if it turns out that, alternatively, linear/isodirectional lettering of the diagram's object – and, accordingly, alphabetical-linear/isodirectional ordering of the letter word – would have been, theoretically, possible, the question arises why the diagrammatic lettering and the letter order in the letter word have not been organized in this way. The most obvious explanation for the deviation from alphabetical linearity and isodirectionality would be that the diagram has been lettered first, and that the text would have had the letters in the letter words ordered on the basis of the diagram's lettering, rather than the other way around. One reason for this could be that the diagrammatic lettering simply opted against linearity or isodirectionality, as there would have been an equally conceivable way of lettering. Another reason could be that the diagram's lettering might have followed a different logic than the text's referencing the objects in the diagram, namely that the objects which were lettered in due order in the diagram were different ones than or did not exactly overlap with those that were referenced in the text. Because of this partial lack of overlap, the linear or isodirectional ordering in the letter words, if done according to the diagram, could turn out to be non-alphabetical, even though the diagram itself was lettered to some extent (alphabetically-)linearly or (alphabetically-)isodirectionally. Whatever the reason for the deviation from (alphabetical-)linear or isodirectional lettering, it is the possibility of an alternative lettering in this way, which would have been available in case that the diagram was lettered according to the diagram-independent letter order of the text, which shall be taken as an initial indication that the diagram preexisted. Subsequently, I shall compare the textual letter order which corresponds to the alternative lettering(s) of the object in the diagram, which would be equally or even more plausible than the transmitted lettering, that is, above all, the alphabetical-linear or -isodirectional letter order matching linear or isodirectional lettering, and compare it to the letter order of the transmitted text with respect to the question which of them would be the one to choose if the diagram had not been there from the start. If, indeed, the transmitted letter order would, thus, be not only in itself elusive, but clearly had less evidence than another variant which has also been proven viable with regard to how the object *could be* lettered in a diagram, this will be a strong indication for the underlying presence of a diagram. However, the possibility that the letter order which appears to be elusive at first sight has been formed in view of the whole geometrical situation, even if the diagram had not been there materially, has to be taken into account, seeing that the configurations, especially in book I, are not so complex that they could not be mentally pictured – particularly by someone who had previously already discovered the proof presented by the proposition.

According to these general deliberations, the instances of non-alphabetical linear and isodirectional lettering would, however, have to be considered individually regarding their indicativeness as to an underlying diagram, because the extent to which the diagrammatic lettering is fixed and to which the letter ordering in the text is elusive without one's having the diagram at hand or rather the most obvious way to do it even on merely picturing the geometrical situation, differs from case to case. Here, one has to consider carefully in each individual case which coming-about of the letter order is more likely: that the letter order must have been 'read' from the diagram or could have been shaped on merely imagining the geometrical situation. Very often, one will not be able to tell. I shall confine myself to discussing some representative examples.

If it appears to be highly probable in certain cases that the diagrammatic lettering of the objects has foregone the ordering of the letters in the textual references to these objects, one may ask further why the letter order has been chosen to be linear or isodirectional, respectively, whereas the underlying lettering is non-linear/-isodirectional. Since the letters have been shown to be ordered on the basis of the viewable diagram, the ordering seems to have been influenced, above all, by how an object was viewed, and linearity and isodirectionality seem to be the principles according to which the text implies doing so. On determining that linearity and isodirectionality are principles which indicate that an object is being viewed, the question arises, why these two of all things? What makes them eligible for organizing the description of geometrical objects in a diagram, and what motivates the viewing of the objects according to them?

Lastly, one would also have to examine whether the application of linearity and isodirectionality in relation to alphabeticality in the letter words differs according to parts of the proposition or according to the letter words' introduction and reuse, in order to see whether the text changes the perspective after the construction of the objects or, literally, 'views' the objects of the diagram throughout. It is the latter which, on the one hand, would point to the diagrams' being present from the beginning, and, on the other hand, to the formal standardization of the letter order in letter words to 'principles of viewing.'

Following these considerations, all instances of four-letter words, that is, all references to quadrangles in *Elements* I have been checked with respect to (1) their adherence to alphabeticality and linearity, (2) the frequency in which each of the four possible combinations of letter ordering occur, and, where indicated, (3) the corresponding letter order in the diagram. The same has been done, respectively, for all occurrences of parallel straight lines in I 27–38 regarding the relation of alphabeticality and isodirectionality.

Linearity and alphabeticality in the lettering and letter order of quadrangles

First, for linearity and alphabeticality in quadrangles: of all 53 four-lettered words in our sample,[333] an overwhelming number (52) had been shown to be linear. Surprisingly, in view of the dominance of the alphabetical principle in our then sample I 1–20, the present sample is alphabetical roughly to the same extent as it is non-alphabetical (26 vs. 27 instances). Accordingly, about half of the letter words are alphabetical and linear (25), and roughly the same number is non-alphabetical and linear, while there is only one occurrence of alphabetical and non-linear ordering, and none which is both non-alphabetical and non-linear. The, in view of the assessment of alphabeticality in I 1–20, unusually high proportion of non-alphabeticality could, indeed, have to do with the comparative complexity of the object kind on which the present sample is focused, i.e., four-lettered letter words referring to quadrangles. For the first 20 propositions did not contain letter words made up of more than three letters. I shall explore this connection shortly. In any case, because of the deviation from the alphabetical principle of ordering letters in roughly half of all cases, however the deviation will turn out to be motivated, the sample will be possibly indicative as to how the letter order in multi-lettered letter words is standardized and as to whether the diagrams were standardly present with the proposition texts.

First of all, the extreme adherence to the linear principle, and more than 50% deviation from the other otherwise dominant principle of letter ordering, alphabeticality, at the same time, show that the principle of linearity in the textual description of multi-lettered objects, as a matter of fact, has been explicitly pursued so as to appear almost completely standardized; by contrast, the alphabetical principle, on the grounds of these overall numbers, does not appear to be standardized at all and to have been applied in the letter words only secondarily if it happened to coincide with 'linearity.' However, these numbers do not give away the reasons for the apparent prioritization of linearity in the letter words for quadrangles. These can only be determined with regard to the letter distribution in the corresponding diagram, which shall be considered shortly with respect to the conclusiveness of diagram regarding the standardization of the diagram.

Second, regarding the question of whether the diagram has been present to begin with or not, it is the group of letter words which are linear but non-alphabetical which is potentially indicative, while the alphabetical-linear occurrences which represent

333 In total, there are 86 references to quadrangles, 33 of which are two-lettered.

roughly the other half of all references to quadrangles are generally non-indicative. As has been pointed out, the non-alphabetical linear instances are not in any case indicative of referring to an already present diagram, but only if they either deviate from the alphabeticality of the corresponding linear lettering of the object in the diagram, or if the corresponding non-linear lettering of the quadrangle can be shown to deviate freely from the principle of (alphabetical-)linear lettering and is not predetermined by the type of diagrammatic figure of which the object forms part or by the order of its construction.

As regards the diagrammatic lettering of the quadrangles, all non-alphabetically linearly ordered letter words correspond to non-(alphabetical-)linear lettering in the diagram, while there is not a single occurrence of (alphabetical-)linear lettering which is referred to by non-alphabetical linear letter ordering in the respective letter word. This suggests that alphabetical-linear lettering (e.g., ΑΒΓΔ), wherever the diagrammatic object has been or must end up to be organized according to it, is by all means adhered to in the description of the diagram, that is, in spite of the possibility of non-alphabetical deviation from it, which is always given: the linearity could be retained in the description of the object by using the alphabetical form mentioned (ΑΒΓΔ) or one of the non-alphabetical but equally linear ones ΒΓΔΑ, ΓΔΑΒ, ΔΑΒΓ, which are never chosen. Regarding the above question of what ordering principle is decisive for the letter order in multi-lettered words, this means that the preferred linear order of description is also alphabetical wherever it can be. The next question, however, concerns those cases where it cannot, and is, why it cannot in those cases.

The assessment of the adjunct diagrams shows that, by and large, the quadrangles of *Elements* I are lettered (alphabetically-)linearly in the diagram if they introduce all of their four letters newly into the text, and may be, consequently, drawn and lettered all at once, and non-(alphabetically-)linearly if they have to integrate letters which have already been introduced as part of other objects. Thus, the non-linearity corresponding to the comparatively high proportion of non-alphabeticality in the four-lettered letter words referring to quadrangles indeed seems to have to do with the aforementioned constraints put on the attachment of the letters to more complex objects and the according increasing probability of necessitated deviation from the alphabetical. As has been pointed out, with more points constituting an object as part of a larger geometrical configuration, this object becomes more likely to be constructed by integration of other, previously constructed multi-lettered objects (rather than just a point), mostly a straight line as one of its sides, or even two straight lines constituting three of the four points of the then constructed triangle, the lettering of which already predetermines to a considerable degree that of the quadrangle. This is the case unless the quadrangle is the first object of the configuration to be constructed. Indeed, the diagrammatic lettering of quadrangles is (alphabetical-)linear in all cases where the object is constructed first, except one, and it deviates from the linear type when the object is introduced later.

I look at three examples of quadrangles from *Elements* I which are referred to by non-alphabetical linear letter words and which are, correspondingly, lettered non-linearly in the diagrams. They differ according to how far their lettering in the diagram is fixed because of the point at which they occur in the order of construction, and according to the degree of elusiveness of the non-alphabetical linear letter order in the references to them. According to the above explained methodological considerations, I shall, in each case, determine first whether there would have been, on the basis of the transmitted lettering, an alternative, more plausible way of ordering the letters in the letter word than the one which occurs in the text. If there is, this way shall be taken as an indication that the text has assumed a certain perspective onto the already lettered object in an already present diagram. Second, I shall consider whether there would have been alternative ways of diagrammatic lettering of the quadrangle which included (alphabetical-)linear lettering to which alphabetical-linear letter ordering in the text would have corresponded. Third, I shall compare the forms eventually evolving from alternative lettering to the transmitted letter order as to which one has more evidence if the text is considered without its diagram. The three examples cover a range from fixed to non-fixed diagrammatic lettering. I shall start with one quadrangle of which all four points have already been referenced by the foregoing construction of the configuration; second, I shall look at a quadrangle which has already had two of its letters fixed by the time of its own introduction into the text; third and finally, I shall consider the only one example from *Elements* I in which the lettering is completely flexible, but which is still non-linear, and turns out a non-alphabetical linear letter order, accordingly.

First, the trapezoid ΑΒΗΔ from I 35 (Fig. 45). Its letter order is non-alphabetical linear, corresponding to non-linear lettering in the diagram. In view of the diagram, it does not appear to be particularly elusive. There is no way how the quadrangle, on the basis of the present lettering, could be referred to in an alphabetical-linear way. The only other way how the letters could be ordered which would be equally conceivable as the transmitted one by its adherence to one of the standardized principles, would be the non-linear alphabetical order ΑΒΔΗ which would trace the quadrangle by mentioning its parallel lines isodirectionally, a type of reference which is attested in the letter word ΑΒΓΔ in the conclusion of I 34 (the only occurrence of alphabetical non-linear letter ordering of quadrangles in our sample).[334] In view of the otherwise complete adherence to linearity in the description of quadrangles, and in view of the other quadrangles described in I 35, which are all referred to linearly, the linear letter order seems to be the one which is more plausible, although *in itself*, it does not suggest a perspective onto the diagram any more than the other letter order which would have been conceivable. However, it is by taking into account the analogical direction which the linear tracing of all the quadrangles in this diagram invariantly takes which seems to reveal that the description takes a certain perspective onto a materially underlying diagram: all four-letter words (ΑΒΓΔ, ΕΒΓΖ, ΑΒΗΔ, ΕΗΓΖ) start,

334 *Elem.* I 34 (vol. 1, p. 48.7 Heiberg & Stamatis).

as it were, at the upper one of the parallel straight lines mentioned in the hypothesis with the upper-left point of the respective object, following it downward to or toward the lower parallel line, then to the right, and finally to the upper parallel line again. The principle adhered to by this description might be called 'isodirectionality,' though Netz himself has used it only of parallel straight lines. I shall elaborate on the possibilities and the necessity of expanding the concept for our own endeavor below. In the present instance, in any case, linearity itself does not indicate viewing, but only by analogy with other, isodirectional instances of linear ordering, that is, as a precondition of a larger-scale isodirectionality. Regarding the indicativeness of the analogical linearity of the letter order ΑΒΗΔ, one must keep in mind that this quadrangle just like ΕΗΓΖ is introduced into the text at a very late point when the diagram, if it were to be constructed from the text, would have been fully drawn and lettered anyway (both quadrangles do not introduce any letters into the text). So the perspective, in this case, could have been chosen on the basis of the simultaneously constructed diagram, not on the basis of one which has had to be underlying from the beginning. Thus, for the question whether the analogical linearity of the quadrangles in I 35 indicates perspective, one needs to consider the other likewise non-alphabetical linear quadrangle ΕΒΓΖ, which is introduced, after the alphabetical-linear ΑΒΓΔ, at the beginning of the proposition, introducing itself two new letters into the text. This will indeed be my second example. It is because of the late occurrence of the presently discussed trapezoid ΕΗΓΖ that its diagrammatic lettering will be unrevealing. As all four of its letters have already been introduced into the text, before the letter word for the trapezoid itself is introduced, one cannot try out different letterings without changing randomly one or several of the previously introduced letters, and one cannot determine on the basis of the lettering of this quadrangle and its 'lettering history,' whether the lettering has followed principles which point to the lettering of the diagram previously to the text, or whether its diagram has been lettered according to the text. Three of its letters have been introduced as part of the quadrangle ΑΒΓΔ, and Η has been introduced and applied to the point at which the two parallelograms cut each other, after the parallelograms themselves have been introduced. This only shows that the diagram is, in any case, lettered in the order of the objects' introduction into the text, but it does not show whether it, regarding the lettering of the objects themselves, follows its own principles, or, conversely, the letter order of the text.

Second, the just mentioned parallelogram ΕΒΓΖ from the same proposition I 35 (Fig. 45). It is introduced as the second object of the proposition in conjunction with another parallelogram, ΑΒΓΔ: Ἔστω παραλληλόγραμμα τὰ ΑΒΓΔ, ΕΒΓΖ ἐπὶ τῆς αὐτῆς βάσεως τῆς ΒΓ καὶ ἐν ταῖς αὐταῖς παραλλήλοις ταῖς ΑΖ, ΒΓ· ("Let ΑΒΓΔ, ΕΒΓΖ be parallelograms on the same base ΒΓ and in the same parallels ΑΖ, ΒΓ").[335] While the letter order in the reference to the first parallelogram, like in almost all others referring to

335 *Elem.* I 35 (vol. 1, p. 48.12–16 Heiberg & Stamatis).

quadrangles which are the first object to be introduced into the respective proposition, is alphabetical-linear, based on a linear lettering of the diagram, the order of EBΓZ, mentioned by the very next word, is already, like the references to almost all other quadrangles referred to *not* as the first object, non-alphabetical and linear, in accordance with its non-linear lettering in the diagram. The non-alphabetical letter order EBΓZ is elusive without the diagram at hand; alternatively, it would have been possible to order the letters in an alphabetical non-linear way, mentioning the two parallel lines BΓ and EZ in an isodirectional way and starting with the line which has already been introduced, before introducing the two new points in alphabetical order. This would make sense if the construction of the diagram and its lettering was to be accomplished on the basis of the text. The letter order which can be found in the text instead becomes easily comprehensible when tracing the letters in the diagram: it is referred to not only linearly, but also analogically with ABΓΔ, in the isodirectional way by which all four quadrangles referred to in this proposition are aligned with each other. In order to stick to this kind of 'isodirectionality,' as I have called it by extension of Netz's concept, the letter order of the second parallelogram must necessarily be non-alphabetical, as two of its letters – the ones which it shares with ABΓΔ– are already fixed, so the order of the letters, if linear by analogy with ABΓΔ, cannot start with B. While the linearity of EBΓZ alone is not indicative, the analogy may entail that the diagram has been viewed while being described. It has been argued that the isodirectionally linear letter order of the previously discussed quadrangle, because of its late introduction into the proposition at a point when all of its letters have already been introduced, may have been formed on the basis of the already constructed diagram and its description up to this point, and is, thus, not necessarily indicative of the underlying presence of the diagram. The reference to the present quadrangle, by contrast, appears at the very beginning of the proposition, and it posits a quadrangle of which only two points so far have been introduced. Rather than mentioning its points chronologically in the order in which they have been and are to be made part of the diagram, the letter order indicates that the quadrangle is being viewed in comparison with another quadrangle and, thus, as part of the whole diagrammatic figure. Taking this perspective at the very beginning of the proposition seems to presuppose the ready configuration of the two quadrangles, especially because the two analogical references to the quadrangles are introduced before the exposition even specifies how ABΓΔ and EBΓZ are to be related. For while one knows from the two-letter words that the two parallelograms must share one straight line BΓ, the relation of AΔ to EZ becomes only clear from the immediately following specification which locates the two parallelograms between BΓ and AZ and, consequently, AΔ and EZ on the same straight line AZ, that is, between AZ. This specification, in turn, also allocates Z to a specific one of the points of the two free points of the parallelogram, which leaves the remaining possible spot for E (as it must lie on AZ, and, thus, between A and Z). Thus, the order in which the objects are mentioned and specified, that is, with the parallelograms being mentioned first, which already presuppose by their letter order the specification accomplished only afterward, suggests that the description is done in view of the ready diagram. However, the figure, thus far consisting of two parallelograms between two parallel lines, is sufficiently manageable to be pictured mentally and hence described in order for it to be constructed accordingly.[336] The letter order could, thus, have been preferred over the alternatively suggested alphabetical-non-linear one even if the diagram was only just to be constructed, as it might have served to bring to mind immediately how the second quadrangle is to be related to the first one. This means that the letter order itself, by comparison with its alternative, does not necessarily indicate the viewing of a diagram. As far as the lettering of the quadrangle in the diagram is concerned, which is non-linear as it is, there would have been, despite the fact that two letters are already fixed, an alternative, namely linear way (ΒΓΕΖ). In order for the lettering of the parallelogram to be linear, Ε and Ζ would have to switch places, and the straight line subsequently introduced as ΑΖ would have to be named ΑΕ. This means that, while the partial overlap of the second parallelogram with the first fixes and, thereby, restricts the choice of the starting point of the linear sequence in the second one, i.e., Β, it does not preclude that this parallelogram, just like ΑΒΓΔ, could be lettered in a linear way. Thus, the diagrammatic lettering of the second parallelogram which corresponds to the non-alphabetically linearly ordered letter word ΕΒΓΖ, in fact, turns out to deviate to some extent freely from the principle of (alphabetical-)linear lettering; the corresponding letter order of the letter word does not actually have to be non-alphabetical, because the corresponding diagrammatic lettering is not forced to deviate from the linear principle by the restrictions imposed by the order of the construction. The alphabetical-linear letter order indeed seems to be at least as conceivable as the transmitted non-alphabetical-linear one in case that the text not had been written on the basis of a diagram, because it would, after the alphabetical-linear introduction of the first parallelogram, first trace the part which both parallelograms share once more, before mentioning the two new points in alphabetical-linear order, and, as the first parallelogram, in anti-clockwise direction. It is, however, not sure which letter order would have been intuitively preferred, if the text had not been written on the basis of a ready diagram. For the analogical situation of the two parallelograms might have been, as has been pointed out, envisaged even without a diagram, and the two new letters might have been arranged

336 It is another question whether the belated specification of many of the objects – the fact that they often cannot be located the exact moment they are being referred to – indicates that they are not actually meant to be constructed, since they are already there, or whether it implies that the objects are not to be constructed in the exact sequence in which they appear, but that the construction is assigned to some extent summarily. I have argued that, as long as the diagram can be constructed by the end of the 'constructing section,' the summary character of the latter, including permutations of the order of construction and specification of letters, is not conclusive of the preexistence of the diagram. In the present case, the permuted order of specification is, in any case, wholly non-indicative regarding the question of the diagram's preexistence, as the specification is accomplished within the same sentence.

in alphabetical order – at the beginning of the word E, at the end of it Z, around the two known ones, which are mentioned by analogy with their middle position in ΑΒΓΔ in second and third position. Thus, as the two ways of letter ordering resulting from the two ways of diagrammatic lettering seem to be equally viable, the theoretical possibility of linear lettering of the diagrammatic object is not indicative of the preexistence of the diagram.

However, the lettering of the present parallelogram sheds some light on how the diagrams might have been lettered before the text had been formulated, and may, in so doing, indicate its own precedence. It has been noted that quadrangles, when introduced into the proposition as the first object, are (almost) always lettered linearly, but (almost) always non-linearly when introduced later. While the previous example had proved that the deviation from linear lettering in the case of 'late' introduction may be due to the situation that all points of the object have been previously introduced, but not *as part* of that object, the present example shows that, even if not all of the letters have yet been applied, and linearity is still an option, the lettering deviates from linearity. The reason for this may be that the lettering of the diagram, though following the order of the introduction of the objects into the proposition, does not consider each of the introduced objects as a whole, but only the points which have not been lettered. That is, the lettering of the diagram is not detached from the described order of the diagram's coming into being – otherwise it could be lettered very differently according to the principles of letter ordering and lettering, e.g., by having the four points on the upper straight line called linearly Α, Β, Γ, Δ, the point where the parallelograms cut each other E, and the two points on the lower parallel straight line isodirectionally with the upper one Z and H. Instead, the first parallelogram is, in accordance with the text, lettered first, before the second one is lettered – however, not *as a parallelogram*. Rather, the lettering seems to consider only the points left to be lettered, that is, the straight line EZ, which is lettered isodirectionally with the parallel straight lines constituting the first parallelogram, ΑΔ and ΒΓ. Thus, the lettering only takes the whole object into account, if it awaits complete lettering: therefore, all but one of the parallelograms which are the respective first objects to be introduced are lettered linearly, while only one of those introduced at some later stage is: in the subsequent proposition I 36, again, two parallelograms are introduced as lying between the same parallel lines, but this time, they do not overlap, so the second parallelogram is, like the first, lettered linearly (ΑΒΓΔ, ΕΖΗΘ). By contrast, the present parallelogram in proposition I 35 seems to be lettered non-linearly, since the text does not seem to be lettered by the already formed letter word of the text, nor in order for it to be alphabetical-linear once the letter word is formed, when the text is being formulated on the basis of the diagram. This means that the lettering, rather than being accomplished on the basis of the fully formulated proposition, might be done on some rough (unlettered) version of it, in which the order of construction is already fixed, but not yet the letter order of the letter words, as in the fully spelled out proposition text. This could explain both the rough adherence of the diagrammatic lettering to the order of construction, and the divergence of the lettering of quadrangles, which deviates from the principle of linearity in half of the cases, and the ordering of the letter words which is governed throughout by the linear viewing of the quadrangles, which, in turn, in many of the cases, seems to be due to an analogical, 'isodirectional' viewing of the quadrangles. Thus, in view of this (arguably generalizable) observation about the way in which the lettering is likely to have been accomplished, the letter order EBΓZ of the present parallelogram seems to indicate that it has been shaped on the basis of the lettered diagram. Yet it is not fully conclusive, as one cannot preclude that the lettering, contrary to our suggestion, might not have been done by taking the full quadrangle into account, that is, as in the case of the second, linearly referenced parallelogram of the subsequent proposition, by analogy with the first quadrangle: in this way, the first letter to be referenced would be the one named E, then, B and Γ would be bypassed as they have already been lettered, and finally, Z would be allocated. In this case, the way the quadrangle has been lettered in the diagram would be the same, whether it is done on or as the basis of the text, and the letter order would, therefore, not be indicative of either way.

The third example of a non-linearly lettered parallelogram is the one constituting the figure of I 34 (Fig. 56), referred to by the non-alphabetical linear letter sequence AΓΔB at the beginning of the proposition,[337] and by the alphabetical non-linear one ABΓΔ in the conclusion.[338] It has already been mentioned above as being the only one quadrangle in *Elements* I which is lettered non-alphabetically despite its being the first object to be introduced into the proposition. It shall be argued that AΓΔB is the only letter sequence referring to a quadrangle in book I in which case the linearity of description, in combination with the corresponding diagrammatic lettering, conclusively points to the preexistence of the diagram. First of all, the letter order of AΓΔB is utterly counterintuitive, especially because no other objects, adding further points and letters to diagram and text, will be introduced: this means that this time, the non-alphabeticality cannot be explained by the analogy with the linearity of other objects. Instead, one would expect the parallelogram, like all others in book I which are named by the first four letters of the alphabet, to be referred to by the alphabetical letter sequence ABΓΔ – as it, indeed, happens, at the end of the proposition, where AΓΔB is permuted to ABΓΔ. The reason why, initially, the text prefers the non-alphabetical letter sequence, must be in order to adhere to the ordering principle of linearity, and since the corresponding lettering is non-linear, this is only possible in a non-alphabetical way. There is no recognizable reason why the letter order switches at the end. The question is why the lettering is non-linear in the first place, unlike in all other instances of quadrangles lettered by the A, B, Γ, and Δ, because just like in these instances, none of the letters has been allocated so far, so that the lettering is completely free. The non-linear lettering may be

337 *Elem.* I 34 (vol. 1, p. 47.1 Heiberg & Stamatis).
338 *Elem.* I 34 (vol. 1, p. 48.7 Heiberg & Stamatis).

due to the analogy with the figure of the immediately preceding I 33 (Fig. 57), which is lettered in the same non-linear way.[339] There, however, the parallelogram is not considered as such, but only in terms of its parallel straight lines and the likewise parallel straight lines connecting them. The present proposition, though it deals with the parallelogram as parallelogram, uses the exact same figure for doing so which simply seems to have been copied from the previous one. The crucial point here is that, while the non-linear lettering of the quadrangle is explicable in the context of the previous proposition, it is by no means compulsive; nothing would have prevented the lettering from being linear like in all other cases of ΑΒΓΔ quadrangles in *Elements* I. Of the kinds of letter order corresponding to each lettering – linear non-alphabetical ΑΓΔΒ or non-linear alphabetical ΑΒΓΔ in the case of the present non-linear lettering, alphabetical-linear ΑΒΓΔ in the case of the alternative lettering – the latter, alternative variant clearly seems to be the preferable one: both in view of the dominant practice in book I, and, particularly, with regard to the preference of linearity expressed by the introductory letter order ΑΓΔΒ (its later permutation to the non-linear form notwithstanding). Therefore, the use of the linear and elusively non-alphabetical reference ΑΓΔΒ at the very beginning of the proof for the first mention of the parallelogram presents strong evidence for the foregoing lettering of the diagram. For the letter sequence ΑΓΔΒ is only conceivable as a description of the non-linear lettering of the underlying diagram; if, by contrast, the diagram had been lettered only on the basis of the text, the letter order in the first reference to the parallelogram in the text would never have been chosen to be ΑΓΔΒ, but most surely ΑΒΓΔ. This letter order could have resulted either in a linear or in a non-linear lettering of the diagram, with regard to which the letter order itself would be alphabetical-linear or alphabetical-non-linear, respectively. In view of the preference of the text for linearity which is indicated by the transmitted letter order, the letter order of ΑΒΓΔ, if it had been applied, would, in any case, be meant to be alphabetical-linear. Crucially, this letter order, which most certainly would have appeared in the text if the diagram did not preexist, is dismissed for the elusively non-alphabetical ΑΓΔΒ. This can only be because the diagram, lettered non-linearly, did preexist. Thus, it is the preferability of an alternative lettering, enabled by the flexibility of the transmitted diagrammatic lettering, and of the corresponding alternative letter ordering, which conclusively points to foregoing lettering in the diagram.

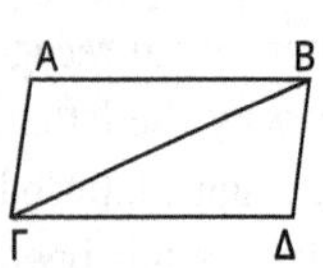

Fig. 56: I 34 according to Heiberg & Stamatis.

339 Acerbi 2007 prints the lettering of the diagram in our two manuscripts, whereas Heiberg & Stamatis present the diagram mirror-invertedly compared with the manuscripts' version.

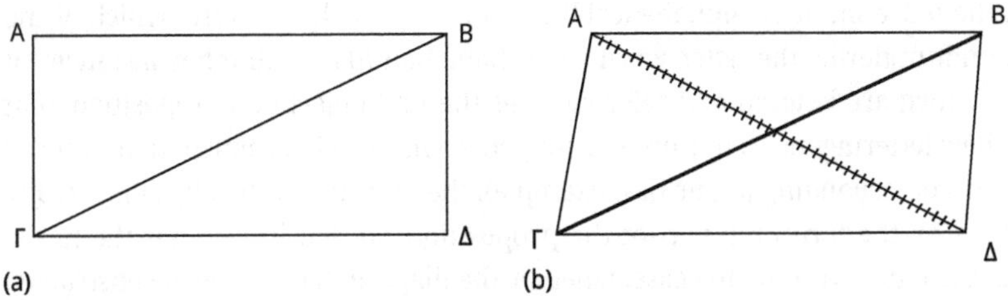

Fig. 57: (a) I 33 according to Vat. gr. 190.1, fol. 30v; (b) I 33 according to MS. D'Orville 301, fol. 24r.

Regarding the question whether linearity, if it indeed indicates a viewing perspective, does so throughout or only after the construction has been accomplished and could have been viewed anyway, it is notable that, while linearity dominates the text throughout, the only one conclusive example is the just discussed quadrangle of I 34 (Fig. 56) which is introduced as the first (and only) object of the proposition in the exposition part. While all other letter words referring to quadrangles are not (fully) conclusive, it is noteworthy that none of them contain any permutation with respect to the introductory form of the letter word but retain a consistent perspective on the diagram throughout the text, if the linearity is formed according to a perspective at all. It is only the ΑΓΔΒ in I 34 which is permuted – however, ironically, from the form which indicates perspective to the form which agrees with the non-linear lettering of the text and does *not* presuppose perspective. This way around, the permutation does not present a counter-indication of the diagram's preexistence; rather, the perspective-indicating introductory form suggests that the underlying form is being viewed from the start, and the concluding form which imitates the lettering of the diagram would do so precisely on the basis of viewing the order of the diagram.

To conclude from the analysis of the relation of alphabeticality and linearity in the letter words referring to quadrangles: from a comparison of the application of linearity in relation to alphabeticality in the letter words and in the diagram, it has emerged that linearity is a principle of letter organization pursued in both realms, but that it is a lot more dominant in the letter words where it is applied almost throughout, even at the expense of alphabeticality. The synoptic investigation of letter words and their corresponding diagrammatic lettering regarding the ramifications of the dominance of linearity has been focused precisely on the non-alphabetical linear letter words as being potentially indicative of the application of linearity as a 'viewing principle,' which would suggest the underlying presence of a diagram. Despite the large portion of such words in our sample, linearity has proved, though many times indicative, in most cases inconclusive with respect to the standardized preexistence of the diagram. There is only one case (our last example) in which the elusive non-alphabetical linearity of the letter word could have been shown to point unequivocally to the underlying presence of the lettered diagram, since the corresponding diagrammatic lettering deviates from linearity without being forced to do so. The lettering could have been linear in accordance with the more intuitive alphabetical-linear form of the letter word, which, if the diagram did not underlie, the letter word could have had like in all other instances of quadrangles which are lettered and referenced as the first object of a proposition, that is, without their lettering and letter order being otherwise predetermined. In a number of other cases (corresponding to our first example), the lettering of the object is already predetermined by the foregoing part of the proposition, so that it precedes the linear formation of the letter word in any case, whether the diagram has just been constructed and lettered according to the text or has preexisted from the start. The linearity, here, could be a way of describing the quadrangle analogically with the other likewise linearly referenced quadrangles of the proposition; or it could indicate a viewing of the quadrangle in relation to these other quadrangle, but the viewing would not be indicative of the preexistence of the diagram. The largest number of letter words referring to quadrangles corresponds to our second example. Here, the quadrangle has only been halfway constructed and lettered by the preceding part of the proposition, and the remaining points could be attributed linearly, but are not.

These cases may be telling because they seem to show how the diagram has been lettered independently of the final, lettered version of the text, that is, following the order in which the objects are constructed, but taking into account only the points yet to be lettered, not the objects as a whole as they correspond to the letter words. While the example (like other comparable ones) in itself is not fully conclusive, as the letters could have been placed the way they are *even if* the whole object had been taken into account, it could provide, alongside with the large number of comparable cases, an explanation for the unusually great amount of non-alphabeticality in the present sample and an indication of the priority of diagrammatic lettering. The lettering of the diagram follows certain other rules than the ordering of the letters in the text which become apparent when objects consisting of many points are considered; it follows the alphabetical-linear principle, but only with regard to the points still awaiting lettering, whereas the letter words follow the linear principle with regard to the whole object to which they refer, but have to dispense with alphabeticality where the lettering is non-linear. If the letter words were not shaped in view of the diagram, they might have been ordered according to alphabeticality or even alphabetical linearity throughout (since the application of alphabeticality is not based on viewing or "visualisation") and would have caused the diagrammatic lettering to be linear in some more cases. However, as it could be argued in each case that its kind of diagrammatic lettering is not necessarily less preferable from the point of view of the text if it were to describe the construction by itself, our third example from I 34 remains the only one which conclusively indicates the precedence of diagrammatic lettering. The 'linearity' of the references to quadrangles itself is, thus, largely inconclusive of the preexistence of the diagram.

However, the examples have given some indication in which way linearity could be shown to be due to a perspective of the text onto the diagram and to be, therefore, conclusive of an underlying diagram: the individual instances of 'linearity,' in fact, seem to be interconnected – both proposition-internally, and, even more remarkably, interpropositionally. That is to say, each reference to a quadrangle is not linearly ordered in a random way, but the linearity seems to be aligned with, first of all, other quadrangles mentioned in the same proposition as far as the starting point, direction, and endpoint of tracing a quadrangle along its contour are concerned. It is in this sense that the analogical linearity of several quadrangles has been called above 'isodirectional,' anticipating the other principle of letter ordering. Indeed, almost all propositions which feature quadrangles at all refer to a number of them which could be subject to isodirectional description. This isodirectionality could indicate an overall perspective of the text onto the diagram and account for the standardized application of linearity which forms the basis of this kind of isodirectionality. With respect to isodirectionality, the letters would not simply be ordered linearly, but linearly in a certain way, most often at the expense of alphabeticality. However, the example of I 35 (Fig. 45) has shown that the proposition-internal isodirectionality of quadrangles itself is not necessarily conclusive of a viewing direction from which an underlying diagram is described. Because of the limited degree of complexity of the diagrammatic figures, linear-isodirectional description could also be a way of making apparent (or: visualizing) the analogical relation of the objects if the diagrams were not there from the start.

More indicatively, the linear description of quadrangles may have been aligned isodirectionally even across propositions, i.e., it seems to trace the contour of the quadrangles mostly from left to right, starting at the top left point, going down to the bottom left point, to the right and up again.[340] This could show that the diagrams are viewed, if at all, not idiosyncratically from a certain viewpoint each, but that the text takes a standardized perspective onto quadrangles (and, as I shall show, on the diagram in general). The absolute number and proportion of objects affected by this seems to indicate strongly such a common perspective. One might still, however, object that this perspective, as we have seen with linearity itself, is also the one which governs the lettering of the diagram as far as possible, since the very first quadrangle of each of the respective propositions is lettered (alphabetically-)linearly precisely in the described way. Again, the case of I 34 (Fig. 56) could be indicative, as its description of a quadrangle is aligned with the dominant direction, even though it does not need to be aligned innerpropositionally with another quadrangle as there is none. The lettering of the quadrangles introduced after the first deviates from this isodirectional linearity – presumably, as I have argued, because the lettering takes into ac-

340 These directions refer to the diagram as it is printed in all our editions on the basis of the Greek Byzantine manuscripts (whereas it is printed mirror-invertedly in the Arabic ones). I shall make repeated reference to these directions in the following; at the end of the section, I shall even put forward the suggestion that the ordering of the letters in multi-lettered letter words in the original version of the text standardly followed the two main directions left-to-right and top-to-bottom, and give an explanation why these two of all, which also explains the inverted look in the Arabic manuscripts.

count only the objects to be lettered, not the whole objects described by the text which include points which have already been lettered. It has been pointed out at the example of the lettering of the second quadrangle of I 35 (ΕΒΓΖ) that the lettering of the as yet unlettered points Ε and Ζ follows the same direction as that of ΑΔ and ΒΓ and, in so doing, causes the lettering of the quadrangle to be non-alphabetical (Fig. 45). In view of the dominant direction of the lettering and of the letter order of quadrangles, the lettering of Ε and Ζ might not only be isodirectional with that of the two straight lines, but follow an overall direction, that is, from left to right, as it were, which emerges from the lettering and description both of the quadrangles and the straight lines of I 35. With respect to the second quadrangle of I 35, this means that the non-alphabeticality of ΕΒΓΖ is indeed inconclusive of the underlying presence of a diagram, as the diagrammatic lettering of Ε and Ζ, if it was done prior to the text, coincides with the description of the quadrangle according to the supposed overall perspective of viewing quadrangles. With respect to the standardization of the letter order according to the viewing of the diagram in general, the question arises whether not only the lettering of different objects, but also the order of the letter words followed an overall direction, and whether this overall isodirectionality of the letter words could be shown to be indicative of the underlying presence of a diagram. This will be investigated in the last part of the section.

At present, it should be noted in conclusion that the practice of linearity in the letter words, although it is extremely standardized compared with the diagrammatic lettering and seems to account for the deviation from alphabeticality in half of the cases, is itself largely inconclusive of the viewing of an underlying diagram. If at all, it is as a precondition for a perspective on the diagram which aligns different quadrangles, or even different objects with one another that linearity could be indirectly indicative. This is not to say that the linearity which characterizes the description of certain objects does not originate from the situation that the diagram was looked at while the text was written. But 'linearity' alone does not suffice to prove that the diagram was present at the time of writing. This is because, first, the scope of linearity is very small: few objects qualify at all for being described in a linear way. A principle which can, if at all, only account for the letter order of specific small groups of objects, can hardly be overall conclusive. Moreover, as only few object kinds are eligible, comparisons of different objects are hardly possible. As regards the object kinds which do qualify, it is not sure whether the conclusions drawn from them regarding the practice of linearity can be generalized or whether they only apply to that particular kind, since each kind differently allows for the application of linearity in the analytically decisive corresponding diagrammatic lettering. In *Elements* I, of the letter words for objects which are indeed eligible (quadrangles), almost all are non-indicative of viewing despite the large portion of potentially indicative non-alphabetically ordered ones. This is because the deviation from the alphabetical-linear principle in the letter ordering in the text and diagram almost always could be explained by the restricted possibilities and conventions of describing quadrangles. Second, linearity only pertains to distinct manageable objects most of which could be described from memory. Therefore, it is never clear whether the letter order is formed on the basis of viewing or simply envisaging the diagram. Linearity, thus, could only be indicative as a principle of viewing if it reached beyond individual distinct objects. Netz's example of a straight line defined by three points, which is, in fact, a conjunction of two straight lines, could provide a starting point for extending the scope of the principle thus. However, in this way, it becomes impossible to find out how standardly the principle is applied, because linearity as a way of viewing different objects as interconnected is only visible where it in fact applies, but it is hard to recognize that it has not been applied where it has not. In addition, referring linearly to any kinds of objects which are mentioned successively often may not be possible to begin with for a number of reasons (let alone lettering them). This means that it is hard to specify a sample which is manageable and at least potentially indicative. Third, and relatedly, because of its restriction to isolated objects, the analysis of linearity alone cannot account for the analogical or 'isodirectional' description of different linear letter words. However, it seems to be this analogical description which is, in many cases, not only based on, but in fact responsible for the application of linearity. This can be seen from the fact that it is only a specific, not any, kind of linearity which is chosen: that which traces the object from a certain point 'isodirectionally' by analogy with the description of other objects of the same kind. Here, it is, therefore, the deviation from a specific *kind* of linearity for the sake of another one fitting the superordinate scheme which is indicative, rather than the deviation from alphabetical description. For the letter order would be in many cases non-alphabetical anyway, even if the corresponding diagrammatic lettering was (alphabetical-)linear, as this linearity would only allow for one particular kind of linear description but not necessarily the one which is linear by analogy. While the standardization of the letter words according to linearity can be explained in this way, it has been pointed out that the analogical description of a number of objects of the same kind would be eventually as little indicative as the linearity of individual objects. For analogical description of such a simple kind could still have been managed without an underlying diagram and might have been the preferred way of referring to objects of the same kind anyway in order to make apparent their similarity. Thus, the concept of linearity will not be analyzed any further, as it has turned out to be in itself little productive with respect to our question, mainly because of its small scope – both in terms of the number of eligible objects and object kinds and in terms of the number of objects concerned at a time (i.e., one). Its analysis has, however, pointed to the relevance of the concept of isodirectionality – and it has, in anticipation of my subsequent introduction of the concept, pointed to the way in which it would have to be extended with regard to our question: namely, by enlarging the scope of kinds of objects and of the number of objects concerned at the same time.

Isodirectionality and alphabeticality in the lettering and letter order of parallel straight lines

The application of isodirectionality in the combined references to parallel straight lines was not quite, but almost as standardized as that of linearity: 52 of 57 mentions of two or three parallel straight lines (91%) in our sample (I 27–38) are isodirectional. Unlike in the case of our linearity sample, the principle of alphabeticality is extremely standardized (again, 52 of 57 mentions are alphabetical). In fact, it is even more dominant than in our general survey of I 1–20, where the amount of alphabetically ordered letter words was 76% as opposed to 91% here. From this, it follows that the order of the majority of letter words referring to parallel straight lines must be alphabetical-linear (48 of 57). This leaves only four non-alphabetical isodirectional combinations, four alphabetical non-isodirectional ones and one which is neither alphabetical nor isodirectional.

The alphabetical-isodirectional letter order necessarily corresponds to an (alphabetical-)isodirectional lettering of the parallel lines in the diagram. Therefore, the majority of the isodirectional occurrences immediately turn out to be non-indicative of the origin of isodirectionality. Because, without necessarily indicating a viewing direction, this letter order could simply follow the alphabetical order of the letters in an underlying diagram; or it could relate the parallel lines analogically by means of alphabeticality even in a 'blind' description of the diagram according to which the lettering would have been subsequently accomplished. (Alphabetical-)isodirectional lettering is in any case the most intuitive way of aligning two parallel lines, and the according alphabetical-isodirectional letter order is the only one which suggests itself for referring to such a lettering, because the alternative non-alphabetical and non-isodirectional way of referring to it is generally less preferable as it does not abide by any of the standardized principles. It is only in a specific geometrical situation that it could make sense and possibly indicate the presence of a diagram. For example, in our sample, the only one instance of a non-alphabetical non-isodirectional letter order is based on alphabetical-isodirectional lettering:[341] the two parallel lines ΑΓ, ΔΒ of the parallelogram ΑΓΔΒ in I 34, cited above as the only example of linearity as a viewing principle (Fig. 56). As the letters constituting these words have already been introduced before, the lettering is in any case fixed. The non-alphabetical non-isodirectional order of the letters could imitate the non-alphabetical linear reference of the quadrangle and indicate the viewing of the diagram in the way of the latter – however, after the non-alphabetical linear introduction of the parallelogram,[342] the two parallel straight lines in question have already been mentioned several times alphabetically-isodirectionally

341 That is, there is no case of theoretically possible non-alphabetical non-isodirectional letter order corresponding to non-(alphabetical-)isodirectional lettering in our sample.
342 *Elem.* I 34 (vol. 1, p. 47.1 Heiberg & Stamatis).

(ΑΓ, ΒΔ)[343] before ΒΔ is permuted to ΔΒ.[344] This happens just before the parallelogram ΑΓΔΒ is referred to differently as ΑΒΓΔ,[345] that is, by its other two (isodirectionally lettered) parallel straight lines ΑΒ and ΓΔ. In this way, it is not clear why the non-isodirectional perspective is taken with regard to ΑΓ and ΒΔ at that moment.

Alphabetical non-isodirectional letter ordering in the references to parallel straight lines is always based on non-(alphabetical-)isodirectional lettering. It is generally non-indicative of the viewing of the diagram, as it deviates from the only one perspective which intuitively suggests itself with regard to parallel straight lines, and does not itself suggest a more conceivable perspective. It could only be by analogy with other pairs of parallel straight lines similarly lettered, that it may point to an overall viewing perspective, which goes in each case back and forth, as it were. In any case, the adherence to alphabeticality (at the expense of any kind of directionality) could always be the kind of letter order chosen if no diagram was present and therefore could not be conclusive. In all four instances of this type in our sample, the lettering has already been accomplished for the most part at the moment at which the references occur. Therefore, since the lettering of the two lines could not have been the result of a choice, we cannot tell from it whether there would have been a more viable alternative way of doing it which would have corresponded to the letter words if their ordering came first. Only in this case, it would be clear that the present lettering was accomplished before the text was.

Three of the four cases of alphabetical non-isodirectional letter ordering occur in I 32 of the same pair ΑΒ and ΓΕ (Fig. 58). ΓΕ, in fact, introduces Ε, but it could not have been any other letter, because all alphabetically preceding letters have already been fixed. The order of ΓΕ is explained by the order of the construction: "Through the point Γ, ΓΕ should be drawn parallel to the straight line ΑΒ."[346] The order of the letters of the straight line ΑΒ could be either isodirectional with ΓΕ but non-alphabetical ΒΑ, or, as it is in our text, alphabetical but non-isodirectional ΑΒ. This alphabetical order is retained in the two other occurrences of the couple in the proof part.[347] There, it is actually more plausible, as ΑΒ is both times mentioned first in line, and the following ΓΕ mentions the remaining letter of the triangle ΑΒΓ first. In this way, not only each reference individually is alphabetical, but also the whole letter sequence of the two lines is. The preference of alphabeticality over isodirectionality in the description is explicable by the geometrical configuration and its order of the construction, not by an overall perspective which would relate ΑΒ and ΓΕ to yet other straight lines: the two parallel lines are constituted by the three points of a triangle and a point which is added to the construction afterward.

---

343 *Elem.* I 34 (vol. 1, p. 47.10, 18 Heiberg & Stamatis).

344 *Elem.* I 34 (vol. 1, p. 48.5 Heiberg & Stamatis).

345 *Elem.* I 34 (vol. 1, p. 48.7 Heiberg & Stamatis).

346 Ἤχθω γὰρ διὰ τοῦ Γ σημείου τῇ ΑΒ εὐθείᾳ παράλληλος ἡ ΓΕ (*Elem.* I 32 [vol. 1, p. 44.17 f. Heiberg & Stamatis]).

347 *Elem.* I 32 (vol. 1, p. 44.19 Heiberg & Stamatis), *Elem.* I 32 (vol. 1, p. 45.1 Heiberg & Stamatis).

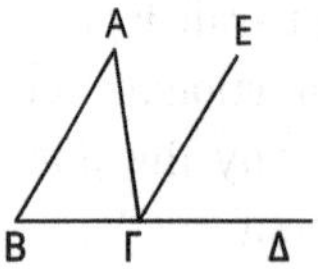

Fig. 58: I 32 according to Heiberg & Stamatis.

The only other occurrence of alphabetical non-isodirectional letter ordering is in I 38 (Fig. 41). Here, two straight lines are to be drawn: ZΘ parallel to the side ΔE of the triangle ΔEZ and BH parallel to the side ΓA of the triangle ABΓ.[348] The triangles and their points have already been introduced, H and Θ are specified just before the construction of BH and ZΘ is assigned. Because of this foregoing order of construction, the resulting pairs of parallel straight lines cannot but be lettered non-(alphabetically-) isodirectionally. Notably, the reference to them is in the case of BH and ΓA isodirectional and non-alphabetical, but in the case of ZΘ and ΔE non-isodirectional and alphabetical. In both cases, the letter ordering of the respective new straight line to be constructed is accounted for by the mention of the respective point of the triangle (B and Z) "through" (διά) which the straight line is to be drawn; this point is, accordingly, mentioned first in the reference to the respective straight line. The description of the respective second straight line in one case (ΓA) follows the direction of the reference to the newly constructed one; in the other (ΔE), it takes the opposite direction. Seeing that the first pair of references appears to deviate from the necessarily non-(alphabetical-) isodirectional lettering for the sake of isodirectionality, the non-analogically non-isodirectional referencing of the second pair is surprising. Here, alphabeticality is preferred presumably because the sequence "ΔE . . . ZΘ" first has the letters of the triangle mentioned in the alphabetical order of its reference ΔEZ, then point Θ which does not belong to the triangle. In addition, the combined letter sequence is also the one of the subsequently introduced parallelogram ΔEZΘ.[349] By contrast, in the case of the other pair, the two references in the order of their being mentioned in the text (ΓA . . . BH) could not represent the letter order of the references to the triangle (ABΓ) and to the parallelogram (HBΓA) anyway. Therefore, the alphabetical non-isodirectional way of letter ordering would not necessarily be the more plausible way of doing so. But an ordering AΓ, BH would have had the advantage of mentioning all three letters of the already constructed triangle as far as possible in alphabetical order, before mentioning the letter H, which has been introduced but is located precisely only by the present construction assignment. The reason why the order AΓ could have been the preferred one if the construction was not described on the basis of the ready diagram is that this line is mentioned first. In order for it to be ΓA, the order of BH, mentioned only afterward, would have to been envisaged. In any case, like in the case of the other three instances from I 32 (Fig. 58), the reason for the preference of alphabeticality by the

348 *Elem.* I 38 (vol. 1, p. 52.3–5 Heiberg & Stamatis).
349 *Elem.* I 38 (vol. 1, p. 52.5 f. Heiberg & Stamatis).

letter order of "ΔΕ . . . ΖΘ" is not a superordinate kind of isodirectionality. Here, ΔΕ and ΖΘ actually defy the possibility of such an overarching viewing of the diagram, because they could have aligned themselves with the isodirectionality of the other pair of parallel straight lines.

This leaves the remaining four cases of non-alphabetical isodirectional letter ordering which are potentially the most indicative ones a viewing of the diagram as they deviate from the alphabetical principle presumably for the sake of mentioning the lines analogically, which could (but does not have to) suggest that the text assumes a certain perspective onto the diagram. Two of the non-alphabetical isodirectional references correspond to non-isodirectional lettering from which they deviate by isodirectionality; the other two correspond to isodirectional lettering, which they retain, though by inverting the alphabetical letter ordering enabled by the lettering and looking at the straight lines non-alphabetically from the other side.

The first instance of the first type is from I 38 (Fig. 41) the couple of isodirectionally non-alphabetically referenced parallel straight lines (ΓΑ, ΒΗ). It has just been referred to in the discussion of the other pair of parallel straight lines of the same proposition (ΔΕ, ΖΘ) which is conversely alphabetically non-isodirectionally referenced. The example is highly indicative of the explicit application of isodirectionality, and of the principle's suggesting the viewing of a preexisting diagram. This is because the pair of letter words already appears in the construction part and involves the construction of one of the lines. As has been argued, the direction of ΒΗ is explained by the mention of Β "through" which the line must be drawn, and the order of ΓΑ could be explained by the analogy.[350] However, ΓΑ is the first of the two lines to be mentioned, and its order suggests that it has been chosen literally *in view of* the order of the letters in the subsequent reference to ΒΗ. Otherwise, an alphabetical non-isodirectional lettering in the manner of ΑΓ, ΒΗ might have been more conceivable. This one would have been analogical with that of ΔΕ, ΖΘ,[351] in the case of which the chosen alphabetical non-isodirectional lettering is in any case a plausible way of referring to the straight lines. The choice of isodirectionality in the case of ΓΑ, ΒΗ is especially indicative in view of the fact that ΔΕ, ΖΘ, which could have been referred to isodirectionally, are not. For this means that the reference to ΓΑ, ΒΗ did not have to be isodirectional either. It presents, thus, an explicit deviation from a kind of letter ordering which would have suggested itself more from the point of view of construction. Thus, both pairs of references seem to have been 'read from' the diagram. While the first pair was viewed isodirectionally, the viewing of the second one is linear as it were, tracing first the triangle and then the new line up to point Θ. A 'blind' description of the diagram, by contrast, would have referred to both pairs analogically.

---

350 *Elem.* I 38 (vol. 1, p. 52.3 f. Heiberg & Stamatis).

351 *Elem.* I 38 (vol. 1, p. 52.5 Heiberg & Stamatis).

An instance of the second of the types of non-alphabetical isodirectional letter ordering occurs in I 36 (Fig. 59) with the parallel lines EB, ΘΓ. The letters have been introduced in the exposition as part of the parallelograms ΑΒΓΔ and ΕΖΗΘ,[352] and the parallel lines have been constructed alphabetically-isodirectionally (ΒΕ, ΓΘ) corresponding to the isodirectional lettering predetermined by the letter order of the parallelograms.[353] In the reference to the pair in the proof part, the isodirectionality of the lettering is retained, but the alphabetical direction of the lettering is permuted. This could have been done for the sake of a larger-scale isodirectionality: as we shall see immediately in the next example, looking down on the diagram seems to have been another preferred direction of viewing the diagram besides the one of looking at it from the left-hand side. This preference could also have influenced the standardized presentation of the quadrangles starting from the upper-left corner, with its first two points indicating a viewing direction from top to bottom. (For the sake of linearity, the view goes up again with the final two letters.) I am going to explore this kind of isodirectionality more closely below. The present case of the pair of parallel lines itself is in any case non-conclusive as regards the question of the preexistence of the diagram, although the second reference to the pair explicitly deviates from the previous alphabetical-isodirectional reference to the straight lines. However, the deviation only occurs in the proof part, when the diagram can be assumed to have been constructed anyway, and the references to the parallelograms which appear from the beginning of the proposition are not conclusive either, as both of them are alphabetical-isodirectional which they might have been even without an underlying diagram. (To be sure, it could be that even with a diagram underlying, the text might describe the construction from the point of view of construction and change the point of view after the constructing section. But in cases like the present in which the viewing direction changes after the construction, the reason could also be as explained, so these cases could not be counted to be fully conclusive.)

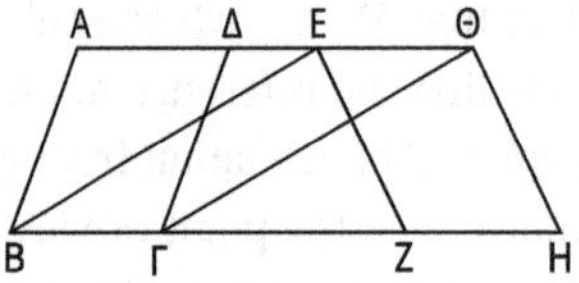

Fig. 59: I 36 according to Heiberg & Stamatis.

The last two instances of potentially indicative isodirectional non-alphabetical letter ordering occur in I 35 (Fig. 45). One of the examples belongs to the first type of this letter order regarding the corresponding lettering, the other one to the second. On the one hand, the straight lines AB, ΔΓ are referred to isodirectionally despite their non-

352 *Elem.* I 36 (vol. 1, p. 49.16 Heiberg & Stamatis).

353 *Elem.* I 36 (vol. 1, p. 50.1 Heiberg & Stamatis).

isodirectional lettering.[354] This is a case which is highly indicative of isodirectionality as a viewing principle because of the inversion of the alphabetical order despite the previous reference to the letters in an alphabetical way as part of the letter word for the parallelogram ΑΒΓΔ, by which the four letters have been introduced and fixed.[355] One might think that the letter order changes to isodirectionality because the diagram has been constructed by now and can be viewed. In this case, the isodirectionality of the lines would not be conclusive of the underlying presence of a diagram from the very beginning. However, the parallelogram is referred to in a linear way throughout the proposition (that is, non-isodirectionally as regards the two straight lines in question), and this consistently linear letter ordering of the parallelogram has itself been suspected of arising from the isodirectional description of the diagram. Therefore, the isodirectional reference to ΑΒ and ΔΓ does not seem to be due to a change of perspective after the construction. Nevertheless, since the isodirectionality *could* have been applied on the basis of the by then constructed diagram, it cannot be fully conclusive of the preexistence of the diagram, although it is in any case strongly indicative of the viewing of the diagram at the moment of its occurrence. It would have to occur already in the construction in order to be proof of the underlying presence of the diagram from the beginning.

On the other hand, also ΕΒ and ΖΓ, the two parallel straight lines constituting the other parallelogram ΕΒΓΖ, are referred to isodirectionally;[356] but this time, the isodirectional letter ordering corresponds to isodirectional diagrammatic lettering – only the direction of the reference is the opposite from the one suggested by the lettering. This case, too, is highly indicative of the application of isodirectionality as a viewing principle; in fact, it is even more indicative than the other pair of parallel lines in the same proposition, because presently, alphabetical-isodirectional letter ordering would actually have been possible on account of the corresponding lettering. As we have seen above, this is the dominant form of lettering parallel straight lines which is almost as dominantly retained in the textual references to them. Here, the alphabetical-isodirectional lettering of the lines in question is the result of the non-(alphabetical-) linear lettering of the parallelogram by which the letters of the parallel lines have been fixed at the beginning of the proposition. As has been argued above, the lettering could have followed the order of construction, with Β and Γ having been applied as part of the parallelogram ΑΒΓΔ, and Ε and Ζ subsequently and maybe alphabetically-isodirectionally with Α and Δ and with Β and Γ; the linear non-alphabetical letter order of the parallelogram, however, seems to have been chosen isodirectionally with the first parallelogram. The letter order of the parallel lines in question deviates not only from their isodirectional lettering arising from the previous attachment of the

354 *Elem.* I 35 (vol. 1, pp. 48.20–49.1 Heiberg & Stamatis).

355 *Elem.* I 35 (vol. 1, p. 48.12 Heiberg & Stamatis).

356 *Elem.* I 35 (vol. 1, p. 49.3 f. Heiberg & Stamatis).

letters, but also from the – albeit non-alphabetical non-isodirectional – order which the linear reference to the parallelogram might suggest (EB, ΓZ). The reason seems to be the larger-scale isodirectionality which aligns both pairs of parallel straight lines. The direction of the view could be predetermined by the alphabetical order of the first of the four straight lines to be mentioned; but it could also be due to a superordinate isodirectional principle which reflects that the diagram is being viewed from top to bottom, because in each case, of the two points constituting the straight line, the one lying on the upper straight line is mentioned first. The lettering of AB could imply that the lettering, too, adheres, if possible, to this particular direction. It is, accordingly, not only where letter order follows this direction but where the lettering at the same time does not that the direction indicates viewing. On a related note, the left-to-right direction has been identified as a potential principle which might reflect a preferred viewing (and lettering) direction. Here, the horizontal parallel straight lines are all referred to according to this direction, but, as all of the letter words are alphabetically ordered in accordance with the lettering, they cannot be conclusive of viewing. But even the vertical straight lines and parallelograms involving them might in fact testify to the combination of both directions: the top-to-bottom direction does not only align the two pairs of parallel straight lines with one another, but also with the two parallelograms: the references to both start, as I pointed out above, from the upper left-hand corner, follow the figure to the right and end at the upper-right corner – not at the lower right, to be sure, but this would only have been possible by dispensing with linearity. Moreover, the four top-to-bottom ordered straight lines are mentioned from left to right, as are the two parallelograms. In light of this overarching isodirectionality, the two pairs of parallel straight lines could be suggestive of the viewing of an underlying diagram. Each pair by itself and even the two pairs combined only indicate that *a* diagram is being viewed. However, this could also have been the diagram as it has just been constructed according to the foregoing instructions, because the references occur only late in the proof part. But the directions determine already the opening references to the parallelograms which precede any construction on the basis of the text – and which themselves appear more indicative of the viewing of a preexisting diagram than has been initially suspected.[357]

To conclude: the analysis of the isodirectionality of parallel straight lines has shown that the great standardization of isodirectionality in the letter order corresponds to an equally great standardization of isodirectional lettering. As a result, the standardized order of pairs of letter words referring to parallel straight lines is alphabetical-isodirectional. Because of this large coincidence of letter ordering and lettering, the isodirectionality of parallel straight lines is not necessarily indicative of the viewing of a diagram. It could be the result of the viewing of a diagram which must

357 See above pp. 363–367 my discussion of the indicativeness of the linearity of EBΓZ.

therefore underlie the description, or it could have been chosen for the sake of analogically alphabetical presentation independently of a diagram. Crucially, isodirectional description has turned out fully indicative of the viewing of a preexisting diagram only under two conditions: (1) if it implies the choice of isodirectionality by deviating from the corresponding alphabetical lettering, i.e., if it is non-alphabetical, and (2) if it is standardized throughout the text, not only after the constructing section, when the ready diagram could have been viewed anyhow in case that the diagram has been lettered simultaneously with the text. There are only four potentially conclusive, i.e., non-alphabetical isodirectional cases at all. Of those, only three are in fact indicative of the viewing of a diagram, but only one is by itself: the pair from I 38 which features the isodirectional presentation as early as the construction part. The two pairs from I 35 are not conclusive themselves but only by indicating their adherence to isodirectionality on a larger scale. As with the quadrangles which we have reviewed with respect to linearity, permutation rarely occurred with pairs of parallel straight lines. Considering the number of permutations in our sample I 1–20, this seems to be remarkable. The stable representation of the objects considered seems to be due precisely to their isodirectional alignment with objects of the same kind: the quadrangles most often appear together with other quadrangles the references to which are isodirectionally linearly ordered; pairs of parallel straight lines are objects mostly being referred to in conjunction with one another, and have been taken into account here only if they are.

Just like in the case of linearity, the analysis of isodirectionality has shown that the objects reviewed are by themselves hardly conclusive of the viewing of a diagram. I now point out the reasons for this, referring constantly to the previous analysis of linearity for underlining the structural similarities of the two concepts and the differences of the results which have to do with the respective object kind. The non-indicativeness of isodirectionality is due mostly to the restriction of the review to instances of two parallel straight lines. First of all, the limitation to letter words referring to one kind of object turns out only a small sample compared with the overall number of letter words. In the present sample, the number of relevant objects has actually been relatively higher than in book I, as it has been chosen precisely for the sake of providing a large sample of references to pairs of parallel straight lines which could be analyzed with regard to isodirectionality. Moreover, if the concept is restricted to parallel straight lines only, it can only say something about whether the letter words imply the viewing of these very objects. Second, the pairs of parallel straight lines have turned out to be both lettered and referred to in most cases alphabetically-isodirectionally, and only few instances in which the letter order deviates from the lettering have been at least potentially indicative. In the case of quadrangles, by contrast, there has been much less coincidence between lettering and letter ordering. This seems to be because isodirectional lettering of objects as simple as parallel straight lines is easier to realize than linear lettering of quadrangles, which are more likely to be constructed around previously constructed objects and consist of four points which would have to appear in a certain order. In order to letter two parallel lines isodirectionally, only two points of one object have to be related to two of another; even the starting point of doing so is often flexible. Third, isodirectionality is in any case the most obvious way of representing two parallel straight lines analogically (it is even more obvious than linearity in the case of parallelograms or squares, where linear presentation is common, although mentioning two of their parallel lines successively is a viable alternative). Therefore, isodirectional referencing of parallel straight lines may always be due either to the direction of viewing or to geometrically analogical representation which does not presuppose viewing. Fourth, and relatedly, because of the restriction to two objects as simple as two parallel straight lines, the letter configuration could be envisaged in any case and arranged 'blindly' in an analogical manner. Fifth, as linearity was confined in each case to one object only, isodirectionality as it has been suggested by Netz a principle of organizing the letter order of references to parallel straight lines only concerns objects in isolation. In this way, it does not take into account isodirectional relations of a pair of parallel straight lines to other (pairs of) parallel straight lines which might determine the letter order of each. However, it has turned out even from the few examples considered more closely that it is, if at all, with respect to such larger isodirectional relations that the letter order of parallel straight lines becomes indicative of viewing. I have pointed out isodirectional relations of pairs of parallel straight lines to other pairs, and even to other kinds of objects, such as parallelograms, which consist of parallel straight lines and may therefore be compared with pairs of parallel straight lines as regards their direction. This is regardless of whether the pairs of straight lines are parallel also to one another or not. (In my examples, the pairs of straight lines which have been compared are parts of parallelograms lying between the same parallel lines and in this way comparable regarding their directions, even if the pairs are not parallel to one another.) The examples have given some indication that the objects which are presented isodirectionally are not only viewed in relation to one another from some direction but from specific standardized directions. These directions seem to not only align parallel lines and objects consisting of parallel lines within a proposition, but to recur across propositions. However, even such large-scale isodirectionality does not necessarily testify to the viewing of the diagram. It could also result from the underlying lettering which, again, also seems to adhere to the preferred directions, or it could coincide with the analogical letter ordering of references to parallel objects. Therefore, it is again only the deviance from the corresponding lettering and from geometrically analogical presentation which proves without doubt that the letter order is due to viewing, but such deviance should become more likely the more objects of a proposition are involved. In any case, if certain specific directions indeed turn out to be standardized in the *Elements* as directions of viewing, it becomes more likely for instances of a certain kind of isodirectionality to indicate the viewing of a diagram.

The larger role of '(iso)directionality'

Like the concept of linearity, that of isodirectionality is relevant to the question of the standardization of the diagram and diagrammatic representation, if at all, only as part of a larger-scale isodirectionality. Unlike linearity, isodirectionality can indeed be applied at a larger scale. It has been shown punctually how the concept may be extended with regard to parallel objects: pairs of parallel straight lines may be regarded in conjunction with other pairs, and with larger objects which consist of pairs of parallel straight lines, such as certain kinds of quadrangles. In my examples, I have considered the isodirectionality of pairs of parallel straight lines and of parallelograms which are not (fully) parallel to each other but which are comparable regarding directions as they lie between parallel straight lines: they unequivocally display isodirectionality or non-isodirectionality in relation to the parallel lines between which they lie, because the reference to, and viewing of them starts either at one of the parallel lines or at the other, and ends at one or the other, respectively. Since, however, the original concept of isodirectionality implies parallelism, isodirectionality in a strict sense does not apply to such cases if the objects themselves (or the parts with regard to which they are compared) are not parallel. To be more precise, one could use the term 'roughly isodirectional' for non-parallel objects or object clusters which nevertheless appear to be described from a common direction, as indicated by their letter order, while reserving 'isodirectionality' in a geometrically correct sense for parallel objects, as it is only by parallelism that objects can imply the same direction. Even more generally, the term "directionality"[358] may be employed for referring to all kinds of object relations established by the letter order which suggest a certain direction of viewing. Stretched in this way, 'directionality' may be used to detect traces of the viewing of an underlying diagram in the description of objects which are materially related to each other in the diagram in various ways, and which seem to be aligned with each other by a more or less common perspective. This allows for tracking (iso)directionality at a larger scale: not only with respect to geometrically coupled objects, but also to object clusters, and eventually to the diagram as a whole. I have already hinted several times at the possibility that there may be standardized perspectives which govern the viewing of the diagram and which determine the letter order unless specific geometrical relations are deviatingly illustrated. It is such an overall (iso)directionality which would indicate the underlying presence of a diagram. This is because it allows (1) for distinguishing between geometrically analogical and aesthetically directional alignment of objects, (2) for recognizing substantial deviation from the underlying lettering (although it may be characterized by a similar directionality to a certain extent), and (3) for aligning objects on such a large scale that their relations cannot be overlooked without a diagram any longer.

---

358 Netz himself uses the more general term alternatively to "isodirectionality" (1999a, 82).

A typology of '(iso)directionality'

Let us now turn to the various forms that directionality can take, and to the ways in which it shapes the view onto the diagram by means of the letter order. I am going to set up a tentative typology of object relations which may be represented isodirectionally. For each type, I am providing small-scale examples which only comprise two objects, and at certain points, I am explaining how these incidents of directionality combine at a larger scale. Having focused on particular objects and object combinations in the two previous two sections, I am not making any restrictions with respect to the geometrical objects considered here. Rather, I am starting from straight lines as the most basic type of multi-lettered objects which, in combination with other straight lines, are capable of indicating direction by their letter order; when looking at more complex object configurations, larger objects are also included even though they tend to indicate directions less unambiguously. For example, circles can be wholly indicative regarding directions if the three points by which they are referenced are not fixed on the circle at points at which the circle is cut by other objects; moreover, the movement is not into one direction only but so as to lead back to its starting point. Similarly, the references to triangles and the linear ones to quadrangles are characterized by such 'circular' directionality. In any case, such letter words may indicate the starting point of viewing and whether the view moves back toward it by taking one direction or another. The larger object configurations I am looking at either consist of objects which are actually connected in the diagram by sharing points with one another, or are solely combined in the text by viewing. Throughout my discussion of types of directionality, I make reference to the overall directions which I suspect to have determined the specific directions which characterize the descriptions of smaller and larger object relations (top-to-bottom, left-to-right). At the end, I attempt to assess the standardization of the preferred directions and evaluate their overall influence on the letter order. In the subsequent part, I shall think about the origin of the standardization of these particular directions and about the ramifications of their standardization with regard to our question of the role of formal standardization in the *Elements*.

In setting up the present typology, I am not aiming for completeness nor for systematization. By contrast, the previous analysis of linearity in the letter words for quadrangles and isodirectionality in references to parallel straight lines has been both systematic and complete with regard to the sample chosen. It has turned out two main results on an abstract level: first that the assessment of individual objects and standardized small-scale configurations is largely inconclusive with regard to our questions; second that the references to objects which are isodirectional by parallelism, are, if at all, only significant in relation to references to other objects parallel to them and to object relations likewise characterized by parallelism, but that even such larger relations are often inconclusive. These results suggest that it would be more productive to look at larger configurations including different kinds of objects and non-parallel object relations for whether they show signs of isodirectionality. However, as has been pointed out, such configurations cannot display isodirectionality in a strict sense, because non-parallel objects could be called, if at all, only 'roughly isodirectional.' This rather vague term suggests that an analysis of letter words with respect to 'rough isodirectionality' can hardly be done on the basis of few objective, unambiguous criteria if these words reference (also) non-parallel objects. Therefore, such an analysis can be neither systematic nor complete.

The purpose of starting with small-scale cases of differently related straight lines before considering larger geometrical configurations is to test which object relations would be perceived as 'roughly isodirectional' by contrast with such that would not. From this, it should emerge how far it is possible to push the boundaries of 'rough isodirectionality.' This means that I look at the letter words and the corresponding objects from a wholly aesthetic, non-mathematical point of view. To be sure, already applying the concept of isodirectionality as it has been put forward by Netz is to approach the text from a non-mathematical stance, because neither the diagram nor the letter words contain directions themselves. It is only by presupposing that the letter order implies the viewpoint of someone looking at the objects that it may be assumed to reflect a direction of viewing. However, to call the direction of two analogically lettered parallel straight lines 'the same' by the word 'isodirectional' is mathematically correct. By contrast, applying vaguely 'rough isodirectionality' to non-parallel objects does not have any mathematical grounding, because non-parallel straight lines in any case point into *different* directions (and if one traces the directions into which they point, they continue to diverge from one another). Nevertheless, I am trying to use the geometrical characteristics, above all, the sizes of the angles made by straight lines, in order to find aesthetic criteria to demarcate tentatively the concept of 'rough isodirectionality.'

The small-scale cases by which I explain what it means for non-parallel straight lines to be roughly isodirectional will not be any more indicative of the viewing of the diagram than the cases of isodirectional parallel straight lines. They will only be so in individual contexts of larger object configurations which adhere to one or both of the preferred directions of description, and by forming part of a multitude of other instances. Therefore, it does not make sense to assess again isodirectionality with respect to the different kinds of relations of individual objects. Instead, I discuss, on the one hand, the individual situation of a small number of larger cases more thoroughly, and the ways in which they apply the supposedly overall directions. On the other hand, I make a statistical count of the directions which seem to be preferred.

The discussion of such larger contexts is complex and for this reason will only be done exemplarily. Crucially, it is precisely where a description is chosen which looks at the figure from a direction which does not coincide with the one which suggests itself the most from a geometrical stance that it indicates viewing. Strictly speaking, there is no kind of description which suggests itself more from a mathematical point of view than any other; there is only a correct description and a wrong one, but not one which is better or more correct than another correct one. Since the letters are permutable within the letter words within certain limits, any letter order which abides by those limits is as good as any other which does so too. However, one could call a description 'geometrical' if it imitates the geometrical situation, mostly by analogical ordering of the letters; it is the one which can also be accomplished from memory as it recalls analogical relations between objects. Further, it is crucial, as has been noted before, that the 'viewing perspective' applies from the very beginning of the proposition, as this would indicate unambiguously that the diagram has been there from the start. With respect to my large-scale examples, I also reassess the role of permutation. Supposedly, it has a greater significance in the references to straight lines which are not part of a couple of parallel straight lines which are, as we have seen, always referenced with respect to one another and change their letter order only rarely if the whole couple is related to other objects for a change.

Lastly, I am going to reassess the significance of the preferred viewing directions as they emerge from the foregoing discussion of the large-scale examples. Here, I shall once again take the individual straight lines into account and make a survey of the directions indicated by their letter order in relation to the directions suggested by the lettering of the diagram.

This section is mainly based on I 47 (Fig. 46), though I am making recourse to the examples previously discussed with respect to isodirectionality. I 47 is particularly apt for studying viewing directions: this is first because it has one of the most complex figures of *Elements* I, which on the one hand displays many different object relations, and on the other hand provides the possibility of observing directionality at a larger scale. Moreover, the figure is – in a non-mathematical sense – 'concentrical' or 'symmetrical,' in that it has a triangle at its center, from which three squares extend which are erected on either of its three sides; additionally, four straight lines connect the squares with the triangle at analogous points on the squares and the triangle, respectively. Finally, one straight line cuts the square on the hypotenuse in two rectangular halves which are the size of the respective square on the legs of the triangle, with the corresponding rectangular halves and squares lying on the same side of the figure.

The lettering of the figure, corresponding with its construction, starts from the triangle at the center, moves on to the remaining points of the three squares, starting with the square on the hypotenuse and continuing with the squares on the legs, and ends with the outer points of the straight lines connecting the squares to the triangle, and of the straight line bisecting the square on the hypotenuse. This means that if the corresponding letter words in the text followed the alphabetical lettering of the diagram, the directions indicated by their letter order would be of the type 'from the inside to the outside,' with the description starting at the points of the triangle and leading on to the outer points of the squares. The letter order would nevertheless imply a slight tendency of left-to-right description, because the lettering, though it proceeds from the center of the figure, starts on the left-hand side as regards the individual objects (e.g., the triangle ΑΒΓ and the square ΒΔΕΓ)[359] and where two analogical objects are introduced at the same time (e.g., the squares ΗΒ, ΘΓ).[360]

If the description was geometrically analogical, it would either likewise start from the triangular center of the figure and look at the other ends of the squares (and thus coincide with the description according to the alphabetical lettering); or it would, conversely, and non-alphabetically, start at the outer ends of the squares and look toward the triangle. To be sure, in either case, the description, whether it leads away from the center, or whether it is directed toward it, also implies some kind of common directionality which applies to all objects described. However, the implied direction is only 'the same' or 'roughly the same' in an analogical sense. 'Analogical isodirectionality,' as I will call it, refers to descriptions which are based on looking at the different objects in terms of their relation to one another; it does not presuppose looking at the diagram from a specific viewpoint from which the description of the objects *literally* takes 'roughly the same direction.' Similarly, a description which is alphabetical wherever it can be, does not betray a view onto the diagram from a specific standpoint; rather, it traces the objects partly from the inside to the outside, and partly, to a very limited extent, from the left-hand to the right-hand side. Accordingly, both an analogical and an alphabetical description of I 47 would be non-indicative of viewing the diagram: both of them could be accomplished on the basis of envisaging the diagram in spite of its comparative complexity because of the 'symmetry' of the arrangement.

By contrast, a letter ordering emerging from a description which follows roughly the same direction in a literal sense, implies a specific viewpoint outside the diagram from which the text refers to it. In this regard, I 47 provides a case in point: because its 'symmetrical' diagrammatic arrangement allows for distinguishing quite clearly a 'roughly isodirectional' from an alphabetical or a geometrically analogical description. It is, thus, potentially more indicative of a larger-scale isodirectionality than the cases of diagrams containing several supposedly isodirectional quadrangles and pairs of parallel straight lines which I have previously cited; because those diagrams are all lettered from the left-hand and to the right-hand side and to some extent also from top to bottom. If the description of I 47 turns out to be accomplished according to what have appeared to be presumably the preferred directions of description, some overlap is still to be expected with a kind of description which is analogical and one which corresponds to the diagrammatic lettering. Accordingly, those cases are the most conclusive ones in which 'roughly isodirectional' description is decidedly preferred over geometrically analogical description.

359 *Elem.* I 47 (vol. 1, p. 63.8, 11 f. Heiberg & Stamatis).

360 *Elem.* I 47 (vol. 1, p. 63.12 Heiberg & Stamatis).

Isodirectional references to parallel objects

Before turning to 'rough isodirectionality,' let us look at different types of cases of isodirectional parallel objects in I 47 (Fig. 46) and see whether they adhere to the supposedly preferred viewing directions, and (if so) how conclusive they are by themselves with respect to the viewing of the diagram: (1) isodirectional parallel straight lines, (2) isodirectional consecutive straight lines, and (3) isodirectional parallel diagonals of parallel squares.

(1) Isodirectional parallel straight lines: I have already mentioned above, when introducing 'the concept of isodirectionality,' the case of the three parallel straight lines ΒΔ, ΓΕ, and ΑΛ, which are introduced in the construction part: καὶ διὰ τοῦ Α ὁποτέρᾳ τῶν ΒΔ, ΓΕ παράλληλος ἤχθω ἡ ΑΛ ("and through A should be drawn ΑΛ parallel to ΒΔ and ΓΕ").[361] All three straight lines are not only isodirectional, but also appear to be looked at from the same point of view, i.e., from above the diagram, as their letter order invariantly implies the direction 'top-to-bottom.' Moreover, the order in which ΒΔ and ΓΕ are mentioned follows the left-to-right direction (whereas ΑΛ is mentioned separately, as it is the straight line only to be drawn). Neither the one nor the other viewing direction implies the viewing of the diagram conclusively, because both the letter order of the three references and the order in which two of the parallel lines are mentioned are alphabetical. Additionally, the letter sequences are also analogical, starting each time with the point of the triangle and ending with that of the square (ΒΔ, ΓΕ) or on the square (ΑΛ). The switch and reswitch of ΒΔ in the further course of the proposition could suggest retrospectively that the quoted part from the construction follows isodirectionality: ΔΒ is used permutedly in the linear description of the straight lines ΔΒ and ΒΑ, analogically with ΖΒ and ΒΓ (I come back to this example below).[362] Afterward, it is used in the alphabetical form again in conjunction with the references to the parallel straight line ΑΛ, and to the triangle ΑΒΔ and the parallelogram ΒΛ which both lie between ΒΔ and ΑΛ.[363] Here, additionally, more parallel objects are aligned isodirectionally with the two parallel straight lines: the parallelogram ΒΛ and the triangle side ΒΔ implied by the reference to the triangle ΑΒΔ. However, even though this passage in which ΒΔ reappears reswitched reinforces the impression of isodirectionality, all objects again abide by alphabeticality and analogy. Therefore, they are again no more indicative of viewing than the initial passage from the construction part; and ΒΔ might be switched back to the alphabetical order simply because it is no longer used in analogical description with ΖΒ, but not for the sake of isodirectionality.

(2) Isodirectional consecutive straight lines: this is a special case of isodirectional parallels. It refers to two straight lines which lie on the same line and share one point,

361 *Elem.* I 47 (vol. 1, p. 63.12 f. Heiberg & Stamatis).
362 *Elem.* I 47 (vol. 1, p. 64.1–3 Heiberg & Stamatis).
363 *Elem.* I 47 (vol. 1, p. 64.8–12 Heiberg & Stamatis).

and which are mentioned in such a way that the second letter of the first line is the same as the first letter of the second one. Isodirectionality here coincided with linearity.[364] I 47 provides two examples within one sentence: ἐπ' εὐθείας ἄρα ἐστὶν ἡ ΓΑ τῇ ΑΗ. διὰ τὰ αὐτὰ δὴ καὶ ἡ ΒΑ τῇ ΑΘ ἐστιν ἐπ' εὐθείας. ("[. . .] therefore, ΓΑ is on a straight line with ΑΗ. For the same reason ΒΑ is also on a straight line with ΑΘ").[365] The explicit isodirectionality of each pair of consecutive straight lines is indicated by the change of the letters in the reference to the straight line ΓΑ which has been introduced and designated ever since alphabetically as ΑΓ. The likewise non-alphabetical straight line ΒΑ, by contrast, has always been called thus for reasons which will be explained below. Each pair of consecutive straight lines is itself isodirectional. The explicitness of the isodirectionality and linearity is recognizable from the switch to a non-alphabetical letter order in the reference to one of the straight lines (ΓΑ). The purpose seems to be to underline stylistically the assertion that the two straight lines lie on the same straight line. Moreover, the linear-isodirectional kind of description suggests itself the most for aligning consecutive straight lines, as it allows for referring consecutively to the three points constituting the two straight lines. Accordingly, all other instances of consecutive straight lines lying on the same straight line in I 47 except one (on which below shortly) abide by isodirectionality (and linearity). Therefore, also in the present case, this mode is preferred to having all straight lines referred to alphabetically. As to the particular direction implied by the letter order, the isodirectionality of the two lines is in each case non-indicative of the viewing of the diagram. On the one hand, they do not follow the preferred directions for the most part, with one pair pointing left, the other one right, and each of them from bottom to top. On the other hand, as a pair, each of them adheres to alphabeticality as far as this is possible for linear straight lines (ΒΑ, ΑΘ instead of ΘΑ, ΑΒ, and ΓΑ, ΑΗ instead of ΗΑ, ΑΓ), so the letter order is in any case the most conceivable one and therefore non-conclusive of a particular directionality which implies viewing the diagram. In combination with one another, the two pairs of consecutive straight lines can serve as an example of 'analogical isodirectionality:' the reference to each starts with a point on the triangle and ends with a point on the respective square. The reason why it is not the other way around could be twofold: on the one hand, the just mentioned alphabeticality of the letter order of the *pairs*; on the other hand, ΒΑ is isodirectional with its own mention at the beginning of the sentence part of which I have quoted above, and with the implication of the sequence ΒΑ as part of the references to the angles ΒΑΓ and ΒΑΗ at this point. Whatever the motivation for the letter order there, it seems to have influenced the order of the sequence ΒΑ, ΑΘ, and the direction of ΓΑ, ΑΗ seems to be chosen by analogy with ΒΑ, ΑΘ. Thus, the pair of the two couples of consecutive

---

364 This is the above-mentioned way of expanding the concept of linearity and making it refer to more than one object.

365 *Elem.* I 47 (vol. 1, p. 63.18 f. Heiberg & Stamatis).

straight lines by itself is non-indicative of the viewing of the diagram, because the letter order can be explained by alphabeticality and by analogy.

Let us have a brief look at the only one instance in this proposition at which consecutive straight lines – one of the two pairs which have just been discussed – are referred to non-isodirectionally. At the beginning of the very sentence I have quoted, ΓΑ is still referred to in the alphabetical form ΑΓ in which it has been introduced into the proposition. The two consecutive straight lines which come to be referred to isodirectionally in the later part of the sentence are referenced here as ΑΓ, ΑΗ: καὶ ἐπεὶ ὀρθή ἐστιν ἑκατέρα τῶν ὑπὸ ΒΑΓ, ΒΑΗ γωνιῶν, πρὸς δή τινι εὐθείᾳ τῇ ΒΑ καὶ τῷ πρὸς αὐτῇ σημείῳ τῷ Α δύο εὐθεῖαι αἱ ΑΓ, ΑΗ μὴ ἐπὶ τὰ αὐτὰ μέρη κείμεναι τὰς ἐφεξῆς γωνίας δυσὶν ὀρθαῖς ἴσας ποιοῦσιν· ("And since each of the angles ΒΑΓ and ΒΑΗ is right, the two straight lines ΑΓ and ΑΗ not lying on the same side make adjacent angles equal to two right ones at the straight line ΒΑ at point Α on it").[366] Here, the text does not emphasize the two lines' lying on the same straight line, but their lying *on different sides* (sc. of ΒΑ). Accordingly, the letter order does not emphasize the continuity, but rather the separateness of the two straight lines. The reason for the particular way of non-isodirectional letter order, beginning in either case with Α, could be alphabeticality, but especially analogy: the proof starts by referring both alphabetically and analogically to the angles ΒΑΓ and ΒΑΗ. These letter 'names' imply the references to the straight lines in question and already suggest the letter order ΑΓ, ΑΗ in the subsequent actual references to them. Further, the text mentions the straight line ΒΑ at which, at point Α, the two straight lines form adjacent angles. Since Α is emphatically mentioned as the point of reference for both straight lines, and the point which demarcates the different sides on which they lie, the references to them most naturally begin in each case from Α and end, analogically, with the respective point at opposite ends of the same line of which both ΑΓ and ΑΗ form part. Therefore, although the letter order of the two straight lines points in opposite directions, they can be called 'analogically isodirectional' in the above defined sense. However, they are clearly not isodirectional in the literal sense which implies their being looked at from a fixed point of view. With respect to the isodirectionality of non-parallel straight lines, which I deal with under the next point, the whole sequence 'ΒΑΓ, ΒΑΗ, ΒΑ, ΑΓ, ΑΗ' might be considered as being potentially indicative of the viewing of the diagram. For the view onto these objects appears to move repeatedly from Β to Α and from there into opposite directions. However, because of the analogy and the alphabeticality of the letter order of the angles, and of the consecutive straight lines, the viewing of the diagram cannot be proved. Lastly, the non-alphabetical order of ΒΑ might give some indication of viewing. It could have been formed by analogy with the alphabetical order of the angles which are mentioned first, but it has already been introduced in this way in the definition of the goal. If the letter order turns to be due to (non-

366 *Elem.* I 47 (vol. 1, p. 63.14–18 Heiberg & Stamatis).

analogical) isodirectionality there, then the whole present sequence (the first three letter names of which start with "BA") could be due to the same view onto the diagram which the text eventually has had from the beginning. I shall come back to this under the next point ('Roughly isodirectional' references to non-parallel objects).

(3) Isodirectional parallel diagonals of parallel squares: this is a special case of parallel straight lines. Squares and rectangles can be referred to not only by four or three of their points but also by two diagonal points. If two parallel squares or rectangles are referenced by their analogically parallel diagonals, the letter order of these references can be compared with respect to their directionality, even though the diagonals themselves do not appear as straight lines in the diagram, and even though the objects that are being viewed are.[367] In I 47, two of the three squares are referred to in this way; throughout the proposition, their references retain the very letter order by which they are introduced in the construction part: Ἀναγεγράφθω γὰρ ἀπὸ μὲν τῆς ΒΓ τετράγωνον τὸ ΒΔΕΓ, ἀπὸ δὲ τῶν ΒΑ, ΑΓ τὰ ΗΒ, ΘΓ [. . .] ("For let there be described on ΒΓ the square ΒΔΕΓ, and on ΒΑ, ΑΓ the squares ΗΒ, ΘΓ [. . .]").[368] ΗΒ and ΘΓ are mentioned isodirectionally and conform to one of the arguably preferred viewing directions, top-to-bottom. This is the first case regarded so far which strongly indicates that the squares are looked at from a specific point of view, i.e., from above. This is because both of their letters are non-alphabetically ordered. The non-alphabetical letter order of both ΗΒ and ΘΓ is suspicious first because Β and Γ have already been introduced into the text and may be expected to be mentioned before Η and Θ which are introduced only by these very references. Second, the analogy with the third square (referred to first in the above construction assignment) suggests that Β and Γ are mentioned first: because these are the points shared with the triangle, and the four-lettered reference ΒΔΕΓ likewise starts with a letter on the triangle. Therefore, the non-alphabetical letter order of ΗΒ and ΘΓ only makes sense as indicating one of the arguably preferred viewing directions. In fact, this could even explain the letter order of the third square ΒΔΕΓ which starts alphabetically, and with a letter of the triangle, by contrast with the other two: at least by its initial part, it implies the same view onto the diagram (going down) as the other two ones – before the view on the square necessarily must go up again. This, however, would mean to compare objects which are analogical (squares on sides of the same triangle), but non-parallel, so I leave a more detailed comparison of all squares and rectangles, parallel or non-parallel, for the next section. Even without taking into account those other references to squares and rectangles, the isodirectionality of ΗΒ and ΘΓ is highly indicative of the viewing according to the top-to-bottom direction. It is, however, not very indicative of the left-to-right direction, as the imagined diagonals are directed only slightly toward the right-hand side in depictions which have ΒΓ as the horizontal line. The order of the joint references to "ΗΒ, ΘΓ,"

367 Netz includes such a case in his short review of the isodirectionality of parallel straight lines (1999a, 81).

368 *Elem.* I 47 (vol. 1, p. 63.11 f. Heiberg & Stamatis).

although it always has the one of the left-hand side first, is non-indicative, because it is also the alphabetically suggested variant.

'Roughly isodirectional' references to non-parallel objects

Two main cases are going to be considered: (1) roughly isodirectional non-parallel consecutive straight lines, and (2) roughly isodirectional non-parallel non-consecutive straight lines.

(1) Roughly isodirectional non-parallel consecutive straight lines. 'Consecutive' means that they share one point and are consequently constituted by three points and referred to by three letters in total. Here, I begin by looking at two main cases: on the one hand, at two consecutive straight lines being referred to by mentioning the letter which they share first in either of the references; on the other hand, at two straight lines being referred to linearly (i.e., in the manner of the isodirectional consecutive straight lines considered above). Then, I am going to compare a larger-scale case which includes the joint reference to two pairs of consecutive non-parallel straight lines of the linear type.

The definition of the goal of I 47 provides us with an idea of either of the two senses in which the text can present non-parallel consecutive straight lines roughly isodirectionally: λέγω, ὅτι τὸ ἀπὸ τῆς ΒΓ τετράγωνον ἴσον ἐστὶ τοῖς ἀπὸ τῶν ΒΑ, ΑΓ τετραγώνοις ("I say that the square on ΒΓ is equal to the squares on ΒΑ and ΑΓ").[369] The three straight lines mentioned here are the ones forming the right triangle, which has been introduced before as ΑΒΓ and as having the angle ΒΑΓ right. The two cases of 'rough isodirectionality' which I recognize are represented in this sentence by the two pairs of successively mentioned straight lines, on the one hand ΒΓ and ΒΑ, and on the other hand ΒΑ, ΑΓ. The conjunction of the latter two straight lines is closer because they represent the triangle's legs, the squares on which are compared with the one on the hypotenuse ΒΓ which appears to be set apart from ΒΑ, ΑΓ accordingly. Therefore, these two are definitely envisaged together. However, the order of the text suggests that also the first leg to be mentioned (ΒΑ) is regarded and referred to in view of the hypotenuse ΒΓ which is mentioned just beforehand.

ΒΓ and ΒΑ: These are references to two non-parallel consecutive straight lines, which are referred to non-linearly, with each of the letter words having the shared letter Β as the first one. They may be called 'roughly isodirectional,' insofar the two straight lines which are presented from the direction of Β make an acute angle and point into a *similar* direction from the point of view of Β. The order of the letter has them point from the left- to the right-hand side of the diagram, or, more precisely, northeast if one divides the diagram into four parts according to the points of the

369 *Elem.* I 47 (vol. 1, p. 63.9 f. Heiberg & Stamatis).

compass. Of course, in a mathematical sense, they are not isodirectional, for only parallels qualify in that sense, and rough isodirectionality is not a geometrically valid category of describing object relations in a diagram. However, in an aesthetic sense (which is the one I am interested in), it does seem possible to conceive of them as taking a similar direction. By contrast, a description of straight lines making an obtuse angle, if it starts from the shared point, is not roughly isodirectional; neither is, even more so, a description of two straight lines lying on the same straight line, if it starts from the shared point (as in the case of AΓ, AH analyzed above). The limiting case seems to be that of two straight lines making a right angle and being described each starting with the point at which they make it. Here, it seems to depend on the context whether such a description still appears isodirectional or not. In the present case in which BΓ and BA make an acute angle, it is the non-alphabetical introduction of BA, and the feasibility of the alternative alphabetical one AB which could indicate an explicit application of rough isodirectionality. To be sure, if one considers only the relation of the two present references, the letter order AB suggests itself only if the word is mentioned as the first one – by analogy with the reference to the triangle which has been introduced as ABΓ. By contrast, with BΓ mentioned first, the sequence 'BΓ . . . AB' is not suggested by any of the preceding letter names, it is non-linear and – by any definition – non-isodirectional. However, with respect to the following letter word AΓ, the likewise alphabetical order AB appears to be a viable alternative to BA. It would present the very case of consecutive straight lines making a right angle and being referred to each from the shared point which, as I have argued, could be regarded as roughly isodirectional in some circumstances. I turn now to this second couple, the joint reference to the two legs of the triangle, and see in which sense, if any, the sequence "BA, AΓ" is roughly isodirectional, and how it compares to the alternatively conceivable variant 'AB, AΓ.'

BA and AΓ: these two-letter words refer to consecutive straight lines in a linear way: that is, the point which is shared by both (A) does not occur, like in the previous case, as the first letter of the two references. Rather, it occurs as the second one in the reference to the first-mentioned straight line, and as the first one in the reference to the second-mentioned straight line. Accordingly, A is presented as the terminal point of the first and as the starting point of the second of the consecutive straight lines. Isodirectionality in the case of consecutive straight lines presupposes, apart from the linearity of the reference, that the two lines lie on the same straight line (as in the example of ΓA, AH discussed above). However, in order for the description to be 'roughly isodirectional' in the more pictorial or aesthetic sense which I am tracing, it suffices that the angle made by the two lines at the shared point be obtuse. By contrast, two consecutive straight lines which make an acute angle and are referred to linearly, are in any case not isodirectional. Again, the case in which they make a right angle, as presently, seems to be ambiguous regarding the question of 'rough isodirectionality,' and to depend on the text. Presently, there does seem to be a sense in which BA and AΓ are roughly isodirectional: although the sequence points at first northeast and then southeast, it points overall into roughly the same direction as the straight line BΓ (left to right), because the sequence shares the starting point (B) and terminal point (Γ) with BΓ. It is, thus, especially in view of the aforementioned line BΓ that the sequence "BA, AΓ" can be perceived to be roughly isodirectional and to conform to the left-to-right direction. In this respect, it should be understood that the indication of a direction in the case of straight lines (εὐθεῖαι γραμμαί) pertains only to the length of the straight lines demarcated by two points, but not beyond this length,[370] although directions are always described in an indeterminate way (left, right, east, west, etc.). The reason why the consecutive straight lines BA and AΓ are presented roughly isodirectionally could be precisely the rough isodirectionality of this sequence with BΓ. As a result, the overall viewing direction emerging from the definition of the goal appears to be from left to right: first, the reference to BΓ points in this direction, then the text goes back to the left-hand side (so as to make BΓ and BA roughly isodirectional) and proceeds, again from left to right, via A again to Γ. To be sure, the ordering of the letters in "BA, AΓ" might simply be analogical with the letter order of the afore-introduced right angle BAΓ. This one might have been ordered itself according to the left-to-right direction, but it could also have been chosen simply by adhering to the alphabetical principle. In this latter case, neither the reference to the angle nor the subsequent reference to "BA, AΓ" would be fully conclusive with respect to the abidance by a specific viewing direction. It is the comparison with alternative possibilities of ordering the letters of "BA, AΓ" which could finally give some conclusive indication of the significance of viewing in this first part of the proposition. The only other way of letter ordering which might be preferable to the present one, is the one "AB, AΓ" which I have pointed out at the end of the preceding paragraph. It might be preferred if the diagram had not been present because it is alphabetical and analogical (both references start at A and end at a point of the hypotenuse, and thus present the analogical legs in the same way). The at least equal viability of this alternative letter ordering in a situation which does not presuppose a specific point of view could indicate that the letter order in the text might have originated from viewing the diagram, and it might even be considered isodirectional in the first sense I have sketched out for consecutive non-parallel straight lines. By contrast, the alternative alphabetical letter ordering would be unlikely to result from a viewing perspective. To be sure, it might even be considered roughly isodirectional in the sense sketched out in the preceding paragraph, and imply the top-to-bottom direction. However, because of the right angle the implication of rough isodirectionality is not clear, and there is no other letter name in this immediate context which displays roughly the same viewing direction and in so doing would confirm that AB and AΓ are roughly isodirectional (in the way that the direction of BΓ confirms the isodirectionality of "BA, AΓ"). For, while the proposition by many of its letter words overall testifies to the top-bottom direction, the exposition and definition of the goal

370 The definition of εὐθεῖα γραμμή is obscure (*Elem.* I, Def. 4), but *Elem.* I, Post. 1 makes it clear that it is to be understood in the sense of the German "Strecke" as opposed to "Gerade" (Thaer 1970, 418 *ad loc.*).

are indicative of viewing the diagram from the left-hand side. This direction is not due to the lettering of the triangle, which displays the left-to-right direction only in part; the letter ordering goes beyond the lettering in pursuing this direction. As a result of these considerations, both pairs of consecutive straight lines ΒΑ and ΑΓ and ΒΓ and ΒΑ are not only explicitly roughly isodirectional, but also indicative of a particular direction of viewing.

ΔΒ, ΒΓ and ΖΒ, ΒΑ; ΔΒ, ΒΑ and ΖΒ, ΒΓ: two corresponding pairs of roughly isodirectional non-parallel consecutive straight lines in context. The following longer passage contains four linear references to consecutive non-parallel straight lines. By itself, it cannot be in any case fully conclusive of the presence of the diagram, because it stems from the proof part in which the diagram must be assumed to have been constructed so as to be able to be viewed anyway. However, the passage is instructive as it serves to sharpen the criteria by which 'rough isodirectionality' and its conclusiveness with respect to the viewing of the diagram can be determined. The part referring to the pairs of consecutive straight lines is marked in italics:[371]

> καὶ ἐπεὶ ἴση ἐστὶν ἡ ὑπὸ ΔΒΓ γωνία τῇ ὑπὸ ΖΒΑ· ὀρθὴ γὰρ ἑκατέρα· κοινὴ προσκείσθω ἡ ὑπὸ ΑΒΓ· ὅλη ἄρα ἡ ὑπὸ ΔΒΑ ὅλῃ τῇ ὑπὸ ΖΒΓ ἐστιν ἴση. *καὶ ἐπεὶ ἴση ἐστὶν ἡ μὲν ΔΒ τῇ ΒΓ, ἡ δὲ ΖΒ τῇ ΒΑ, δύο δὴ αἱ ΔΒ, ΒΑ δύο ταῖς ΖΒ, ΒΓ ἴσαι εἰσὶν ἑκατέρα ἑκατέρᾳ·*
>
> And, since the angle ΔΒΓ is equal to the angle ΖΒΑ: for each is right: let the angle ΑΒΓ be added to each; therefore, the whole angle ΔΒΑ is equal to the whole angle ΖΒΓ. *And, since ΔΒ is equal to ΒΓ, and ΖΒ to ΒΑ, the two sides ΔΒ, ΒΑ are equal to the two sides ΖΒ, ΒΓ respectively;*

The passage begins by stating the equality of two analogous right angles (ΔΒΓ, ΖΒΑ) made in Β on different sides of Β and proceeds to deduce from this the equality of two other analogous obtuse angles, likewise made in Β and on different sides of Β (ΔΒΑ, ΖΒΓ). Then it asserts the equality of two consecutive non-parallel straight lines (ΔΒ, ΒΓ) and the analogous equality of two other consecutive non-parallel straight lines (ΖΒ, ΒΑ). These are the straight lines which make the right angle mentioned at the beginning of the passage. Both pairs of consecutive straight lines are analogously two sides of a square on the triangle: ΔΒ, ΒΓ of ΒΔΕΓ, and ΖΒ, ΒΑ of ΗΒ. The respective first-mentioned line is not on the triangle, the second line, respectively, is the line which the respective square shares with the triangle. All four lines have in common point Β which is part of the triangle and of both squares. The correspondence of the two sets of straight lines is implied by their analogous linear description: in each case, it starts from the square erected on the adjacent sides of the triangle, more precisely, from the respective external point (Δ and Ζ) of those sides of the square which face each other, lying on opposite sides of Β; then, in both cases, the description runs along these sides of the squares toward their common point Β of the triangle, and from here runs apart again, following the side which the respective square shares with the trian-

371 *Elem.* I 47 (vol. 1, pp. 63.20–64.3 Heiberg & Stamatis).

gle (as it is the base line on which the respective square is erected), and stops at the end of these lines in points Γ and A, respectively. Subsequently, from the established respective equality of the two sides of the two squares, the corresponding equality (ἴσαι εἰσὶν ἑκατέρα ἑκατέρᾳ) of the sides ΔB, BA and ZB, BΓ is deduced. The sentence deduces plainly that the respective corresponding lines are equal to each other, that is, on the one hand ΔB and BΓ, and on the other hand BA and ZB, but in so doing, it also implies that the respective two lines combined equal one another.[372] Again, the description is linear in an analogous way: it begins with Δ and Z, respectively (even in the same order as before), and traces the opposing sides of the squares toward B on the triangle. This time, the description runs in crosswise direction and continues on the base line of the triangle on which the respective *other* square is erected: that is, the description which has started from Δ continues toward A, and the one from Z toward Γ. The text forms again two linear sequences Δ, B, A and Z, B, Γ, and their geometrical correspondence is again mirrored by the analogy of the textual exposition.

Let us consider the four pairs of linearly presented consecutive straight lines mentioned in this passage individually: the first two each make a right angle (ΔB, BΓ and ZB, BA), and the subsequent two an obtuse one (ΔB, BA and ZB, BΓ). Each of those which make an obtuse angle are in any case roughly isodirectional in the sense we have defined at the beginning (e.g., of the consecutive lines 'ZB, BΓ,' ZB leans toward the bottom right, BΓ further to the right, so their rough direction is toward the right, and slightly downward). By contrast, those of the consecutive lines which make a right angle are not by themselves indicative of rough isodirectionality (e.g., of the consecutive lines ΔB, BΓ, the letter order of the first one points only upward, that of the second one only to the right). In the specific case of ΔB, BΓ, it is the switch of the letters of the straight line ΔB which has been introduced in the construction part as BΔ which could indicate an explicit directionality of the pair of straight lines. However, the switch occurs only in the proof part and is, therefore, in any case, not indica-

372 The expression ἑκατέρα ἑκατέρᾳ makes it clear that the equality of the respective corresponding lines is meant (LSJ s.v.), but the sentence must mean these lines equal one another *as part of* the conjunction of the two pairs of straight lines, which therefore also equal each other, because the equality of the lines individually has already been established by the previous sentence. The problem has been felt that the straight lines do not appear in the order in which they correspond to one another. Thaer tries to solve this problem by adding "(überkreuz)" to his translation of the sentence ("so sind zwei Seiten DB, BA zwei Seiten FB, BC (überkreuz) entsprechend gleich;" 1970, 32); Heath (2014, 350 *ad.loc.*) finds himself constrained to change the transmitted order of the letters of the first couple of straight lines, that is, both the order of the two letter words and the order of the letters within (in his transcription from DB, BA to AB, BD), so as to make the corresponding lines and letters occur in the according order. However, this does not seem necessary since the sentence ultimately wants to assert the equality of the two pairs of consecutive lines, and since from the previous sentence it is already clear which of the lines equal one another. Changing the transmitted letter order is not only unnecessary, but also changes the analogy with the previous and subsequent presentation of corresponding objects.

tive of viewing. But it is above all the analogous presentation of the whole passage which suggests that the pairs of consecutive straight lines are presented isodirectionally both individually and with the pairs of consecutive straight lines – and with the angles – which start from the same point and lie on the same side of the diagram. For example, the consecutive straight lines ΔB and BΓ can be held to be isodirectional in spite of their making a right angle because of their analogous presentation with their respective counterparts ZB and BA: these lie on the other side of B and likewise appear to be isodirectional from the comparison. Moreover, the analogy of ΔB, BΓ and ZB, BA is framed by the foregoing juxtaposition of two pairs of opposite angles (ΔBΓ and ZBA, and ΔBA and ZBΓ). The comparison of the letter words for the angles and consecutive straight lines is enabled by the conventionalized linear presentation of the angles: in the present case, they involve the same letters, in the same order, as the pairs of consecutive straight lines (as these are the very straight lines making the angles). In this way, the first pair of opposite angles is composed of the same letter combination (presented in the same order) which is taken up by the first analogy of consecutive straight lines: ΔB, BΓ and ZB, BA. In view of the foregoing opposition of the same letter combination and order, the order of the letters in the references to the present consecutive straight lines appears to be explicitly chosen so as to repeat the directions implied by the references to the angles: the two consecutive straight lines ΔB, BΓ are isodirectional with ΔBΓ; and the (rough) isodirectionality of ΔB, BΓ themselves is implied by the repeated opposition with the direction of the analogously presented letter sequence ZBA. Moreover, the second pair of angles (ΔBA and ZBΓ) implicitly aligns the letter sequence ΔBA with the foregoing one ΔBΓ by the analogy with the opposite sequences. The sequences ΔBΓ and ΔBA appear as roughly isodirectional, especially by comparison with the sequences formed in reference to the angles on the other side of B; and in light of this rough isodirectionality, the sequence ΔBΓ, as it subsequently appears in the form of the consecutive straight lines ΔB and BΓ, appears to be roughly isodirectional itself. Finally, the order of the letters in the references to the second couple of consecutive straight lines (ΔB, BA) takes up the letter order of the earlier reference to the angle ΔBA and recalls its rough isodirectionality with the angle ΔBΓ. Accordingly, the order of ΔB, BA (which is itself roughly isodirectional) confirms the rough isodirectionality of ΔB, BA with ΔB, BΓ, and, retrospectively, also that of the sequence ΔB, BΓ itself. The analysis has focused on the mutual impact of the description of different objects onto the perceived direction of specific ones of these objects with special regard to the directionality of the consecutive straight lines making a right angle. It confirms the initial intuition that the perception of isodirectionality as regards consecutive non-parallel lines depends on the context rather than the exact angle between the two lines. Having, thus, established that the consecutive straight lines and angles on either side of the common point B are isodirectional above all by the constant analogy with the objects on the respective other side, the question is whether the description of the geometrical situation as a whole comprising all these objects constitutes a large-scale case of 'rough isodirectionality.' In a sense, it does – that is, if one looks at the overall directionality of the passage. The rough direction of description is from southwest to northeast, and there are no components which contradict this classification completely, and only very few which contradict it partly, such as the letter sequence ZB (some behave neutrally by being indifferent to either the east/west or the north/south direction). However, the internal bipartite structure of the description is more dominant than this overall direction: in fact, the description in terms of analogous pairs of (pairs of) objects turns out two strings of roughly isodirectional objects on either side of point B. Although the two strings partially overlap (with points A and Γ belonging sometimes to the one and sometimes to the other string), the two directions evolving from the analogous description remain visible: the viewing direction starts from two distant points, runs together in one and drifts apart again toward two distant points. This means that the two consecutive lines are not aesthetically isodirectional to one another, but only analogically. Their description presupposes two instead of one common viewpoint, and it is therefore, in an aesthetic sense, not even *roughly* isodirectional. In a purely aesthetic, non-geometrical sense, how they behave to each other is mirror or point symmetrically rather than isodirectionally. In this sense, such an analogous exposition of the geometrical situation is no less visualizing than 'aesthetical' isodirectionality; the point, however, is, that analogous description is what one might go for even without having the diagram at hand and, therefore, does not betray any viewing of the diagram. To be sure, the fact that analogous description emerges as a predominant feature in I 47 does not indicate that the proposition was not dependent on a diagram; but we have to look for deviations from this standardized practice in the manner of aesthetically isodirectional description to find clear evidence of diagrammatization.

(2) Roughly isodirectional non-parallel non-consecutive straight lines: apart from non-parallel consecutive objects, even geometrical objects which do not have any material connection in the diagram (as do consecutive straight lines) may be described 'roughly isodirectionally.' Here, two examples are going to be considered: first, the references to two non-consecutive straight lines in their immediate context in the construction and in the proof part, and second, the combined consideration of all squares and rectangles in I 47.

ΑΔ, ΖΓ: the first conjoint occurrence of these two straight lines in the construction which assigns that they are (to be) drawn: καὶ ἐπεζεύχθωσαν αἱ ΑΔ, ΖΓ ("and let ΑΔ, ΖΓ be joined").[373] Their isodirectionality, if any, is not evident, because their respective letter order has them point individually in different directions (both are directed south, ΑΔ more clearly than ΖΓ; ΑΔ points slightly west and ΖΓ clearly east). However, the view on the *complex* of the two straight lines is from top to bottom and from left

373 *Elem.* I 47 (vol. 1, p. 63.13 f. Heiberg & Stamatis).

to right, for the view from the starting points A and Z onto the terminal points Δ and Γ is in this direction. Unlike in the previously discussed case of analogically isodirectional consecutive lines, assessing isodirectionality on the basis of their conjoint appearance seems to be possible. For the angle which the two lines make by cutting each another and which separates them before and after doing so, is slightly below 90 degrees (which, in the case of consecutive straight lines, had been singled out as a landmark for determining 'rough isodirectionality'). Moreover, both lines have a common starting point from which they could be viewed roughly isodirectionally: their description starts in each case from the square HB. A strong indication for the explicit isodirectionality of the description of the two lines is presented by the non-alphabetical introduction of ZΓ by contrast with the alphabetical one of AΔ. At the point of the assignment of the construction of the two lines, all letters which they involve have already been attributed to the respective objects, though Z only implicitly (by the introduction of the square HB). The alphabetical letter order ΓZ would have suggested itself also with respect to the geometrical analogy of the two straight lines, because it would have the description start, analogically with AΔ, at the respective point on the triangle, and have it end at the point of the respective square. (The alternative manner of presenting the lines analogically – starting from the squares and ending at the triangle – would result in non-alphabetical letter ordering both times and therefore would not be more conceivable. The letter order AΔ suggests itself especially because of the preceding one AΛ which is alphabetical and indicative of the same direction as AΔ.) Notably, the text explicitly deviates from such a both alphabetical and analogous exposition at the end of the construction part in favor of presenting both straight lines roughly isodirectionally from the perspective of the square HB. This seems to reflect the viewing of an underlying diagram rather than an attempt to facilitate the construction of the two lines by means of visualization. The foregoing part of the construction confirms this, as some of the preceding letter words deviate likewise from alphabetical and geometrically analogical presentation (my previous examples of the letter words for the consecutive straight lines BA, AΓ and for the squares HB, ΘΓ), and prefigure the left-to-right direction (BA, AΓ) and the top-to-bottom one (BΔEΓ; HB, ΘΓ; BΔ, ΓE; AΛ; on the references to the squares, see my next and final example) of the present AΔ, ZΓ.

The conjoint reference to our two non-parallel non-consecutive straight lines is repeated in the proof part, and the letter order of both is exactly the same. Pointedly, the reference is made just after the longer passage from the beginning of the proof part which I have quoted earlier and which presents two pairs of angles and of consecutive straight lines in an analogically isodirectional way. The passage quoted goes on this way: καὶ γωνία ἡ ὑπὸ ΔΒΑ γωνίᾳ τῇ ὑπὸ ΖΒΓ ἴση· βάσις ἄρα ἡ ΑΔ βάσει τῇ ΖΓ [ἐστιν] ἴση, καὶ τὸ ΑΒΔ τρίγωνον τῷ ΖΒΓ τριγώνῳ ἐστὶν ἴσον· ("and the angle ΔBA is equal to the angle ZBΓ; therefore the base AΔ is equal to the base ZΓ, and the triangle

ABΔ is equal to the triangle ZBΓ").[374] At first, the equality of the two analogous angles is repeated which had already been deduced in the preceding passage; the angles are again presented in analogous non-alphabetical fashion. From this, the equality of the two bases (i.e., of the subsequently mentioned triangles), our AΔ and ZΓ, is deduced. Abruptly, the chain of analogous presentation ends (it began at the beginning of the proof part after the roughly isodirectional, top-to-bottom construction), and the two *analogical* bases are described again, like at the end of the construction part, in the roughly isodirectional, *non-analogical* way, starting both times at the square HB. The subsequent deduction of the equality of the according triangles ABΔ and ZBΓ sticks to this roughly isodirectional, non-analogical presentation. Accordingly, the two pairs of non-parallel non-consecutive objects form another larger-scale case of 'rough isodirectionality' involving two different types of objects. Both in the case of the bases and of the triangles, it is the deviation from the foregoing analogically non-alphabetical presentation which is indicative of isodirectionality: AΔ and ABΔ would be expected to appear in the form ΔA and ΔBA, respectively, that is, analogically with ZΓ and ZBΓ (in the construction part, by contrast, the indicative letter order had been that of ZΓ, which would have been expected alphabetically there, if it were not for rough isodirectionality). This example shows that it is not in any case the switches of the letter order of particular words which is meaningful, e.g., regarding isodirectionality, but rather in general the deviation from the foregoing alphabetical or non-alphabetical, analogical or non-analogical manner of presentation. The reason why in the present case the description suddenly deviates from the analogous presentation and is instead roughly isodirectional by showing the alphabetical letter order AΔ can hardly be for the sake of repeating the form of the earlier appearance of the joint reference AΔ, ZΓ. It is too far away, and the multitude of the preceding references ΔBΓ, ΔBA, ΔB, ΔB, ΔBA[375] would have suggested to start the present one with Δ, too. Rather, the sudden change seems to indicate that at this moment, the diagram is being viewed 'non-geometrically:' that is, without making an effort of presenting the objects according to their relations; the objects are 'read from' it according to an apparently standardized way of reading the diagram which applies whenever the description does not orientate itself toward specific object relations. In the present case, it would have been more intuitive to adhere to the analogical mode until the end of the passage which is concerned with analogical objects. But apparently, the more intuitive (because standardized) mode of describing an underlying diagram is the roughly isodirectional one, so the change toward it here might have occurred inadvertently. In any case, it is adhered to also in the subsequent reference to the triangles which, again, confirms that the order AΔ is not chosen because of the earlier appearance of the word but more likely reflect a change in the

374 *Elem.* I 47 (vol. 1, p. 64.3–7 Heiberg & Stamatis).

375 I.e., in the passage *Elem.* I 47 (vol. 1, pp. 63.20–64.5 Heiberg & Stamatis).

way of viewing.[376] Although the two occurrences of ΑΔ, ΖΓ indicate the viewing of the diagram quite prominently, neither is fully conclusive with respect to viewing: the latter, as it occurs only in the proof part; the former as it occurs at a point of the construction at which the relevant letters have already been attributed.

ΒΔΕΓ; ΗΒ, ΘΓ; ΓΛ, ΒΛ: my last example for the rough isodirectionality of non-parallel objects comprises the references to the squares and rectangles of I 47 all of which stick to the same letter order throughout the proposition. Some of these are parallel objects – on the one hand, the squares ΗΒ, ΘΓ, the isodirectionality of which I have already discussed above; on the other hand, the square ΒΔΕΓ and the rectangles ΓΛ, ΒΛ which have emerged from cutting the square into two parts. However, not only the parallel objects are described isodirectionally, but indeed all of the quadrilateral objects are roughly isodirectional: they display the top-to-bottom direction throughout. Thus, they constitute a case of large-scale isodirectionality, and it is only by this overall context of isodirectionality that the direction of the parallel quadrangles becomes explicable. The main objects of this array are the three squares erected on the three sides of the triangle. Since they are also three of the central objects of the proposition, they are the first objects to be constructed and provide strong evidence for an underlying diagram, as their construction is already indicative of its being viewed. I have discussed above the isodirectionality of the parallel references to the squares ΗΒ and ΘΓ which are indicative of viewing because they are non-alphabetical and non-analogical with the reference to the third (but first-mentioned) square of the proposition ΒΔΕΓ: they start from the points not on the triangle, whereas ΒΔΕΓ starts with Β of the triangle and at least *begins* alphabetically by its first two letters (although it continues back to the triangle non-alphabetically).[377] I have pointed out that the order of ΗΒ, ΘΓ could be explained by the standardized top-to-bottom direction and that their non-analogical description with ΒΔΕΓ could be explained by the 'rough isodirectionality' of the three objects. For, while ΒΔΕΓ is not parallel with the other two squares, its description seems to indicate (at least by its first letters) roughly the same direction as does theirs, namely from top to bottom. To be sure, the three references are not really comparable because of the differing number of letters, and strictly, ΒΔΕΓ is not even capable of displaying one and the same direction throughout because of its linearity. In fact, it displays three directions instead of one, and can be compared with respect to directionality only with same-lettered references.[378] If one wants to extract one direction from it nevertheless,

---

376 This explanation for the order of the letters in the reference to the triangle ΑΒΔ is more likely than supposing – with respect to the role which disambiguation can have in determining the letter order – that it is chosen in contradistinction to the letter order in the reference to the angle ΔΒΑ which occurs shortly before.

377 *Elem.* I 47 (vol. 1, p. 63.11 f. Heiberg & Stamatis).

378 A four-lettered reference to a square could only be in itself isodirectional by mentioning the letters in the order of two of its parallel lines, which, as have seen, is uncommon in *Elements* I (one occurrence in total), even though, as in the present case, many are lettered in the diagram in this way.

one has to decide which of the three is the decisive one. Crucially, ΒΔΕΓ is introduced first; this means that, if at all, ΗΒ, ΘΓ are modelled on it by adhering to its direction, rather than the other way around. By itself, the letter order of ΗΒ, ΘΓ is unambiguously top-to-bottom but counter-intuitively so, and cannot be explained in any other way but by its adhering to the standardized viewing direction and by its being potentially isodirectional with yet another object in doing so. Accordingly, this object, if any, would be the immediately preceding square ΒΔΕΓ; ΗΒ, ΘΓ could indicate retrospectively that the order ΒΔΕΓ by its initial letters likewise displays the top-to-bottom direction, as the viewing of themselves and their letter order would have been influenced by it. By itself, ΒΔΕΓ is probably not indicative of its being viewed according to the top-bottom (and left-right) directions, like most other quadrangles in *Elements* I: the lettering of the square in the diagram could suggest that it has been done beforehand (Β and Γ as part of the lettering of the triangle, and Δ and Ε according to left-to-right lettering afterward), because a diagrammatic lettering accomplished according to a linear letter order might have been alphabetical-linear (in anti-clockwise direction), if the letters in the text had not been lettered according to the diagram,[379] but an alphabetical-linear anti-clockwise letter order would not be necessarily preferable to the present non-alphabetical linear one. While the quadrangles lettered thus are themselves non-indicative of viewing, it is precisely their alignment with other, different objects (or, as presently, the same kind of objects referenced by a different kind of letter words) which shows that the way of the letter order is explained by viewing. Presently, the construction of the straight line ΑΛ (Λ being a point on the square's side ΔΕ) parallel to two sides ΒΔ, ΓΕ of ΒΔΕΓ, which follows directly upon the construction of the three squares, gives further indication of the viewing direction of ΒΔΕΓ.[380] All three lines, which either form part of ΒΔΕΓ (ΒΔ, ΓΕ) or appear in close proximity to it (ΑΛ, which runs through it) are lettered isodirectionally from top to bottom. Even though they are not conclusive regarding their directionality as they are lettered alphabetically and thus expectably, they could suggest that ΒΔΕΓ, introduced just before, is viewed in the same top-to bottom manner; the reference to ΒΔ recalls that of ΒΔΕΓ by citing its first two letters, and the sequence ΒΔ, ΓΕ suggests how ΒΔΕΓ would be ordered if it were not for the linearity of references to quadrangles which is standardized in *Elements* I.

More conclusively, the top-to-bottom directionality of ΒΔΕΓ is indicated by the letter order of the references to the rectangles ΒΛ and ΓΛ (here called, more indeterminately, "parallelograms") which emerge as two parts of that square ΒΔΕΓ produced by the straight line ΑΛ. Because of their alphabeticality, ΒΛ and ΓΛ do not give indication of a viewing direction themselves. But the explicit analogy of ΒΛ, ΓΛ and ΗΒ, ΘΓ toward the end of the proof displays the non-analogical presentation of the upper squares and the lower one more clearly than the introduction of ΗΒ, ΘΓ in compari-

379 The case of ΒΔΕΓ is similar to that of ΕΒΓΖ discussed in detail above pp. 379 f.
380 *Elem.* I 47 (vol. 1, p. 63.12 f. Heiberg & Stamatis).

son with ΒΔΕΓ; and it underscores that the non-analogical presentation is due to rough top-to-bottom isodirectionality. The text runs as follows: ἴσον ἄρα ἐστὶ καὶ τὸ ΒΛ παραλληλόγραμμον τῷ ΗΒ τετραγώνῳ. ὁμοίως δὴ ἐπιζευγνυμένων τῶν ΑΕ, ΒΚ δειχθήσεται καὶ τὸ ΓΛ παραλληλόγραμμον ἴσον τῷ ΘΓ τετραγώνῳ· ("Therefore the parallelogram ΒΛ is also equal to the square ΗΒ. Similarly, if ΑΕ, ΒΚ be joined, the parallelogram ΓΛ can also be proved equal to the square ΘΓ").[381] With respect to the four quadrangles mentioned, this passage presents a large-scale case of rough isodirectionality as it aligns two non-parallel pairs of parallel objects. All four objects are presented by their diagonals; while in the case of ΗΒ and ΘΓ, the implied diagonals themselves are parallel, ΒΛ and ΓΛ are represented by non-parallel imaginary straight lines, but the directions of the four straight lines are in any case comparable and are regarded with respect to the direction of the square they represent. All four references indicate that the squares and parallelograms are viewed from top to bottom. They are aligned formally by rough isodirectionality rather than by analogy, even though both times the respective two objects are said to be equal. The case is thus comparable to the preceding example in which opposite angles and pairs of consecutive straight lines are said to be equal but in which 'analogical isodirectionality' is used to highlight the correspondence. Presently, the references to the squares and corresponding parallelograms would even start with the same letter in both cases if they were presented analogically (ΒΛ, ΒΗ; ΓΛ, ΓΘ), which would underline their correspondence not only in terms of size but also in terms of their lying on the same side of the diagram (i.e., on the respective side of ΑΛ). In comparison with the alphabetical presentation of ΒΛ and ΓΛ, the invariant non-alphabeticality of the references to ΗΒ and ΘΓ in I 47 and the resulting non-analogical manner of presentation become all the more conspicuous. Conversely, in light of the form of ΗΒ and ΘΓ, also the letter order of ΒΛ and ΓΛ, which are introduced into the text fairly late, becomes indicative of the viewing of the diagram. For with respect to the letter order of ΗΒ and ΘΓ, they could be ordered non-alphabetically for the sake of analogical presentation. But from the reason why ΗΒ and ΘΓ are non-alphabetically ordered (top-to-bottom isodirectionality) it becomes clear why ΒΛ and ΓΛ are not presented analogically non-alphabetically: they are alphabetically ordered in order to be roughly isodirectionally with the two squares. Finally, from the equality of ΗΒ and ΘΓ and the respective parallelograms, the equality of the sum of ΗΒ and ΘΓ and the sum of the parallelograms, which is the square ΒΔΕΓ, is deduced: ὅλον ἄρα τὸ ΒΔΕΓ τετράγωνον δυσὶ τοῖς ΗΒ, ΘΓ τετραγώνοις ἴσον ἐστίν ("therefore the whole square ΒΔΕΓ is equal to the two squares ΗΒ and ΘΓ").[382] By taking the place of ΒΛ and ΓΛ in relation to ΗΒ, ΘΓ (compared with the preceding part of the sentence), ΒΔΕΓ is once more aligned with the top-to-bottom direction of the other four quadrilateral objects.

381 *Elem.* I 47 (vol. 1, p. 64.16–19 Heiberg & Stamatis).
382 *Elem.* I 47 (vol. 1, p. 65.1 f. Heiberg & Stamatis).

Nevertheless, ΒΔΕΓ stands out from the other four references by its four-lettered form. While this is the dominant way of referring to quadrangles in *Elements* I, it is the exception in I 47 and therefore in need of explanation. To be sure, the square ΒΔΕΓ is the first of them to be mentioned in the proposition, so one might suspect it is for this reason that it is introduced in the more exhaustive form, while the other two squares, as well as the parallelograms into which the square ΒΔΕΓ is divided later on, take the abbreviated form. In the case of ΒΛ and ΓΛ, the four-lettered form is not possible, because the point at which ΑΛ cuts ΒΓ, which is the respective fourth point of the two rectangles, never gets named. But these two objects are introduced late and the letter words referring to them cannot have influenced the form of ΗΒ, ΘΓ. The four letters constituting ΗΒ and ΘΓ are not four consecutive letters of the alphabet (Α, Β, Ζ, Η; Α, Γ, Θ, Κ) as in the case of ΒΔΕΓ (Β, Γ, Δ, Ε), but this has not prevented other parallelograms in book I from being referenced in a four-lettered way. Therefore I suggest that the varying form of the references to the quadrangles has a point: the square ΒΔΕΓ stands out from the other two squares ΗΒ and ΘΓ as it is their sum; this is nothing less than the theorem proved by the proposition. Likewise, the parallelograms ΒΛ and ΓΛ, being of the same area as ΗΒ and ΘΓ, respectively, amount to the area of ΒΔΕΓ and are referred to in the way of the two smaller squares equaling one of them. This means that the use of synonyms here (which is generally rare in the *Elements*), on the one hand, serves to underscore by means of form the analogous equality of the two smaller squares and the corresponding rectangles by contrast with the single bigger square; on the other hand, the fact that this bigger square is the sum of the two smaller ones, and of the rectangles, respectively, could be represented by the number of the letters, which in the case of ΒΔΕΓ is twice as high as that of each of the smaller objects, so that both times, two two-lettered objects correspond to one four-lettered one.

Of the five letter words for squares and parallelograms considered here, only ΗΒ, ΘΓ are conclusive with respect to the presence of a diagram from the beginning, as these references already indicate their being viewed at the moment when their construction is assigned. However, the five references together indicate quite clearly that the objects to which they refer are viewed roughly from one direction, that is, from top to bottom. While not all of the references are equally indicative, their pertaining to roughly the same direction is suggested on three grounds: first, by the number of other quadrangles which indicate roughly the same direction; second, by their mutually influencing how they are perceived; and third, by their standardized form which they keep throughout the proposition. In the present case, it is precisely the non-permutation of the individual letter 'names' which is especially indicative of their indicating viewing. The example confirms what has been suspected with respect to another example of rough isodirectionality (the straight lines ΑΔ, ΖΓ): that it is often the deviation in terms of alphabeticality from an alphabetical or non-alphabetical 'setting' rather than the switches of individual names which is indicative of the description's abidance by a viewing direction.

The indicativeness of the present example goes beyond that of the former ones: those have shown that not only two parallel objects may be presented isodirectionally, but also larger complexes of different parallel and non-parallel objects, including straight lines, squares, parallelograms, angles, triangles; the context of their rough isodirectionality many times only just sheds light on the letter order of individual references. The present example, however, is the only one to even indicate unequivocally that a complex of objects is viewed in a preexisting diagram, not just in the one drawn on the basis of the text, because it includes references indicative of viewing already as early as the construction. These references are the ones to the parallel squares HB and ΘΓ, which are indicative of an underlying diagram as early as their introduction; but it is only their subsequent standardized use in relation to the other three quadrangles which confirms conclusively that the description of all five implies viewing and presupposes a diagram. Since the five objects account for a large and constitutive portion of the diagram and are present in longer coherent parts of the text, their rough isodirectionality implies that not only a particular object combination, but indeed the whole diagram is viewed while described, and that it is viewed largely according to a particular direction. Additionally, other objects are also aligned with the squares and parallelograms by their top-to-bottom directionality: the most closely related ones are the three straight lines AΛ, BΔ, ΓE which have been mentioned above. Besides, also the couple AΔ, ZΓ which has been shown to display the top-to-bottom and especially the left-to-right direction could be adduced; its left-to-right direction could be connected to the order in which the two smaller squares and the two parallelograms are mentioned – though themselves non-indicatively because of their alphabeticality – from left to right.

The standardization of two main viewing directions

The analysis of comprehensive cases of 'rough isodirectionality' in I 47 has confirmed the assumption that the two directions top-to-bottom and left-to-right shape the description of the diagram. These two directions of description have been shown to be indicative of viewing where they presuppose a ready diagram and do not overlap completely with the directions of lettering of this diagram which appears to have followed similar rules. This is the case wherever the directions cannot be explained by the alphabetical principle of letter ordering or by analogical description.

In the preceding section, my aim was to find indicative cases of 'rough isodirectionality' in view of the apparent prevalence of the two viewing directions. For this purpose, it was necessary to analyze particular larger passages regarding the complex interaction of the application of different letter ordering principles in the interplay of the lettering and letter ordering of words mentioned together. While the analysis has turned out a number of indicative examples, it has also shown that many cases display the said directions, however, non-indicatively, and that some do not display them, either because they cannot because of the constraints imposed by the diagrammatic lettering, or because they deviate from them purposefully for the sake of analogy. Therefore, although the examples of large-scale isodirectionality display both the left-to-right and the top-to-bottom direction prominently, it is difficult to quantify their standardization on the basis of them, especially in relation to the principle of analogical letter ordering.

This analogical way of relating letter words has emerged from the foregoing analysis as a prominent means of structuring the description of the diagram which partly overlaps with (aesthetic) isodirectionality and partly competes with it. However, its prominence also has to do with the choice of the sample passages. 'Passages' are by definition coherent parts of text, and coherent parts of the *Elements* always contain such objects which are interrelated closely by the text, no matter their kind or relation in the diagram. While analogy appears to be particularly apt for relating such objects stylistically, it is also confined to characterizing object relations which the text establishes anyway in specific coherent parts of it. Beyond such passages, analogies could be repeated within the same proposition (but this would not count as a large-scale case of analogy), or further objects could be aligned to the previous ones, again on the basis of the geometrical relations to be established between the earlier analogical relation and the present one. But it is not possible to find such objects described analogically which the text does not relate analogically explicitly; and many objects of a diagram are not analogically relatable in the first place, because they either do not have comparable properties or they do not compare within a particular diagrammatic configuration (e.g., the triangle and any one of the squares and parallelograms in I 47: the proposition does not establish any analogy between them, and if they were described in a comparable way, i.e., linearly-isodirectionally, their description would be classified as roughly isodirectional rather than analogical because of the lack of comparability of the objects). Accordingly, analogical description is mostly confined to coherent parts of text dealing with a particular geometrical situation of the diagram, and in any case cannot be observed across propositions.

By contrast, rough isodirectionality in an aesthetic sense can connect all kinds of objects whether they are diagrammatically or deductively connected or not, because it does not imitate object relations stylistically but orientates itself toward apparently standardized directions which do not have any intrinsic connection with the geometrical situation.[383] For this reason, isodirectionality does not only apply beyond particular passages, but even beyond individual propositions. However, if isodirectionality is traced as above within coherent passages, or in different passages of a proposition which relate the same objects which have been mentioned isodirectionally before with more, likewise isodirectional objects – then isodirectionality, just like analogy,

383 To be sure, there are certain cases in which the diagrammatic configuration does not allow certain objects to be presented roughly isodirectionally; but 'rough isodirectionality' in the wide sense in which we have defined it, does not have any geometrical conditions attached to it.

ties up stylistically such objects which are being geometrically related at that very moment anyway. This means that within particular passages, isodirectionality is always bound up with geometrical analogy. This is true whether isodirectionality underscores the analogical relation – when coinciding with a more obviously analogical way of description and being therefore non-indicative of viewing – or whether it obliterates this analogy to a certain extent instead of emphasizing it – this is where isodirectionality has been deemed indicative of viewing as it occurs in the place of a description which imitates the geometrical analogy. In fact, if confined to certain passages or recurring object configurations, isodirectionality could be seen as just another kind of analogical description, as 'isodirectional analogy,' as it were.[384] For instance, the rough isodirectionality of the squares and parallelograms of I 47 which have been traced as part of different combinations of them in two larger passages of the proof could be understood as a way of emphasizing the analogous relations of all five of them (rather than as an indication of their being viewed in accordance with a standardized direction of viewing). Thus, it is because of the analogical nature of the object descriptions in the *Elements* that analogy and isodirectionality appear to be similarly frequent in coherent passages of text and that they often coincide.

The close reading of larger or related passages of text with regard to large-scale isodirectionality remains the only way to detect cases of 'rough isodirectionality' which are indicative of the viewing of a diagram according to the main directions. However, such an analysis does not reveal the overarching significance of isodirectionality both within and beyond individual propositions, nor the great degree of its standardization. At the end of my analysis, I have related the individually indicative larger-scale cases of 'rough isodirectionality,' that is, the case of the five squares and the straight lines ΑΛ, ΒΔ, ΓΕ, as well as that of the straight lines ΑΔ, ΖΓ. To this, one could add the introduction of the sides of the triangle ΒΓ, ΒΑ, ΑΓ. Such a synopsis proves that the description of the diagram strongly adheres to the two viewing directions left-right and top-bottom independently of specific, even larger, object configurations. However, in this way, one can only accumulate indicative large-scale cases (and likewise, such large-scale cases which are non-indicative or even display contrary directions); but one cannot count them systematically in relation to one another. If one seeks to assess the standardization of the viewing directions systematically, one can only do so by counting the directions displayed by the letter words individually.

To be sure, such a count could not distinguish between cases of rough isodirectionality which are indicative of the viewing of a diagram and cases which are not – but which display the directions merely because of analogical or alphabetical description. Such a distinction can only be done by means of the previous kind of close readings of larger-scale cases. However, once it has been established that the description

384 Above I had coined the term 'analogical isodirectionality' for referring to combinations of letter words which are analogically described, and are only isodirectional in this regard.

of a diagram in certain cases does indicate conclusively the viewing of an underlying diagram, as it has been done for I 47, it is clear that this diagram at least would have been viewed throughout rather than that some objects of it would have been but others not, or that some parts would be based on it but others not. Thus, it can be assumed to have been viewed in any case. But even a description based on viewing can follow a certain viewpoint regardless of the object configuration, or trace the objects along the alphabetical lettering or according to analogical relations. The count of the directions displayed by the description alone will not allow for such distinctions, but can only serve to quantify the directions of *description*. Further, for the specificity of I 47, it may be possible to conclude from these on the standardization of the *viewing* directions. Generally, it appears from our examples that the lettering of the objects of the diagram in the order of their construction followed the same directions as the description (if the lettering indeed preceded the text). Thus, it is often not clear whether the description according to the supposedly preferred directions is due to the viewing of the diagram, or simply coincides with its lettering. In the case of I 47, however, the concentrical and 'symmetrical' shape of the diagram may allow for a possibly indicative assessment of the directions of descriptions because it has been lettered from the inside to the outside. As far as the individual objects are concerned, a left-to-right direction is visible, but an overall adherence to the left-to-right and top-to-bottom directions in the description would be distinguishable from the overall direction of the diagrammatic lettering.

As for the method and the results of the evaluation of the letter words of I 47: I have counted all letter words of I 47 which consist of two or more letters. I have only taken into account two orientations (the horizontal and the perpendicular) and four directions, that is, left-to-right, right-to-left, top-to-bottom, and bottom-to-top, and I have regarded the diagram as it is printed in all modern editions and as I have done it so far, with the first straight line to be mentioned (ΒΓ) drawn horizontally with Β on its left and Γ on its right. As far as references consisting of more than two letters are concerned, which 'make turns' and do not follow one and the same direction throughout, I have considered in each case (a combination of) the first and the last point for determining their direction. Of all 71 multi-lettered references in this proposition, 53 point from left to right, while only 10 take the opposite direction right to left and 8 do not show any inclination to either side, with all, or at least their beginning and end points being on the same perpendicular. Similarly, the majority of multi-lettered references presents the diagram from top to bottom (41), whereas only 10 are viewed the other way around from bottom to top, and 20 remain on the same horizontal line and are, thus, neutral with regard to these two sides. This means that the directions which have been suspected to dominate the description of the diagram – left-to-right and top-to-bottom – are not 100% but still very much standardized. Accordingly, if the directions of description are indeed based on the directions of viewing, a general tendency of looking into one and the same direction throughout the proposition emerges, namely, from the upper left-hand to the lower right-hand corner. Indeed, the dominance of the two directions left-to-right and top-to-bottom in the present case seems to indicate the viewing of the diagram according to these directions, because it is ascribable neither to the directions of the lettering of the diagram nor to the underlying analogical relations. For both of them are oriented overall not linearly from left to right or top to bottom, but concentrically from the center of the figure away or toward the center. This conclusion, however, can only be drawn on the basis of the foregoing close analysis of particular examples with respect to the relation of the diagrammatic lettering and the textual letter ordering, but the directions of the lettering of the objects in the diagram cannot be counted in any useful way in order to be compared with those implied by the letter words.

Three questions arise from the assessment of the letter words of I 47 regarding the directions of viewing and descriptions: first, how representative is the result of the assessment of I 47 that the diagram is viewed and described mainly from two main directions? Second, why are the two viewing directions (in I 47, but presumably also beyond) greatly, but not fully standardized, and what does the alternative mode of analogical description imply regarding the question of the viewing of the diagram? And third, what if the diagrams are turned upside-down: is it really the top-to-bottom and the left-to-right directions by which the viewing of the diagram appears to be oriented, and if so, why these two directions of all?

First, as to whether I 47 can be held representative at all with respect to the standardization of the directionality of the letter order: even though it is a small sample, the results it has yielded may be generalizable because of their great indicativeness. There are two reasons for this indicativeness: on the one hand, I 47 belongs to those propositions in *Elements* I which involve the most letters and therefore are more likely to display conclusive large-scale cases of rough isodirectionality than many of those propositions which contain diagrams with fewer letters. On the other hand, the viewing directions are clearly distinguishable because of the particular 'symmetrical' and concentrical diagrammatic configuration, on which I have just elaborated. The viewing directions which the letter words of I 47 imply on account of the size and shape of the diagram may be generalized precisely because they do not emerge from the particular configuration and shape of the diagram and do not, unlike the bits of analogical description, imitate and help to explain the object relations especially well. Rather, they must be motivated by something else and be generally applicable. As nothing indicates why the viewing directions would characterize the description of this of all diagrams, it is not unlikely that they are standardized overall for some reason. A rough review of book I indeed confirms that two directions mentioned seem to characterize the letter ordering throughout; I have made recourse to various possible instances of ordering the letters according to the left-to-right and top-to-bottom directions in the previous sections on the linearity of the letter words for quadrangles and the isodirectionality of parallel lines. If one wanted to analyze systematically the directions implied by the letter order *as viewing directions* for a more comprehensive sample, one would have to proceed as in the case of I 47 per proposition: first look for indicative large-scale cases for the viewing of the diagram and then count the viewing directions as displayed by all the letter words of the proposition. However, the problem will be to find a larger number of indicative cases in book I: for one thing, some diagrams, as those involving circles, do not lend themselves to isodirectionality to begin with, let alone to descriptions leaning toward one specific direction such as left-to-right. For another thing, of those which do, many propositions, at least of *Elements* I, contain only very few objects and letters. While their letter words do indicate the above dominant directions of descriptions, they mostly do not provide themselves evidence for the viewing of the diagram, so one cannot be sure of the origin of the directionality. Thus, I have analyzed the viewing directions exemplarily in I 47; the reason which I suggest shortly for why the diagram was viewed in this way might point to their standardization beyond this proposition.

Second, as to the parts of I 47 which do not abide by the standardized viewing directions. In fact, while these directions do not characterize the letter words necessarily whenever the geometrical situation allows it, there are no letter words in this proposition which deviate from the ordering according to the two main directions randomly. The reasons for the deviation are either that the respective object complex cannot be represented isodirectionally or that it is to be analogically described. To be sure, analogical and isodirectional description do not necessarily exclude one another. In some cases, analogical description coincides with rough isodirectionality (by which isodirectionality becomes non-indicative of viewing); e.g. the letter order in the references to the straight lines AE, BK, though one of them points upward, the other one downward, both are directed to the left-hand side, that is, are roughly isodirectional, but they are also alphabetical and analogical, beginning each time with the point on the triangle. In other cases, analogical description has to deviate from the left-to-right and top-to-bottom directions in order to be analogical in the intended way: e.g., the joint reference to the straight lines ΑΓ, ΑΗ, and the correlated references to the pairs of straight lines ΓΑ, ΑΗ and ΒΑ, ΑΘ. These analogical non-isodirectional parts[385] do not present any counter-indication for the viewing of the diagram. Rather, they show that the description on the basis of viewing might not have followed certain standardized directions mechanically but that the objects also could have been described analogically on the basis of the underlying diagram. Analogical description could serve to imitate the geometrical situation by means of form for didactic reasons – but also for aesthetic ones. In fact, both types of letter ordering – the isodirectional and the analogical way – are aesthetic in the sense that there is nothing geometrical about them. To be sure, analogical letter ordering seems to suggest itself as a way of displaying object relations on the level of text, whereas (non-analogical) isodirectional lettering appears to run counter to the textual relations which are to be established. Accordingly, the former type of lettering might be regarded as informative, the latter as aesthetic. But indeed, just as there is nothing geometrical about looking at differ-

385 Above, I have referred to the phenomenon as 'analogical isodirectionality.'

ent geometrical objects or even at the whole diagram from one and the same standardized direction, there is nothing geometrical either about presenting objects analogically by means of the letter order. It does not make the deduction any more valid and is essentially a matter of form. In this way, isodirectionality and analogy are both part of the same practice: the standardized application of aesthetic, non-mathematical form to the order of the letters, that is, to a constitutive part of the proof.

Third, as to the kind of viewing directions which appear to be standardized in our diagrams. One might object that we cannot know *which* the dominant direction was, or whether there was a dominant direction at all, since some or all of the diagrams might have been drawn the other way around, and turning them upside down would result in the opposite directions. That the ancient Greek editions should have featured the diagrams according to the same orientation which they have in our modern editions, is suggested by the manuscripts: in them, the orientation of the diagrams is largely standardized,[386] and changes only with regard to what is the left-hand and what is the right-hand side in accordance with the direction in which the respective language is written and read. That is, the Byzantine manuscripts invariantly show them in the way of our modern editions,[387] whereas the Arabic manuscripts show them mirror-invertedly.[388] From this one might infer that the direction of drawing (and lettering) corresponds with the culturally dominant direction of writing, which is, in Greek, from top to bottom and from left to right (in Arabic, by contrast, from right to left). Thus, it could be that (part of) the reason why the Euclidean diagrams were drawn, lettered and read from left to right and top to bottom was that texts, too, at least by the time of Euclid, were written and read according to these directions. In this respect, the explicit reflection about the directions of drawing by the first-century AD geographer Ptolemy are noteworthy, which I am going to consider now.

4.5 Drawing directions in Ptolemy

In the prologue to the second book of his *Γεωγραφικὴ ὑφήγησις* Ptolemy claims that he has arranged the objects of his work, that is, the descriptions of all the regions of the world, according to the map to be drawn from them, proceeding from left to right (ἐπὶ δεξιά), as the drawing (καταγραφή) may be accomplished easily (εὐχρήστου) in such a way, because the hand of the drawer may move on from those objects which

386 There are a small number of deviations as far as orientation is concerned (cf. p. 368, n. 339), but overall, the directions of drawing are standardized.

387 Cf., e.g., the diagram in our two sample manuscripts B and P.

388 Cf. the diagram for I 47 in the *Commentaries on Euclid* of the Muslim Persian mathematician Nasir al-Din al-Tusi (1201–1274) from Khorasan (https://www.loc.gov/resource/gdcwdl.wdl_07465/?sp=49).

have already been drawn up to those which have not (ἀπὸ τῶν ἤδη κατατεταγμένων ἐπὶ τὰ μηδέπω).[389] After stating the general principle of his τάξις, Ptolemy explains how he is going to accomplish proceeding from left to right: namely – and this is notable in the face of the preferred directions of the description of the diagrams in Euclid – by drawing the northern objects before the southern ones, and the western ones before the eastern ones; he adds that from the drawer's and beholder's point of view, "above" means "northern," and "right" "eastern."[390] In more concrete terms, Ptolemy concludes (II 1.6 f.) that this means starting with Europe, proceeding with Africa, and ending with Asia. Strikingly, thus, Ptolemy recommends the very directions for drawing the map which we have detected in the description of the diagram in I 47, and, for the same reason, has proceeded according to them in his own description of the map. Even though Ptolemy does not explicitly say it, the map, thus, should be drawn like a text should be written: from left to right, and from top to bottom. The reason he himself gives is a practical one: the hand should move towards the empty space. It may be that he recommended doing so in order for the hand not to become dirty or simply in order for the drawer to be able look at what he has already drawn while going on drawing. While those explanations are based on the assumption that the Greeks drew with their right hand,[391] the most natural explanation would be that drawing in this direction imitates (Greek) writing, which has always been (at least from a certain, in any case, pre-Euclidean time on) from left to right – and which, indeed, benefits in the above ways from being done with the right hand.

For our purposes, it is important to understand the order of writing and drawing in Ptolemy. The different stages of the process are indicated by Ptolemy's use of the tenses in his meta-remarks at the beginning of books II and VIII.[392] In the prologue to book II he refers to both describing and drawing the maps in the future tense, but suggests that the drawing of the maps is to be accomplished only *after* or *in the course of* their description, and in any case not beforehand. Accordingly, he plunges right into the detailed description of the maps, which is a tabular listing of all topographical items to be registered in the map which provides their exact location in a coordinate system by means of degree values (II 2–VII 4). It is only afterward that he lets follow

389 Ptolemy, *Geography* II 1.4: Προειλόμεθα δὲ τάξιν τοῦ περὶ τὴν καταγραφὴν εὐχρήστου πανταχῆ ποιούμενοι πρόνοιαν, τουτέστι καθ' ἣν ἐπὶ δεξιὰ ποιησόμεθα τὰς μεταβάσεις, ἀπὸ τῶν ἤδη κατατεταγμένων ἐπὶ τὰ μηδέπω τῆς χειρὸς ἐκλαμβανομένης. I owe the reference to Ptolemy's reflection on drawing directions to Renate Burri.

390 II 1.5: Τοῦτο δὲ γένοιτ' ἄν, εἰ γράφοιτο τά τε βορειότερα πρότερα τῶν νοτιωτέρων καὶ τὰ δυσμικώτερα τῶν ἀπηλιωτικωτέρων, ὅτι πρὸς τὰς τῶν ἐγγραφόντων ἢ ἐντυγχανόντων ὄψεις ἄνω μὲν ἡμῖν ὑπόκειται τὰ βορειότερα, δεξιὰ δὲ τὰ ἀπηλιωτικώτερα τῆς οἰκουμένης ἐπί τε τῆς σφαίρας καὶ τοῦ πίνακος.

391 Throughout Greco-Roman antiquity, right-handedness was the stereotypical ideal, while left-handedness, for general negative associations with the "left," was regarded as practically and morally inferior. For this reason, even though there is pictorial and literary evidence that left-handers (even famous ones) did exist, monumental art avoided depicting it, and children were trained and even retrained so as to use their right hands predominantly (Humer 2005, 318–327).

392 Stückelberger & Graßhoff 2006, 20.

an outline of the map of the ecumene (VII 5), the guideline to the construction of an armillary sphere (VII 6 f.), and, most importantly, summary descriptions of the individual maps of parts of the ecumene (VIII) to which the numbers in the foregoing detailed tabular description refer and to whose production, therefore, Ptolemy refers as lying in the past (VIII 2.1 ἐποιησάμεθα πίνακες).[393] That is, not only were the maps included most probably in this final part of the work,[394] but they were also drawn as "working maps" according to the outlines in book VIII, that is, as a first version presupposed by the numbering, with the order of the parts of the work, thus having (at least partly) inverted that of their production. If we take into account Ptolemy's explanation for the direction of the description, the completion of the maps according to their exact description has taken place after or simultaneously with that description. In any case, later readers will be able to draw both the first and the version of the map within the coordinate system which Ptolemy has specifically designed as a precaution against the loss of the maps in the course of the transmission, that is, as a strategy of securing the communicative success.[395] Because of the numbers it uses, the relation between the text and the external reference system is more unambiguous than that between the Euclidean propositions and the adjunct diagrams.

Despite the supposedly different respective origination of text and diagram in each case, the parallel between Ptolemy and Euclid in terms of the directions of the text referring to an imagined or real image is striking. The explicit reflection of Ptolemy may serve as evidence that a scientific drawing was described from one of its sides to the other and even in those directions which correspond to the reading and writing of a text. However, while it has been argued so far that the directions indicated by the order of the letters are proof of the prior existence of the diagram, from the comparison with Ptolemy the question arises whether the Euclidean diagrams were also to be drawn only afterward from or simultaneously with the textual description, and whether, consequently, its drawing paid attention to the very directions which the description follows. For, the very reason Ptolemy explicitly gives for choosing the directions which we also find in I 47 is that they facilitate the production of the drawing. I continue to argue that the diagram must have been included in the Euclidean text: namely, on the grounds that its description differs partly from the supposed drawing and lettering in terms of direction, whereas in Ptolemy, both processes are causally connected and, thus, invariantly share the direction.

The crucial difference is that Ptolemy is talking about a geographical map, not a geometrical diagram. This difference results in different ways of, on the one hand, drawing, and, on the other hand, reading it. First, on the aspect of drawing: while a map may be

393 Similarly, the construction guide for diagrams at the end of book VII representing an armillary sphere employs, like the Euclidean diagrams, the perfect passive imperative.

394 This is, additionally, suggested by the inclusion of maps in 17 of the Byzantine manuscripts (Stückelberger & Graßhoff 2006, 27).

395 Stückelberger & Graßhoff 2006, 13 on I 18.2.

drawn from whatever direction within a given framework, there are certain limits to choosing directions when drawing a diagram, because the (technical) construction of geometrical objects must proceed from or at least be directly related to objects which must be drawn first; e.g., in I 47, the triangle cannot but precede the three squares. However, what we might conclude from Ptolemy's reflection on the drawing of the map for the drawing of the diagram is that wherever two points were to be joined in order to construct a straight line, this might have been done preferably in the directions mentioned by Ptolemy,[396] and that the objects on the left-hand side tend to be drawn before those on the right-hand side in case where the order does not matter.

Second, on the difference of reading the map in Ptolemy and the diagram in Euclid: since all Ptolemy does is to describe by means of a list, he may sustain a given direction throughout; a mathematical proposition, by contrast, must first describe the construction in a certain order and, thus, in a somehow predetermined direction, and then perform the deduction with reference to the objects relevant for the respective step, so the direction of referring to the diagram cannot be linear from one side of the picture to the other. Accordingly, the mathematical text cannot be as rigorous with regard to directions to begin with. It is all the more surprising, then, that the direction of description does not change considerably between the first part of the proposition, including exposition and construction, and the proof part (which may have been done from the viewer's perspective in any case).

The particular directions of viewing the drawing are the same in Ptolemy and Euclid; however, while in Ptolemy, it is said to be so because of the direction of drawing, in Euclid it seems to be so in spite of it. Whereas Ptolemy specifically chooses the direction in order to accomplish the goal of his work, namely drawing a map of the ecumene, in Euclid the overall viewing direction implied by the letter words at times even counterimitates the geometrical situation (even though it is not able to contradict it as the letter order outside certain rules does not have any influence on the validity of the construction and deduction). In any case, the monodirectional depiction in itself does nothing for securing the communication of the diagram in the case of its loss (a danger which seems to have been real in the face of Ptolemy's comment), so the reliance on both media may risk indeterminacy. In any case, while the preference of a certain direction is not (always) non-informative, there is, as we have pointed out, nothing geometrical about looking at geometrical objects from the same or even at the whole diagram from one standardized direction. It is in this sense that the sensitivity to direction might be called 'aesthetic.' Ptolemy's explicit reflection on the direction of drawing implicitly testifies to the shared direction of producing (and reading) diagram and text – and therefore seems to be crucial for understanding the non-geometrical layer in Euclid.

396 That is, provided that the geometrical objects of the Euclidean diagrams have indeed not been constructed properly but simply been drawn from point to point (cf. Section 4.3.3 on the silence of the text on how to accomplish the construction).

5 The relation of form and clarity in the *Elements*

The analysis of the features of the textual form of the *Elements*, including the diagrammatic letters, in Chapters 3 and 4 has shown that each of them is standardized; in fact, many of the features are extremely standardized. However, most of them are not one hundred percent standardized, that is, they are not 'unambiguous' in the sense in which I used the term in the preceding two parts of the study. Nevertheless, the overall impression is that the textual and diagrammatic standardization is extreme and that the text is on account of this utterly unambiguous.

In Chapters 3 and 4, the determination of the relation between standardization and unambiguousness has been very technical and arguably exact with regard to every single feature. However, as regards the overall unambiguousness of the text of the *Elements*, the analysis has only been able to convey an impression or a tendency: as it is, it does not yet serve to settle the relation of standardization and unambiguousness more globally; it does not answer the question whether the *Elements* is clear or not. Nor does it serve to determine the function of standardization and the according role of clarity: whether standardization produces clarity, and if so, whether it is used specifically to do so or motivated by something else. Nor does it tell us generally how many, if any, and which readers perceive the standardization as clear.

I now attempt to determine the exact relation between standardization and unambiguousness in order ultimately to be able to define the role of clarity and form in the work. In doing so, I focus on each of the two aspects of unambiguousness, that is, on its logical and communicative side. I consider first how standardization is related to the 'logic,' or 'logical' unambiguousness of the proof, and second how it is related to the communication of the proof.

5.1 Standardization and 'logical' clarity

First, I consider unambiguousness as a textual feature which takes account of the 'logic,' mathematical precision, and consistency which is required in order for the single proof and the whole system of proofs ('Satzsystem') to be valid. This function of the form comes to mind immediately in the context of mathematics. By scrutinizing the exact relation of form and 'logic' and by testing the limits of their correspondence, this section intends to forestall the thought that the standardized form of elements texts follows inevitably from the proof content or 'logic' and functions solely in the service of proof. In order to show that the relation between formal standardization and consistency of proof is not self-evident, I compare the actual degree of standardization of the *Elements* with the degree of standardization which is required by mathematical unambiguousness. However, it proves difficult to name criteria for determining a logically necessary degree of standardization, as the text, even though it makes use of logical

https://doi.org/10.1515/9783110670462-005

argument, does not abide by the formal rules of (e.g., Aristotelian) logic. One cannot but approach the question tentatively starting from the 'logical' method which the text applies, i.e., the axiomatic-deductive one.

The amount of standardization which is necessitated by the deduction can only be tested on a trial basis. The method which I use for distinguishing necessary from unnecessary standardization is what I call the 'method of subtraction;' it consists of 'subtracting' the standardization of textual features from the text in order to consider whether the proof remains consistent and unambiguous even afterward and can, therefore, dispense with the standardization subtracted, or not. I developed this method from a passage in Aristotle's *Topics* in which he defines the superfluous parts of definitions: ἁπλῶς δ' εἰπεῖν ἅπαν περίεργον οὗ ἀφαιρεθέντος τὸ λοιπὸν δῆλον ποιεῖ τὸ ὁριζόμενον ("Speaking generally, everything is superfluous upon whose removal the remainder still makes the term that is being defined clear.").[1] I call 'necessary standardization' that sort whose absence or deficiency which will cause the proposition to be inconsistent or unsuccessful and/or the system of propositions to be incoherent. Conversely, 'unnecessary standardization' is that which can be 'subtracted' from the text without any consequences for the 'logic' of the proof. The 'subtraction' will be done as a thought experiment with regard to the logic of the Euclidean proofs in general, not with regard to any particular propositions.

The distinction of necessary and unnecessary standardization alone can only tell us whether the text is standardized *beyond* the requirements of the logic of the proof, i.e., by containing 'logically' unnecessary standardization. It cannot tell us whether it is sufficiently standardized in order for the proof to be logically unambiguous, because it is only possible to 'subtract' what is there: with the help of this method, I consider whether the standardization which characterizes the propositions of *Elements* I is necessary (or not) from a logical point of view. It is only after having distinguished 'logically' necessary and unnecessary standardization that I can consider whether the features whose standardization has been deemed necessary are indeed sufficiently standardized in the *Elements*. As the textual features of the *Elements* are for the most part not completely standardized, the necessity of their standardization is a matter of degree, and the question is whether the degree to which the necessary features are standardized is sufficient. On the basis of this, it is possible to say not only whether

1 *Top.* VI 3.140b1 f. Similarly, he formulates b33 f.: Ἢ εἰ ἔστι μὲν ἴδιον τὸ προσκείμενον, ἀφαιρεθέντος δὲ τούτου καὶ ὁ λοιπὸς λόγος ἴδιός ἐστι καὶ δηλοῖ τὴν οὐσίαν. ("Or see if, though the additional matter is proper to the given term, yet when it is removed the rest of the account too is proper and makes clear the substance of the term.") Transl. Pickard-Cambridge, modified. I only use these passages heuristically; on the one hand, because Aristotle is talking specifically about definitions which are not my subject here, on the other hand, because I generally do not use his remarks about form and clarity in order to interpret the Euclidean form. As I have argued above, the reason is that he regards the formal elements of mathematical texts from his own perspective (this can be seen precisely from the present passage about how definitions should be written, to which the Euclidean definitions do not conform).

the standardization exceeds the requirements of 'logical' unambiguousness (on account of the unnecessary features), but also whether the standardization is sufficient from the perspective of proof logic. What will not be considered is whether any qualitatively different textual standardization would be necessary in order for the proof to be logically unambiguous; I assume that the textual form – as it is – is capable of sufficient standardization with regard to the 'logic' of the proof. – Here it is important to stress the following point: by saying that I examine whether the textual form is sufficiently standardized with regard to 'logical' unambiguousness or proof 'logic,' I mean precisely that I consider whether the textual form is sufficiently standardized in order for the proof to be clear *as far as the textual form is able to contribute to 'logical' unambiguousness*. For I do not look at other (i.e., logical-argumentative) preconditions for the unambiguousness of the proof.[2]

Technically, all textual features are in line for being 'subtracted' on a trial basis: e.g., lexical, grammatical, and diagrammatic features. None of them are considered as 'given;' there is no core of textual standardization which is presupposed as necessary. What I presuppose is only the axiomatic-deductive method, because otherwise, one could think of a different method coming with different kinds of textual standardization altogether. It is difficult not to take any textual features for granted, but if certain features are presupposed, certain standardization will be presupposed too, and this would somehow predetermine the picture of textual standardization. I do not systematically go through all textual features again; rather, the aim is to show based on examples what is necessary and what is not.

'Logically' necessary and unnecessary standardization

'Logically' necessary standardization includes, e.g., the standardized use of the diagrammatic letters and of the geometrical lexicon. The reference to a lettered diagram by means of letters has been shown to be necessary from a certain degree of proof complexity onward. Indeed, the *Elements* uses letters in every single proof – even in the simplest ones.[3]

The letters must be introduced into the text and connected with the diagram[4] – unless a diagram is applied. The comprehensive analysis of Chapter 4 has shown that the inclusion of a diagram in many of the propositions is certain, but that its standard-

2 As has been pointed out in the introduction, the present study is not about how far textual form is responsible for proof 'logic,' but about whether the textual form is sufficiently explained by the needs for 'logical' unambiguousness.

3 By contrast, Aristotle paraphrased very simple diagrams without using letters (Asper 2007, 121).

4 This is an element of logical standardization which is only partly accomplished by textual standardization; I consider it here in listing necessary and unnecessary textual standardization nevertheless, because the necessity of the feature of the diagram can only be discussed with regard to it.

ized inclusion, though it seems highly obvious, cannot be proved. As the diagram is probably standardly applied, and as most of the letters are introduced into the text, a 'double standardization' emerges which is unnecessary from a logical point of view.

The geometrical terms must be completely standardized, too, in order for the deduction to be unambiguous; that is, they must not only be used in the same way within the unit of a proposition, but throughout the work, so that the propositions and their proofs can be presupposed by subsequent ones and referred to simply by their exact wording being repeated. As has been shown in Section 3.2.1, the lexicon is almost completely standardized insofar as there are no homonyms and only few synonyms.

'Logically' unnecessary standardization could include, e.g., the "second-order" vocabulary,[5] that is, the 'logical' language of the proposition: As we have seen in Section 3.3.3, the use of ἔστω is fairly standardized, but ἔστω can also be substituted by other present imperatives and is therefore technically dispensable. Moreover, from the analysis of the 'textual construction' (Section 4.3) the question has arisen whether the exposition imperatives and the construction imperatives necessarily have to be formalized in different ways, because neither assigns an actual construction; both effectively merely posit the objects. That the two corresponding parts of the proposition are separated by the definition of the goal has to do with the logical structure of the proposition – exposition and definition of the goal repeat the general πρότασις in particular terms – but their strict formal distinction seems to contain an element of unnecessary textual standardization.

As for the introductory forms of the definition of the goal, λέγω ὅτι/ δεῖ δή, these have been shown to be one hundred percent standardized in theorem and problem proposition; however, one may ask what would happen to the validity of the proof if, e.g., λέγω was substituted by another 'verb of saying.' Probably, it would not change anything as long as all other features of textual standardization remained in place. But it is hard to say whether it would affect the validity of the proof *in the eyes of readers* who are used to the form.

The construction imperatives seem to contain an element of unnecessary standardization, too: it is the characteristic ending of the verb forms which seems to be extremely repetitive, as several of these peculiar forms generally show up in close proximity. I shall consider the general possibility of abbreviation shortly.

The element of the "second-order language" which seems to be the most unnecessary one is the concluding formula ὅπερ ἔδει δεῖξαι/ ποιῆσαι. It confirms that the proof has been accomplished after the result of the proof has already been stated both in particular and in general terms. As I have pointed out, the formula is uncommon outside Euclid's *Elements* and, even here, it is not applied throughout; some scholars think it has been standardly applied only by later editors. However, as it is,

5 Netz 1999a, 62.

its standardization fits the general style; it is only one of several elements of the textual form which seem to be unnecessary from the point of view of 'logical' unambiguousness. Modern translators print the abbreviated form "Q.E.D.," presumably because it is the standardized way to end a proof today. The preceding συμπέρασμα likewise has been suspected, because it is rare outside Euclid and often not applied in the *Elements*. It does have a logical justification, because it generalizes from the particular result of the proof and points back to the beginning of the proof. Nevertheless, the verbatim repetition of the πρότασις (in theorem propositions), which occurs in many successive proofs, is somewhat awkward. The reason why modern editors abbreviate the συμπέρασμα seems to be precisely that they deem it superfluous, not that they think it is spurious, in which case they would have left it out completely.[6]

Other cases of logically unnecessary standardization include such in which the text is elliptic some of the time, indicating that an element can be left out at other times as well, e.g., terms such as τὸ Α (σημεῖον) which refer to particular objects in the diagram and from which the noun denoting the abstract geometrical object can be left out (cf. Section 3.2). A similar case is the one of ἐστίν as a predicate noun: it occurs, for instance, many times as part of the phrase ἴσον ἐστίν, although the text shows that it can also be left out. Such cases display a tendency toward textual standardization which results in more text than would be strictly necessary from a logical point of view. The resulting amount of superfluous text is remarkable in view of the economy which otherwise characterizes the text, as it dispenses with everything except for the bare proofs.

In Section 4.4.4, I have shown that the order of the letters in the 'letter words' is standardized according to various principles. These are all cases of logically unnecessary standardization. Even the form of the diagram itself, in case it was included in the text, could have been unnecessarily standardized, as I have argued in Sections 4.1 and 4.2: namely if it was drawn with the help of a ruler and compass, although the use of these instruments is not necessitated by the proof.

Overall, the analysis of the standardization of necessary and unnecessary features with regard to mathematical unambiguousness shows first that the text is sufficiently standardized with regard to 'logical' unambiguousness, and second that many features are unnecessarily standardized. The text even overfulfills the requirements for 'logically' necessary standardization; it is overstandardized.

This dichotomy of necessary and unnecessary standardization does not make any claim to definitiveness, and the necessity of the individual features remains debatable in every single case. This is because it proves difficult to consider the features of textual standardization in isolation, but they must be isolated in order to be subtracted. On the one hand, the scope of the features unavoidably predetermines the relation of their standardization to unambiguousness, and accordingly the necessity of their stan-

6 Heath prints "Therefore etc." throughout; Thaer " – S.".

dardization. On the other hand, the textual features are interdependent regarding their standardization. Individual features entail either the necessity or the non-necessity of other features of textual standardization (either-or-necessity). Thus, necessary standardization can be compensated for by other forms of standardization (so that which is actually necessary turns out not to be necessary in every case, and the necessity can then only be determined on a case-by-case basis).

Stylistic indications for unnecessary standardization

The question arises as to whether logically unnecessary standardization can be stylistically identified. I suggest that there are two main – paradoxically contrary – stylistic indications for unnecessary standardization.

The first is the imperfectness of standardization, i.e., that a feature is less than one hundred percent standardized. Stylistically, imperfectness is recognizable as variation or ellipsis: variation means that the otherwise standardized feature is substituted by another one or that its form is changed; ellipsis means that it or part of it is left out. The deficiency of the standardization of a feature indicates that the proof can do without its standardization (or in any case without its invariable use). For if its standardization were 'logically' required, the standardization would have to be assumed to be one hundred percent. This is true unless the deviation from standardization was a mistake, but the deviation from one hundred percent is usually too strong for it to be explicable in this way. Netz has argued that the deviation from full standardization suggests that standardization cannot be for the sake of disambiguation.[7] Whatever the motivation for (somewhat deficient) standardization (more on which shortly), it points to a surplus of standardization with respect to unambiguousness because its pervasive use was not necessary in the first place.

The second indication for unnecessary standardization is perfect, one hundred percent standardization which results in mechanical repetition and uneconomic redundancy. For while, according to my preceding argument, imperfectness of standardization suggests that standardization is unnecessary, the perfectness of standardization seems to be a necessary, but not a sufficient condition for the necessity of standardization. Perfect standardization many times lead to more text being written than is actually necessary from a logical point of view. The superfluous part of the text could have been left out altogether or abbreviated. In fact, one might go as far as to say that it is the very exactness of repetition or perfectness of standardization which opens up the possibility for some sort of cancelling, that is, either by ellipsis or abbreviation;[8] and as soon as this possibility opens up, standardization or detailedness is no longer necessary. This does not

7 Netz 1999a *passim.*

8 Cf. Netz 1999a, 145 on the mutually related processes of repetition and ellipsis.

contradict the above statement that certain types of (perfect) standardization in the *Elements* are indeed mathematically necessary. But while standardization may be logically necessary in the first place, the resulting standardized detailedness in some cases enables ellipsis and abbreviation and, in doing so, turns out to be superfluous. In these cases, the standardization of the resulting abbreviated phrases would be again necessary.

The problem is how to identify cases of 'unnecessary' standardization where standardization is perfect, that is, by what criteria to choose textual phenomena for subtraction on a trial basis. In cases of imperfect standardization, one may subtract precisely that which is imperfectly standardized; by contrast, in instances of perfect standardization, the text itself does not suggest whether the feature and how much of it is logically superfluous. One might conclude, however, from the general possibility of ellipsis to which the *Elements* testifies, that perfect standardization presents a license to abbreviate. The abbreviation of verb forms is unusual in the *Elements*, but it is practiced in later mathematical texts. Such evidence for mathematical abbreviations elsewhere shows that they are, though not a genre tradition, nevertheless possible from the 'logical'-mathematical perspective. In any case, while the imperfectness of standardization could serve to identify cases of logically unnecessary standardization, features which are completely standardized do not unequivocally indicate 'logically' unnecessary standardization: they could suggest just the opposite.

Clarity of proof and excess of form

The standardization of the *Elements*, even though it is not one hundred percent, is, in any case, sufficient from a 'logical' point of view so as to supply the text with utter unambiguousness. For it is next to perfect wherever necessary. What is more, it has turned out that even in the cases of 'necessary' standardization, one hundred percent standardization is not always strictly necessary. This is because a lack of perfection in one case of 'necessary' standardization will be compensated for by strict consistency in another. Therefore, at the end of the day, every letter, word, phrase, verb ending, mode, tense, and structural element will be clear (that is, unambiguous) from the text itself. Even more compellingly, the Euclidean text contains a lot of standardization which is logically 'unnecessary' according to my classification. In this way, the text is not only sufficiently standardized, but overfulfills the requirements for 'logical' unambiguousness by means of standardization, and produces an excess of textual form as regards the mathematical contents and 'logic.'

The division of standardization into a 'necessary' and an 'unnecessary' type and further subtypes can serve as a preliminary model for tentatively quantifying the relation of form and content, for grasping the surplus of form, and, thus, for determining an aesthetic dimension of the mathematical text. However, it has turned out that there are not any decisive criteria for determining what is merely formal. The stylistic indications for 'overunambiguousness' – imperfection of standardization and physical superfluousness – have emerged from the analysis only retrospectively as reflexes of the excess of the form's exceeding the necessary degree for consistency at the cognitive level; they cannot serve to unambiguously define the borders of content-related form. Thus, the trial-and-error method of subtraction remains the only way to approach the point where form transcends proof, although it only allows for testing standardization on a case-by-case basis, and provides a fluid rather than a definitive picture.

The purpose of the conceptualization of 'necessary' and 'unnecessary' standardization with regard to mathematical unambiguousness, however, has not been a definitive localization of the border between the two among the standardized features of Euclid's *Elements*. Rather, it is a thought experiment which, for one thing, provides a way to conceive of unambiguousness as a feature which the *Elements* objectively contain in the form of fulfilling the requirements for 'necessary' standardization; the experiment, thus, shows that the work is autonomous with regard to clarity. For another thing, the conceptualization gives an idea of there being an enormous amount of form (that is, standardization) which does not naturally emerge from geometrical proof and its logical needs. Moreover, the bipartite scheme, precisely by its lack of definitiveness, points to the lack of objectiveness which characterizes unambiguousness in the *Elements* and, as I shall contend, beyond. The limits of investigation of unambiguousness from a logical perspective are due to the fact that what is necessary about the text's standardization cannot be determined unless by including the perspective of a recipient and, thus, by looking at the communicative side of unambiguousness. This is true especially as far as the compensation of certain types of standardization by certain others is concerned.

The analysis of the form of the *Elements* has shown that the text is self-contained with regard to unambiguousness, that its unambiguousness is *technically* objective. The text itself, through its standardization, offers the instruments for its own clarification; it includes everything necessary to comprehend it (and even more). It might not in any case be obvious from the start, but ultimately, there is only one possible way to understand the meaning of a particular word, phrase, sentence, or proposition. The text, when taken into account in its entirety, seems to be unmistakable. However, this characterization indicates the communicative aspect of clarity (clarity-as-comprehensibility, the ideal of which is unmistakability). It points to the decisive role of reception in determining how clear and unambiguous a text actually is: It is only ever clear to *someone*. 'Objective clarity' from a cognitive-communicative stance, that is, 'unmistakability,' is a kind of clarity which pertains to any reader one can think of, across time and cultures.

5.2 Standardization and communicative clarity

The focus of this section is on the recognizability of the textual unambiguousness produced by (over)standardization. I discuss how much standardization is necessary for the comprehensibility of the work, and how much is again, from this point of view, too, unnecessary because it exceeds unmistakability; and how far logically unnecessary standardization can be explained by communicatively necessary standardization. For the purpose of distinguishing necessary and unnecessary standardization from a communicative point of view, I apply again the 'method of subtraction.' However, in order to determine the relation between standardization and *communicative* clarity, the method can only be applied with regard to a specific readership, i.e., with regard to what they would have found necessary or unnecessary. This is because it cannot be presupposed that the *Elements* was universally clear: in the brief survey of the extant remarks on the form of the *Elements* and of the genre in general which I have provided above in the general introduction, it has already been shown that there seem to have been diverging perceptions of the language and style of elements texts. On the one hand, Euclid is emphatically called clear by his commentators; on the other hand, there are indications that the work was by no means universally clear in antiquity, although there is no explicit ascription of obscurity.

My focus is on the intended readership of Euclid, that is, on the readers of Euclid's original communicative context, and their perception of the extreme standardization of the *Elements*. We lack any explicit statements by Euclid's readership and any information about Euclid's communicative context and for whom the work was written. Accordingly, we will not be able to know whether Euclid's original readers found the *Elements* clear. However, it is likely that they would have done so, in view of the increase and stabilization of the textual features in the history of the genre, and the ongoing standardization of the form in Euclid's aftermath. In any case, the intended readers, if any, would have found the work and its form clear, because it is with regard to them that the form was originally applied. In order to determine the relation of standardization and communicative clarity with regard to the original readers, 'clarity' will be regarded as comprehensibility: the question is how much of the standardization would have been necessary in order for Euclid's original readers to understand the work.[9]

Before it is possible to apply the 'method of subtraction,' it must be determined first who Euclid's intended readers were, and second what the purpose and function of the work were and in what kind of communicative situation the work was received, because the *Elements* itself is silent on these questions. First, in order to determine the original readership, the explicit and implicit reception of the *Elements* and its form will be recalled, as well as, *e contrario*, the information we have about later reading contexts. It has been pointed out in reviewing the few extant remarks on the form of the *Elements*

9 The cognitive-communicative dimension of excessive standardization with regard to recipients contemporary to ourselves could be analyzed productively with the help of the cognitive sciences such as neurolinguistics. However, since my primary interest will be in the contemporary readers of Euclid, whose cognitive abilities and processes we can access no longer, it will not be possible to determine what Euclid's readership would have thought sufficiently standardized by this method (for the application of "psycholinguistics" to ancient texts cf. Vatri 2016).

that they indicate a perception of obscurity rather than clarity.[10] They point out, above all, the strangeness and 'overprecision' of the language (ἀκριβολογία).[11] This does not have to imply necessarily that the language was perceived as obscure, but it suggests at least that this language was not taken for granted, and that it is not the language of those who make these judgments nor of those who read them. Moreover, these passages have a critical, pejorative undertone.[12] They stem from non-mathematical authors (Aristotle, Galen) and reflect their non-mathematical readership.[13] From this, one may conclude that the standardized form conveyed *obscurity* to readers outside the generic context and that the form would have been perceived as clear, if at all, within a mathematical context of reception. The Galenic passage seems to support this view, because Galen claims that no one except the mathematician "understands" a certain mathematical term.

As has been pointed out in the review of the explicit remarks, the only explicit attributions of clarity to Euclid are by his late antique commentators;[14] that is, they stem from a mathematical (or, in the case of Proclus, mathematical-philosophical) context, but from one which is presumably very different from the one in which the *Elements* originated, because they are from the period in which the *Elements* was already canonized and used in the context of (philosophical) teaching. This means that ascriptions of clarity are made from an 'etic' perspective; they might reflect a traditional view on Euclid, but more probably the status of the canonized work and in any case will have fulfilled a purpose for the commentators. I have suggested possible functions of emphasizing Euclid's clarity both with regard to the experts and the students whom Proclus' commentary seems to have addressed simultaneously.[15] Proclus' attribution of clarity is consistent with his commentary practice insofar as he does not really clarify nor present his work as clarification.[16] A main practice of his, however, is to explicate things which the Euclidean text leaves implicit because of its strict standardization.[17] We have pointed out that the implicitness of the *Elements*, a corollary of its standardized style, could be a potential source of obscurity for readers who are not acquainted with the style.[18] In this way, Proclus' commentary could implicitly testify to a perception of obscurity among his student readership.[19]

---

10 See Section 1.3.
11 See Section 1.3, 35 f.
12 Ibid.
13 Ibid.
14 See Section 1.3, 34.
15 See Section 2.3.2.
16 See Section 2.3.1, 100 f.
17 See Section 2.3.1, 100.
18 See Section 2.3.1, 100.
19 This hypothesis is put forward by Saito 2009. To be sure, as Netz 1998, 263–273 has pointed out, systematization, standardization and unfolding things which are left implicit are typical practices of 'deuteronomic' texts; in this way, Proclus' commentary practices do not necessarily indicate that his

The development of the text by its editors and commentators in late ancient times suggests something similar. As I have pointed out, the ancient editors, above all Theon, who have engaged in 'improving' the Euclidean text as part of its ongoing canonization, have tried to make the text more explicit in various ways: on the one hand, in logical terms, by adding obvious arguments to the deduction; on the other hand, by changing the textual form toward greater explicitness: by spelling out earlier proved propositions, and in so doing, disrupting the linear deductive structure or by making explicit the book structure and separating the units of the proposition by numbering the propositions.[20] Further, there has been a drive to systematize and explicate rules on the basis of implicit standardization, for instance, for the proposition structure.[21] These rules, in turn, seem to have been applied to the text in order to increase the less-than-one-hundred percent implicit standardization even further. Even the perpetuation of the implicit standardization could be explained by the endeavor to explicate the text: once the implicit standardization, e.g., of the internal proposition structure, is complete, there is again something explicit about it, because what I have called above the 'set frames,' the lexical-syntactical markers of the proposition parts, then work like (explicit) headings.

Theon, whose recension has most decisively shaped our text, was a teacher of mathematics; this suggests that the effort to make the text more explicit was due to the fact that his students found the text obscure as it is. Both the late ancient commentators and editors appear to try to compensate for the implicitness of the Euclidean text in order to make it more comprehensible for (part of) their readership, i.e., learners of mathematics. In so doing, the commentaries and editions indicate that the work – implicit as it is – that is, without any didactic framing, additional arguments, explicit quotations of preceding proofs, explanations, adjunct commentary or information as to its purpose and use, cannot have been directed to mathematical beginners.

Thus, two kinds of readers have been ruled out as intended readers: first, non-mathematical readers, on the grounds of the extant remarks on the form of mathe-

readers found the text obscure. However, by contrast with other texts, the *Elements* provides ample opportunity for making things explicit; its commentators do not have to insinuate its obscurity.

20 This seems to have happened fairly early; there is no evidence of proposition numbering for the time of Euclid, but already for the second century and third century AD (P. Oxy. i 29 and P. [inv.] Berol. 17469 (Asper 2007, 177, n. 568 and 189, 647 with reference to Althoff 1995, 271).

21 The scheme which Proclus describes in his commentary and which we still use for describing the (internal) proposition structure has first been deemed an invention by Diophantes (Knorr 1993, cf. Doxiadis & Sialaros 2013, 381, n. 85); Netz has emphasized that it is in any case a later systematization, maybe by Proclus himself (see Section 1.2 of the introduction on the formal features of the *Elements* p. 23, n. 134); Doxiadis & Sialaros 2013, 381 with n. 85 point out that the scheme could even have been Euclidean, mainly with reference to Hero, *Def.* 137.1.1 who provides a description of the structure which "gives a better fit to Euclid's practice." The difficulty of safely attributing the scheme to anyone in part has to do with the extreme implicit standardization of the proposition structure: anyone who attempts to infer a rule from it will end up with a similar one.

matics which seem to indicate that the form was received as obscure outside mathematics; and learners of mathematics, especially beginners. This may surprise us considering that the textual form contains the rules for deciphering itself and in this way also for deciphering the content, and considering that the proofs toward the beginning of the work are rather basic and could have provided a suitable introduction to geometry and to the whole work. However, from the pejorative reflections of the form, it seems that non-experts may be driven away by the inaccessible form before they are able to profit from its unambiguousness. The teachers of the late ancient schools and institutions prepare for this by writing additional explanatory works and by changing the original style of the text. Thus, it turns out that the unambiguousness of the work does not automatically pertain to any reader even though they technically would be able to understand it.

Having excluded lay readers and mathematical beginners, this leaves mathematical experts as the intended recipients of the work and its particular kind of clarity. That is, the *Elements* appear to be a work of expert communication within the mathematical group context in which elements texts were the usual text form. The group of mathematicians is extremely small,[22] though it does not only consist of individual mathematicians who have to write letters to each other for the sake of long-distance communication with other solitary mathematicians, as the case of Archimedes of Syracuse indicates. There seem to have been a few centers where mathematicians were based.[23] In any case, these groups were small, socially homogeneous (aristocratic) and inaccessible to people outside the group.[24] This explains why the standardized form of elements texts was perceived as strange in non-mathematical contexts of reception: the readership of Aristotle and of Galen, who make the remarks about the unusualness, seems to be remotely acquainted with the text form, because the two authors count on the approval of their readers; but non-mathematical readers will hardly ever have been confronted with elements texts before the *Elements* became canonized and institutionalized in Euclid's aftermath.

Having determined the intended readers and, as far as possible, the communicative context, I now recall what we know from earlier studies about the conditions of the reception and the presumed function of the *Elements*, and square it with the implications of the preceding analysis of the textual form. Since it has been concluded *ex negativo* from comparing the reception of the *Elements* in the late ancient school context that the intended readers are not mathematical beginners, some functions which have been suggested for the *Elements* can be excluded immediately. The *Elements* cannot have been written as a 'course book' or 'course material' for the use of students in class. As such, it would have to be accompanied by a teacher who simultaneously sup-

22 Netz 2002b.

23 Asper 2007, 168 *pace* Netz.

24 Cf. Netz 2002b, 216: "If Greek mathematics involved few people, it means that the very access to mathematics was extremely complicated."

plied the information in the way later commentaries do. While the late ancient commentaries and editions are written in view of the ready work by Euclid, the question prompts itself as to why the work was not prepared for being used for teaching from the beginning. As it is, the *Elements* does not make any reference to a teaching context. It is equally unlikely that the work would have been useful to mathematical students as a textbook which they were to study on their own. This use could be attested by the papyrus evidence which we have surveyed, but it comes up only after the canonization of the work shortly after Euclid's own time, and the use of the work in school contexts. Finally, one might consider whether mathematicians were supposed to use the work for teaching their students; whether the text either was a script written for a lecture or whether it was written out and expanded after each lecture from what were previously 'lecture notes.' Either scenario can be excluded because of the extreme standardization of the *Elements* which indicates that the work must have a written origin. It must also have been received as a written work, that is, by being read according to the linear structure which results from the strict focus on and standardization of the deduction.

The *Elements* is completely decontextualized by its extreme textual standardization, which has been shown to include, most probably, the standardized application of a lettered diagram. It is standardized so as to be read and understood out of context; and it does not contain any reference to a teaching situation or any other oral context of reception, precisely because there was no institution which provided such a context. The mathematicians communicated within the group independently from any intellectual or educational institution, and they seem to have compensated for this by using a decontextualizing text form.[25] The generic standardization and impersonality seem to indicate that there was no competition among the group members.[26]

This is how the function of the *Elements* can be described most generally: it communicates mathematical knowledge to other non-institutionalized mathematicians in writing. The specific function of this written knowledge communication can be concluded by comparing the content and scope of the text to that of what I have called the 'second,' Archimedean type of elements texts. These texts present newly discovered proofs (or rather proof chains) to individual fellow mathematicians whom they address in introductory letters. The *Elements*, by contrast, systematically exposes the foundational knowledge of geometry and arithmetic without delimiting its readership. In view of the rather basic and comprehensive content which the text communicates to other expert mathematicians, one may conclude that it aims at accumulating approved, consensual knowledge, and serves as a reference work and as a readily accessible basis for further research of both Euclid himself and of (present and future) colleagues.

---

25 Asper 2007, 147–156.

26 Asper 2007, 161–163.

Finally, with regard to this communicative situation and function of the *Elements*, I reassess the necessity of standardization with respect to cognitive-communicative clarity and with respect to the related socio-cognitive function of building consensus. Again, I apply the 'method of subtraction,' but in the way of a more general thought experiment. I ask whether the extreme standardization, which has been shown to cause 'over-unambiguousness' from a logical perspective, is necessary from a communicative stance or whether lesser standardization would suffice.

First, it must be considered how much standardization is necessary in view of the danger of misunderstanding posed by decontextualization in the non-institutionalized communicative context: whether the *Elements*, if some of its standardization is 'subtracted,' that is, if it is standardized to a lesser degree, runs the risk of being misunderstood. I argue that, in view of the content of the work and the readership which we have determined, additional standardization beyond the requirements of 'logical' unambiguousness for the sake of communication is not necessary. The knowledge to be communicated by the *Elements* is, in part, foundational and for the most part not new. Accumulating such knowledge to make it available and accessible to oneself in the future as well as to other contemporary or later mathematicians does not require great effort in terms of securing understanding. Considering the examples provided in the previous part for logically unnecessary standardization, these could be safely subtracted from the text without making the text less comprehensible: e.g., the standardization of the concluding formula or even that of the συμπέρασμα or the standardized application of ἐστί as a predicate noun. It appears that these features of standardization are equally unnecessary from a communicative point of view as they have been shown to be from a logical one. One might even question whether all of the 'logically' necessary standardization is still necessary in view of the subject of the work and the expert readership, for Euclid's fellow mathematicians would have been able to compensate for missing elements of necessary standardization, e.g., the diagrammatization of the text of easy propositions.

Second, and on a related note, the socio-cognitive function of building consensus (which is about convincing the readers of the truth of the text so as to make them agree) has been attributed to the form of the *Elements*, since it has an important role in modern mathematics which does not consider a proof valid unless its truth has been acknowledged by the mathematical community.[27] It has been argued that it is with regard to building consensus that the extreme formal standardization works as 'technology of trust' in the place of personal authority or a consolidating institution, mostly

27 Heintz 2000, 178. An example for this is the supposed proof of the 'abc conjecture' by the Japanese mathematician Shin'ichi Mochizuki; first published in 2015 in the form of pre-prints amounting to 500 pages, the proof is so obscure, not least because of its non-standard, flowery language, that no one so far has understood it and acknowledged its truth and existence (http://www.sueddeutsche.de/wissen/mathematik-das-grosse-abc-1.3141614-2). I owe the reference to this proof to Simon Renner.

through objectification which is created by standardization and impersonality.[28] I should question that additional standardization (i.e., beyond the unambiguousness of the proofs) is needed with regard to such a function. This is because of the lack of competition among mathematicians which the form seems to indicate: it makes the author disappear completely. The lack of competition diminishes the importance of the supposed function of consensus building. This is confirmed by the contents of the work and by the longstanding textual tradition up to the time of Euclid of adding to it: the *Elements* primarily seems to fix consensual knowledge rather than build consensus concerning its contents. The lack of competitiveness is what distinguishes the communicative context decisively from that of modern academia, and the function of the *Elements* differs from that of modern science papers with regard to the role of consensus. The standardized form of the latter, the IMRAD scheme, which has been compared to the standardized form of elements texts,[29] has the function of building consensus in a competitive milieu in the first place. Considering the supposedly non-competitive context in which the *Elements* emerged, the extreme standardization beyond 'logical unambiguousness' does not seem to be explicable by the function of consensus building. It does not appear to serve to persuade the recipient of the truth and validity of the proofs.

5.3 The meaning of the form and the role of clarity

Young man, in mathematics you don't understand things. You just get used to them.
John von Neumann

Concluding from my analysis of the relation of standardization, on the one hand, to logical unambiguousness, and on the other hand, to communicative clarity, there remains a considerable part of standardization in the *Elements* which does not seem to have any cognitive-communicative function. There seems to be an amount of extra standardization which presents an excessive layer of form.

The need for unambiguousness and consensus formation will have accounted for the adoption of the text's standardized form and its development at the time when theoretical mathematics came into being. However, by the time of Euclid, the *Elements* displays a degree of standardization which exceeds by far the requirements for unambiguousness. Therefore, I argue that, in Euclid's *Elements*, unambiguousness/clarity pursued for the sake of guaranteeing the transmission of knowledge is no longer the primary objective or driving force of standardization. I argue that historically, the communicative motivation for standardization will have been replaced by another motivation as the predominant one. Nevertheless, the standardization of the textual form is to some extent indispensable to proof logic and communication, but

28 Asper 2007, 185–188.
29 Asper 2007, 118.

these functions will in any case be fulfilled by the extreme standardization, as the analysis of the textual standardization has shown. As regards the text of the *Elements* itself, the point of distinguishing between necessary and unnecessary standardization was to be able to discern whether the textual form is fully explicable by the communicative function of the work. The distinction has served as an analytical tool. However, this is not to say that there is any point *within* the text from which point on standardization stops serving the *objective* of unambiguousness in order to serve something else instead. I argue that in the *Elements*, the drive to excessive standardization as a whole is non-functional regarding knowledge transmission and codification, and in this sense 'aesthetic.' It is only retrospectively that one can distinguish overall a communicatively functional and an aesthetic dimension of the textual form, without being able to demarcate these dimensions clearly within the text. I refer to the aesthetic dimension as 'surplus, excessive, or aesthetic (layer of) form.'

It is at least part of the point of excessive, aesthetic form that it does not explicitly pronounce a certain point and that it cannot be reduced to some point but ultimately remains open to differing interpretations. To a certain extent, the point of form is form itself.[30] However, to find out what motivates the excessiveness of standardization in the first place, I now consider what agenda brings it about and creates an aesthetic dimension, and which 'message' is carried by the surplus layer of form. Two main lines of interpretation have been followed on this question: both of them have emphasized the theoretical character of the *Elements*. One of these lines has been to explain the standardized form within the Platonist context which we can assume for Euclid. In this regard, the exclusion of all subjectivity by standardization would have been a Platonist statement about mathematics: 'Not only the geometrical objects, but also the proofs preexist and are only *discovered* by the mathematician.' Considering Plato's above-cited critique of the procedural language of, supposedly, the tradition of practical mathematics in view of the status of mathematical objects and object relations, Euclid and other writers of elements texts might use the objectifying language to foreground the objective and merely theoretical character of their own occupation.[31] Another line of interpretation of the extreme standardization similarly highlights the theoretical inclination of the *Elements*, but from a sociological rather than an ideological point of view. The standardized form has been differently explained in terms of a distinction from 'the practical:' on the one hand, as the endeavor of the mathematicians as part of the otiose intellectual elite to distinguish themselves from the practical life within the polis and from the practical professions of those who work for money;[32] on the other hand, as an attempt of the mathematicians of the ele-

30 Cf. Lowrie 2005 and Hamilton 2003 on obscurity itself being the point of obscurity.
31 Cf. Asper 2007, 125–127.
32 Netz 1999a, 271–312.

ments tradition to position themselves in decided contradistinction to the longstanding *practical* tradition of mathematics.[33]

Before going into these interpretations, the 'practical' dimension of excessive standardization must be emphasized. In view of it, there would be a certain irony in a self-fashioning in terms of a 'theoretical life' or 'theoretical' mathematics. Euclid makes himself vanish behind what appear to be the actions of mathematics itself. But in fact, the impersonality and seeming objectivity of the text are the result of Euclid's own *practices*. I argue that he, at closer inspection, more than anything emerges as a mathematical *practitioner* – as one who 'does' *words*, not mathematical operations, to be sure.

I claim that Greek theoretical mathematics is in some way more about doing something as one is used to do it (according to an established form) than about 'understanding' or 'discovering' something. The reason is that writing the proof – as opposed to discovering it – takes up such an important part of supposedly 'theoretical' mathematics. As I have argued, it is only the description of the proof which accomplishes it in the first place, and this description is diligently done in every single case. It has been emphasized that the description of the proof, though it is the description of the result rather than of the process of proof discovery, is itself a process. The process of proof description is accomplished by practicing the standardized text form. This entails performing a set number of standardized steps by applying the standardized textual features, and Euclid takes great care to 'do' the standardized things in the case of each proposition, even though many of the easier proofs could dispense with some of the standardization. At least to some extent, writing the proof consists of non-cognitive practices: of literally doing the same things again and again. In this spirit, John von Neumann famously said the above-quoted to a fellow mathematician who expressed his concern about not to understand a certain mathematical method: "Young man, in mathematics you don't understand things. You just get used to them."

The impression that the *Elements*, by its textual form, witnesses 'doing' besides 'understanding' is not only based on the schematic and repetitive nature of the texts, which implies that the mathematician is constantly working according to a given scheme. It is also grounded on an observation about the cognitive space in which the proofs take place, and the cognitive processes which are by convention at the author's disposal. In the *Elements*, 'understanding' is confined to an abstract (or 'theoretical') level, even though the proof itself is accomplished with regard to a concrete diagrammatic case. Cognition takes place exclusively in the closed system of the logic of object relations, which the author never has to leave. Even within that system, thinking is restricted to *doing* 'since-then' operations on the basis of assumptions and proved propositions. Furthermore, already proven propositions will no longer be questioned but implemented into these 'since-then' operations as givens, so that a proof never

33 Asper 2007, 197–210.

needs to be grasped completely, but only with respect to those bits that are actually proved in the proposition at hand.

I argue, therefore, that the *Elements* is marked by a 'poetics of doing:' it 'does' according to standardization and in this way also '*does* standardization.' With regard to its stylistic effects, one might also call it a 'poetics of repetition.' Excessive standardization comes from a drive on the part of the mathematicians to do things, again and again, as they are used to doing them – to practice the proof, which is to practice the expected text form.

It is hard to imagine that such excessive indulgence in formalization to which the textual form of the *Elements* testifies is the direct consequence of a certain philosophical viewpoint or a desire for social or generic distinction. On the one hand, the extreme standardization cannot be the result of a Platonist agenda, because the formal surplus, at least at a closer look, produces a materiality of text which overlays the transparent objectiveness at which a 'Platonist standardization' would aim. On the other hand, assuming that the 'theoretical mathematicians' endeavored to distinguish themselves as a group from other groups by using the peculiar textual form makes sense only for the early period of the genre, in which the formal features were developed or adopted. A sense of social distinction, besides the above-discussed communicative motivations and a Platonist understanding of mathematics, could have influenced the formation of the formal features. However, as standardization became gradually more and more dominant and extreme during the long development of the genre (and even after Euclid in the textual transmission of the *Elements*), I find it hard to believe that social distinction would have remained a major driving force for standardization throughout.

My own suggestion is that the standardized form, through its constant reuse, became increasingly aestheticized until standardization was to be practiced for its own sake; at some point, the mathematicians practiced standardization because they enjoyed doing so. This is especially the case as readership and authorship coincide in the case of the *Elements* and will have shared formal preferences and expectations just as the aesthetic pleasure caused by their being met: Euclid will have been a reader of elements texts just as his readers will have composed elements propositions themselves. Standardization will be expected by and aesthetically appeal to the readership, Euclid's (author) colleagues and the reader Euclid himself; each of them will respond to this expectation by practicing standardization likewise, and, in so doing, will all the more increase the expectations and the aesthetic pleasure arising from proliferating standardization. By abiding especially meticulously by what would have become conventions in his own time, Euclid was to push them to another level. I argue that the formal features of elements texts, by the time of Euclid, are the result of a process which one might call 'l'art pour l'art'[34] – a process in which the formal

---

34 Asper 2007, 130 grants that theoretical mathematics is "economically l'art pour l'art," by not being in any way profitable.

features would have been expressly pursued for their own sake rather than having emerged from cognitive processes as a "self-regulating conventionality."

It is only *by* practicing standardization for its own sake that the text can be taken to deliver some kind of message. I argue, tentatively, that the extreme standardization could mirror what has come to be known as 'the mathematical life,' including, that is, a complete devotion to mathematical problems, cherishing a peculiar lifestyle and befriending only a few like-minded people. In doing so, the extreme use of standardization could have contributed to cultivating this image. Both in terms of the indulgence in a practice of standardization for its own sake and in terms of the obsessiveness of the practitioners in so doing and their according otherworldly image in the eyes of the rest of the society, a social group nowadays resembling that of Euclid and his fellow mathematicians seems to be that of computer programming nerds. By implication, such an inward-looking message about the identity of mathematicians would also make a point about the relation of the mathematicians to their surroundings: the mathematical elite can afford to spend their time idly with problems that are irrelevant to both society and to practical mathematicians (who, fittingly, practice a less seclusive style). As I pointed out, however, I do not believe that an endeavor for distinction was still the motivating force for standardization in Euclid's time, and so cannot account for the extreme manifestation of the genre features in his work.

With regard to the intended readership of Euclid, I now ask what role clarity has in the *Elements* in relation to the aesthetic practice of standardization. I argue that, although cognitive-communicative clarity is not the aim of the form, it is a direct effect of standardization. By aiming past clarity at something else, that is, at meeting the readership's formal expectations, standardization also produces clarity as unambiguousness. It does so by its very orientation toward the formal expectations of the readership. As I have tried to show, the *Elements* produces a greater amount of standardization than necessary from the point of view of 'logical' and 'communicative' unambiguousness. I argue that its overstandardization does not reverse the effect of clarity to cause obscurity instead. To the contrary, one might go as far as to say that the use of excessive standardization in the *Elements* creates the expectation of a new kind of clarity, in the sense that such an overdone standardization comes to be expected as part of 'communicative clarity' in elements texts in the aftermath of the work. I refer to it as 'aesthetic clarity,' in order to distinguish it from and add it to my previous dichotomic classification of 'logical' and 'communicative' clarity.

Finally, how specific, or how universal is the clarity of the *Elements*? To Euclid's original readers, the *Elements* would have been clear. Beyond the context of theoretical mathematics, it would have been clear to some and unclear to others. Arguably, the use of the very formal features which cause clarity in the communicative context of elements texts would not have had the same effect in other genres of ancient Greek science. Clarity seems to be context-bound; it seems to be achieved by adhering to the generic features. In the context of elements texts, the *Elements* uses the generic form most rigorously and seems to have been clearer than the earlier exponents of the genre which it surpassed in form and clarity. Archimedes, by contrast, plays with the strict 'Euclidean' standardization, and, in so doing, testifies to it.[35] This playfulness may have caused obscurity even within the context of the same genre,[36] but again, the form is geared to the intended readers. In Archimedes' case, these were individual mathematicians who arguably would have shared his taste and enjoyed the slight deviation from the generic practice of 'doing as one is used to;' they would have been able to understand the Archimedean play with structure and turn obscurity into clarity.

Clarity comes from (over)fulfilling the reader's or group's expectations by adhering to the generic form, obscurity from breaking them. Thus, the present approach can be used for analyzing the relation between form and clarity and the meaning of the form in the same way. It remains to ask whether the clarity of the *Elements* is any different from that of the clearest works of other genres, considering that the work can still be understood today. The reason for its long-term clarity seems to be that the *Elements* fulfills the generic expectations precisely through standardization; it standardizes standardization, as it were – and for this reason may be universally understood, even if not everybody finds it clear.

---

35 Netz 2009. Rather than evolving from the zeitgeist as part of an Alexandrian aesthetic, the Archimedean play could be a reaction to the established generic form (although it is not sure whether Archimedes knew the Euclidean work).

36 Netz 2009 *passim* repeatedly mentions the "opacity" resulting from Archimedes' and Apollonius' play.

Bibliography

Texts Quoted

Burnet, J. (ed.) 1901. *Platonis opera, Vol. 2: Tetralogiae III–IV*. Oxford.

Duke, E. A., and W. F. Hicken *et al.* (eds.) 1995. *Platonis Opera, vol. 1: Tetralogiae I–II*. Oxford.

Friedlein, G. (ed.) 1873. *Procli Diadochi in primum Euclidis Elementorum librum commentarii*. Leipzig.

Heiberg, J. L., and E. S. Stamatis (eds.) 1969–1973. *Euclidis Elementa*. 5 vols. Leipzig.

Heiberg, J. L. (ed.) 1893. *Apollonii Pergaei Conica, quae graecae exstant: cum commentariis antiquis*. 2 vols. Leipzig.

Heiberg, J. L. (ed.) 1972–1975. *Archimedis opera omnia: cum commentariis Eutocii*. 4 vols. 2nd ed., Stuttgart.

Heiberg, J. L. (ed.) 1895. *Euclidis optica, Opticorum recensio Theonis, Catoptrica, cum scholiis antiquis*. Leipzig.

Heiberg, J. L. (ed.) 1912. *Heronis Alexandrini opera quae supersunt omnia. Vol. IV: Heronis Definitiones cum variis collectionibus. Heronis quae feruntur Geometrica*. Leipzig.

Helmreich, G. (ed.) 1923. *Galeni De alimentorum facultatibus*. Corp. Med. Graec. V 4.2. Leipzig, 199–386.

Kranz, W. (ed.) 1951–1952. *Die Fragmente der Vorsokratiker*. Griechisch und Deutsch von H. Diels. 3 vols. 6th ed., Berlin.

Kühn, C. G. (ed.) 1964–1965. *Claudii Galeni opera omnia*. 20 vols. Repr., Hildesheim.

Marcovich, M. (ed.) 1999–2002. *Diogenis Laertii Vitae philosophorum*. 3 vols. Stuttgart.

Meyerhof, M., and J. Schacht (eds.) 1931. *Galen. Über die medizinischen Namen. De vocibus in arte medica usitatis*. Arab. und dt. Berlin.

Moraux, P. (ed.) 1965. *Aristote. Du ciel. Texte établi et traduit*. Paris.

Müller, I. (ed.) 1891. *Claudii Galeni scripta minora*. Bd. 2. Leipzig, 91–124.

Reinhardt, T. (ed.) 2022. *M. Tulli Ciceronis Academicus primus, fragmenta et testimonia Academicorum librorum, Lucullus*. Oxford.

Reynolds, L. D. (ed.) 1998. *M. Tulli Ciceronis De finibus bonorum et malorum libri quinque*. Oxford.

Ross, W. D., and L. Minio-Paluello (ed.) 1964. *Aristotelis Analytica priora et posteriora*. Oxford.

Ross, W. D. (ed.) 1959. *Aristotelis Ars rhetorica*. Oxford.

Ross, W. D. (ed.) 1958. *Aristotelis Topica et Sophistici elenchi*. Oxford.

Ross, W. D. (ed.). *Aristotle's Metaphysics: A Revised Text with Introduction and Commentary*. Oxford.

Slings, S. R. (ed.) 2003. *Plato: Respublica*. Oxford.

Stückelberger, A., and G. Graßhoff (eds.) 2006. *Klaudios Ptolemaios. Handbuch der Geographie. Griechisch–Deutsch*. Basel.

Thomson, W., and G. Junge (eds.) 1930. *The Commentary of Pappus on Book X of Euclid's Elements. Arabic Text and Translation*. Cambridge, Mass.

Toomer, G. J. (ed.) 1990. *Apollonius: Conics, Books V to VII. The Arabic Translation of the Lost Greek Original in the Version of the Banû Mûsâ*. New York.

Usener, H., and L. Radermacher (eds.) 1899. *Dionysii Halicarnasei opuscula*. Vol 1. Leipzig.

Wehrli, F. (ed.) 1969. *Die Schule des Aristoteles. Heft 8*. 2nd ed., Basel.

Winterbottom, M. (ed.) 1970. *M. Fabi Quintiliani Institutionis oratoriae libri duodecim*. Vol. 2. Oxford.

Works Quoted

Acerbi, F. 2003. "Drowning by Multiples. Remarks on the Fifth Book of Euclid's *Elements*, with Special Emphasis on Prop. 8." In: *Archive for History of Exact Sciences* 57.3, 175–242.

Acerbi, F. 2007 (ed.). *Euclide. Tutte le opere. Testo greco a fronte*. Milan.

Acerbi, F. 2008. "Euclid's *Pseudaria*." In: *Archive for History of Exact Sciences* 62.5, 511–551.

https://doi.org/10.1515/9783110670462-006

Acerbi, F. 2010. *Il silenzio delle sirene. La matematica greca antica.* 1st ed., Rome.
Acerbi, F. 2011a. "The Language of the 'Givens:' its Forms and its Use as a Deductive Tool in Greek Mathematics." In: *Archive for History of Exact Sciences* 65.2, 119–153.
Acerbi, F. 2011b. *La sintassi logica della matematica greca.* Pisa.
Acerbi, F. 2012. "I codici stilistichi della matematica greca: dimostrazioni, procedure, algoritmi." In: *Quaderni Urbinati di Cultura Classica Nuova Serie* 101.2, 167–214.
Acerbi, F. 2013. "Aristotle and Euclid's Postulates." In: *Classical Quarterly* 63.2, 680–685.
Acerbi, F. 2016. "Byzantine Recensions of Greek Mathematical and Astronomical Texts: A Survey." In: *Estudios Bizantinos* 4, 133–213.
Acerbi, F. 2018. "Hellenistic Mathematics." In: Keyser, P. T. and J. Scarborough (eds.). *Oxford Handbook of Science and Medicine in the Classical World.* Oxford, 269–292.
Acerbi, F. 2021. *The Logical Syntax of Greek Mathematics.* Cham.
Allwein, G., and J. Barwise (eds.) 1996. *Logical Reasoning with Diagrams.* New York.
Althoff, J. 1995. *Studien zu den Anfängen der wissenschaftlichen Literatur bei den Griechen.* Habilitationsschrift. Freiburg im Breisgau.
Angeli, A., and T. Dorandi 1987. "Il pensiero matematico di Demetrio Lacone." In: *Cronache Ercolanesi* 17, 89–103.
Artmann, B. 1985. "Über voreuklidische 'Elemente', deren Autor Propositionen vermied." In: *Archive for History of Exact Sciences* 4, 291–306.
Artmann, B. 1991. "Euclid's Elements and its Prehistory." In: Mueller, I. (ed.). *Peri tōn Mathēmatōn.* Edmonton, Alberta, 1–47.
Artmann, B. 1999. *Euclid – The Creation of Mathematics.* New York.
Asmuth, B. 2003. Art. "Perspicuitas." In: *Historisches Wörterbuch der Rhetorik* 6, 814–874.
Asper, M. 2001. "Stoicheia und Gesetze. Spekulationen zur Entstehung mathematischer Textformen in Griechenland." In: *Antike Naturwissenschaft und ihre Rezeption* 11, 73–106.
Asper, M. 2003a. "Mathematik, Milieu, Text. Die frühgriechische(n) Mathematik(en) und ihr Umfeld." In: *Sudhoffs Archiv* 87, 1–31.
Asper, M. 2003b. "Rev. Netz 1999." In: *Gnomon* 75.1, 7–12.
Asper, M. 2007. *Griechische Wissenschaftstexte. Formen, Funktionen, Differenzierungsgeschichten.* Stuttgart.
Asper, M. 2009. "The Two Cultures of Mathematics in Ancient Greece." In: Robson, E., and J. Stedall (eds.). *The Oxford Handbook of the History of Mathematics.* Oxford, 107–132.
Asper, M. 2011a. *Erzählungen in der (griechischen) Mathematik? Ein Survey.* Antrittsvorlesung 5. Mai 2011. Berlin.
Asper, M. 2011b. "'Frame Tales' in Ancient Greek Science Writing." In: Pohl, K.-H., and G. Wöhrle (eds.). *Form und Gehalt in Texten der griechischen und der chinesischen Philosophie.* Stuttgart, 91–112.
Asper, M. 2011c. "Hoppe 100. Zehn Bemerkungen zu Mathematik und Astronomie im klassischen Altertum (erster Teil, I-VII)." In: Schwindt, J. P. (ed.). *Edmund Hoppe: Mathematik und Astronomie im klassischen Altertum.* Bd. 1. Heidelberg, 277–314.
Asper, M. 2012a. "Hoppe 101. Zehn Bemerkungen zu Mathematik und Astronomie im klassischen Altertum (zweiter Teil, VIII-X)." In: Schwindt, J. P. (ed.). *Edmund Hoppe: Mathematik und Astronomie im klassischen Altertum.* Bd. 2. Heidelberg, 5–27.
Asper, M. 2012b. "'True' and 'False' Errors in Ancient (Greek) Computation." In: Geller, M., and K. Geus (eds.). *Productive Errors: Scientific Concepts in Antiquity.* Berlin, 47–66.
Asper, M. 2013a (ed.). *Writing Science. Medical and Mathematical Authorship in Ancient Greece.* Berlin.
Asper, M. 2013b. "Introduction." In: Asper, M. (ed.). *Writing Science. Medical and Mathematical Authorship in Ancient Greece.* Berlin, 1–13.
Asper, M. 2013c. "Making Up Progress." In: Asper, M. (ed.). *Writing Science. Medical and Mathematical Authorship in Ancient Greece.* Berlin, 411–430.
Asper 2013d. "Rev. Netz 2009." In: *Classical Review* 63, 75–77.

Asper, M. 2014. “Mathematik.” In: Zimmermann, B., and A. Rengakos (eds.). *Handbuch der Altertumswissenschaft VII 2. Handbuch der griechischen Literatur der Antike*. Bd. 2: Die Literatur der klassischen und hellenistischen Zeit. München, 453–459.

Asper, M., and S. Föllinger 2014. “Fachliteratur. Einleitung.” In: Zimmermann, B., and A. Rengakos (eds.). *Handbuch der Altertumswissenschaft VII 2. Handbuch der griechischen Literatur der Antike*. Bd. 2: Die Literatur der klassischen und hellenistischen Zeit. München, 459–481.

Asper, M. 2016. “On Systematicity. How to Write (Ancient Greek) Science.” In: *Trends in Classics* 8.1, 103–123.

Barnes, J. 1992. “Metacommentary.” In: *Oxford Studies in Ancient Philosophy* 10, 267–281.

Becker, O. 1936a. “Die Lehre vom Geraden und Ungeraden im Neunten Buch der Euklidischen Elemente.” In: *Quellen und Studien zur Geschichte der Mathematik, Astronomie und Physik, Abt. B: Studien* 3, 533–553.

Becker, O. 1936b. “Zur Textgestaltung des eudemischen Berichts über die Quadratur der Möndchen durch Hippokrates von Chios.” In: *Quellen und Studien zur Geschichte der Mathematik, Astronomie und Physik, Abt. B: Studien* 3, 411–419.

Brashear, W. M. 1994. “Vier neue Texte zum antiken Bildungsweg.” In: *Archiv für Papyrusforschung* 40, 29–35.

Burkert, W. 1959. “Στοιχεῖον. Eine semasiologische Studie.” In: *Philologus* 103, 167–197.

Burkert, W. 1982. “Konstruktion und Seinsstruktur: Praxis und Platonismus in der griechischen Mathematik.” In: *Abhandlungen der Braunschweigischen Wissenschaftlichen Gesellschaft* 34, 125–141.

Burnyeat, M. F. 1987. “Platonism and Mathematics: A Prelude to Discussion.” In: Graeser, A. (ed.). *Mathematics and Metaphysics in Aristotle*. Xth Symposium Aristotelicum. Bern, 213–240.

Burnyeat, M. F. 1999. “Knowledge is Perception: ‘Theaetetus’ 151d-184a.” In: Fine, G. (ed.). *Plato 1: Metaphysics and Epistemology*. Oxford, 320–354.

Cambiano, G. 1992. “La nascita dei trattati e dei manuali.” In: Cambiano, G., L. Canfora *et al.* (eds.). *Lo spazio letterario della Grecia antica. Vol. 1: La produzione e la circolazione del testo. T. 1: La ‘Polis.’* Rome, 525–553.

Carter, L. B. 1986. *The Quiet Athenian*. Oxford.

Cleary, J. J. 1995. *Aristotle and Mathematics: Aporetic Method in Cosmology and Methaphysics*. Leiden.

Cuomo, S. 2001. *Ancient Mathematics*. London.

Detel, W. 1993. *Aristoteles. Analytica Posteriora. Übersetzt und erläutert*. 2 vols. Berlin.

Doxiadis, A., and M. Sialaros 2013. “Sing, Muse, of the Hypotenuse: Influences of Poetry and Rhetoric on the Formation of Greek Mathematics.” In: Asper, M. (ed.). *Writing Science. Medical and Mathematical Authorship in Ancient Greece*. Berlin.

van der Eijk, Ph. J. 1997. “Towards a Rhetoric of Ancient Scientific Discourse.” In: Bakker, E. J. (ed.). *Grammar as Interpretation. Greek Literature in its Linguistic Contexts*. Leiden, 77–129.

Federspiel, M. 1992. “Sur locution ἐφ’ οὗ/ἐφ’ ᾧ servant à designer des êtres géométriques par des lettres.” In: Guillaumin, J.-Y. (ed.). *Mathématiques dans l’antiquité*. Saint-Étienne, 9–25.

Federspiel, M. 1995. “Sur l’opposition défini/indéfini dans la langue des mathématiques grecques.” In: *Les Études Classiques* 63, 249–293.

Federspiel, M. 2015. “Sur une partie spécifique de la démonstration dans les textes géométriques grecs classiques: l’anaphore.” In: *Pallas* 97, 33–50.

Fowler, D. H. 1999. *The Mathematics of Plato’s Academy. A New Reconstruction*. 2nd ed., Oxford.

v. Fritz, K. 1955. “Die ἀρχαί in der griechischen Mathematik.” In: *Archiv für Begriffsgeschichte* 1, 13–103.

Fuhrmann, M. 1966. “Obscuritas: das Problem der Dunkelheit in der rhetorischen und literarästhetischen Theorie der Antike.” In: Iser, W. (ed.): *Immanente Ästhetik, ästhetische Reflexion: Lyrik als Paradigma der Moderne. Kolloquium Köln 1964; Vorlagen und Verhandlungen*. Köln, 47–72.

Gasser, A.-M. 2024. “Form, Terminology, and Clarity in Aristotle.” In: Asper, M. (ed.). *Coming to Terms. Approaches to (Ancient) Terminologies*. Berlin, 131–181.

Grenfell, B., and A. Hunt 1898 (eds.). *The Oxyrhynchus Papyri. Part 1 and 2*. London.

Grenfell, B., A. Hunt, and D. Hogarth 1900. *Fayûm Towns and Their Papyri*. London.
Groebner, V. 2012. *Wissenschaftssprache. Eine Gebrauchsanweisung*. Konstanz.
Hamilton, J. 2003. *Soliciting Darkness. Pindar, Obscurity, and the Classical Tradition*. Cambridge, MA.
Heath, T. L. 1926. *The Thirteen Books of Euclid's Elements*. Vols. 1–3. Translated from the Text of Heiberg with Introduction and Commentary. Repr. of the 2nd ed., Cambridge.
Heiberg, J. L. 1882. Litterargeschichtliche Studien über Euklid. Leipzig.
Heiberg, J. L. 1885. "Ein Palimpsest der Elemente Euklids." In: *Philologus* 44, 353–366.
Heiberg, J. L. 1900. "Quelques papyrus traitant de mathématiques." In: *Extrait du Bulletin de l'Académie royale des Sciences et des Lettres de Danemark, Copenhague* 2, 1–25.
Heiberg, J. L. 1903a. "Paralipomena zu Euklid." In: *Hermes* 38, 46–74.
Heiberg, J. L. 1903b. "Paralipomena zu Euklid." In: *Hermes* 38, 161–201.
Heiberg, J. L. 1903c. "Paralipomena zu Euklid." In: *Hermes* 38, 321–356.
Heintz, B. 2000. *Die Innenwelt der Mathematik. Zur Kultur und Praxis einer beweisenden Disziplin*. Wien.
Hilbert, D. 1899. "Grundlagen der Geometrie." In: *Festschrift zur Feier der Enthüllung des Gauss-Weber Denkmals in Göttingen*. Leipzig.
Holmes, F. L. 1987. "Scientific Writing and Scientific Discovery." In: *Isis* 78, 220–235.
Inwood, B., and L. P. Gerson (eds.) 2008. *The Stoics Reader*. Indianapolis.
Iser, W. 1994. *Der Akt des Lesens, Theorie ästhetischer Wirkung*. 4th ed., München.
Jackson, H. 1920. "Aristotle's Lecture-Room and Lectures." In: *Journal of Philosophy* 35, 191–200.
Kahn, Ch. H. 1981. "The Role of Nous in the Cognition of First Principles in Posterior Analytics II 19." In: Berti, E. (ed.). Aristotle on Science. The Posterior Analytics. Padua, 385–414.
Kanthak, A.-M. 2013. "Obscuritas – eine Strategie griechischer Wissenschaftsliteratur?" In: Schmitzer, U. (ed.). *Enzyklopädie der Philologie. Themen und Methoden der Klassischen Philologie heute*. Göttingen, 157–183.
Kennedy, G. A. 2003. *Progymnasmata: Greek Textbooks of Prose Composition and Rhetoric*. Leiden.
Kidd, S. E. 2014. *Nonsense and Meaning in Ancient Greek Comedy*. Cambridge.
Knorr, W. R. 1975. *The Evolution of the Euclidean Elements. A Study of the Theory of Incommensurable Magnitudes and Its Significance for Early Greek Geometry*. Dordrecht.
Knorr, W. R. 1978. "Archimedes and the *Elements*. Proposal for a Revised Chronological Ordering of the Archimedean Corpus." In: *Archive for History of Exact Sciences* 19, 211–290.
Knorr, W. R. 1981. "On the Early History of Axiomatics: The Interaction of Mathematics and Philosophy in Greek Antiquity." In: Hintikka, J., D. Gruender *et al.* (eds.). *Theory Change, Ancient Axiomatics, and Galileo's Methodology*. Dordrecht, 145–186.
Knorr, W. R. 1982. "Infinity and Continuity: The Interaction of Mathematics and Philosophy in Antiquity." In: N. Kretzmann (ed.). *Infinity and Continuity in Ancient and Medieval Thought*. Ithaca, 112–145.
Knorr, W. R. 1993. "*Arithmêtikê stoicheiôsis*. On Diophantus and Hero of Alexandria." In: *History of Mathematics* 20, 180–192.
Knorr, W. R. 1996. "The Wrong Text of Euclid: On Heiberg's Text and its Alternatives." In: *Centaurus* 38, 208–276.
Krämer, S. 2016. *Figuration, Anschauung, Erkenntnis. Grundlinien einer Diagrammatologie*. Berlin.
Kullmann, W. 1974. *Wissenschaft und Methode. Interpretationen zur aristotelischen Theorie der Naturwissenschaft*. Berlin.
Kullmann, W. 1981. "Die Funktion der Mathematischen Beispiele in Aristoteles' Analytica Posteriora." In: Berti, E. (ed.). *Aristotle on Science. The Posterior Analytics*. Padua, 245-270.
Lachtermann, D. R. 1989. *The Ethics of Geometry. A Genealogy of Modernity*. New York.
Laks, A., and G. Most (eds.) 2016. *Early Greek Philosophy. Vol. 5: Western Greek Thinkers, Part 2*. Cambridge, Mass.
Lehoux, D. 2000. "Rev. Netz 1999." In: *Bryn Mawr Classical Review* 11.2, 1–5.
Lengen, R. 2002. *Form und Funktion der aristotelischen Pragmatie. Die Kommunikation mit dem Rezipienten*. Stuttgart.

Lewis, O. 2017. *Praxagoras of Cos on Arteries, Pulse and Pneuma*. Leiden.
Lowrie, M. 2005. "Inside out: In Defense of Form." In: *Transactions of the American Philological Association* 135, 35–48.
Luhmann, N. 1981. "Die Unwahrscheinlichkeit der Kommunikation." In: Luhmann, N. (ed.). *Soziologische Aufklärung 3. Soziales System, Gesellschaft, Organisation*. Opladen, 25–34.
Luhmann, N. 2000. "Lesen lernen." In: Gente, P., H. Paris *et al.* (eds.). *Short Cuts 1.* Frankfurt am Main, 150–157.
Mainberger, S. 2003. *Die Kunst des Aufzählens. Elemente zu einer Poetik des Enumerativen*. Berlin.
Manders, K. 2008. "The Euclidean Diagram." In: Mancosu, P. (ed.). *The Philosophy of Mathematical Practice*. Oxford, 80–133.
Mansfeld, J. 1994. *Prolegomena. Questions to Be Settled Before the Study of an Author, or a Text*. Leiden.
Mau, J., and W. Müller 1982. "Mathematische Ostraka aus der Berliner Sammlung." In: *Archiv für Papyrusforschung* 17, 1–10.
Morrow, G. R. 1992. *Proclus: A Commentary on the First Book of Euclid's 'Elements.' Translated with Introduction and Notes*. Princeton.
Mueller, I. 1969. "Euclid's *Elements* and the Axiomatic Method." In: *The British Journal for the Philosophy of Science* 20, 289–309.
Mueller, I. 1979. "The Completeness of Stoic Propositional Logic." In: *Notre Dame Journal of Formal Logic* 20, 201–215.
Mueller, I. 1981. *Philosophy of Mathematics and Deductive Structure in Euclid's Elements*. Cambridge, Mass.
Mueller, I. 1991. "On the Notion of a Mathematical Starting Point in Plato, Aristotle, and Euclid." In: Bowen, A. C. (ed.). *Science and Philosophy in Classical Greece*. New York, 59–97.
Mugler, Ch. 1958. *Dictionnaire historique de la terminologie géométrique des grecs*. 2 vols. Paris.
Mumma, J. 2010. "Proofs, Pictures and Euclid." *Synthese* 175, 255–287.
Mumma, J. 2012. "Constructive Geometrical Reasoning and Diagrams." In: *Synthese* 186, 103–119.
Mumma, J., and M. Panza 2012. "Diagrams in Mathematics: History and Philosophy." In: *Synthese* 186, 1–5.
Netz, R. 1998. "Deuteronomic Texts: Late Antiquity and the History of Mathematics." In: *Revue d'histoire des mathématiques* 4, 261–288.
Netz, R. 1999a. *The Shaping of Deduction in Greek Mathematics. A Study in Cognitive History*. Cambridge.
Netz, R. 1999b. "Proclus' Division of the Mathematical Proposition into Parts: How and Why Was It Formulated?" In: *Classical Quarterly* 49, 282–303.
Netz, R. 2002a. "Counter Culture: Towards a History of Greek Numeracy." In: *History of Science* 40, 321–352.
Netz, R. 2002b. "Greek Mathematicians: A Group Picture." In: Tuplin, C. J., and T. E. Rihll (eds.). *Science and Mathematics in Ancient Greek Culture*. Oxford, 196–216.
Netz, R. 2004a. "Eudemus of Rhodes, Hippocrates of Chios and the Earliest Form of Greek Mathematical Text." In: *Centaurus* 46, 243–286.
Netz, R. 2004b. *The Transformation of Mathematics in the Early Mediterranean World: From Problems to Equations*. Cambridge.
Netz, R. 2009. *Ludic Proof. Greek Mathematics and the Alexandrian Aesthetic*. Cambridge.
Netz, R. 2013. "Authorial Presence in the Ancient Exact Sciences." In: Asper, M. (ed.). *Writing Science. Mathematical and Medical Authorship in Ancient Greece*. Berlin, 217–253.
Neuenschwander, E. 1972/73. "Die ersten vier Bücher der Elemente Euklids. Untersuchungen über den mathematischen Aufbau, die Zitierweise und die Entstehungsgeschichte." In: *Archive for History of Exact Sciences* 9, 325–380.
Panza, M. 2012. "The Twofold Role of Diagrams in Euclid's Plane Geometry." In: *Synthese* 186, 55–102.
Petrucci, F. M. 2016. "Rev. Acerbi 2010." In: *Journal of Hellenic Studies* 136, 294–295.
Piaget, J. 1972. *Die Entwicklung des Erkennens I. Das mathematische Denken* (orig. Paris 1950). Mit einer Einführung von H. Aebli, übersetzt von F. Kubli. Stuttgart 1972.

Porter, J. E. 1995. "Content and Form in Philodemus: The History of an Evasion." In: Obbink, D. (ed.). *Philodemus and Poetry: Poetic Theory and Practice in Lucretius, Philodemus and Horace*. Oxford, 97–147.

Rapp, C. 2013. "Sprachliche Gestaltung und philosophische Klarheit bei Aristoteles." In: Erler, M., and J. E. Heßler (eds.). *Argument und literarische Form in antiker Philosophie*. Berlin, 283–303.

Rommevaux, S., A. Djebbar, and B. Vitrac 2001. "Remarques sur l'histoire du texte des Éléments d'Euclide." *Archive for History of Exact Sciences* 55, 221–295.

Rudio, F. 1907. *Der Bericht des Simplicius über die Quadraturen des Antiphon und des Hippokrates*. Leipzig.

Saito, K. 2006. "A Preliminary Study in the Critical Assessment of Diagrams in Greek Mathematical Works." In: *SCIAMVS* 7, 81–144.

Saito, K. 2009. "Reading Ancient Greek Mathematics." In: Robson, E., and J. Stendall (eds.). *The Oxford Handbook of the History of Mathematics*. Oxford, 801–829.

Saito, K. 2012. "Traditions of the Diagram, Tradition of the Text: A Case Study." In: *Synthese* 186, 7–20.

Saito K., and N. Sidoli 2012. "Diagrams and Arguments in Ancient Greek Mathematics: Lessons Drawn from Comparisons of the Manuscript Diagrams with Those in Modern Critical Editions." In: Chemla, K. (ed.). *The History of Mathematical Proof in Ancient Traditions*. Cambridge, 135–162.

Schneider, I. 1979. *Archimedes. Ingenieur, Naturwissenschaftler und Mathematiker*. Darmstadt.

Schnur, H. 2010. *Schreiben: Eine lebensnahe Anleitung für die Geistes- und Sozialwissenschaften*. Wiesbaden.

Schütrumpf, E. 1987. "Form und Stil aristotelischer Pragmatien." In: *Philologus* 133, 177–191.

Seidenberg, A. 1977/78. "The Origin of Mathematics." In: *Archive for History of Exact Sciences* 18, 301–342.

Shin, S.-J. 2012. "The Forgotten Individual: Diagrammatic Reasoning in Mathematics." In: *Synthese* 186, 149–168.

Sluiter, I. 2016. "Obscurity." In: Grafton, A., and. G. W. Most (eds.). *Canonical Texts and Scholarly Practices. A Global Comparative Approach*. Cambridge, 34–51.

Smyth, H. W. 1956. *Greek Grammar*. Rev. by G. M. Messing. Cambridge, Mass.

Strunk, W., and E. B. White 2000. The Elements of Style. With Revisions, and Introduction, and a Chapter on Writing. 4th ed., Boston.

Stückelberger, A., and G. Graßhoff (eds.) 2006. *Klaudios Ptolemaios. Handbuch der Geographie. Griechisch–Deutsch*. Basel.

Szábo, Á. 1962. "Der älteste Versuch einer definitorisch-axiomatischen Grundlegung der Mathematik." In: *Osiris* 14, 308–369.

Szábo, Á. 1994. *Die Entfaltung der griechischen Mathematik*. Mannheim.

Szlezák, T. A. (ed.) 2001. *Platonisches Philosophieren. Zehn Vorträge zu Ehren von Hans Joachim Krämer*. Hildesheim.

Thaer, C. (eds.) 1970. *Euklid. Die Elemente. Buch I – XIII. Nach Heibergs Text aus dem Griechischen übersetzt*. Darmstadt.

Turner, E. G., D. H. Fowler, L. Koenen, and L. C. Youtie 1985. "Euclid, *Elements* I, Definitions 1–10 (P. Mich III 143)." In: *Yale Classical Studies* 28, 13–24.

Unguru, S. 1975. "On the Need to Rewrite the History of Greek Mathematics." In: *Archive for History of Exact Sciences* 15, 67–114.

Vatri, A. 2016. *Orality and Performance in Classical Attic Prose. A Linguistic Approach*. Oxford.

Vitrac B. 2001. "La transmission des textes mathématiques: l'exemple des *Éléments* d'Euclide." In: Girard, L., and C. Jacob (eds.). *Des Alexandries I. Du livre au texte*. Paris, 339–355.

Vitrac, B. 2004. "A propos des démonstrations alternatives et autres substitutions de preuves dans les Éléments d'Euclide." In: *Archive for History of Exact Sciences* 59, 1–44.

Vitrac, B. 2008. "Promenade dans les préfaces des textes mathématiques grecs anciens." In: P. Radelet-de-Grave (ed.). *Liber amicorum Jean Dhombres*. Turnhout, 9–46.

Vitrac, B. 2012. "The Euclidean Ideal of Proof in the Elements and Philological Uncertainties of Heiberg's Edition of the Text." In: Chemla, K. (ed.). *The History of Mathematical Proof in Ancient Traditions*. Cambridge, 69–134.

van der Waerden, B. L. 1976. "Defence of a 'Shocking' Point of View." In: *Archive for History of Exact Sciences* 15, 199–210.

van der Waerden, B. L. 1977/78. "Die Postulate und Konstruktionen in der frühgriechischen Geometrie." *Archive for History of Exact Sciences* 18, 343–357.

Waschkies, H.-J. 1989. *Anfänge der Arithmetik im alten Orient und bei den Griechen*. Amsterdam.

Waschkies, H.-J. 1995. "Prinzipien der griechischen Mathematik: Platon, Aristoteles, Proklos und Euklids *Elemente*." In: Döhring, K., B. Herzhoff *et al.* (eds.). *Antike Naturwissenschaft und ihre Rezeption. V*. Trier, 91–153.

Waschkies, H.-J. 1998. "Mathematische Schriftsteller." In: H. Flashar (ed.), *Grundriss der Geschichte der Philosophie. Die Philosophie der Antike, Bd. 2/1: Sophistik, Sokrates, Sokratik, Mathematik, Medizin*. Basel, 365–453.

Webb, R. H., and P. Weller 2012. "Enargeia." In: Greene, R. (ed.). *The Princeton Encyclopedia of Poetry and Poetics*. 4th ed., Princeton, 409.

Zanker, G. 1981. "Enargeia in the Ancient Criticism of Poetry." In: *Rheinisches Museum für Philologie* 124, 297–311.

Digital Resources

Vat. gr. 190.1 and Vat. gr. 190.2 (Vatican Library)
http://digi.vatlib.it/view/MSS_Vat.gr.190.pt.1, https://digi.vatlib.it/view/MSS_Vat.gr.190.pt.2 (Digital Vatican Library, last accessed 7 May 2024).

MS. D'Orville 301 (Bodleian Library, Oxford)
https://digital.bodleian.ox.ac.uk/objects/d4a23501-0b98-4aff-acd6-fe06fe9b62e3/ (Digital Bodleian, last accessed 7 May 2024).

Nasir al-Din al-Tusi, Commentaries on Euclid (Library of Congress)
https://www.loc.gov/resource/gdcwdl.wdl_07465/?sp=49 (last accessed 7 May 2024)

http://www.sueddeutsche.de/wissen/mathematik-das-grosse-abc-1.3141614-2 (last accessed 7 May 2024).

General index

https://doi.org/10.1515/9783110670462-007

Index locorum

https://doi.org/10.1515/9783110670462-008